IET MATERIALS, CIRCUITS AND DEVICES SERIES 58

Magnetorheological Materials and their Applications

Other volumes in this series:

Volume 2 **Analogue IC Design: The current-mode approach** C. Toumazou, F.J. Lidgey and D.G. Haigh (Editors)
Volume 3 **Analogue–Digital ASICs: Circuit techniques, design tools and applications** R.S. Soin, F. Maloberti and J. France (Editors)
Volume 4 **Algorithmic and Knowledge-Based CAD for VLSI** G.E. Taylor and G. Russell (Editors)
Volume 5 **Switched Currents: An analogue technique for digital technology** C. Toumazou, J.B.C. Hughes and N.C. Battersby (Editors)
Volume 6 **High-Frequency Circuit Engineering** F. Nibler *et al.*
Volume 8 **Low-Power High-Frequency Microelectronics: A unified approach** G. Machado (Editor)
Volume 9 **VLSI Testing: Digital and mixed analogue/digital techniques** S.L. Hurst
Volume 10 **Distributed Feedback Semiconductor Lasers** J.E. Carroll, J.E.A. Whiteaway and R.G. S. Plumb
Volume 11 **Selected Topics in Advanced Solid State and Fibre Optic Sensors** S.M. Vaezi-Nejad (Editor)
Volume 12 **Strained Silicon Heterostructures: Materials and devices** C.K. Maiti, N.B. Chakrabarti and S.K. Ray
Volume 13 **RFIC and MMIC Design and Technology** I.D. Robertson and S. Lucyzyn (Editors)
Volume 14 **Design of High Frequency Integrated Analogue Filters** Y. Sun (Editor)
Volume 15 **Foundations of Digital Signal Processing: Theory, algorithms and hardware design** P. Gaydecki
Volume 16 **Wireless Communications Circuits and Systems** Y. Sun (Editor)
Volume 17 **The Switching Function: Analysis of power electronic circuits** C. Marouchos
Volume 18 **System on Chip: Next generation electronics** B. Al-Hashimi (Editor)
Volume 19 **Test and Diagnosis of Analogue, Mixed-Signal and RF Integrated Circuits: The system on chip approach** Y. Sun (Editor)
Volume 20 **Low Power and Low Voltage Circuit Design With the FGMOS Transistor** E. Rodriguez-Villegas
Volume 21 **Technology Computer Aided Design for Si, SiGe and GaAs Integrated Circuits** C.K. Maiti and G.A. Armstrong
Volume 22 **Nanotechnologies** M. Wautelet *et al.*
Volume 23 **Understandable Electric Circuits** M. Wang
Volume 24 **Fundamentals of Electromagnetic Levitation: Engineering sustainability through efficiency** A.J. Sangster
Volume 25 **Optical MEMS for Chemical Analysis and Biomedicine** H. Jiang (Editor)
Volume 26 **High Speed Data Converters** A.M.A. Ali
Volume 27 **Nano-Scaled Semiconductor Devices** E.A. Gutiérrez-D (Editor)
Volume 29 **Nano-CMOS and Post-CMOS Electronics: Devices and modelling** Saraju P. Mohanty and Ashok Srivastava
Volume 30 **Nano-CMOS and Post-CMOS Electronics: Circuits and design** Saraju P. Mohanty and Ashok Srivastava
Volume 32 **Oscillator Circuits: Frontiers in design, analysis and applications** Y. Nishio (Editor)
Volume 33 **High Frequency MOSFET Gate Drivers** Z. Zhang and Y. Liu
Volume 38 **System Design with Memristor Technologies** L. Guckert and E.E. Swartzlander Jr.
Volume 39 **Functionality-Enhanced Devices: An alternative to Moore's law** P.-E. Gaillardon (Editor)
Volume 40 **Digitally Enhanced Mixed Signal Systems** C. Jabbour, P. Desgreys and D. Dallett (Editors)
Volume 42 **Radio Frequency and Microwave Power Amplifiers: Theory, design and applications** A. Grebennikov (Editor)
Volume 43 **Negative Group Delay Devices: From concepts to applications** B. Ravelo (Editor)
Volume 47 **Understandable Electric Circuits: Key concepts, 2nd Edition** M. Wang
Volume 60 **IP Core Protection and Hardware-Assisted Security for Consumer Electronics** A. Sengupta and S. Mohanty
Volume 68 **High Quality Liquid Crystal Displays and Smart Devices, Vol. 1 and Vol. 2** S. Ishihara, S. Kobayashi and Y. Ukai (Editors)
Volume 69 **Fibre Bragg Gratings in Harsh and Space Environments: Principles and applications** B. Aïssa, E.I. Haddad, R.V. Kruzelecky, W.R. Jamroz

Magnetorheological Materials and their Applications

Edited by
Seung-Bok Choi and Weihua Li

The Institution of Engineering and Technology

Published by The Institution of Engineering and Technology, London, United Kingdom

The Institution of Engineering and Technology is registered as a Charity in England & Wales (no. 211014) and Scotland (no. SC038698).

First published 2019

The Institution of Engineering and Technology
Michael Faraday House
Six Hills Way, Stevenage
Herts, SG1 2AY, United Kingdom

www.theiet.org

British Library Cataloguing in Publication Data
A catalogue record for this product is available from the British Library

ISBN 978-1-78561-770-6 (hardback)
ISBN 978-1-78561-771-3 (PDF)

Typeset in India by MPS Limited
Printed in the UK by CPI Group (UK) Ltd, Croydon

Contents

Preface

New generations of smart materials featuring sensing or/and actuating capabilities by external stimuli have produced many innovative control devices or control systems in diverse industries for the last two decades. Currently, some of control devices using smart material actuators such as vehicle shock absorber are commercially available, but some characteristics of smart materials are still far from meeting the unstructured environment in which practical requirements such as reliability and cost-effectiveness are severe. Among numerous smart materials developed or invented so far, magnetorheological materials whose rheological and mechanical properties are controlled by the intensity of the magnetic field are most actively researched in recent years due to the robustness under variable service conditions in unstructured environmental operations. Thus, it is timely and appropriate to present most recent research works on the magnetorheological materials in a single book. This book consists of 14 chapters, and its scope includes magnetorheological fluids (MRFs), magnetorheological elastomers (MREs) and magnetorheological plastomers (MRPs) from material characterizations to application devices and systems.

Chapter 1 discusses on the redispersibility of MRF subjected to magnetic dispersed phase and carrier liquid domain considering surfactant and thixotropic additives to develop advanced physical models.

Chapter 2 introduces the discrete element method to calculate the field-induced magnetostatic interaction due to long-ranged dipolar one and the finite element method to calculate the field distribution and the yield stress as a function of the strain.

Chapter 3 presents a new hybrid optimal controller by modifying conventional sliding mode and H-infinity controllers to enhance the robustness and stability of MR seat damper system subjected to uncertainties and external disturbances.

Chapter 4 discusses how the models and methods can be used in solving engineering and research problems directly related to MR actuators which call for the expertise in solid body mechanics, fluid mechanics, heat transfer, electromagnetics, power electronics, chemistry and control theory.

Chapter 5 aims to analyze the effects of combined working coils of MR absorber on the shock mitigation performance by adopting a physical scale model of the field gun designed with four-stage parallel electromagnetic coils and annular flow channels.

Chapter 6 proposes a principle and configuration of a novel reciprocating rotary damper based on MR fluid from which high controllable torque range is possible through a piecewise control strategy with respect to vehicle body roll angular velocity.

Chapter 7 presents a hybrid controller to achieve several advantages of a semi-active MR seat suspension for heavy duty vehicles in terms of the fail-safe function, low power consumption and high vibration control performance showing excellent ride comfort.

Chapter 8 focuses on the modeling, optimal design and evaluation of different configurations of side-coil MRB such as single side-coil, double side-coil and triple side-coil MR brakes.

Chapter 9 provides a comprehensive knowledge of MRE by reviewing its material preparation, test characterization, mechanism investigation, engineering applications as well as the prospects of future development.

Chapter 10 presents micromechanical modelling solutions investigating the relationship among tensile modulus, magnetization and particle volume fraction of the structured MRE under a uniform magnetic field by considering the macroscopic constitutive laws and the anisotropy of these particular materials.

Chapter 11 develops a new kind of MRE with Brominated Isoprene Isobutylene Rubber (BIIR) matrix, summaries its difference with MRE based on NR matrix and models its mechanical property by using the magneto-viscoelasticity parametric model.

Chapter 12 explores the existing application of MRE-based devices by introducing the working principle, parameter design method and prototype performance.

Chapter 13 proposes a hybrid MRE absorber with nonlinearity and high adaptability and experimentally evaluates and verifies its widened effective frequency bandwidth, absorption effectiveness and variable stiffness property.

Chapter 14 details the development of a MRE base isolation system with controllable stiffness for seismic protection of civil structures, including isolator development, control design and shake table testing, aiming to resolve the inherent shortcomings of both traditional and hybrid base isolation systems.

Chapter 15 provides a detailed introduction to MRPs, including its synthetic methods, mechanical and electrical properties, evolution of microstructure, mechanism of the magnetic field-induced properties and the applications.

The editors trust that this book will stimulate the readership to cogitate upon the significant role of magnetorheological materials in various fields. In addition, all contributing authors of this book sincerely hope that the readers of this book will get some novel insights for more advanced material characteristics and creative applications. This book can be used as a text book for graduate students who are doing research on smart materials and a reference book to anyone who wants to invent new devices or systems utilizing the field-dependent magnetorheological materials.

Finally, the editors would like to sincerely express gratitude to all contributing authors for high-quality articles and the editorial staff of this book for their great effort in the process of edition.

Professor Seung-Bok Choi
Inha University, Korea
Professor Weihua Li
University of Wollongong, Australia

Chapter 1

Redispersibility and its relevance in the formulation of magnetorheological fluids

Sara R. Gomes de Sousa[1] *and Antonio J. Faria Bombard*[1]

In the literature about magnetorheological fluids (MRFs), with both theoretical and experimental results of physical models to explain the magnetorheological (MR) effect, often the MRF is prepared with only two components, a magnetic dispersed phase and a carrier liquid, while attempting to keep the MRF formulation as simple as possible. On the other hand, many patents of MRF include three, four or more components, such as some surfactant and thixotropic additives. In order to formulate a good and reliable MRF for different applications such as MR shock absorbers, clutches and brakes, the MRF redispersibility is a challenge, but necessary key property for out-of-lab real-world applications. In this chapter, we show how to measure and evaluate the MRF redispersibility.

1.1 Introduction

According to the Merriam-Webster Dictionary, dispersibility is defined as "the quality or state of being dispersible." Dispersible is an adjective related to the transitive verb to disperse, which, among other meanings, has the following meaning in the context of chemistry: "to distribute (something, such as fine particles) more or less evenly throughout a medium." Therefore, one can understand redispersibility as the ability of a dispersion or suspension (in the colloidal physical–chemical context) to be redispersed.

In this chapter, we will be more interested in how easy or difficult it is to redisperse a sample of MRF after some time (days, weeks, months, years, etc.) under rest; usually, under normal gravity, 1*g*. Sometimes, one can be interested in measuring the redispersibility of MRF after subjecting it to centrifugal force. In this case, one must specify how many "*g*"s have been applied to the sample and how much time the sample was centrifuged as this directly affects the sediment.

Quite commonly, especially when beginners start to work in magnetorheology, is there concern with the MRF "stability." It is important to understand the

[1]Physics and Chemistry Institute, Federal University of Itajubá, Brazil

difference between "sedimentation stability," related to Stokes' Law and the large density mismatch of the dispersed phase (usually dense, magnetic solid particles, with a size range of ~0.1–10 μm, as iron microparticles) and the liquid-dispersing medium (whose densities ranges from 786 to 1,890 kg/m^3 at temperatures ~25 °C). By far, the most common magnetic material investigated and employed to formulate MRF is carbonyl iron powder. Its density is ~7,860 kg/m^3.

On the one hand, sometimes one wants to keep the MRF as simple as possible to get theoretical, physical models to explain and fit experimental results well. These MRFs were prepared only with oil (perhaps other liquid at times) and the magnetic phase. Usually, these MRF samples settle very fast, and the only sedimentation retarders are the parameters of Stokes' Law (liquid viscosity, particle size and density phase mismatch).

On the other hand, even those commercial MRFs that are well formulated, with proprietary additives (dispersing, thixotropic, anti-wear, etc.), and that are sold by large industries, such as LORD Inc. or BASF SE, are prone to show phase separation. Additionally, anyone who has bought a commercial MRF or prepared their own MRF knows well that this phase separation is simply a matter of time.

However, much more serious than the old problem of settling (or sedimentation and phase separation), intrinsic to MRF, is the problem of colloidal stability. Related "cousins" of MRF, the ferrofluids, do not show phase separation since they are true colloids; their particles are in nanometer size range, usually 1–10 nm. Ferrofluid's, if well chemically prepared, are very stable. Besides not settling, surfactants and Brownian motion keep their particles apart, in such a way that they are not subject to coagulation.

This phenomenon, coagulation, in the context of colloids, is the more serious problem when formulating a good MRF. Without dispersing additives, the only thing that can hinder the iron particles to coagulate is if the van der Waals forces are null. As these forces are always omnipresent in almost any dispersion, the only way to cancel them is to match the Hamaker constant in the magnetic phase (iron powder or other) with the Hamaker constant of the liquid phase [1]. Besides van der Waals attractive forces, magnetic materials also can aggregate due to remnant magnetization and affect the redispersibility of MRF [2].

Several approaches were tested, including coating the iron core microparticles with thin, nanometric shells from different polymers. This can be a solution, although converting the carbonyl iron particles—expensive themselves—is even more expensive.

Another possible way that is easier and more efficient regarding time and money is testing (even by trial-and-error) different dispersing additives and finding one that modifies the surface of the magnetic phase, matching the "physical–chemical" nature of the dispersed phase with the carrier liquid Hamaker constant. Following one of the recipes described in US patent # 5,645,752: "Thixotropic Magnetorheological Materials" assigned to LORD Corp., invented by Weiss *et al.* [3], one of the authors of this chapter prepared a lot of MRFs using carbonyl iron powder, a silicone oil (different branch of that in the patent, but still silicone oil) and the same dispersing additive "SILWET L-7500" (Witco/Momentive chemicals) described in the patent. This batch of MRF was prepared in 1999. After years under rest, settling and phase separation of iron powder have been observed several times since then. However, just shaking the bottle containing the MRF by hand was enough to add energy to redisperse the entire dispersion completely. We think that this is a good example of this approach.

Throughout this chapter will be described in a clear and objective way one method of analysis for the redispersibility tests of MRFs with different formulations. This can contribute with different manner to analyses the quality of MRF after long period of resting and sedimentation.

1.1.1 Methodology of the redispersibility test

Although one can always use a simple spatula, according to ASTM D869-85, or the so-called "Instinctive hardness test" [4], we will describe here the approach first employed to test the redispersibility of an MRF (to the best of our knowledge) by Kieburg *et al.* [5].

This is a modified rheometer penetration test version of the ASTM D869-85: "Standard Test Method for Evaluating Degree of Settling of Paint" [6]. Table 1.1 is an adapted short, partial reproduction of the recommended Rating in ASTM D869-85. For a more detailed, full rating, refer to ASTM D869-85.

Table 1.1 is used just to show how subjective the "hand spatula" method can be in quantifying the redispersibility. However, if the other way is not available, using a steel spatula and human tactile feeling is a good tool to compare (at least qualitatively) MRF formulations after sedimentation.

Instead, a subjective, only qualitative rating evaluation based on a tactile feeling of a human hand, the normal force, capacitive sensor built-in some commercial rheometers, is employed to measure how easy/difficult it is to penetrate the iron powder cake sediment. The test can be run in an MRF sample container like a test tube (15 mL volume, or with ~100 mm height). If available, a rheometer equipped with normal force sensor, vane rotor and vertical downward movement with controlled, constant speed is preferable.

Table 1.1 Quality of paint after 6 months, according ASTM D869 (adapted)

Grade	Paint condition
10	Perfect suspension. No change from the original condition
8	A feel of settling and slight deposit. No significant resistance to sidewise movement of spatula
6	Definite cake of settled pigment. Definite resistance to sidewise movement of spatula
4	Difficult to move spatula through cake sidewise and slight edgewise resistance
2	If spatula has been forced through the cake, it is very difficult to move sidewise
0	Very firm cake that cannot be reincorporated with the liquid to form a smooth paint by stirring manually

*Intermediate conditions, among those above, are given the appropriate odd number.

In order to accomplish the test with a rheometer, one test tube containing the MRF sample is left under rest in order, to settle under normal gravity. After the time scheduled (24 hours, 1 week, 1 month, etc.), the test tube is put under the center aligned with the rheometer head, and a penetration test is run. One can employ a needle or a Vane tool, or even a steel blade (e.g. 50 mm height × 10 mm width × 1 mm thickness) attached to the rheometer coupling. Then the head is moved downward at a constant speed. Usually the movement speed is set on the software to 1 mm/s and makes use of the rheometer normal force readings and registers points every 0.05 s (or 50 μm penetration). In addition, an event control can be programmed in the rheometer software, so that the test is automatically interrupted if the normal force reaches the maximum that the capacitive sensor of the rheometer supports (50 N in our case).

Another possibility, as demonstrated by Iglesias *et al.* [7] and Portillo and Iglesias [4], is to use an analytical scale and a mechanism employing a stepper motor to move the tube containing an MRF to be tested under the load cell of electronic scale. Different approaches to measure or improve the redispersibility of MRF are possible too. See additional references [8–19], for details.

Figure 1.1 shows a steel blade tool attached to the rheometer head (Anton Paar MCR 301) and a tube containing the MRF sample in position to run the penetration

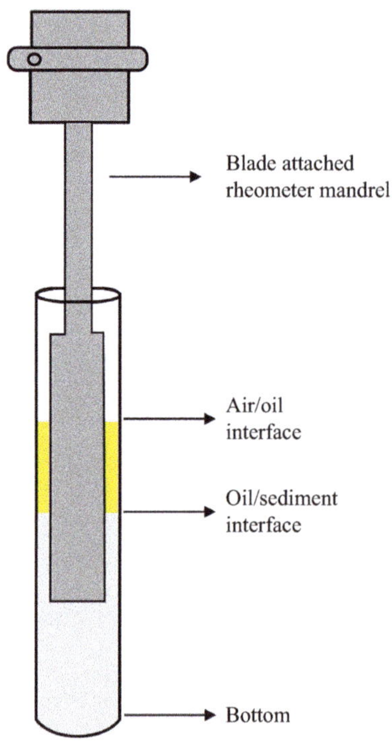

Figure 1.1 The steel blade attached to the rheometer head and a tube with MRF to run the penetration redispersibility test. A vane rotor can also be used

redispersibility test. The head started at the top position (arbitrary) and is moved downward, penetrating the MRF and its iron powder content (or any other magnetic solid sediment). All of the tests here were performed with MRF sedimenting at room temperature, over months or even years. But the variation of temperature oscillated between 15 and 35 °C considering winter and summer in the place of the experiments.

1.2 Results and discussion

1.2.1 Effect of sedimentation time

The next figures present redispersibility test curves in which normal force (*F*) in newtons (N) is a function of the depth (*d*) in millimeters (mm). Although the *X*-axis can also be presented as the gap, simple subtraction can be used to invert it. Instead of the value of the gap (distance between the lower tip of the blade or vane) to the bottom of the tube, one can show the depth as the distance the tool that penetrates the sediment. We prefer this way of presenting our results, but both (gap or depth) give the same information.

Figure 1.2 shows the results for the same MRF, with 80% by weight of carbonyl iron powder and additives. These were tested after 1 week, 1 month, 5 months and 10 months of sedimentation in normal gravity. The normal force resolution of the capacitance sensor of the rheometer used for the tests is 1 mN.

In Figure 1.2, it can be observed that even after almost a year under rest, the sediment did not exceed 0.3 N. In addition to the fact that the curves did not change much from 1 week to 1 month or 10 months, they indicate that this formulation can be considered "easy" to redisperse, or that it has good redispersibility.

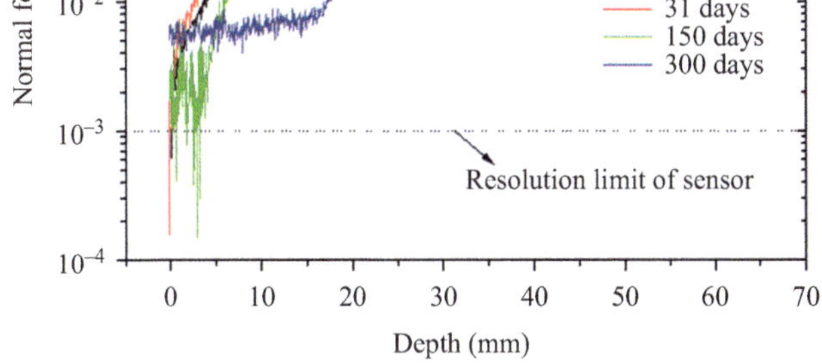

Figure 1.2 The normal force as a function of depth. MRF with 80 wt.% carbonyl iron powder in Poly(alpha-olefine) (PAO) oil

Integrating the area under the normal force × depth curves gives the mechanical work (energy) exerted by the blade when it penetrates the sediment. The higher value of the work shows that it is more compact and hard to penetrate. However, it must be considered that as this is a mathematical product, $w = F \times d$, we should not analyze only the value of the work.

There may be cases where only 1 mm of penetration of the normal force reaches a peak of 40 or 50 N (50 N is the operating limit of the MCR-301 rheometer). In this case even if the work results in a value of 40 mJ, it is an indication of a very hard sediment (*hard cake*) in the formation. Thus, an MRF of redispersibility becomes difficult or impossible after a certain point. In other words, it is considered a poorly formulated MRF.

Accordingly, we have to consider both the *normal force* × *depth* curves and the *work* × *depth* curves in the evaluation of redispersibility. Figure 1.3 corresponds to the integration of the areas (work) under the curves of Figure 1.2.

It can be observed in Figure 1.3 that the work is below 10 mJ, even after 300 days of sedimentation, confirming that this can be considered a well-formulated MRF. Even after 300 days, an MRF remains "easy" to redisperse. The results of Figure 1.3 are summarized in Table 1.2.

Figures 1.2 and 1.3 and Table 1.2 are examples of good MRF formulations in terms of redispersibility since the maximum force of blade penetration into the sediment, even after 300 days of settling under rest, was below 0.3 N, and the total work of each test was always below 10 mJ for 50 mm (or more) penetration.

It is important to understand that although the work as mechanical energy spent by the blade in order to penetrate the solid powder sediment can be useful in comparing different MRF formulations, this is merely secondary information and normal force peak together with the depth penetration is much more significant.

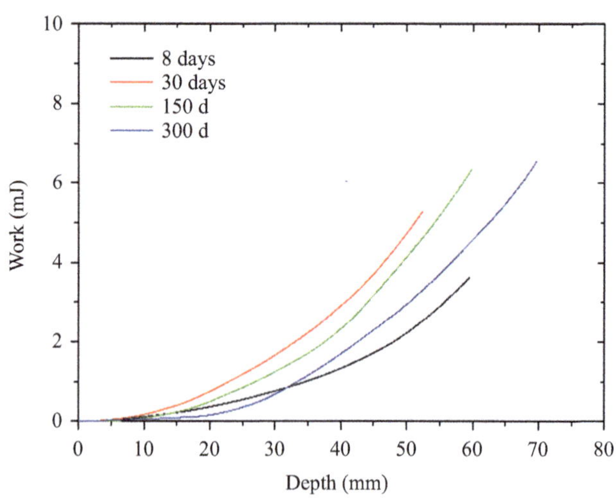

Figure 1.3 Work spent on penetration test, by integration of the curves in Figure 1.2

Table 1.2 Work from Figures 1.2 and 1.3 with a sample of 80 wt.% MRF in PAO oil

Time (days)	Max normal force (N)	Depth (mm)	Work (mJ)
8	0.15	60	3.6
30	0.20	52.6	5.3
150	0.25	60	6.4
300	0.28	70	6.6

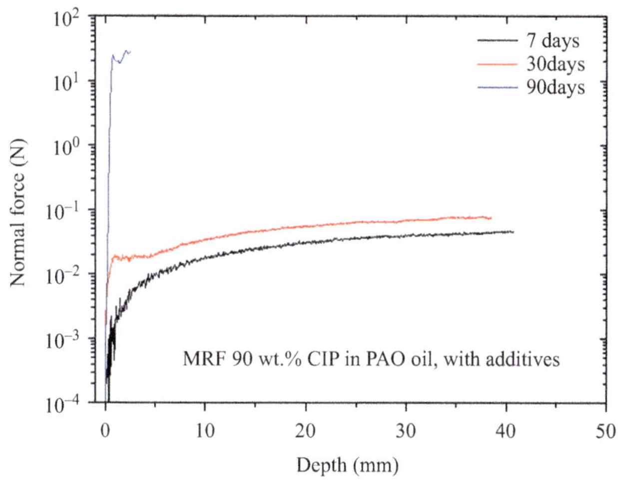

Figure 1.4 Redispersibility curves under normal gravity

For instance, let us consider the hypothetical examples: an MRF "A" which presents maximum normal force of 0.5 N, a depth of 50 mm and penetration work of 25 mJ and then a second MRF "B" with the same working value of 25 mJ that reaches 25 N of normal force with only 1 mm of depth.

It would be possible to think that the two formulations of MRF "A" and "B" are good considering only the parameter "work" since both presented the value of 25 mJ. However, during the penetration test, MRF "B" had a depth of 1 mm and a normal force of 25 N indicating that there is a formation of hard cake, and therefore, it is a bad MRF formulation. Different from MRF "B," MRF "A" reached a depth of 50 mm reaching only 0.5 N of the normal force that would indicate a good MRF formulation.

As an example of the description above, we may consider the example of Figure 1.4 with normal force × depth in log scale. In sequence, Table 1.3 shows the same results present in Figure 1.4.

The redispersibility curve after 7 days shows that the formulation of this MRF appears to be excellent since the normal force did not exceed 0.05 N at a depth of 40 mm. The same can be said after a month of sedimentation with the force below

Table 1.3 Redispersibility results from Figure 1.4 after sedimentation

Time (days)	Max normal force (N)	Depth (mm)	Work (mJ)
7	0.05	40.7	1.15
30	0.075	38.6	1.94
90	29.5	2.68	48.8

0.10 N at practically the same depth. After 3 months of rest, the test indicated a force of 30 N at only 2.7 mm depth, clearly demonstrating the formation of a hard cake. Therefore, the MRF in question does not have good redispersibility, although it may appear in the first 30 days.

In this case, the work and the normal force peak agree, demonstrating that at times before 30 days the formulation seems to be acceptable, but after 90 days, it changed and showed hard cake sediment.

1.2.2 Effect of additives in MRF formulation

To make it clear that raw data of normal force versus penetration depth in the redispersibility test are more important than just the mechanical working value, the example in Figure 1.5 illustrates three MRF formulations that have all been assayed after 1 month of sedimentation. The iron powder and the concentration were the same in the three formulations. The only difference between the three MRFs was the dispersant additive employed in each. The test was programmed (via event control in the software) to abort the test if the force reached 30 N. Table 1.4 summarizes the values of normal force, depth and work for each MRF tested.

If one takes into account only the work spent on the penetration of the blade, Additive #1 seems to be the best since this formulation has a lower working value (78 mJ). However, since in the experiment the normal force was limited to maximum 30 N, all formulations have a peak force value of ~30 N. In this case, what is more relevant is the distance that the blade has been able to penetrate. The smaller the distance—independent of the area under the curve (i.e. the mechanical work)—the more compact, harder and also more difficult it is to redisperse the sediment formed in the sample composition.

Therefore, in the case of the samples of Figure 1.5 and Table 1.4, Additive #1 was the worst of the three, Additive #3 was regular and Additive #2 was the best since it presented the longest distance of penetration which also indicates a less compacted sediment.

This shows that the value of the parameter "work" is relative and must be used with caution, and always in conjunction with the force × depth curves. In this case, none of the three formulations showed a good result after 1 month. But it is a case that illustrates just how mechanical work—alone—should not be taken as a criterion for evaluating the redispersibility of an well-formulated MRF, at least not in the penetration test. Thus, the parameter "work" is only relevant when comparing the same depth of sediment.

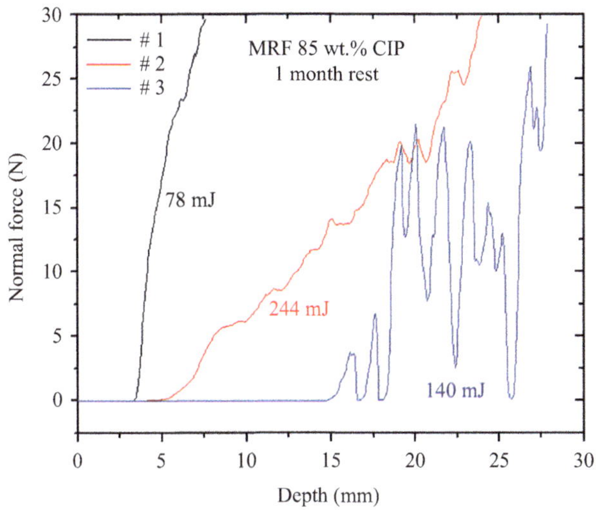

Figure 1.5 Redispersibility curves for MRF in PAO oil, with three different additives

Table 1.4 Redispersibility results from Figure 1.5 after 30 days with different additives

MRF with additive	Normal force peak (N)	Maximum depth * (mm)	Work (mJ)
#1	29.67	4.27	78
#2	29.9	20.0	244
#3	27.9	13.56	140

*The depth here considered as the distance traveled by the blade between the first layer of iron powder sediment (mud line) and the 30 N limit reach.

To demonstrate three completely different additives (K, Y, M) from those first used in the preparation of new MRF formulations, see Figure 1.6. After 1 year of rest and sedimentation, the fluids were analyzed in the redispersibility test. All were subjected to the same penetration distance while keeping the *event control* when the depth reached 1 mm before the bottom of the container tube. The force applied was limited by 50 N, which did not occur for any of the three formulations.

In Figure 1.6, we can see that Additive “K” is the best of the three additives used with an excellent redispersibility. Considering that after 1 year the work did not get to 0.05 mJ and the peak force 0.018 N, Additive “Y” is acceptable with a work of ~3 mJ and a peak force of 1.93 N. However, Additive “W” is the worst with almost 24 mJ of work and a peak force of 18 N after the same amount of time.

Figures 1.7 and 1.8 show the redispersibility test curves for MRF formulated with 85 wt.% Carbonyl iron powder (CIP), but with two different dispersing additives D-1 and D-2, respectively.

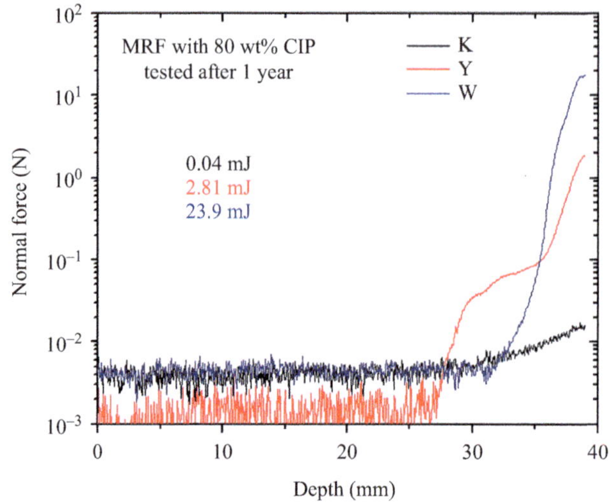

Figure 1.6 Redispersibility curves for MRF in PAO oil, with three different additives, after 1 year

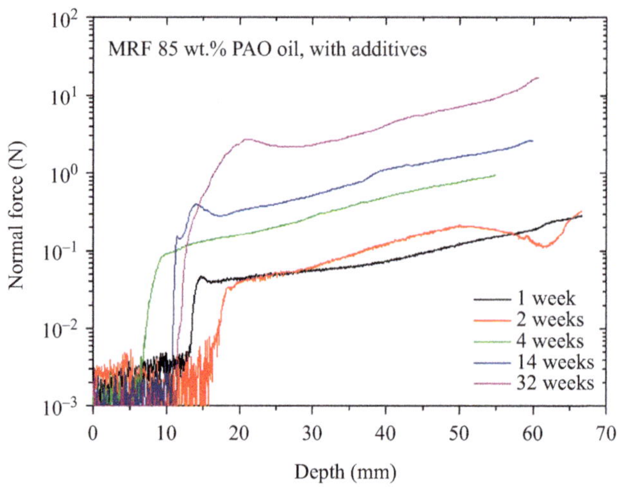

Figure 1.7 Redispersibility curves with additives dispersing "D-1" and thixotropic "T"

As can be seen from Figure 1.7 and Table 1.5, it is not so difficult to formulate an MRF whose redispersibility, as evaluated in short time intervals, is good and acceptable. After 3 weeks or longer, it is not easy to get a low value of normal force with its associated work. Table 1.6 summarizes the work, spent while the blade penetrates the sediment at different times of settling.

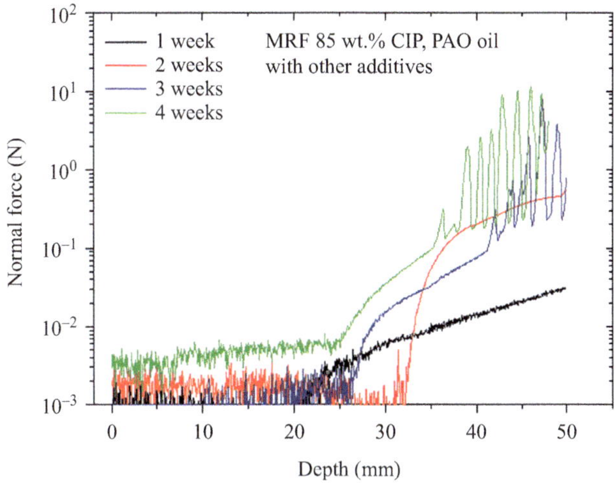

Figure 1.8 Redispersibility curves with additives dispersing "D-2" and thixotropic "T"

Table 1.5 Redispersibility results from Figure 1.7 with additive dispersing D-1

Weeks	Work (mJ)
1	5.5
2	6.2
4	18
14	48
32	232

Table 1.6 Redispersibility results from Figure 1.8 with additive dispersing D-2

Weeks	Work (mJ)
1	0.4
2	4.3
3	9.1
4	24.6

On the other hand, employing the same CIP and poly(alpha-olefine) oil, at same volume fraction, but with different dispersing additive "D-2," the redispersibility behavior can change completely according to Figure 1.8.

Again, if only the redispersibility work values are compared, one might think that Additive D-2 is good because the total work after 4 weeks of resting is relatively low. However, from Figure 1.8, the pattern of peaks and valleys, which can occur with some formulations, is still poor after only 3 weeks, reaching 10 N and more which is a bad result. This is another example in which one must be careful to not evaluate only the values of work, but also the peak of normal force measured in the redispersibility test.

1.2.3 Effect of centrifuging the MRF

Figures 1.9 and 1.10 show, respectively, normal force × depth and work × depth, for a commercial MRF sample, after 1 week and 18 weeks, tested under normal

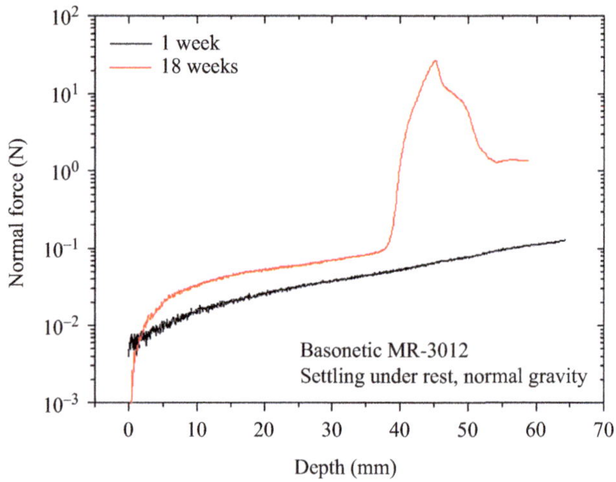

Figure 1.9 Redispersibility curves of a commercial MRF

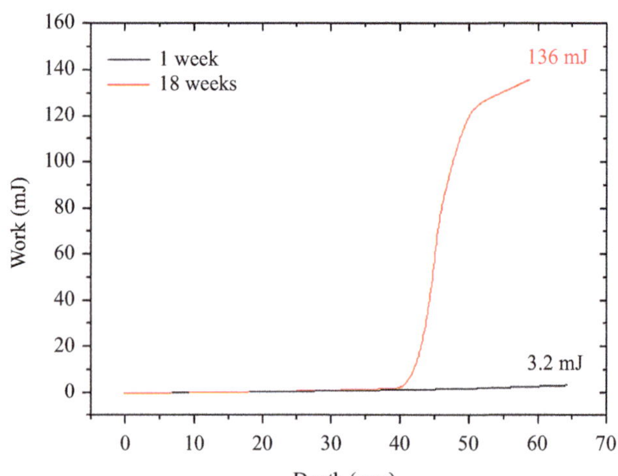

Figure 1.10 Work by integration, of the area under curves of Figure 1.9

gravity. The same MRF, but after centrifuged under 2,000*g* by 15 min, was also tested, and the result is presented in Figure 1.11.

We can see at Figure 1.11 that at the first 30–35 mm of depth, the blade was moving downward in the air, or just starting to penetrate the liquid supernatant. Note the rapid increase of the registered normal force at depth > 40 mm.

Figure 1.12 shows the effect of centrifuging on three different MRFs: one commercial (BASF MR-3012) and two of our MRF formulations. One was prepared with 90 wt.% one size CIP. The other sample is a blend of two sizes.

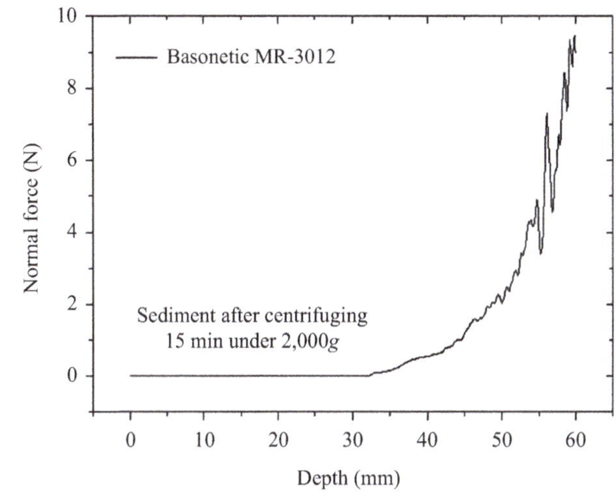

Figure 1.11 Redispersibility of commercial MRF after centrifugation

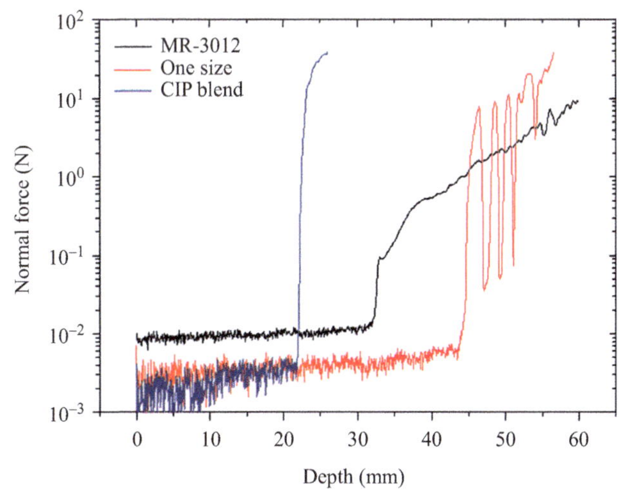

Figure 1.12 Redispersibility of different MRF after centrifugation: 2,000g by 15 min

In Figure 1.12, the region with "noise" corresponds to the region where the blade attached to the rheometer is moving downward, but in the air or inside the upper liquid supernatant of MRF sample. Note that for the MRF with blend sizes, the blade penetrated only 4.3 mm and reached 40 N of normal force very quickly, an indication that its sediment is very hard to penetrate, thus making it a bad result. The sediment for the MRF prepared with only one size also reached 40 N, but the blade penetrated ~13 mm, and with a succession of increase and decrease of the normal force. This suggests that the sediment is formed in layers. The best result, in this case, was obtained with the commercial MRF sample. In general, mixing sizes of iron powders can improve the redispersibility [20], but not always, as Figure 1.12 illustrates.

1.2.4 Long-time settling (1-year) redispersibility

A well-formulated MRF, like that one shown in Figure 1.13, resulted in the redispersibility test with a normal force lower than 0.1 N and work around 1 mJ, for ~20 mm blade penetration into iron sediment after 1 month of settling under rest. Even after more than 1 year of settling under rest, the normal force reached less than 0.13 N and work of ~1.8 mJ.

A priori it can be apparent to get a good result in the redispersibility test when one is testing several different dispersing additives for iron powder and seeking to formulate a good MRF. It is simple, considering only short (7 days) times of settling under rest. This is shown in Figure 1.14, for MRFs all formulated with 85 wt.% carbonyl iron powder but with different additives after 1 week of sedimentation.

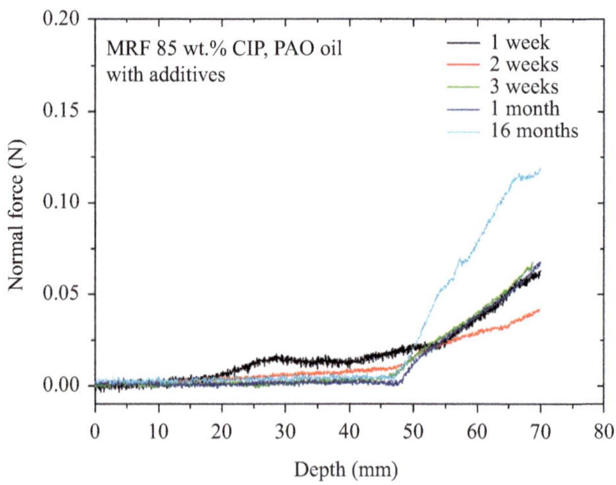

Figure 1.13 A well-formulated MRF, after 1 month of sedimentation

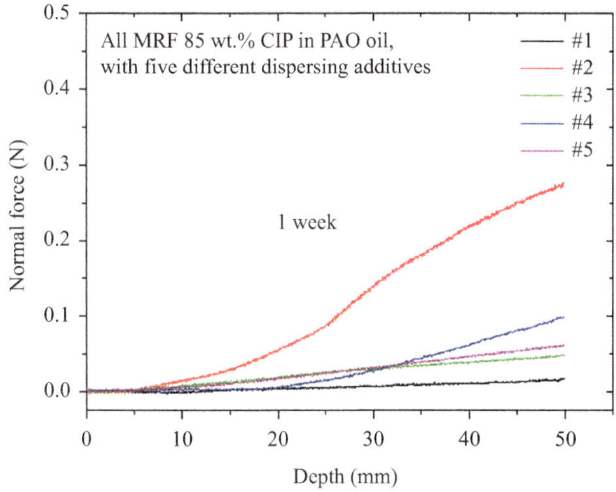

Figure 1.14 Redispersibility test for five formulations of MRF with different dispersing additives

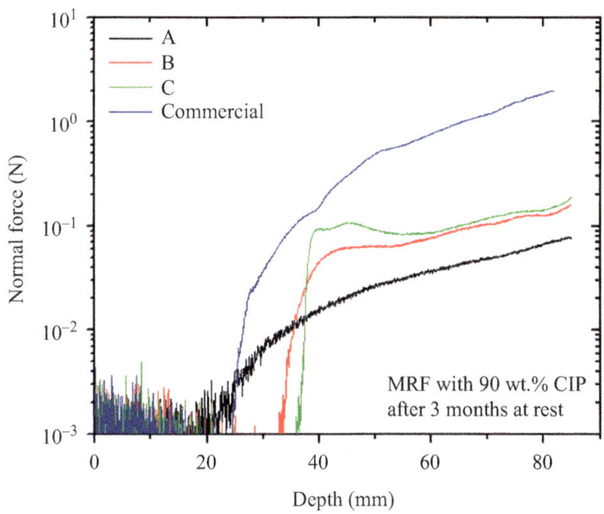

Figure 1.15 Redispersibility curves of MRF, with different dispersing additives. Blue line is a commercial MRF

1.2.5 Redispersibility of MRF with 48 vol.%

Figures 1.15–1.17 show the redispersibility test results for MRF with high-volume concentration ~48 vol.% (or 90 wt.%) of iron powder, after 3 months under rest. The relevance of the additives employed is evident.

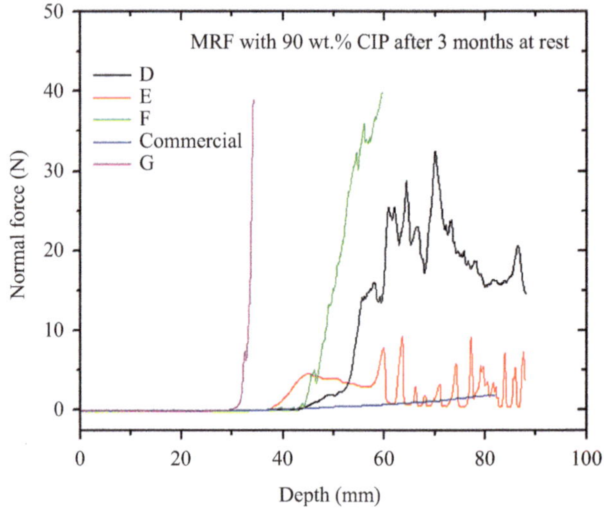

Figure 1.16 Redispersibility curves with different dispersing additives. Blue line is a commercial

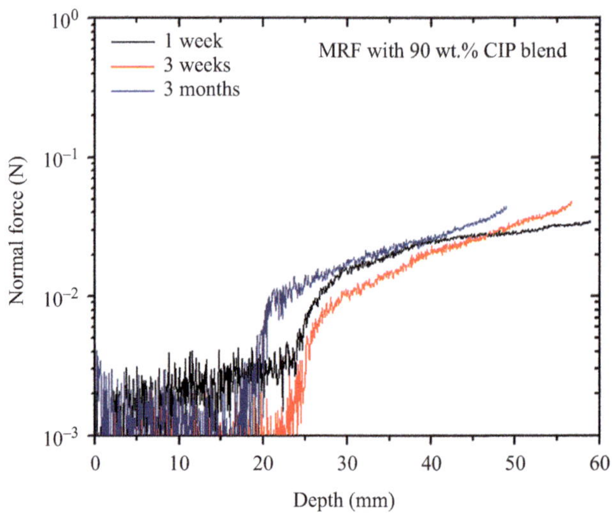

Figure 1.17 Redispersibility curves of one MRF formulation with ~48 vol.% CIP size blend

1.3 Conclusions

Redispersibility in MRFs is determined to be "good" when after a long period of time the MRF does not require much work to disperse while maintaining a penetration depth equal or close to the initial one.

It is also possible to analyze different redispersibility using different dispersing additives where each one will act and respond in a different way in obtaining the curves of normal force versus depth and consequently in the obtained values of work.

We should be careful when initially analyzing an MRF as soon as it has been prepared because it can be very good and, after a short period of time, it starts to show "hard cake" as a "good" fluid.

In other words, the redispersibility of an MRF, in particular, cannot be analyzed and concluded to be "good" with a single analysis of response work, but rather as a function of the depth and normal force curves after long periods of rest time.

Acknowledgments

Mrs. Sara R.G. de Sousa acknowledges CAPES by her PhD student fellowship. AJFB acknowledges FULBRIGHT (US Dept. of State, Bureau of Educational and Cultural Affairs) financial support during his 3-month research at the University of Wisconsin–Madison. We also thank FAPEMIG Grant APQ-01824-17, and CNPq.

References

[1] Israelachvili, Jacob N. *Intermolecular and Surface Forces*, 3rd edition; Amsterdam, The Netherlands: Academic Press; 2011; *Ch. 13.*

[2] Phulé, Pradeep P.; Mihalcin, Matthew P.; Genç, Seval. "The role of the dispersed-phase remnant magnetization on the redispersibility of magnetorheological fluids." *J. Mater. Res.* 1999; 14(7), 3037–3041.

[3] Weiss, Keith D.; Nixon, Donald A.; Carlson, J. David; Margida, Anthony J. "Thixotropic magnetorheological materials." US Patent 5,645,752. July 8th, 1997.

[4] Portillo, M. Aguillera; Iglesias, G.R. "Magnetic nanoparticles as a redispersing additive in magnetorheological fluid." *Journal of Nanomaterials.* 2017; 1–8. Article ID 9026219, DOI: 10.1155/2017/9026219

[5] Kieburg, C.; Oetter, G.; Lochtman, R., *et al.* "High performance magnetorheological fluids tailored for a 700 Nm automotive 4-wheel-drive clutch." In: *Proceedings of the 10th Intl. Conf. on Electrorheological Fluids and Magnetorheological Suspensions—ERMR'2006*, Lake Tahoe, USA; Edited by F. Gordaninejad *et al.*; World Sci.; Singapore; ISBN: 978-981-277-119-3; page 105.

[6] ASTM D869 - 85 (Reapproved 2015); "Standard test method for evaluating degree of settling of paint." DOI: 10.1520/D0869-85R15.

[7] Iglesias, G.R.; López-López, M.T.; Durán, J.D.G.; González-Caballero, F.; Delgado, A.V. "Dynamic characterization of extremely bidisperse magnetorheological fluids." *J. Colloid and Interface Sci.* 2012; 377, 153–159.

[8] Lopez-Lopez, M.T.; de Vicente, J.; Bossis, G.; Gonzalez-Caballero, F.; Duran, J.D.G. "Preparation of stable magnetorheological fluids based on extremely bimodal iron-magnetite suspensions." *Journal of Materials Research.* 2005; 20(4): 874–881.

[9] Lopez-Lopez, M.T.; Zugaldia, A.; Gonzalez-Caballero, F.; Duran, J.D.G. "Sedimentation and redispersion phenomena in iron-based magnetorheological fluids." *Journal of Rheology*. 2006; 50(4), 543–560.

[10] Phule, P.P.; Ginder, J.M. "Synthesis and properties of novel magnetorheological fluids having improved stability and redispersibility." *International Journal of Modern Physics B*. 1999; 13(14–16), 2019–2027.

[11] Ngatu, G.T.; Wereley, N.M.; Karli, J.O.; Bell, R.C. "Dimorphic magnetorheological fluids: exploiting partial substitution of microspheres by nanowires." *Smart Materials & Structures*. 2008; 17(4), 1–8.

[12] Jun, J.B.; Uhm, S.Y.; Ryu, J.H.; Suh, K.D. "Synthesis and characterization of monodisperse magnetic composite particles for magnetorheological fluid materials." *Colloids and Surface A - Physicochemical and Engineering Aspects*. 2005; 260(1–3), 157–164.

[13] Shah, K.; Oh, J.S.; Choi, S.B.; Upadhyay, R.B. "Plate-like iron particles based bidisperse magnetorheological fluid." *Journal of Applied Physics*. 2013; 114(21).

[14] Machovsky, M., *et al.* "The enhanced magnetorheological performance of carbonyl iron suspensions using magnetic Fe3O4/ZHS hybrid composite sheets." *RSC Advances*. 2015; 5(24), 19213–19219.

[15] Shah, K.; Phu, D.X.; Choi, S.B. "Rheological properties of bi-dispersed magnetorheological fluids based on plate-like iron particles with application to a small-sized damper." *Journal of Applied Physics*. 2014; 115(20), 1–10.

[16] Cvek, Martin; Mrlik, Miroslav; Moucka, Robert; Sedlacik, Michal. "A systematical study of the overall influence of carbon allotrope additives on performance, stability and redispersibility of magnetorheological fluids." *Colloids and Surfaces A - Physicochemical and Engineering Aspects*. 2018; 543, 83–92.

[17] Lijesh, K.P.; Muzakkir, S.M.; Hirani, Harish. "Rheological measurement of redispersibility and settling to analyze the effect of surfactants on MR particles." *Tribology - Materials Surfaces & Interfaces*. 2016; 10(1), 53–62.

[18] Nin, Fang-Hao; Hu, Zhi-De; Yan, Hua; Yang, Jian-Jian; Zhang, Han-Song. "Rheological response of magnetorheological suspensions sediments and its implications for redispersibility." *Journal of Magnetism and Magnetic Materials*. 2018; 465, 421–429.

[19] Niu, Fang-Hao; Hu, Zhi-De; Yan, Hua; Zhang, Han-Song; Yang, Jian-Jian. "Sedimentation and redispersibility of magnetorheological suspensions while settling under gravity." *Smart Materials and Structures*. 2018; 27(9), 1–12.

[20] Bombard, Antonio J. F.; Teodoro, Joáo V.R. "Magnetorheological fluids with carbonyl and water atomized iron powders." *International Journal of Modern Physics B*. 2011; 25(7): 943–946.

Chapter 2

DEM and FEM simulations in magnetorheology: aggregation kinetics and yield stress

Jose R. Morillas[1,2] and Juan de Vicente[1,2]

Magnetorheological (MR) fluids are magnetic field-responsive two-phase systems. Typically, they are constituted by magnetizable particles suspended in a (non-magnetic) liquid carrier. The most striking characteristic of an MR fluid is the dramatic change in its rheological properties when subjected to an external magnetic field. In particular, MR fluids exhibit a yield stress for sufficiently large field strengths and concentrations. The yield stress is the minimum stress that needs to be overcome for the MR fluid to flow. To be of interest in real applications, the number of particles in dispersion is typically large (concentrations above 10 vol.%), and this complicates the modeling of these kinds of materials [1–3].

Several approaches have been followed in the past to model MR fluids. Some of them follow a continuum medium approach [4–6] while others focus on a microstructural particle-level approach [7–11]. Undoubtedly, particle-level simulations based on the discrete element method (DEM) and finite element method (FEM) have been the most frequently used in the specialized literature.

2.1 Background

2.1.1 Discrete element method

DEMs were originally developed by Cundall and Strack to model granular systems where contact forces are predominant [12]. In these simulations, the motion of individual particles is followed by solving Newton's second law once all interparticle (contact) forces are known [13,14]. In DEM simulations, a MR fluid under the presence of an applied external magnetic field is simulated as an ensemble of N spherical particles interacting with each other, as well as with the simulation box boundaries and the background carrier fluid (see Figure 2.1). Interparticle interactions consist of at least magnetostatic and repulsive forces.

[1]Department of Applied Physics, Faculty of Sciences, University of Granada, Spain
[2]Excellence Research Unit "Modeling Nature" (MNat), University of Granada, Spain

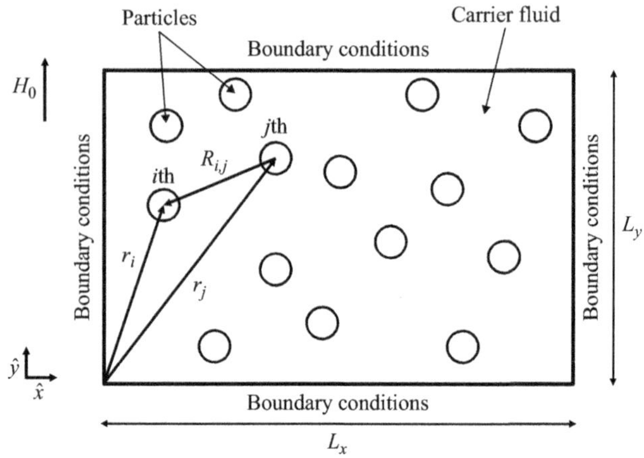

Figure 2.1 Two-dimensional simulation box traditionally used in DEM. An ensemble of N particles is randomly prepositioned inside the computational box. The particles interact with their neighbors, carrier fluid and confining walls through magnetostatic, repulsive and Stokes' drag forces

Traditionally, these interactions are supposed to be known analytically. In this sense, the field-induced magnetostatic interaction is approximated as a long-ranged dipolar one (i.e. neglecting higher-order multipoles). Repulsive forces are incorporated in the model to avoid particle–particle and particle–wall interpenetration. They are frequently modeled as exponential functions (despite the fact that there is not a physical reason for this). Finally, the interaction between the particles and the carrier fluid is accounted for using the Stokes' drag law. In most cases, this is done in only "one way" (i.e. fluid motion is not coupled with particles) [7,15–22].

Knowing, a priori, the total force acting on the particles facilitates the integration of Newton's second law. Moreover, in the particular case of MR fluids the inertial term can be neglected because of its small value in comparison to the rest of interactions [7,23]. As an output of the simulations, the dynamics of the whole ensemble (particle position, velocity and acceleration) is recorded in order to post-analyze it computing (averaged over all particles) the stress, pair correlation function, mean cluster size, etc.

Previous assumptions are strong and difficult to fulfill in magnetorheology. The reason for this is that, in experiments, the particle volume fraction is typically large; this invalidates the use of the dipolar and Stokes' drag approximations that are strictly valid for isolated spheres in homogeneous magnetic fields and fluid flows, respectively. However, such a simple model provides first-order useful information because a large number of particles N can be simulated and statistical averages can be performed [24]. Second, and more important, is the fact that these simplified simulations are capable to reproduce the phenomena observed in MR fluids under static (at rest) and dynamic (in presence of fluid flow) experiments.

Available DEM simulations in the absence of flow (i.e. at rest) typically focus in the understanding of the short-time aggregation process in dilute MR fluids upon the application of a magnetic field. Investigations reported in the literature concern the calculation of aggregation times and structure formation kinetics under different conditions such as the particular kind of repulsive forces [7] or the confinement level in the field direction [19,22,25,26].

DEM simulations for MR fluids in the presence of flow are extensive. They cover canonical flows: squeeze flow [27,28], pressure-driven flow [29,30] and simple shear flow [18,31–34]. Undoubtedly, steady shear flow has received the widest attention [17]. Recently, a very good agreement has been reported between DEM simulations in steady shear flow and both experiments [10] and the structural viscosity model [35] for a wide range of particle concentrations. A master curve has also been proposed in terms of a reduced Mason number Mn (defined as the ratio between hydrodynamic Stokes' drag force and magnetostatic dipolar force) [36].

The good agreement between in-flow DEM simulations and experiments cannot be extended however to the preyield regime (i.e. Mn $\rightarrow$ 0), that is, when hydrodynamic forces are negligible in comparison to magnetostatic ones. In this regime, depending on the sample volume fraction ϕ, one can find either isolated single-particle chains (at very small volume fractions), column-like structures (at intermediate volume fractions) and tightly packed gap-spanning aggregates (at high concentrations).

In the preyield regime, relevant forces are traditionally reduced to magnetostatic and repulsive, being both typically very large as the particles are in close contact-forming aggregates. In these conditions, the simulation time step, which is controlled by the maximum force at the step, must be reduced and simulations become computationally very expensive [17]. In addition, for small interparticle distances, magnetostatic forces are also strongly influenced by multipolar terms and the dipolar approximation becomes inappropriate [37]. Under this scenario, other kinds of computational methods are needed to model MR fluids [38,39].

2.1.2 Finite element method

FEM magnetostatic simulations are preferred to model MR fluids in quasi-static conditions [38,40,41]. In FEM simulations, the exact magnetic field distribution in a bounded portion (computational domain) inside the MR fluid is calculated by directly solving Maxwell's equations and taking into consideration the constitutive equation of both particles and fluid.

In FEM simulations, the microstructure is prescribed at every strain and the field distribution calculated. With this, magnetic interparticle forces are computed via integration of Maxwell's stress tensor. Finally, the yield stress is identified with the maximum in the force (or stress) as a function of the strain.

Inherent to the method are the imposition of some boundary conditions and the construction of an appropriate mesh. The application of boundary conditions is equivalent to fixing the solution at the boundary. Ideally, the boundary conditions should reflect the influence that the surrounding media has over the solution in the simulated computational domain. The construction of a mesh constitutes a critical step where the nodes, over which the solution is going to be computed, are defined. The density of nodes must be as high as possible in those regions of the computational domain where magnetic field gradients are expected to be large. Ideally, the finer the mesh, the better the solution. However, this is at the expense of a higher computational cost. In addition, depending on the particular problem to be solved, associated with each node there can be more than one unknown quantities (i.e. degrees of freedom). This increases the computational effort as the solver has to run over all nodes as many times as needed to satisfy the required accuracy for each degree of freedom.

In the case of MR fluids, magnetic field gradients are mainly located in the space existing between the particles (i.e. the gaps) due to the focusing effect created by their proximity. Consequently, each gap needs to be densely meshed. However, this constitutes an unaffordable task taking into consideration the large number of particles and the computational power of current computers.

To circumvent this issue, two main strategies have been followed in the past. On the one hand, some works study the interaction between (only) two isolated particles [42–45]. The computational domain in these works usually contains one or two particles (see Figure 2.2(a)), and boundary conditions are imposed far away from them (excepting mirror symmetries if they apply) to properly simulate an isolated doublet. The motivation behind these works is to ascertain how interparticle magnetic forces change with particle shape, surface roughness or interparticle distance. Of course, these results cannot be extrapolated to high concentrated MR fluids.

On the other hand, a second group of works approximates real structures in the MR fluids by periodic lattices or infinite chains [38,40,41,46,47]. In this case, the computational domain includes one particle or a sector of a particle (depending on the symmetries that are present in the simulated structure) (see Figure 2.2(b) and (c)). Although, a priori, these simulations are appropriate for highly concentrated MR fluids, difficulties appear when imposing the boundary conditions (see discussion in the following). Nevertheless, results obtained using this kind of simulations show trends that are in good agreement with experiments.

2.1.3 *More complex approaches*

DEM technique has been extended by including magnetic multipolar terms for a doublet [48,49], inelastic collisions to see a yield stress [50] and magnetic interactions via a fast multipole method (FMM) [23]; allowing particles with non-spherical shapes [11,51,52] or nonmagnetic behavior [53], polydispersity [54–56], Brownian motion (i.e. Brownian dynamics [28,57]), body forces [58]; introducing a non-Newtonian carrier fluid [59] solving rotational, in addition to translational, degrees of freedom [27,60]; and giving a physical meaning to the interaction between particles and walls via contact forces [61] and hydrodynamic ones [62].

Especial effort has also been made to couple particle and fluid motion giving place to different combinations of DEM with other methods. Probably, the simplest and most computationally efficient two-way scheme is based on interchanging

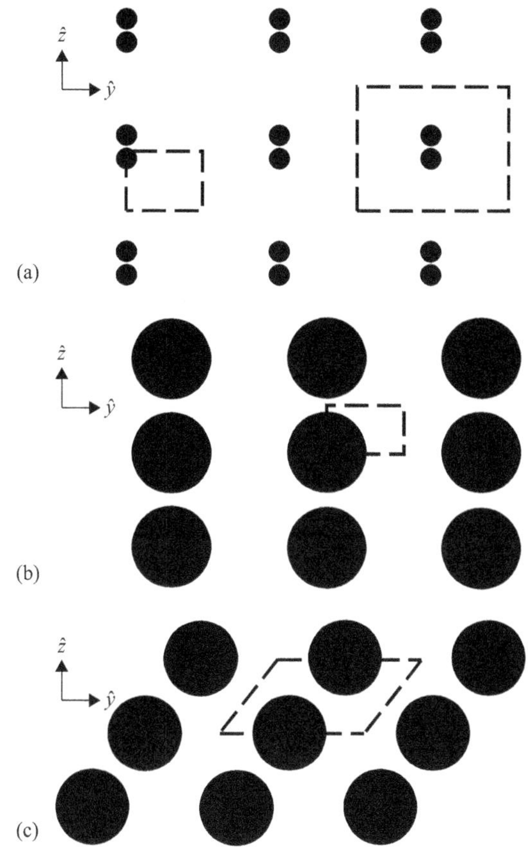

Figure 2.2 Most representative particle lattices simulated with FEM. Dashed lines correspond to the computational domains. (a) Axisymmetric cell including one half of the particle used in [42,43] and two-dimensional (2D) cell including two disks used in [44,45]. (b) Axisymmetric cell including one quarter of particle used in [41,46]. A similar cell including 1/8 of particle is used in [38,40] for three-dimensional (3D) simulations. (c) Lattice and unit cell used in [47] for 3D simulations

fluid–particle force. Once the drag force is computed on each particle (usually from Stokes' law), these forces are included (but with opposite sign to fulfill Newton's third law) as body ones in the governing equation of the fluid flow. These governing equations depend on the way to model the fluid phase; for example, if it is regarded from a micro-/mesoscopic point of view, smooth particle hydrodynamics (DEM-SPH) [63] or lattice Boltzmann methods (DEM-LBM) [30] are commonly used.

Previously described techniques allow both phases coupling in a "coarse" way as fluid flow is not fully resolved at the particle scale. To overcome this issue, smaller "fluid particles" can be used in SPH or finer lattices in LBM (small enough in comparison to the particle size solved in DEM) [64]. In this case, no coupling force between the particles and the fluid is needed because the flow around each particle is solved and thus the force on a particle can be calculated directly from the fluid pressure and stress on its surface.

In another macroscopic approach, the carrier fluid phase is simulated by solving Stokes' equation. In this case, the particles are taken into consideration including again a force coupling term [25] or solving Stokes' equations around each particle via Stokesian dynamics [65], its simplified versions (neglecting hydrodynamics short- [62,66] or long-range [67,68] interactions) or Computational Fluid Dynamics (CFD) [69]. These methods are not restrictive on the particles' interactions (especially in [65] or [69], where multipole and multibody magnetic interactions are also accounted for). Unfortunately, these methods are very expensive from a computational point of view and are usually limited to 2D and a very small number of particles.

In the present work, we will only focus our attention on the application of DEM to solve particle aggregation/structuration in dilute MR fluids and the use of FEM to ascertain the yield stress in concentrated MR fluids.

2.2 Methodology: numerical methods

2.2.1 Discrete element method

The most basic DEM simulations are based on supposing dipolar magnetic interaction, "hard spheres" repulsion and Stokes' drag. As it was mentioned in the previous section, these simulations are strictly valid in the dilute regime, where each particle can be seen as an isolated one. In this section, we will show how this simple model can be used to study the aggregation kinetics of a dilute 2D system of monodisperse magnetic particles at rest under the presence of a homogeneous magnetic field $\boldsymbol{H}_0$.

Let us consider an ensemble of N monodisperse and buoyant spherical particles (see Figure 2.1) subjected to drag (*drag*), magnetostatic (*mag*) and repulsive (*rep*) forces. Hence, the motion of the ith particle in the suspension is given by its Newton's second law:

$$m\ddot{\boldsymbol{r}}_i = \boldsymbol{F}_i^{drag} + \sum_{j \neq i}^{N} \boldsymbol{F}_{i,j}^{mag} + \sum_{j \neq i}^{N} \boldsymbol{F}_{i,j}^{rep} \tag{2.1}$$

where m is the particle mass and $\boldsymbol{r}_i$ is its position vector.

The hydrodynamic force is modeled using Stokes' law for one isolated particle:

$$\boldsymbol{F}_i^{drag} = -\frac{18\eta}{\rho d^2} m\dot{\boldsymbol{r}}_i \tag{2.2}$$

Here η is the fluid shear viscosity, ρ is the particle density and d is the particle diameter.

The magnetic interparticle force is implemented in dipolar approximation neglecting local field effects. Note that in this section, we are solely interested in dilute suspensions. In 2D, supposing that the dipoles are oriented in the $\hat{y}$ direction, the magnetostatic interparticle force between spheres i and j reads as follows:

$$\boldsymbol{F}_{i,j}^{mag} = F_0 \left[\frac{\boldsymbol{R}_{i,j}}{R_{i,j}^5} + \frac{2(\boldsymbol{R}_{i,j} \cdot \hat{\boldsymbol{y}})}{R_{i,j}^5} \hat{\boldsymbol{y}} - 5\boldsymbol{R}_{i,j} \frac{(\boldsymbol{R}_{i,j} \cdot \hat{\boldsymbol{y}})^2}{R_{i,j}^7} \right] \tag{2.3}$$

where $\boldsymbol{R}_{i,j} = \boldsymbol{r}_i - \boldsymbol{r}_j$ and $F_0 = 3\pi\mu_f\mu_0(\beta H_0)^2 d^6/16$ is the magnetic force scale with μ_0 the vacuum permeability, μ_f the fluid relative permeability, $\beta = (\mu_p - \mu_f)/(\mu_p + 2\mu_f)$ the contrast factor and μ_p the particles relative permeability.

Finally, the particles are not allowed to interpenetrate. With this in mind, short-range repulsive interparticle forces are considered. Repulsive forces are modeled here using an exponential form:

$$\boldsymbol{F}_{i,j}^{rep} = \frac{F_0}{d^4} \exp\left[-100 \frac{(R_{i,j} - d)}{d} \right] \frac{\boldsymbol{R}_{i,j}}{R_{i,j}} \tag{2.4}$$

In our simulations, initially, the particles are randomly distributed within a box. No-slip boundary conditions are applied on the walls of the simulation cell (i.e. if a particle touches a boundary, it stays at this place during the rest of the simulation). Particle motion is tracked by solving Newton's second law using a second-order implicit method (generalized-α method with parameter $\alpha = 0.75$) implemented in the FEM package Comsol Multiphysics. Note that the acceleration of the particle is solved in contrast to most of the previous works reported in the literature. During the iteration, the time step is changed accordingly to the maximum force experienced by the particles. Typical values of this time step are around 10 ns.

2.2.2 Finite element method

FEM simulations are carried out in quasi-static regime (no time dependence). As a result, the steady form of Maxwell's equations is used. In addition, in the absence of free currents the Ampère's law is reduced to

$$\boldsymbol{H} = -\nabla V \tag{2.5}$$

where V is the scalar potential. For the problem to be solved, the previous fundamental equation is coupled with Gauss law through the constitutive equation for each subdomain (magnetizable particle and nonmagnetizable fluid):

$$\nabla \cdot \boldsymbol{B} = \mu_0 \nabla \cdot (\mu_r \boldsymbol{H}) = 0 \tag{2.6}$$

Here μ_r is the relative permeability of each subdomain. On the one hand, the carrier fluid is nonmagnetic and therefore $\mu_r = \mu_f = 1$. On the other hand, the particles are nonlinear magnetic materials with a field-dependent $\mu_r = \mu_p > 1$. In this work, we introduced the experimental magnetization curve of the particles directly in the model through the permeability curve $\mu_p = \mu_p(H)$. Note that, due to the absence of free currents and time dependence, the only unknown quantity in the model is a scalar function. Thus, there is only one degree of freedom per node and the computational cost is significantly reduced.

Among the different techniques that have been proposed in the literature to solve (2.5) and (2.6) using FEM, here we will follow the approach by Morillas *et al.* [47] because it can appropriately simulate high volume fractions and true strains thanks to the use of a reduced field formulation and periodic boundary conditions.

A reduced field formulation allows splitting the (total) magnetic field $\boldsymbol{H}$ in two contributions: one contribution corresponds to the externally applied magnetic field $\boldsymbol{H}_e$ and another contribution comes from the presence of the magnetic particles $\boldsymbol{H}_p$. As a result, (2.5) can be substituted by

$$\boldsymbol{H} = \boldsymbol{H}_e + \boldsymbol{H}_p = \boldsymbol{H}_e - \nabla V_p = -\nabla V \tag{2.7}$$

Here V_p is the scalar potential due to the particles. Consequently, only V_p needs to be solved and, more importantly, $\boldsymbol{H}_e$ is not introduced in the model by constraining its magnitude or direction in any boundary of the computational domain.

In order to simulate high volume fractions without a huge computational cost, the particles are supposed to be arranged in a periodic lattice. As a consequence, both the magnetic field and the lattice should share the same periodicity. In this way, computing the magnetic field inside the whole lattice is equivalent to compute it only in the lattice unit cell by imposing periodic boundary conditions. With respect to the particular kind of lattice, taking into consideration the anisotropy induced by the external field in the structure, lattices with a preferred direction (e.g. tetragonal or hexagonal ones) seem to better mimic the actual arrangement in a magnetized MR fluid in the non-strained state.

When a macroscopic deformation is applied to the magnetized MR fluid, the field-induced aggregates change their shape/orientation and therefore particles' positions change as well. In the particular case of a shearing deformation, a common approximation to particles' movement is the so-called affine deformation. In an affine deformation, each particle is displaced in the direction of the shear a distance that is proportional to its vertical position. Despite being a strong simplification, this approximation is of special interest because some of the lattice symmetries are preserved (e.g. tetragonal/hexagonal lattices become monoclinic/triclinic ones). As a result, the new particle arrangement under strain can be also simulated only by changing the computational domain (using the corresponding unit cell) and keeping periodic boundary conditions. Moreover, using the methodology by Morillas *et al.* [47], a wider casuistic can be studied if different bases (such as simple, body/base/face-centered) are considered.

It is worth to note here that most of the previous publications in this field impose mirror-like symmetries hence reducing the applicability of the simulations to non-strained structures (tetragonal lattices) where only elongations are properly mimicked [38,40,41,46].

In this work, we focus in the simplest case: a monoclinic lattice with simple basis. Supposing that $\boldsymbol{H}_e = H_e\hat{\boldsymbol{z}}$, lattice vectors are be given by

$$\boldsymbol{a}_1 = w\hat{\boldsymbol{x}}, \quad \boldsymbol{a}_2 = w\hat{\boldsymbol{y}}, \quad \boldsymbol{a}_3 = d(\gamma\hat{\boldsymbol{y}} + \hat{\boldsymbol{z}}) \tag{2.8}$$

where d is the particle diameter, $w = \sqrt{\pi d^2/6\phi}$ is the lateral distance between chains, ϕ is the particle volume fraction and γ is the strain (applied in the $\hat{\boldsymbol{y}}$ direction).

The magnetic field distribution in the computational domain is solved using Comsol Multiphysics for different applied strains and magnetic fields. Then, the shear stress in the lattice is computed using Maxwell's stress tensor. As a result, a shear stress versus shear strain curve is obtained for each field. Finally, the yield stress is identified as the maximum of each curve.

2.3 Results and discussion

2.3.1 Discrete element method

In this section, we describe a particular application of DEM. Concretely, we simulate the aggregation kinetics in the absence of Brownian motion (i.e. the deterministic aggregation process) in 2D diluted MR fluids under a homogeneous magnetic field.

Initially, $N = 216$ particles of diameter $d = 1$ μm are randomly distributed in a box of base $L_x = 90$ μm and height $L_y = 75$ μm (area fraction $\phi_{2D} = 0.02$, see Figure 2.1). Particle properties are chosen to model real iron microparticles ($\rho = 7.8$ g/cm^3, $\beta = 0.9$). The carrier fluid is supposed to be water ($\eta = 1$ mPa·s, $\mu_f = 1$). At $t = 0$, a magnetic field of strength 113 kA/m is applied and oriented in the $\hat{\boldsymbol{y}}$ direction. The magnetic moments of the particles are supposed to be collinear with the field. During the simulated time (0.03 s), the particles' position is tracked and an identification algorithm is employed [70] to detect and perform statistical analysis on the field-induced clusters.

The mean cluster size S is probably the most relevant statistical parameter to characterize the cluster growth [25,71–73]:

$$S(t) = \sum_s s^2 n_s(t) \Big/ \sum_s s n_s(t) \tag{2.9}$$

Here n_s is the number of clusters with s particles. The time evolution of the mean cluster size provides useful information on the aggregation process. Indeed, by using an appropriate time scale t_s, S collapses in a master curve [7] following a power law with time:

$$S(t) = 1 + (t/t_s)^z \tag{2.10}$$

Here z is the dynamic exponent dictating the aggregation rate and t_s is the time taken by two particles to get in contact. Several values have been found for these two parameters. On the one hand, numerical studies [70,74,75] have yielded a value of $z = 0.5$. However, experiments give z values in a range from 0.4 to 0.6 depending on the experimental conditions (sample kinetic stability, sedimentation, interactions with box, etc.) [76].

On the other hand, theoretical and numerical works have shown that t_s depends on two dimensionless numbers: the so called lambda number $\lambda = F_0/3d^3k_BT$ (i.e. the ratio between magnetostatic and Brownian energies) and the area fraction ϕ_{2D} in the following way:

$$t_s = \frac{d^2}{AD\lambda\phi_{2D}^{5/2}} \tag{2.11}$$

being $D = k_BT/3\pi\eta d$ the diffusion coefficient and being A a prefactor dependent on the particular reference: [16]: 55, [77]: 110, [70]: 662 (this last value will be assumed in the following).

In Figure 2.3, the aggregation master curve ((2.10) using the parameters given in [70]: $z = 0.55$, $A = 662$) is plotted with a solid line together with results from DEM simulations (square symbols). As it can be seen, a reasonably good agreement is found. The small deviations could be a consequence of the different boundary conditions used. In contrast to our simulations, Shahrivar *et al.* [70] considered periodic boundary conditions in the vertical walls.

From the inspection of Figure 2.3, the deterministic aggregation can be described as a two-stage process. Initially, cluster formation takes place between isolated particles to yield small chains of a few particles. In a second stage (which

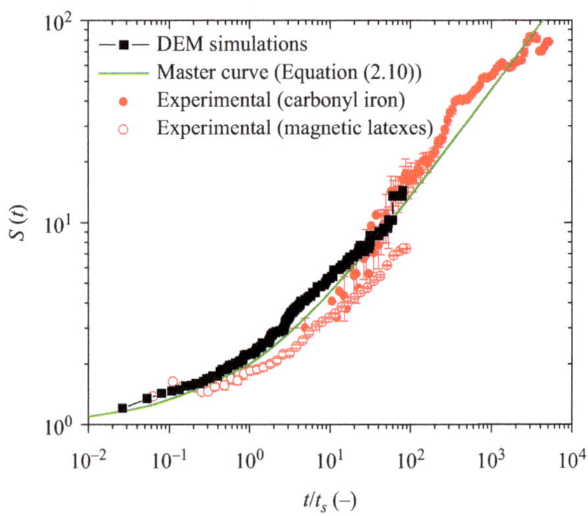

Figure 2.3 Mean cluster size $S(t)$ as a function of the scaled time t/t_s for DEM simulations described in this work (filled squares), master curve from [70] (line), experiments from [70] using carbonyl iron microparticles (filled circles) and magnetic latexes (empty circles)

starts at t_s), chains begin to stick in a tip-to-tip fashion giving place to longer and longer chains aggregates.

As it was pointed out in previous sections, simplified DEM simulations, as they are presented here, suffer from several limitations—most of them related to the proper way to model true existing interactions. On the one hand, hydrodynamic interactions are not expected to play a crucial role in the aggregation process [64], especially in the dilute regime and at short times. On the other hand, multipolar terms could be important because of the close proximity of the particles in the chaining process. In order to elucidate the importance of multipolar interactions, in Figure 2.3, we also show experimental results for magnetic latex suspensions [70]. In magnetic latex suspensions, interactions are weak and only dipolar interactions are important. As observed in Figure 2.3, a good agreement is found between the different sets of data, pointing out that for the experimental applied field (30 kA/m) multipoles are not important hence validating the dipolar approximation employed in the simulations. In summary, multipoles have a deep effect on the interparticle force (see in the following subsection) but seem not to affect that much the aggregation kinetics and the final structure [7,64].

2.3.2 Finite element method

In this section, we discuss the magnetic field strength dependence of the yield stress in concentrated MR fluids. FEM simulations are also compared to experimental results.

The experimental system consists in carbonyl iron (HQ grade, BASF) suspensions in silicone oil (20 mPa·s, Sigma-Aldrich) at different particle volume fractions ϕ; for more details, see [78]. For the carbonyl iron particles used in this work, the diameter is around $d = 1$ μm and their magnetization curve is plotted in Figure 2.4 in terms of relative permeability μ_p. The magnetization curve at room temperature is characterized by a small initial permeability $\mu_i = 4.75$ and a saturation magnetization of $M_S = 1{,}600$ kA/m. Note that, once the kind of lattice is chosen (monoclinic in this case according to (2.8)), only particles' magnetic properties (magnetization curve), volume fraction and diameter are needed to implement the FEM simulation method.

As an example, the normalized shear stress versus shear strain curves are shown in Figure 2.5 for $\phi = 0.2$ under different external magnetic fields. Here, two different normalizations are used following [41] depending on whether the system operates in the linear (low-field) or saturation (high-field) regime. On the one hand, in the linear regime the particle permeability is supposed not to change a lot with the applied magnetic field (see solid line in Figure 2.4). Therefore, the interparticle force is proportional to the applied magnetic field squared: $\tau/\left(\mu_0 H_0^2/2\right)$ (Figure 2.5(a)). On the other hand, for high magnetic fields, particles are fully saturated and consequently their interaction (and resultant yield stress) is bounded by their saturation magnetization: $100\tau/\left(\mu_0 M_s^2/2\right)$ (Figure 2.5(b)).

A closer observation of Figure 2.5 shows several differences with stress versus strain curves previously published in the literature on similar carbonyl iron-based MR fluids [41]. First, the normalized shear stress in the linear regime is one order

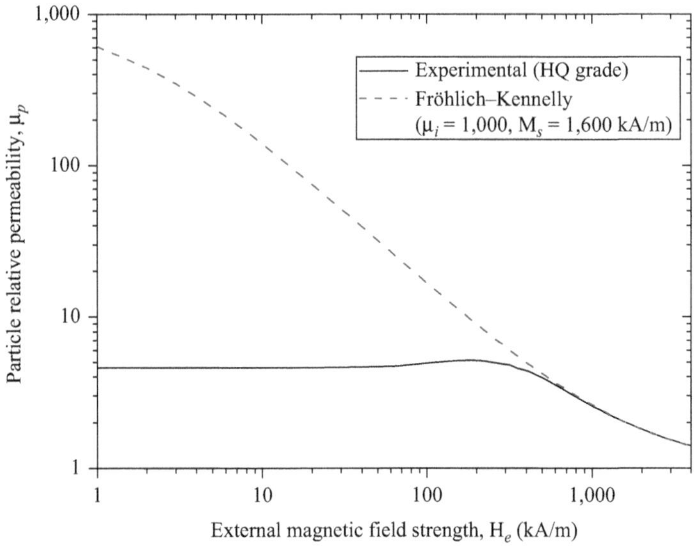

Figure 2.4 Relative permeability μ_p as a function of applied magnetic field H_e according to experimental measurements (carbonyl iron microparticles are HQ grade from BASF) and according to Fröhlich–Kennelly model supposing an initial permeability of μ_i =1,000 and saturation magnetization of M_s =1,600 kA/m (typical values used in the literature)

of magnitude smaller than that shown in Figure 2 of [41]. The reason for this is the different constitutive equation employed. In the paper by Ginder and Davis [41], the magnetization curve of the particles corresponds to a Fröhlich–Kennelly constitutive equation with initial permeability $\mu_i = 1{,}000$ and $M_S = 1{,}600$ kA/m. As it can be seen, this particular value for the initial permeability greatly overestimates our experimental one by three orders of magnitude (see Figure 2.4). As a consequence, the contrast factor in [41] is $\beta \sim 1$ while in our experimental system the contrast factor is of only $\beta \sim 0.55$. As a result, multipolar interactions are much more stronger in [41], and this contributes to a larger difference in the shear stress level [37,64]. Note, in contrast, that in the saturated regime a reasonably good agreement is found because the saturation magnetization employed in the Fröhlich-Kennelly constitutive equation and the experimental one are very similar.

The importance of the particular constitutive model together with multipolar interactions is also manifested in the success of the normalization employed in Figure 2.5. Stress versus strain curves in [41] do not collapse even at small magnetic fields (see Figure 2 in their paper). The reason for this is the field enhancement in the polar regions of the particles because multipoles saturate/reduce particle magnetization/permeability. On the contrary, Figure 2.5 in this work shows a good collapse in two particular field ranges: 4–80 and 160–320 kA/m. These field

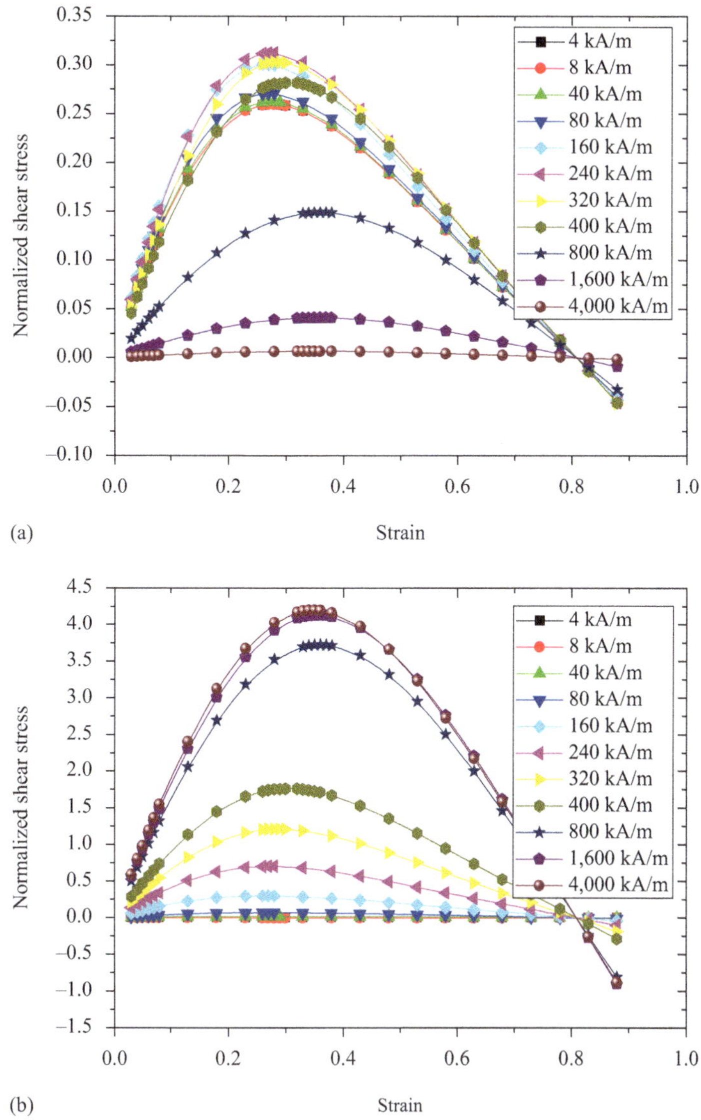

Figure 2.5 Normalized shear stress versus shear strain curves for an MR fluid of $\phi = 0.2$ under different external magnetic fields. (a) Normalization according to linear regime. (b) Normalization according to saturated regime

intervals correlate very well with field intervals where the particle permeability is kept constant in the magnetization curve (see Figure 2.4). This explains the collapse between curves at different fields (even when multipoles increase the local field in the polar regions of the particles).

Apart from the normalization, it is also important to comment on the shape of the curves contained in Figure 2.5. First, the strain corresponding to the maximum in the stress (so-called yield strain) does not significantly change with the field strength in contrast to the predictions by Ginder and Davis [41] in the linear regime. In addition, for large strains, independently of the regime considered (linear or saturation), the stress reaches a negative value again in striking contrast to [41]. The origin of this discrepancy is found in the way the structures are strained. In this work, the lattice experiences a real strain and therefore, under shear, upstream chains get closer to the probe chain and eventually the force reverses its sign becoming negative. However, in [41] chains are not truly sheared following an affine deformation. Instead, they are first elongated in the field direction and later rotated multiplying the stress by a positive corrective factor. This gives place to a stress that asymptotically approaches zero always keeping its sign.

It is important to remark here that the discussion above is not entirely applicable to MR fluids. In particular, negative stresses are never observed because the structures become unstable and break at the yield point. The previous discussion on the shape of stress versus strain curves aims to highlight the use of both periodic boundary conditions and a reduced field formulation allows us to properly model the affine deformation of the lattice.

Similar stress–strain curves were computed for MR fluids having different particle volume fractions in order to identify their yield stress. Results are summarized in Figure 2.6. In Figure 2.6, the yield stress is plotted against the external

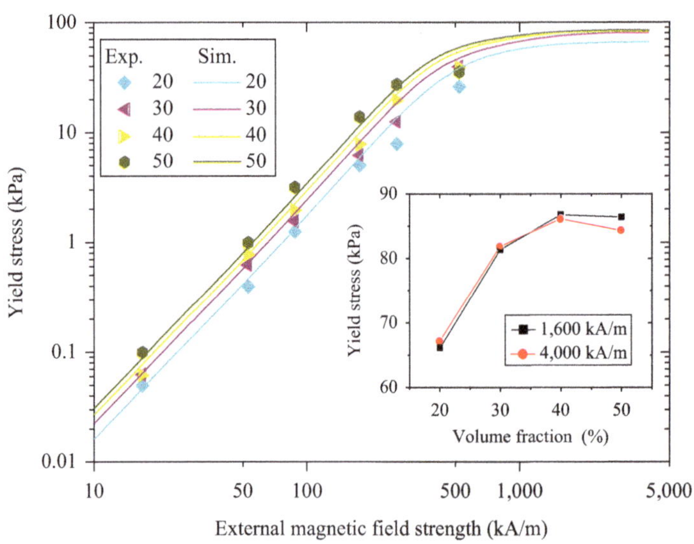

Figure 2.6 Yield stress as a function of the external magnetic field strength for MR fluids with different concentrations. Lines: FEM calculations. Points: experimental data [78]. Inset: simulated results in the saturation regime

magnetic field. As expected, the yield stress increases with H_e^2 for small magnetic field strengths and becomes independent on H_e for sufficiently large magnetic fields close to saturation. It is also worth to remark that in the saturation regime the yield stress does not steadily increase with volume fraction but instead levels of around 40 vol.%. In Figure 2.6, we also include experimental data. Taking into consideration that there is not any free parameter, experiments are in surprisingly good agreement with the FEM simulations despite the fact that particles are not necessarily expected to aggregate in monoclinic lattices for so high particle concentrations. The agreement between experiments and simulations suggests that the yield stress does not seem to be particularly sensitive on the exact microstructural details.

2.4 Conclusions

Modeling MR fluids is a challenging task that has been undertaken from different points of view and using different computational methods. In particular, the use of microscopic approaches is of outstanding interest today.

Among the different microscopic approaches reported in the literature, DEM simulations have proved to be strong candidates to model MR fluids because they are able to gather the most basic features under static and flow conditions [10,17,18,28,29]. DEM simulations are shown here to appropriately capture the aggregation kinetics of dilute MR fluids as well.

On the other hand, FEM simulations are a complementary tool to accurately ascertain magnetostatic interparticle forces [38,40,41,46,47]. Although this type of simulation is hard to be implemented in dynamic systems with a large number of particles due to its high computational cost, it is demonstrated here that it is still very useful in the precise determination of the yield stress in concentrated MR fluids.

Acknowledgments

This work was supported by MINECO MAT 2016-78778-R and PCIN-2015-051 projects (Spain), and European Regional Development Fund (ERDF). J.R. Morillas acknowledges FPU14/01576 fellowship.

References

[1] de Vicente J., Klingenberg D. J., Hidalgo-Álvarez R. "Magnetorheological fluids: a review." *Soft Matter.* 2011; **7**: 3701–10.

[2] Wereley N. (ed.). *Magnetorheology: Advances and Applications.* Cambridge: RSC Publishing; 2013. Vol. 6.

[3] Ghaffari A., Hashemabadi S. H., Ashtiani M. "A review on the simulation and modeling of magnetorheological fluids." *Journal of Intelligent Material Systems and Structures.* 2015; **26**(8): 881–904.

[4] Bossis G., Lemaire E., Volkova O., Clercx H. "Yield stress in magnetorheological and electrorheological fluids: a comparison between microscopic and macroscopic structural models." *Journal of Rheology*. 1997; **41**(3): 687–704.

[5] Shkel Y. M., Klingenberg D. J. "A continuum approach to electrorheology." *Journal of Rheology*. 1999; **43**(5): 1307–22.

[6] von Pfeil K., Graham M. D., Klingenberg D. J., Morris J. F. "A two-fluid model for electro-and magnetorheological suspensions." *International Journal of Modern Physics B*. 2002; **16**(17–18): 2669–75.

[7] Klingenberg D. J., Van Swol F., Zukoski C. F. "Dynamic simulation of electrorheological suspensions." *The Journal of Chemical Physics*. 1989; **91** (12): 7888–95.

[8] Martin J. E., Anderson, R. A. "Chain model of electrorheology." *The Journal of Chemical Physics*. 1996; **104**(12): 4814–27.

[9] Heine M. C., de Vicente J., Klingenberg D. J. "Thermal transport in sheared electro- and magnetorheological fluids." *Physics of Fluids*. 2006; **18**(2): 023301 pp. 11.

[10] Ruiz-López J. A., Fernández-Toledano J. C., Klingenberg D. J., Hidalgo-Álvarez R., de Vicente J. "Model magnetorheology: a direct comparative study between theories, particle-level simulations and experiments, in steady and dynamic oscillatory shear." *Journal of Rheology*. 2016; **60**(1): 61–74.

[11] Satoh A. *Modeling of Magnetic Particle Suspensions for Simulations*. Boca Raton, FL: CRC Press; 2017.

[12] Cundall P. A., Strack O. D. "A discrete numerical model for granular assemblies." *Geotechnique*. 1979; **29**(1): 47–65.

[13] Zhu H. P., Zhou Z. Y., Yang R. Y., Yu A. B. "Discrete particle simulation of particulate systems: theoretical developments." *Chemical Engineering Science*. 2007; **62**(13): 3378–96.

[14] Li S., Marshall J. S., Liu G., Yao Q. "Adhesive particulate flow: the discrete-element method and its application in energy and environmental engineering." *Progress in Energy and Combustion Science*. 2011; **37**(6): 633–68.

[15] Whittle M. "Computer simulation of an electrorheological fluid." *Journal of Non-Newtonian Fluid Mechanics*. 1990; **37**(2-3): 233–63.

[16] See H., Doi M. "Aggregation kinetics in electro-rheological fluids." *Journal of the Physical Society of Japan*. 1991; **60**(8): 2778–82.

[17] Klingenberg D. J., Van Swol F., Zukoski C. F. "The small shear rate response of electrorheological suspensions. I. Simulation in the point–dipole limit." *The Journal of Chemical Physics*. 1991; **94**(9): 6160–9.

[18] See H., Doi M. "Shear resistance of electrorheological fluids under time-varying electric fields." *Journal of Rheology*. 1992; **36**(6): 1143–63.

[19] Mohebi M., Jamasbi N., Liu J. "Simulation of the formation of nonequilibrium structures in magnetorheological fluids subject to an external magnetic field." *Physical review E*. 1996; **54**(5): 5407–13.

[20] Martin J. E. "Thermal chain model of electrorheology and magnetorheology." *Physical Review E*. 2000; **63**(1): 011406 pp. 9.

[21] Martin J. E. "Field-induced rheology in uniaxial and biaxial fields." *International Journal of Modern Physics B*. 2001; **15**(6–7): 574–95.
[22] Ukai T., Maekawa T. "Patterns formed by paramagnetic particles in a horizontal layer of a magnetorheological fluid subjected to a dc magnetic field." *Physical Review E*. 2004; **69**(3): 032501 pp. 4.
[23] Ly H. V., Reitich F., Jolly M. R., Banks H. T., Ito K. "Simulations of particle dynamics in magnetorheological fluids." *Journal of Computational Physics*. 1999; **155**: 160–77.
[24] Sherman S. G., Paley D. A., Wereley N. M. "Parallel simulation of transient magnetorheological direct shear flows using millions of particles." *IEEE Transactions on Magnetics*. 2012; **48**(11): 3517–20.
[25] Climent E., Maxey M. R., Karniadakis G. E. "Dynamics of self-assembled chaining in magnetorheological fluids." *Langmuir*. 2004; **20**(2): 507–13.
[26] Shahrivar K., Carreón-González E., de Vicente J. "Effect of confinement on the aggregation kinetics of dilute magnetorheological fluids." *Smart Materials and Structures*. 2017; **26**(10): 105031 pp. 10.
[27] Liu T., Gu R., Gong X., Xuan S., Wu H., Zhang Z. "Structural and rheological study of magnetic fluids using molecular dynamics." *Magnetohydrodynamics*. 2010; **46**(3): 257–69.
[28] Ruiz-López J. A., Wang Z. W., Hidalgo-Álvarez R., de Vicente J. "Simulations of model magnetorheological fluids in squeeze flow mode." *Journal of Rheology*. 2017; **61**(5): 871–81.
[29] Pappas Y., Klingenberg D. J. "Simulations of magnetorheological suspensions in Poiseuille flow." *Rheologica Acta*. 2006; **45**(5): 621–29.
[30] Gusenbauer M., Schrefl T. "Simulation of magnetic particles in microfluidic channels." *Journal of Magnetism and Magnetic Materials*. 2018; **446**: 185–91.
[31] Klingenberg D. J. "Simulation of the dynamic oscillatory response of electrorheological suspensions: demonstration of a relaxation mechanism." *Journal of Rheology*. 1993; **37**(2): 199–214.
[32] Parthasarathy M., Klingenberg D. J. "A microstructural investigation of the nonlinear response of electrorheological suspensions." *Rheologica Acta*. 1995; **34**(5): 417–29.
[33] Parthasarathy M., Klingenberg D. J. "Large amplitude oscillatory shear of ER suspensions." *Journal of Non-Newtonian Fluid Mechanics*. 1999; **81**(1–2): 83–104.
[34] Sim H. G., Ahn K. H., Lee S. J. "Three-dimensional dynamics simulation of electrorheological fluids under large amplitude oscillatory shear flow." *Journal of Rheology*. 2003; **47**(4): 879–95.
[35] Berli C. L., de Vicente J. "A structural viscosity model for magnetorheology." *Applied Physics Letters*. 2012; **101**(2): 021903 pp. 4.
[36] Ruiz-López J. A., Hidalgo-Álvarez R., de Vicente J. "Towards a universal master curve in magnetorheology." *Smart Materials and Structures*. 2017; **26**(5): 054001 pp. 7.

[37] Keaveny E. E., Maxey M. R. "Modeling the magnetic interactions between paramagnetic beads in magnetorheological fluids." *Journal of Computational Physics*. 2008; **227**(22): 9554–71.

[38] Davis L. C. "Finite-element analysis of particle-particle forces in electrorheological fluids." *Applied Physics Letters*. 1992; **60**(3): 319–21.

[39] Clercx H. J. H., Bossis G. "Many-body electrostatic interactions in electrorheological fluids." *Physical Review E*. 1993; **48**(4): 2721–38.

[40] Davis L. C. "Polarization forces and conductivity effects in electrorheological fluids." *Journal of Applied Physics*. 1992; **72**(4): 1334–40.

[41] Ginder J. M., Davis L. C. "Shear stresses in magnetorheological fluids: role of magnetic saturation." *Applied Physics Letters*. 1994; **65**(26): 3410–2.

[42] Klingenberg D. J., Ulicny J. C., Golden M. A. "Mason numbers for magnetorheology." *Journal of Rheology*. 2007; **51**(5): 883–93.

[43] Ocalan M. *Magnetorheological fluids for extreme environments: stronger, lighter, hotter*, PhD thesis, Massachusetts Institute of Technology, 2011.

[44] Vereda F., Segovia-Gutiérrez J. P., de Vicente J., Hidalgo-Álvarez R. "Particle roughness in magnetorheology: effect on the strength of the field-induced structures." *Journal of Physics D: Applied Physics.* 2014; **48**(1): *015309* pp. 11.

[45] Vereda F., de Vicente J., Hidalgo-Álvarez R. "Effect of surface roughness on the magnetic interaction between micron-sized ferromagnetic particles: finite element method calculations." *Journal of Intelligent Material Systems and Structures*. 2017; **28**(8): 992–8.

[46] Ginder J. M., Davis L. C., Elie L. D. "Rheology of magnetorheological fluids: models and measurements." *International Journal of Modern Physics B*. 1996; **10**(23–24): 3293–303.

[47] Morillas J. R., Bombard A., de Vicente J. "Enhancing magnetorheological effect using bimodal suspensions in the single-multidomain limit." *Smart Materials and Structures*. 2018; **27**(7): 1–7.

[48] Klingenberg D. J., Van Swol F., Zukoski C. F. "The small shear rate response of electrorheological suspensions. II. Extension beyond the point–dipole limit." *The Journal of Chemical Physics*. 1991; **94**(9): 6170–8.

[49] Pei L., Pang H., Ruan X., Gong X., Xuan S. "Magnetorheology of a magnetic fluid based on Fe_3O_4 immobilized SiO_2 core–shell nanospheres: experiments and molecular dynamics simulations." *RSC Advances*. 2017; **7** (14): 8142–50.

[50] Vågberg D., Tighe B. P. "On the apparent yield stress in non-Brownian magnetorheological fluids." *Soft Matter*. 2017; **13**(39): 7207–21.

[51] Goško D., Lācis S. "Magnetorheological suspension composed of fiber particles: numerical simulation of anisotropic behavior." *Proceedings of the International Scientific Colloquium Modelling for Material Processing*; Riga, Latvia, Sept. 2010. Riga: Laboratory for Mathematical Modelling of Environmental and Technological Processes, Univ. of Latvia; 2010. pp. 335–40.

[52] Kor Y. K., See H. "The electrorheological response of elongated particles." *Rheologica Acta*. 2010; **49**(7): 741–56.

[53] Wilson B. T., Klingenberg D. J. "A jamming-like mechanism of yield-stress increase caused by addition of nonmagnetizable particles to magnetorheological suspensions." *Journal of Rheology*. 2017; **61**(4): 601–11.

[54] Kittipoomwong D., Klingenberg D. J., Ulicny J. C. "Dynamic yield stress enhancement in bidisperse magnetorheological fluids." *Journal of Rheology*. 2005; **49**(6): 1521–38.

[55] Sherman S. G., Wereley N. M. "Effect of particle size distribution on chain structures in magnetorheological fluids." *IEEE Transactions on Magnetics*. 2013; **49**(7): 3430–3.

[56] Fernández-Toledano J. C., Ruiz-López J. A., Hidalgo-Álvarez R., de Vicente J. "Simulations of polydisperse magnetorheological fluids: a structural and kinetic investigation." *Journal of Rheology*. 2015; **59**(2): 475–98.

[57] Segovia-Gutiérrez J. P., de Vicente J., Puertas A. M., Hidalgo-Álvarez R. "Describing magnetorheology under a colloidal glass approach." *Physical Review E*. 2017; **95**(5): 052601 pp. 9.

[58] Klingenberg D. J., Ulicny J. C., Smith A. L. "Effects of body forces on the structure and rheology of ER and MR fluids." *International Journal of Modern Physics B*. 2007; **21**(28–29): 4841–48.

[59] Pang H., Pei L., Sun C., Gong X. "Normal stress in magnetorheological polymer gel under large amplitude oscillatory shear." *Journal of Rheology*. 2018; **62**(6): 1409–18.

[60] Kargulewicz M., Iordanoff I., Marrero V., Tichy J. "Modeling of magnetorheological fluids by the discrete element method." *Journal of Tribology*. 2012; **134**(3): 031706 pp. 10.

[61] Lagger H. G., Bierwisch C., Korvink J. G., Moseler M. "Discrete element study of viscous flow in magnetorheological fluids." *Rheologica Acta*. 2014; **53**(5-6): 417–43.

[62] Joung C. G., See H. "The influence of wall interaction on dynamic particle modelling of magneto-rheological suspensions between shearing plates." *Rheologica Acta*. 2008; **47**(8): 917–27.

[63] Lagger H. G., Breinlinger T., Korvink J. G., *et al*. "Influence of hydrodynamic drag model on shear stress in the simulation of magnetorheological fluids." *Journal of Non-Newtonian Fluid Mechanics*. 2015; **218**: 16–26.

[64] Han K., Feng Y. T., Owen D. R. J. "Three-dimensional modelling and simulation of magnetorheological fluids." *International Journal for Numerical Methods in Engineering*. 2010; **84**(11): 1273–302.

[65] Bonnecaze R. T., Brady J. F. "Dynamic simulation of an electrorheological fluid." *The Journal of Chemical Physics*. 1992; **96**(3): 2183–202.

[66] Joung C., See H. "Simulation of magneto-rheological fluids incorporating hydrodynamic effects." *Journal of Central South University of Technology*. 2007; **14**(1): 271–4.

[67] Melrose J. R., Itoh S., Ball R. C. "Simulations of electrorheological fluids with hydrodynamic lubrication." *International Journal of Modern Physics B*. 1996; **10**(23–24): 3211–7.

[68] Satoh A., Chantrell R. W., Coverdale G. N., Kamiyama S. I. "Stokesian dynamics simulations of ferromagnetic colloidal dispersions in a simple shear flow." *Journal of Colloid and Interface Science*. 1998; **203**(2): 233–48.
[69] Kang T. G., Hulsen M. A., den Toonder J. M. "Dynamics of magnetic chains in a shear flow under the influence of a uniform magnetic field." *Physics of Fluids*. 2012; **24**(4): 042001 pp. 19.
[70] Shahrivar K., Carreón-González E., Morillas J. R., de Vicente J. "Aggregation kinetics of carbonyl iron based magnetic suspensions in 2D." *Soft Matter*. 2017; **13**(14): 2677–85.
[71] Promislow J. H., Gast A. P., Fermigier M. "Aggregation kinetics of paramagnetic colloidal particles." *The Journal of Chemical Physics*. 1995; **102** (13): 5492–8.
[72] Domínguez-García P., Melle S., Pastor J. M., Rubio M. A. "Scaling in the aggregation dynamics of a magnetorheological fluid." *Physical Review E*. 2007; **76**(5): 051403 pp. 13.
[73] Erb R. M., Krebs M. D., Alsberg E., Samanta B., Rotello V. M., Yellen B. B. "Beyond diffusion-limited aggregation kinetics in microparticle suspensions." *Physical Review E*. 2009; **80**(5): 051402 pp. 7.
[74] Meakin P., Vicsek T., Family F. "Dynamic cluster-size distribution in cluster-cluster aggregation: effects of cluster diffusivity." *Physical Review B*. 1985; **31**(1): 564–9.
[75] Miyazima S., Meakin P., Family F. "Aggregation of oriented anisotropic particles." *Physical Review A*. 1987; **36**(3): 1421–7.
[76] Shahrivar K., Carreón-González E., de Vicente J. "On the importance of carrier fluid viscosity and particle–wall interactions in magnetic-guided assembly of quasi-2D systems." *Microfluidics and Nanofluidics*. 2017; **21**(7): 120 pp. 9.
[77] Liu D., Maxey M. R., Karniadakis G. E. "Simulations of dynamic self-assembly of paramagnetic microspheres in confined microgeometries." *Journal of Micromechanics and Microengineering*. 2005; **15**(12): 2298–308.
[78] Ruiz-López J. A., Fernández-Toledano J. C., Hidalgo-Álvarez R., de Vicente J. "Testing the mean magnetization approximation, dimensionless and scaling numbers in magnetorheology." *Soft Matter*. 2016; **12**: 1468–76.

Chapter 3

A new novel of composite adaptive optimal control for MR damper system subjected to mixed disturbances

Do Xuan Phu[1]*, Le Dai Hiep*[2] *and Seung-Bok Choi*[3]

This study presents a new hybrid optimal controller to enhance the robustness and stability of dynamic systems subjected to uncertainties and external disturbances. In the formulation of the proposed control system, the Bolza–Meyer (BM) criterion of the optimal controller is modified to meet the system variation, and the sliding-mode controller is modified to obtain the classical control property with the prescribed performance. The proposed controller also includes the *H*-infinity technique as a bridge for connecting the sub-controls and improving the robust performance of the system. Fuzzy neural networks are used as filters to choose the optimal values for the next calculation. Hence, many advantages of fuzzy neural networks are acquired, related to optimal control, sliding mode control, prescribed performance, and *H*-infinity techniques. To demonstrate these advantages, the proposed hybrid controller is applied to a vehicle seat suspension for vibration control. Simulation results show that the proposed control obtains good performances than the compared controllers.

3.1 Introduction

Recently, various hybrid controllers have been developed to cope with parameter uncertainties and severe disturbances in both nonlinear and linear systems. This control issue becomes significant in practical environments, where complex and accurate systems are subjected to real uncertainties and excitations. Hybrid controllers are normally built by adopting the salient benefits or advantages of each control technique. However, control performance of the hybrid controller depends on the combination sequence of several different controllers. Therefore,

[1]Department of Mechatronics and Sensor Systems Technology, Vietnamese-German University, Vietnam
[2]Department of Computational Engineering, Vietnamese-German University, Vietnam
[3]Department of Mechanical Engineering, Inha University, South Korea

the formulation procedure of a new hybrid controller showing high robust control performances is not straightforward. Owing to this difficulty, since the beginning of the research, relatively moderate controllers have been used to optimally combine different control algorithms into a new model. For example, a modified optimal controller with adaptation laws was developed by applying an iteration policy [1]. This policy was connected to the weight of the neural networks and the factors of the Hamilton–Jacobi–Bellman (HJB) equation of optimal control. However, this technique exhibited slow convergence times, adversely effecting robust performance and system stability. Despite this disadvantage, the HJB equation was used in several works because of its easy adaptation to the optimal controller [2–5]. It has also been shown that a hybrid optimal controller using the Hamilton–Jacobi–Isaacs equation exhibits similar disadvantages as HJB for convergence time [6]. To shorten convergence time, the modified Riccati equation has been used [7]. The off-policy method has also been used to reduce the convergence time in the formulation of hybrid optimal controllers [8]. However, the modified BM criterion has been frequently adopted [9–15]. This criterion concentrates on the enhancement of the cost calculation function of the optimal controller, focusing on a constant group, to improve robust performance. In [9–15], to guarantee robustness, a sliding-mode control method was used by adopting a sign function. However, these previous works used matrices of BM as 1 or a unit matrix. This caused disadvantages, such as slow convergence times and large tracking errors.

Recently, several types of hybrid controllers have been developed that do not use the optimal controller as the basis control method. For example, the authors of this work formulated several hybrid controllers using an adaptive control with a fuzzy sliding mode controller-combined *H*-infinity technique [16–21]. In these studies, the application of *H*-infinity, connected with the Riccati-like equation, is the key to formulating the hybrid controller, because it guarantees high control robustness against parameter uncertainties and severe disturbances. These hybrid controllers provide high robust control performance, but the formulation sequence procedure is complex. Furthermore, the choice of optimal control gains is very difficult, because there are no objective or constraint functions. This leads to new work combining several robust control schemes with optimal controllers to eliminate disadvantages. The development of a new hybrid optimal controller is very helpful when dealing with complicated or complex control systems subjected to uncertainties and disturbances in practical environments.

The main technical contribution of this work is (a) the development of a new modification of optimal control based on BM criterion and (b) use the new development of optimal control for design of new adaptive controller including sliding mode control, *H*-infinity technique and fuzzy neural networks model. Some important notes of new controller are also presented in this study. After formulating the hybrid optimal controller, the vibration control system of the vehicle seat using magnetorheological (MR) damper is adopted, and simulations is undertaken.

3.2 New novel of optimal control for nonlinear system

3.2.1 Proposed control law

The previous section shows the efficiency of the value α in optimal control system given by

$$\dot{x}(t) = f_\Sigma(x,t) + B_\Sigma(t)u(t) \tag{3.1}$$

where $f_\Sigma(x,t)$ is a polynomial function. In this section, the function $f_\Sigma(x,t)$ is replaced by the continuous functions of time $A_\Sigma(t)$, and the value α is also proposed to solve the following time-varying system with the linear control input given by

$$\dot{x}(t) = A_\Sigma(t)x(t) + B_\Sigma(t)u(t) \tag{3.2}$$

In the above, $\dot{x}(t) \in R^n$ is the state of system, $u(t) \in R^m$ is the input of the control, $A_\Sigma(t)$ and $B_\Sigma(t)$ are continuous functions of time. In the classical linear optimal control problem, the criterion to be minimized is determined using the BM function:

$$J_1 = \frac{1}{2}[x(T)]^T \psi_\Sigma [x(T)] + \frac{1}{2}\int_{to}^{T} \left(u^T(s)R_\Sigma(s)u(s) + x^T(s)L_\Sigma(s)x(s)\right)ds \tag{3.3}$$

The original cost function is expressed as follows:

$$J_2 = \sum_{i=1}^{n} \psi_{\Sigma ii}|x_i(T)| + \frac{1}{2}\int_{to}^{T} \left(u^T(s)R_\Sigma(s)u(s) + x^T(s)L_\Sigma(s)x(s)\right)ds \tag{3.4}$$

The cost function of the optimal controller proposed in this study is proposed with the value α as follows:

$$J_3 = \sum_{i=1}^{n} \frac{\psi_{\Sigma ii}|x_i(T)|}{\alpha} + \frac{1}{2}\int_{to}^{T} \left(u^T(s)R_\Sigma(s)u(s) + x^T(s)L_\Sigma(s)x(s)\right)ds \tag{3.5}$$

The optimal control law for the system (3.2) is then determined using the classical method as follows:

$$u^*(t) = -(R_\Sigma(t))^{-1}B_\Sigma^T(t)q(t) \tag{3.6}$$

where $q(t) = -Q_\Sigma(t)\text{sign}[x(t)]$. Hence, (3.6) is rewritten as follows:

$$u^*(t) = (R_\Sigma(t))^{-1}B_\Sigma^T(t)Q_\Sigma(t)\text{sign}[x(t)] \tag{3.7}$$

The matrix function $Q_\Sigma(t)$ is determined as follows:

$$Q_\Sigma^*(t) = Q_\Sigma(t)|x(t)| \tag{3.8}$$

Thus, the function $u^*(t)$ of (3.7) is expressed by

$$u^*(t) = (R_\Sigma(t))^{-1}B_\Sigma^T(t){Q_\Sigma}^*(t)\text{sign}[x(t)] \tag{3.9}$$

In this study, a new function $Q_\Sigma^{**}(t)$ defined as $Q_\Sigma^{**}(t) = Q_\Sigma(t)\alpha|x(t)|$ is adopted which satisfies the following the matrix equation:

$$\begin{aligned}
\dot{Q}_\Sigma^{**}(t) &= \frac{d(Q_\Sigma(t)\alpha|x(t)|)}{dt} = \frac{dQ_\Sigma(t)}{dt}\alpha|x(t)| + Q_\Sigma(t)\alpha\frac{d(|x(t)|)}{dt} \\
&= (-A_\Sigma^T(t)Q_\Sigma(t) - Q_\Sigma(t)A_\Sigma(t) + L(t) - Q_\Sigma(t)B_\Sigma(t)R_\Sigma^{-1}(t)B_\Sigma^T(t)Q_\Sigma(t))\alpha|x(t)| \\
&\quad + Q_\Sigma(t)\alpha(A_\Sigma(t)|x(t)| + B_\Sigma(t)R_\Sigma^{-1}(t)B_\Sigma^T(t)Q_\Sigma(t)|x(t)|) \\
&= L_\Sigma(t)\alpha|x(t)| - A_\Sigma^T(t)Q_\Sigma(t)\alpha|x(t)| = L_\Sigma(t)\alpha|x(t)| - A_\Sigma^T Q_\Sigma^{**}
\end{aligned} \tag{3.10}$$

The detailed derivation of the above result is provided in Appendix A. Hence, the function $u^*(t)$ of (3.7) is rewritten by

$$u^*(t) = (R_\Sigma(t))^{-1}B_\Sigma^T(t)Q_\Sigma{}^{**}\text{sign}[x(t)] \tag{3.11}$$

Substituting (3.11) into (3.2) yields the following equation:

$$\dot{x}(t) = A_\Sigma(t)x(t) + B_\Sigma(t)(R_\Sigma(t))^{-1}B_\Sigma^T(t)Q_\Sigma^{**}\text{sign}[x(t)] \tag{3.12}$$

Theorem 3.1. *The optimal regulator for the linear system (3.2) with respect to the criterion (3.5) is given by the control law (3.11) and the gain matrix differential equation* (3.10). *Hence, the optimal control state of the linear system (3.2) is reexpressed by* (3.12).

Proof. The Hamiltonian function of the system is defined as follows:

$$H(x, u, q, t) = \frac{1}{2}\left(u^T R_\Sigma(t)u + x^T L_\Sigma(t)x\right) + q^T\dot{x}(t) \tag{3.13}$$

The optimal control law is then found by taking the partial derivative $\partial H/\partial u = 0$:

$$R_\Sigma(t)u(t) + B_\Sigma^T(t)q(t) = 0 \tag{3.14}$$

Hence, the optima control law is obtained by

$$u^*(t) = -R_\Sigma^{-1}(t)B_\Sigma^T(t)q(t) \tag{3.15}$$

The function $q(t)$ is defined as follows:

$$q(t) = -Q_\Sigma(t)\alpha\ \text{sign}[x(t)] \tag{3.16}$$

where $Q_\Sigma(t)$ is the square symmetric matrix $(n \times n)$, α is the positive constant which is determined by the user, *sign*$[x(t)]$ is the signum (sign) function related to the sliding surface of the sliding mode control. Now, substituting (3.16) into (3.15) gives the following equation:

$$u^*(t) = R_\Sigma^{-1}(t)B_\Sigma^T(t)Q_\Sigma(t)\alpha\ \text{sign}[x(t)] \tag{3.17}$$

Using the definition of costate equation given by $dq(t)/dt = -(\partial H/\partial x)$, the following equation is obtained:

$$-\frac{dq(t)}{dt} = L_\Sigma(t)x(t) + A^T(t)q(t) \tag{3.18}$$

Substituting (3.18) into (3.17) yields the following equation:

$$\dot{Q}_\Sigma(t)\alpha \,\mathrm{sign}[x(t)] + Q_\Sigma(t)\alpha\frac{d(\mathrm{sign}[x(t)])}{dt} = L_\Sigma(t)x(t) - A^T(t)Q_\Sigma(t)\alpha \,\mathrm{sign}[x(t)] \tag{3.19}$$

Then, using the property of $d(\mathrm{sign}[x(t)])/dt = 0$, (3.19) is rewritten as follows:

$$\dot{Q}_\Sigma(t)\alpha \,\mathrm{sign}[x(t)] = L_\Sigma(t)x(t) - A^T(t)Q_\Sigma(t)\alpha \,\mathrm{sign}[x(t)] \tag{3.20}$$

If $x(t) = 0$, then $u(t) = 0$. Thus, the value of $Q_\Sigma(t)$ is not required. If $Q_\Sigma(t)$ is a solution of (3.10), (3.20) is satisfied. At time $t = T$, (3.16) is then rewritten as follows:

$$q(T) = -Q_\Sigma(T)\alpha \,\mathrm{sign}[x(T)] = \frac{\partial J}{\partial x(T)} = \frac{\psi_\Sigma}{\alpha}\mathrm{sign}[x(T)] \tag{3.21}$$

Hence, the value $Q_\Sigma(T)$ is found as follows:

$$Q_\Sigma(T) = -\frac{\psi_\Sigma}{\alpha^2} \tag{3.22}$$

This completes the proof of Theorem 3.1.

3.2.2 Simulation of the proposed optimal control and discussion

To show the advantage of the proposed optimal controller, the simulation system used in [12] is adopted with $R_\Sigma(t) = L_\Sigma(t) = 1$ as follows:

$$\dot{x}(t) = x(t) + u(t)\,, \quad x(0) = 1 \tag{3.23}$$

$$J_1 = 50|x(T)| + \frac{1}{2}\int_0^T \left(u^2(t) + x^2(t)\right)dt \tag{3.24}$$

The solution is obtained from Basin's proposal as follows [12]:

$$\dot{x}(t) = x(t) + Q_\Sigma^*(t)\mathrm{sign}[x(t)] \tag{3.25}$$

$$Q_\Sigma^* = |x(t)| - Q_\Sigma^*(t) \tag{3.26}$$

On the other hand, in this approach, the solution of this problem is determined by (3.20)–(3.22) as follows:

$$\dot{x}(t) = x(t) + Q_\Sigma^{**}(t)\mathrm{sign}[x(t)] \tag{3.27}$$

$$Q_\Sigma^{**} = \alpha|x(t)| - Q_\Sigma^{**}(t) \tag{3.28}$$

Simulation results are presented in Figures 3.1–3.4. Parameters for the simulation are setup as $R_\Sigma(t) = L_\Sigma(t) = 1$, $\alpha = 1$, $q = -2.4142$. The simulation is carried out

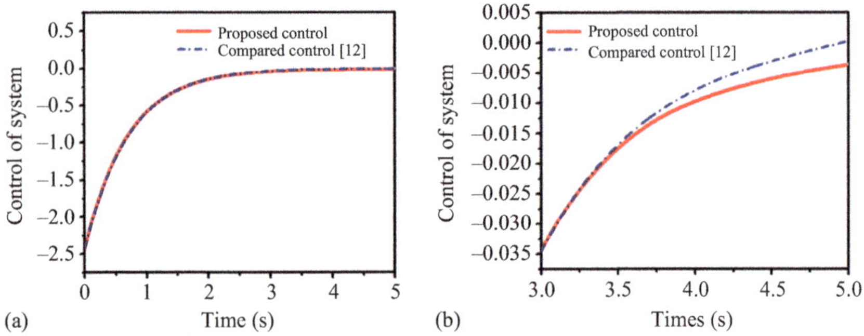

Figure 3.1 Control results when $R_\Sigma(t) = L_\Sigma(t) = 1$: (a) general view, (b) large-scale of the general view

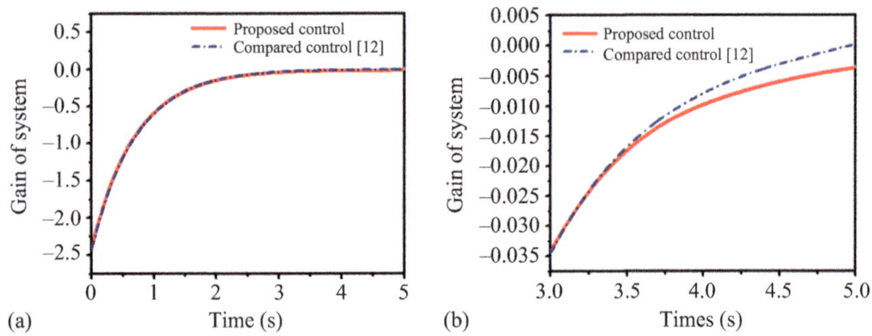

Figure 3.2 Control gains when $R_\Sigma(t) = L_\Sigma(t) = 1$: (a) general view, (b) large-scale of the general view

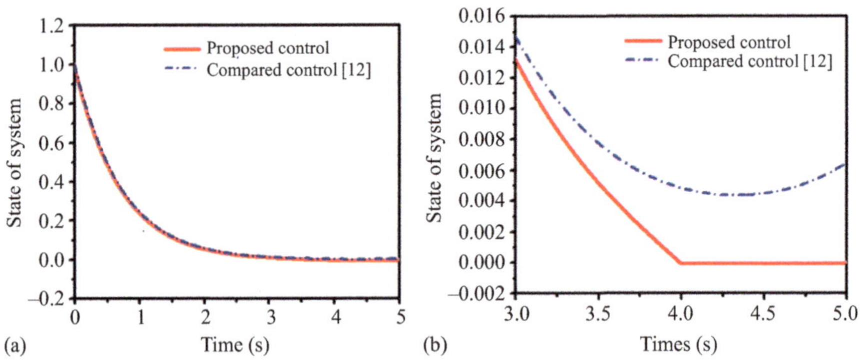

Figure 3.3 State results when $R_\Sigma(t) = L_\Sigma(t) = 1$: (a) general view, (b) large-scale of the general view

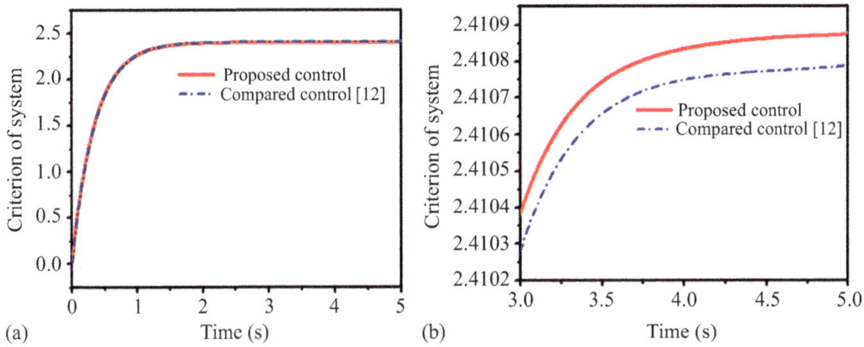

Figure 3.4 Criterion results when $R_\Sigma(t) = L_\Sigma(t) = 1$: (a) general view, (b) large-scale of the general view

with step 0.001 and uses the Runge–Kutta method. The comparative controller used in [12] is adopted with the values of $\alpha = 1$. The value of the Meyer term and the final value of state of system $x(T)$ are used to find the cost of calculation of the comparative controller, and it is determined by $2.4108 + 0.0065 \times 50 = 2.7355$. For the proposed control law, $R_\Sigma(t) = L_\Sigma(t) = 1$, $\alpha = 1.02$, $q = -2.4213$ are used and the cost of calculation is identified by $2.4109 + \left(\left(6.7616 \times 10^{-7} \times 50\right)/\alpha\right) = 2.4109$. From this result, it is understood that the application of α makes the system to decrease the cost of calculation, and the energy for control as shown in Figures 3.1–3.4. In Figure 3.1, the level of energy consumption of two control laws is presented. From the result, the index energy at 5 s is determined by $2.88074 \cdot 10^{-4}$ and -0.00355, respectively, for the comparative controller and the proposed controller. This, of course, directly indicates that the proposed optimal control law is better in saving energy for control action. In Figure 3.2, the variation of gain system is shown for both controllers. The index gain is determined by $2.8807 \cdot 10^{-4}$ and -0.00355 at time 5 s for the comparative controller and the proposed controller, respectively. The convergence of state system for both controller are presented in Figure 3.3 from which the index of state system is identified by 0.0065 and $6.7616 \cdot 10^{-7}$ at 5 s for the comparative controller and the proposed controller validating the higher stability of the proposed optimal control law. The criterion index of two controller is shown in Figure 3.4. The index criterion is determined by 2.41079 and 2.41088 at time 5 s, respectively, for the comparative controller and the proposed controller.

3.3 Design of a new composite controller

In the second proposed controller, a new composite control is designed based on results of the above optimal control. In this design, the grouping of optimal control, *H*-infinity technique, classical sliding mode control, and prescribed sliding mode control with the prescribed performance provide the principal components for the new proposed control.

3.3.1 Fuzzy model

In this study, the interval type-2 fuzzy model is used for the control of uncertain system properties. The granular clustering method is combined with the fuzzy model to reduce computation costs. The applied fuzzy model has been presented in several works [22–31]. Generally, the fuzzy model is expressed as follows:

$$R^j_{\Delta f} \text{ If } h_{\Delta 1} \text{ is } \boldsymbol{H}^j_{\Delta f1} \text{ and} \ldots \text{ and } \boldsymbol{H}^j_{\Delta fn} \text{ is Then } g \text{ is } a^j_0 + \sum_{i=1}^{n} a^j_i h_{\Delta i} \tag{3.29}$$

where $\boldsymbol{H}^j_{\Delta fi}(i = 1,\ldots,n; j = 1,\ldots,m)$ are fuzzy sets, m is the number of rules, and a^j_i are the interval sets. The defuzzified output is expressed by

$$g_{\Delta f} = \frac{g_{\Delta l} + g_{\Delta r}}{2} = \frac{\boldsymbol{\theta}^T_{\Delta l}\xi^f_{\Delta l} + \boldsymbol{\theta}^T_{\Delta r}\xi^f_{\Delta r}}{2} \tag{3.30}$$

where $\boldsymbol{\theta}^T_{\Delta l} = \left[w^l_1 w^l_2 w^l_3 \cdots w^l_n\right]$ and $\boldsymbol{\theta}^T_{\Delta r} = \left[w^r_1 w^r_2 w^r_3 \cdots w^r_n\right]$ are the weighting vectors, and the weighted firing strength vectors are determined by

$$\xi^f_{\Delta l} = \left[\frac{\underline{f}_1}{\sum_{i=1}^{n}\underline{f}_i}\frac{\underline{f}_2}{\sum_{i=1}^{n}\underline{f}_i}\frac{\underline{f}_3}{\sum_{i=1}^{n}\underline{f}_i}\cdots\frac{\underline{f}_n}{\sum_{i=1}^{n}\underline{f}_i}\right]^T,$$

$$\xi^f_{\Delta r} = \left[\frac{\overline{f}_1}{\sum_{i=1}^{n}\overline{f}_i}\frac{\overline{f}_2}{\sum_{i=1}^{n}\overline{f}_i}\frac{\overline{f}_3}{\sum_{i=1}^{n}\overline{f}_i}\cdots\frac{\overline{f}_n}{\sum_{i=1}^{n}\overline{f}_i}\right]^T.$$

3.3.2 Adaptive optimal control

The first step is to define the model of a nonlinear uncertain system. The model of a single-input–single-output nonlinear system, governed by the following equation, is considered:

$$\dot{\mathbf{x}}_\Delta = \mathbf{f}_\Delta(\mathbf{x}_\Delta) + \mathbf{g}_\Delta(\mathbf{x}_\Delta)u(t) + \mathbf{d}_\Delta(\mathbf{t}) \tag{3.31}$$

where $\mathbf{f}_\Delta(\mathbf{x}_\Delta) \in R^n$ and $\mathbf{g}_\Delta(\mathbf{x}_\Delta) \in R^n$ are two unknown nonlinear function vectors; $u(t) \in R^1$ is the control function; $\mathbf{d}_\Delta(\mathbf{t}) \in R^n$ is an external disturbance vector, $|\boldsymbol{d}_\Delta(\mathbf{t})| \le \delta\boldsymbol{d}_\Delta$, where $\delta\mathbf{d}_\Delta \in R^n$ is the upper bound of $\mathbf{d}_\Delta(\mathbf{t})$, and $\mathbf{x}_\Delta = \left[x_{\Delta 1}, x_{\Delta 2}, \ldots, x_{\Delta n}\right]^T = \left[x_1, \dot{x}_1, \ldots, x_1{}^{(n-1)}\right]^T \in R^n$ is the state vector of the system. The classical sliding surface, $s_{\Sigma s}$, is defined as follows:

$$s_{\Delta s} = k_1 x_{\Delta 1} + k_2 x_{\Delta 2} + k_3 x_{\Sigma 3} + \cdots + k_n x_{\Delta n} = \sum_{i=1}^{n} k_i x_{\Delta i} \tag{3.32}$$

where $\mathbf{K} = [k_n, k_{n-1}, k_{n-2}, \ldots, k_1]$ are the coefficients, such that all roots of the polynomial, $\sigma^n + k_{n-1}\sigma^{n-1} + k_{n-2}\sigma^{n-2} + \cdots + k_1$, are in the open left-half complex plane. A new vector, $\tilde{\boldsymbol{x}}_\Delta$, is defined by $\tilde{\boldsymbol{x}}_\Delta = [x_{\Delta 1} x_{\Delta 2} x_{\Delta 3} \cdots x_{\Delta n-1}]^T$, and (3.31) is rewritten as follows:

$$\dot{\tilde{\boldsymbol{x}}}_\Delta = \boldsymbol{S}_{\Delta 1}\tilde{\boldsymbol{x}}_\Delta + \boldsymbol{S}^T_{\Delta 2} s_{\Delta s} \tag{3.33}$$

where

$$S_{\Delta 1} = \begin{bmatrix} 0 & 1 & 0 & \cdots & 0 \\ 0 & 0 & 1 & \cdots & 0 \\ \cdot & \cdot & \cdot & \cdots & \cdot \\ -k_1 & -k_2 & -k_3 & \cdots & -k_{n-1} \end{bmatrix}, S_{\Delta 2} = \begin{bmatrix} 0 \\ 0 \\ \cdot \\ 1 \end{bmatrix}.$$

The tracking error is determined as $e_\Delta(t) = x_{\Delta 1} - x_{\Delta d}$, where x_d is the desired value. $x_{\Delta 1}$ is also x_Δ in (3.31). Thus, the error performance function is defined as follows:

$$\lambda_\Delta(t) = (\lambda_\Delta(0) - \lambda_{\Delta\infty})e^{-lt} + \lambda_{\Delta\infty} \tag{3.34}$$

where $l > 0, 0 < |e_\Delta(0)| < \lambda_\Delta(0), \lambda_{\Delta\infty} > 0, \lambda_{\Delta\infty} < \lambda_\Delta(0)$, then $\lambda_\Delta(t) > 0$, and $\lambda_\Delta(t)$ tend toward $\lambda_{\Delta\infty}$, exponentially. The tracking error of the prescribed performance is also set as follows:

$$e_\Delta(t) = \lambda_\Delta(t)S_\Delta(\varphi) \tag{3.35}$$

From (3.35), function $S_\Delta(\varphi)$ is found as

$$S_\Delta(\varphi) = \frac{e_\Delta(t)}{\lambda_\Delta(t)}. \tag{3.36}$$

From the above conditions, the function, $S_\Delta(\varphi)$, can be developed as follows:

$$S_\Delta(\varphi) = \frac{e^\varphi - e^{-\varphi}}{e^\varphi + e^{-\varphi}} \tag{3.37}$$

The tracking error of the prescribed performance is obtained as

$$-\lambda_\Delta(t) < e_\Delta(t) < \lambda_\Delta(t) \tag{3.38}$$

The inverse function of (3.37) is found by

$$\varphi = \frac{1}{2}\, ln\, \frac{1 + S_\Delta}{1 - S_\Delta} = \frac{1}{2}[ln(\lambda_\Delta + e_\Delta(t)) - \ln(\lambda_\Delta - e_\Delta(t))] \tag{3.39}$$

and the derivatives of (3.39) are found as

$$\dot{\varphi} = \frac{1}{2}\left[\frac{\dot{\lambda}_\Delta + \dot{e}_\Delta(t)}{\lambda_\Delta + e_\Delta(t)} - \frac{\dot{\lambda}_\Delta - \dot{e}_\Delta(t)}{\lambda_\Delta - e_\Delta(t)}\right] \tag{3.40}$$

$$\ddot{\varphi} = M_{\Delta 1} + M_{\Delta 2} + M_{\Delta 3}\ddot{e}_\Delta(t) \tag{3.41}$$

where

$$M_{\Delta 1} = \frac{\ddot{\lambda}_\Delta(\lambda_\Delta + e_\Delta(t)) - \left(\dot{\lambda}_\Delta + \dot{e}_\Delta(t)\right)^2}{2(\lambda_\Delta + e_\Delta(t))^2}, \quad M_{\Delta 2} = -\frac{\ddot{\lambda}_\Delta(\lambda_\Delta - e_\Delta(t)) - \left(\dot{\lambda}_\Delta - \dot{e}_\Delta(t)\right)^2}{2(\lambda_\Delta - e_\Delta(t))^2}$$

$$M_{\Delta 3} = \left(\frac{\lambda_\Delta + e_\Delta(t)}{2(\lambda_\Delta + e_\Delta(t))^2} + \frac{\lambda_\Delta - e_\Delta(t)}{2(\lambda_\Delta - e_\Delta(t))^2}\right).$$

Next, to realize the convergence of $\varphi \to 0$, the prescribed sliding surface is defined as

$$\sigma_{\Delta s} = \dot{\varphi} + c_{\Delta s}\varphi \tag{3.42}$$

where $c_{\Delta s} > 0$. The derivative of (3.42) is found as

$$\dot{\sigma}_{\Delta s} = M_{\Delta 1} + M_{\Delta 2} + M_{\Delta 3}(f_{\Delta}(x_{\Delta}) + g_{\Delta}(x_{\Delta})u(t) + d_{\Delta}(t) - \ddot{x}_{\Delta d}) + c_{\Delta s}\dot{\varphi} \tag{3.43}$$

The lumped uncertainty of the system is then obtained by

$$w_{\Sigma} = M_{\Sigma 3}\tilde{\gamma}_f\xi_{f\Sigma} + M_{\Sigma 3}\tilde{\gamma}_g\xi_{g\Sigma}u + M_{\Sigma 3}d_{\Sigma}(t), \tag{3.44}$$

where $\gamma_f = f_{\Sigma}(x_{\Sigma}) - f_{\Sigma}^*(x_{\Sigma}), \gamma_g = g_{\Sigma}(x_{\Sigma}) - g_{\Sigma}^*(x_{\Sigma})$. Using (3.43) and (3.44), the derivative of (3.43) is reexpressed as follows:

$$\dot{\sigma}_{\Delta s} = M_{\Delta 1} + M_{\Delta 2} + M_{\Delta 3}f_{\Delta}^*(x_{\Delta}) + M_{\Delta 3}g^*(x_{\Delta})u(t) - M_{\Delta 3}\ddot{x}_{\Delta d} + c_{\Delta s}\dot{\varphi} + w_{\Delta} \tag{3.45}$$

Using (3.30), (3.45) is expressed as

$$\dot{\sigma}_{\Delta s} = M_{\Delta 1} + M_{\Delta 2} + M_{\Delta 3}\theta_{f\Delta}^*\xi_{f\Delta} + M_{\Delta 3}\theta_{g\Delta}^*\xi_{g\Delta}u - M_{\Delta 3}\ddot{x}_{\Delta d} + c_{\Delta s}\dot{\varphi} + w_{\Delta} \tag{3.46}$$

where

$$\begin{aligned}
\theta_{f\Delta}^* &= arg\ \min_{\theta_{f\Delta}\in\Delta\theta_{f\Delta}}\left[sup_{x_{\Delta}\in\Delta x_{\Delta}}\left|f_{\Delta}(x_{\Delta}) - f_{\Delta}^*(x_{\Delta})\right|\right],\\
\theta_{g\Delta}^* &= arg\ \min_{\theta_{g\Delta}\in\Delta\theta_{g\Delta}}\left[sup_{x_{\Delta}\in\Delta x_{\Delta}}\left|g_{\Delta}(x_{\Delta}) - g_{\Delta}^*(x_{\Delta})\right|\right],\\
\Delta\theta_{f\Delta} &= \left\{\theta_{f\Delta}\in R^n, \|\theta_{f\Delta}\| \le \Theta_{f\Delta}\right\}, \quad \Delta\theta_{g\Delta} = \left\{\theta_{g\Delta}\in R^n, \|\theta_{g\Delta}\| \le \Theta_{g\Delta}\right\}\\
\Delta x_{\Delta} &= \{x_{\Delta}\in R^n, \|x_{\Delta}\| \le \Theta_{x\Delta}\}
\end{aligned}$$

$\Theta_{f\Delta}, \Theta_{g\Delta}, \Theta_{x\Delta}$ are constant boundaries. From (3.46), an equivalent control is found as follows, based on the assumption, $\dot{\sigma}_{\Delta s} \approx 0$:

$$u_1 = \frac{1}{M_{\Delta 3}\hat{\theta}_{g\Delta}\xi_{g\Delta}}(-M_{\Delta 1} - M_{\Delta 2} - M_{\Delta 3}\hat{\theta}_{f\Delta}\xi_{f\Delta} + M_{\Delta 3}\ddot{x}_{\Delta d} - c_{\Delta s}\dot{\varphi}) \tag{3.47}$$

For the next procedure, the second control, u_2, must be designed to combine the equivalent control u_1 to overcome the error from fuzzy approximation. Thus, the total control, u, is found as follows:

$$u = u_1 + u_2 \tag{3.48}$$

In this study, the design of a new robust controller, u_2, is proposed using the optimal controller combined with several robust control schemes. In the classical linear optimal control problem [11], the criterion to be minimized is determined by BM function, as follows:

$$J_1 = \frac{1}{2}[\mathbf{x}_{\Delta}(T)]^T\boldsymbol{\psi}_{\Delta}[x_{\Delta}(T)] + \frac{1}{2}\int_{to}^{T}\left(\mathbf{u}^T(s)\mathbf{R}_{\Delta}(s)\mathbf{u}(s) + \mathbf{x}_{\Delta}^T(s)\mathbf{L}_{\Delta}(s)\mathbf{x}_{\Delta}(s)\right)ds \tag{3.49}$$

where $\mathbf{R}_\Delta(t)$ is the positive definite symmetric matrix function and $\boldsymbol{\psi}_\Delta, \mathbf{L}_\Delta(t)$ are the nonnegative definite symmetric matrix functions. The original cost function is then expressed as

$$J_2 = \sum_{i=1}^{n} \psi_{\Delta ii}|x_{\Delta i}(T)| + \frac{1}{2}\int_{to}^{T} \left(\mathbf{u}^T(s)\mathbf{R}_\Delta(s)\mathbf{u}(s) + \mathbf{x}_\Delta^T(s)\mathbf{L}_\Delta(s)\mathbf{x}_\Delta(s)\right)ds \quad (3.50)$$

In this study, the proposed cost function is chosen by

$$J_3 = \sum_{i=1}^{n} \frac{\psi_{\Delta ii}|x_{\Delta i}(T)|}{\alpha} + \frac{1}{2}\int_{to}^{T} \left(\mathbf{u}^T(s)\mathbf{R}_\Delta(s)\mathbf{u}(s) + \mathbf{x}_\Delta^T(s)\mathbf{L}_\Delta(s)\mathbf{x}_\Delta(s)\right)ds \quad (3.51)$$

The optimal control law for (3.29) is then determined from the following classical linear system:

$$u_2(t) = -\frac{\sigma_{\Delta s}\Gamma\xi_{\Delta z}}{M_{\Delta 3}\hat{\theta}_{g\Delta}\xi_{g\Delta}}\mathbf{R}_\Delta^{-1}\mathbf{B}_\Delta^T q(t), \quad (3.52)$$

where $q(t) = -\mathbf{P}_\Delta(t)\text{sign}[e_\Delta(t)]$. Hence, (3.52) is rewritten as follows:

$$u_2(t) = -\frac{\sigma_{\Delta s}\Gamma\xi_{\Delta z}}{M_{\Delta 3}\hat{\theta}_g\xi_g}\mathbf{R}_\Delta^{-1}\mathbf{B}_\Delta^T\mathbf{P}_\Delta\text{sign}[e_\Delta(t)] \quad (3.53)$$

From (3.53), the controller, u_2, include two sliding surfaces (i.e., $s_{\Delta s}$ and $\sigma_{\Delta s}$) and components of the optimal control. $P_\Delta(t)$ is found by solving the following equation:

$$\dot{\mathbf{P}}_\Delta = \mathbf{PS_1} + \mathbf{S_1^T P} + \mathbf{L} - \sigma_{\Delta s}\Gamma\xi_{\Delta z}\mathbf{PS_2S_2^TP} + \rho\mathbf{PS_2S_2^TP} \quad (3.54)$$

where $\rho \geq \sigma_{\Delta s}\Gamma\xi_{\Delta z}$; ρ is the prescribed attenuation level; $\mathbf{P} = \mathbf{P^T} \geq 0$; and $\xi_{\Delta z}$ is a the consequent membership value of OIT2F. When the value, $\rho = \sigma_{\Delta s}\Gamma\xi_{\Delta z}$, the Riccati-like equation (3.53) is rewritten as

$$\dot{\mathbf{P}}_\Delta = \mathbf{PS_1} + \mathbf{S_1^T P} + \mathbf{L} \quad (3.55)$$

The optimal quadratic performance index is found as follows:

$$J^* = \frac{1}{2}\frac{[e_\Delta(T)]^T\psi[e_\Delta(T)]}{\alpha} \quad (3.56)$$

From (3.47) and (3.53), the total control, u, of the system is determined as follows:

$$u = \frac{1}{M_{\Delta 3}\hat{\theta}_g\xi_g}\left(-M_{\Delta 1} - M_{\Delta 2} - M_{\Delta 3}\hat{\theta}_{f\Delta}\xi_{f\Delta} + M_{\Delta 3}\ddot{x}_{\Delta d} - c_{\Delta s}\dot{\varphi}\right) + \frac{1}{M_{\Delta 3}\hat{\theta}_{g\Delta}}\xi_{g\Delta}\left(\sigma_{\Delta s}\Gamma\xi_{\Delta z}\mathbf{R}_\Delta^{-1}\mathbf{B}_\Delta^T\mathbf{P}_\Delta\text{sign}[e_\Delta(t)]\right) \quad (3.57)$$

The value, Γ, is the adaptive parameter of (3.54), where its boundary is given by $\Delta\Gamma = \{\Gamma \in R, \|\Gamma\| \leq \Theta_\Gamma, \sigma_{\Delta s}\Gamma\xi_{\Delta z} \leq \rho\}$, and Θ_Γ is the constant boundary. In the case of $P_\Delta(t) = 0$, (3.54) is rewritten as follows:

$$\mathbf{PS}_{\Delta 1} + \mathbf{S}_{\Delta 1}^{\mathbf{T}}\mathbf{P} + \mathbf{L} - \sigma_{\Delta s}\Gamma\xi_{\Delta z}\mathbf{PS}_{\Delta 2}\mathbf{S}_{\Delta 2}^{\mathbf{T}}\mathbf{P} + \rho\mathbf{PS}_{\Delta 2}\mathbf{S}_{\Delta 2}^{\mathbf{T}}\mathbf{P} = 0 \quad (3.58)$$

When $\rho = \sigma_{\Delta s}\Gamma\xi_{\Delta z}$, the Riccati-like function (3.58) is rewritten as:

$$\mathbf{PS}_{\Delta 1} + \mathbf{S}_{\Delta 1}^{\mathbf{T}}\mathbf{P} + \mathbf{L} = 0 \tag{3.59}$$

Based on the controller, u, (3.46) can be derived as follows:

$$\begin{aligned}\dot{\sigma}_{\Delta s} &= M_{\Delta 1} + M_{\Delta 2} + M_{\Delta 3}\tilde{\gamma}_f\xi_{f\Delta} + M_{\Delta 3}\tilde{\gamma}_g\xi_{g\Delta}u - M_{\Delta 3}\ddot{x}_{\Delta d} + c_{\Delta s}\dot{\varphi} + w_\Delta \\ &\quad + \left[M_{\Delta 3}\hat{\theta}_{f\Delta}\xi_{f\Delta} + M_{\Delta 3}\hat{\theta}_{g\Delta}\xi_{g\Delta}u\right]\end{aligned} \tag{3.60}$$

where $\tilde{\gamma}_f = \theta_{f\Delta}^* - \hat{\theta}_{f\Delta}, \tilde{\gamma}_g = \theta_{g\Delta}^* - \hat{\theta}_{g\Delta}$. Using (3.57) and (3.60), the following equation is obtained:

$$\dot{\sigma}_{\Delta s} = \left[M_{\Delta 3}\tilde{\gamma}_f\xi_{f\Delta} + M_{\Delta 3}\tilde{\gamma}_g\xi_{g\Delta}u + w_\Delta\right] + \sigma_{\Delta s}\Gamma\xi_{\Delta z}\mathbf{R}_\Delta^{-1}\mathbf{B}_\Delta^T\mathbf{P}_\Delta \mathrm{sign}[e_\Delta(t)] \tag{3.61}$$

Theorem 3.2. *The stability and robustness of the proposed control system are guaranteed if the main control function is designed as given in (3.47), (3.52), and (3.57), and adaptation laws are proposed as follows*:

$$\begin{aligned}\dot{\tilde{\gamma}}_f &= -\mu_1 M_{\Delta 3}\sigma_{\Delta s}\xi_{f\Delta} \\ \dot{\tilde{\gamma}}_g &= -\mu_2 M_{\Delta 3}\sigma_{\Delta s}\xi_{g\Delta}u \\ \dot{\Gamma} &= -\mu_3\sigma_{\Delta s}\Gamma\xi_{\Delta z}\left(\frac{1}{2}\tilde{\boldsymbol{x}}\boldsymbol{P}\boldsymbol{S}_{\Delta 2}\boldsymbol{S}_{\Delta 2}^T\boldsymbol{P}\tilde{\boldsymbol{x}}^T + \boldsymbol{R}_\Delta^{-1}\boldsymbol{B}_\Delta^T\boldsymbol{P}_\Delta \mathrm{sign}(e_\Delta(t))\right)\end{aligned} \tag{3.62}$$

Proof: To proof Theorem 3.2, a Lyapunov function candidate is proposed as follows:

$$L_v = \frac{1}{2}\sigma_{\Delta s}^2 + \frac{1}{2}\tilde{\mathbf{x}}_\Delta\mathbf{P}\tilde{\mathbf{x}}_\Delta T + \frac{1}{2\mu_1}\tilde{\gamma}_f^2 + \frac{1}{2\mu_2}\tilde{\gamma}_g^2 + \frac{1}{2\mu_3}\Gamma^2. \tag{3.P1}$$

The derivative of (3.P1) is obtained as

$$\dot{L}_v = \sigma_{\Delta s}\dot{\sigma}_{\Delta s} + \frac{1}{2}\dot{\tilde{\mathbf{x}}}_\Delta\mathbf{P}\tilde{\mathbf{x}}_\Delta T + \frac{1}{2}\tilde{\mathbf{x}}_\Delta\mathbf{P}\dot{\tilde{\mathbf{x}}}_\Delta^T + \frac{1}{\mu_1}\tilde{\gamma}_f\dot{\tilde{\gamma}}_f + \frac{1}{\mu_2}\tilde{\gamma}_g\dot{\tilde{\gamma}}_g + \frac{1}{\mu_3}\Gamma\dot{\Gamma}. \tag{3.P2}$$

Substituting (3.61) into (3.P2) yields the following:

$$\begin{aligned}\dot{L}_v &= \sigma_{\Delta s}\left[\left(M_{\Delta 3}\tilde{\gamma}_f\xi_{f\Delta} + M_{\Delta 3}\tilde{\gamma}_g\xi_{g\Delta}u + w_\Delta\right) + \sigma_{\Delta s}\Gamma\xi_{\Delta z}\mathbf{R}_\Delta^{-1}\mathbf{B}_\Delta^T\mathbf{P}_\Delta \mathrm{sign}(e_\Delta)\right] \\ &\quad + \frac{1}{2}\dot{\tilde{\mathbf{x}}}_\Delta\mathbf{P}\tilde{\mathbf{x}}_\Delta T + \frac{1}{2}\tilde{\mathbf{x}}_\Delta\mathbf{P}\dot{\tilde{\mathbf{x}}}_\Delta^T + \left[\frac{1}{\mu_1}\tilde{\gamma}_f\dot{\tilde{\gamma}}_f + \frac{1}{\mu_2}\tilde{\gamma}_g\dot{\tilde{\gamma}}_g + \frac{1}{\mu_3}\Gamma\dot{\Gamma}\right] \\ &\Leftrightarrow \dot{L}_v = \left[M_{\Delta 3}\sigma_{\Delta s}\tilde{\gamma}_f\xi_{f\Delta} + \frac{1}{\mu_1}\tilde{\gamma}_f\dot{\tilde{\gamma}}_f\right] + \left[M_{\Delta 3}\sigma_{\Delta s}\tilde{\gamma}_g\xi_{g\Delta}u + \frac{1}{\mu_2}\tilde{\gamma}_g\dot{\tilde{\gamma}}_g\right] \\ &\quad + \left[\frac{1}{2}\sigma_{\Delta s}\Gamma\xi_{\Delta z}\tilde{\mathbf{x}}_\Delta\mathbf{PS}_2\mathbf{S}_2^{\mathbf{T}}\mathbf{P}\tilde{\mathbf{x}}_\Delta T + \sigma_{\Delta s}\Gamma\xi_{\Delta z}\mathbf{R}_\Delta^{-1}\mathbf{B}_\Delta^T\mathbf{P}_\Delta \mathrm{sign}(e_\Delta)\frac{1}{\mu_3}\Gamma\dot{\Gamma}\right] \\ &\quad + \left[\sigma_{\Delta s}w_\Delta - \frac{1}{2}\rho\tilde{\mathbf{x}}_\Delta\mathbf{PS}_2\mathbf{S}_2^{\mathbf{T}}\mathbf{P}\tilde{\mathbf{x}}_\Sigma T - \frac{1}{2}\tilde{\mathbf{x}}_\Delta T\mathbf{L}\tilde{\mathbf{x}}_\Delta T\right]\end{aligned} \tag{3.P3}$$

Equation (3.58) is used to find (3.P3). Next, substituting (3.62) into (3.P3) yields the following:

$$\dot{L}_v = \sigma_{\Delta s} w_\Delta - \frac{1}{2}\rho \tilde{\boldsymbol{x}}_\Delta \boldsymbol{P}\boldsymbol{S}_2\boldsymbol{S}_2^T \boldsymbol{P}\tilde{\boldsymbol{x}}_\Delta T - \frac{1}{2}\tilde{\boldsymbol{x}}_\Delta T \boldsymbol{L}\tilde{\boldsymbol{x}}_\Delta T$$

$$\Leftrightarrow \dot{L}_v = \left[-\frac{1}{2}\tilde{\boldsymbol{x}}_\Delta T\boldsymbol{L}\tilde{\boldsymbol{x}}_\Delta T - \frac{1}{2}(\sigma_{\Delta s} - w_\Delta)^2 + \frac{1}{2}\sigma_{\Delta s}^2 + \frac{1}{2}{w_\Delta}^2\right]$$

$$-\frac{1}{2}\rho \boldsymbol{P}\boldsymbol{S}_2\boldsymbol{S}_2^T\boldsymbol{P} \le -\frac{1}{2}\tilde{\boldsymbol{x}}_\Delta T\boldsymbol{L}\tilde{\boldsymbol{x}}_\Delta T + \frac{1}{2}\sigma_{\Delta s}^2 + \frac{1}{2}w_\Delta^2$$

$$\Leftrightarrow \dot{L}_v = -\frac{1}{2}\tilde{\boldsymbol{x}}_\Delta T\boldsymbol{L}\tilde{\boldsymbol{x}}_\Delta T + \frac{1}{2}\sigma_{\Delta s}^2 + \frac{1}{2}w_\Delta^2 \le -\frac{1}{2}\tilde{\boldsymbol{x}}_\Delta T\boldsymbol{L}\tilde{\boldsymbol{x}}_\Delta T + \frac{1}{2}w_\Delta^2. \quad (3.P4)$$

The above equation cannot be used for stability conclusion. Hence, it needs to be integrated from $t = 0$ to $t = T$;

$$L_v(0) - L_v(T) + \frac{1}{2}\int_0^T w_\Delta^2 dt \ge \frac{1}{2}\int_0^T \tilde{\boldsymbol{x}}_\Delta \boldsymbol{L}\tilde{\boldsymbol{x}}_\Delta T dt \quad (3.P5)$$

where $L_v(0) = (1/2)\sigma_{\Delta s}^2(0) + (1/2)\tilde{\boldsymbol{x}}_\Delta(0)\boldsymbol{P}\tilde{\boldsymbol{x}}_\Delta T(0) + (1/2\mu_1)\tilde{\gamma}^2{}_f(0) + (1/2\mu_2)$ $\tilde{\gamma}^2{}_g(0) + (1/2\mu_3)\Gamma^2(0)$. The value $L_v(T)$ is always positive. Thus, (3.P5) becomes

$$L_v(0) + \frac{1}{2}\int_0^T w_\Delta^2 dt \ge \frac{1}{2}\int_0^T \tilde{\mathbf{x}}_\Delta \mathbf{L}\tilde{\mathbf{x}}_\Delta^T dt \ge 0 \quad (3.P6)$$

Thus, the proof is complete. The boundedness of the parameters, $\tilde{\gamma}_f$ and $\tilde{\gamma}_g$, are defined by closed sets:

$$\Xi_1 = \left\{\tilde{\gamma}_f \middle| \|\tilde{\gamma}_f\| \le \aleph_{f\Delta}\right\}, \quad \Xi_2 = \left\{\tilde{\gamma}_g \middle| \|\tilde{\gamma}_g\| \le \aleph_{g\Delta}\right\},$$

$$\Xi_{\delta 1} = \left\{\tilde{\gamma}_f \middle| \|\tilde{\gamma}_f\| \le \aleph_{f\Delta} + \delta_1\right\}, \quad \Xi_{\delta 2} = \left\{\tilde{\gamma}_g \middle| \|\tilde{\gamma}_g\| \le \aleph_{g\Delta} + \delta_2\right\},$$

where $\aleph_{f\Delta}$, $\aleph_{g\Delta}$, δ_1, and δ_2 are the choosing parameters. Using the projection algorithm, the above adaptation laws are rewritten as follows:

$$\dot{\tilde{\gamma}}_f = \begin{cases} -\mu_1 M_{\Delta 3}\sigma_{\Delta s}\xi_{f\Delta} & \text{if } \|\tilde{\gamma}_{\mathrm{f}}\| < \aleph_{f\Delta} \text{ or} \left(\|\tilde{\gamma}_{\mathrm{f}}\| = \aleph_{f\Delta} \text{ and } M_{\Delta 3}\sigma_{\Delta s}\xi_{f\Delta}\tilde{\gamma}_{\mathrm{f}} \ge 0\right) \\ -\mu_1 M_{\Delta 3}\sigma_{\Delta s}\xi_{f\Sigma} + \mu_1 \dfrac{\left(\|\tilde{\gamma}_f\|^2 - \aleph_{f\Delta}\right) M_{\Delta 3}\sigma_{\Delta s}\xi_{f\Delta}\tilde{\gamma}_f}{\delta_1 \|\tilde{\gamma}_f\|^2} & \text{if } \|\tilde{\gamma}_{\mathrm{f}}\| = \aleph_{f\Delta} \text{ and } M_{\Delta 3}\sigma_{\Delta s}\xi_{f\Delta}\tilde{\gamma}_{\mathrm{f}} < 0 \end{cases} \quad (3.63)$$

$$\dot{\tilde{\gamma}}_g = \begin{cases} -\mu_2 M_{\Delta 3}\sigma_{\Delta s}\xi_{g\Delta}u & \text{if } \|\tilde{\gamma}_{\mathrm{g}}\| < \aleph_{g\Delta} \text{ or } \left(\|\tilde{\gamma}_{\mathrm{g}}\| = \aleph_{g\Delta} \text{ and } M_{\Delta 3}\sigma_{\Delta s}\xi_{g\Delta}u\tilde{\gamma}_{\mathrm{g}} \ge 0\right) \\ -\mu_2 M_{\Delta 3}\sigma_{\Delta s}\xi_{g\Delta}u + \mu_2 \dfrac{\left(\|\tilde{\gamma}_g\|^2 - \aleph_{g\Delta}\right) M_{\Delta 3}\sigma_{\Delta s}\xi_{g\Delta}u\tilde{\gamma}_g}{\delta_2 \|\tilde{\gamma}_g\|^2} & \text{if } \|\tilde{\gamma}_{\mathrm{g}}\| = \aleph_{g\Delta} \text{ and } M_{\Delta 3}\sigma_{\Delta s}\xi_{g\Delta}u\tilde{\gamma}_{\mathrm{g}} < 0 \end{cases} \quad (3.64)$$

$$\dot{\Gamma} = \begin{cases} & \text{if} \|\Gamma\| < \Theta_\Gamma \\ -\mu_3 \sigma_{\Delta s} \Gamma \xi_{\Delta z} \begin{pmatrix} \frac{1}{2} \tilde{\mathbf{x}}_\Delta \mathbf{P} \mathbf{S}_{\Delta 2} \mathbf{S}_{\Delta 2}^T \mathbf{P} \tilde{\mathbf{x}}_\Sigma T + \\ + \mathbf{R}_\Delta^{-1} \mathbf{B}_\Delta^T \mathbf{P}_\Delta sign(e_\Delta) \end{pmatrix} & \text{or} \begin{pmatrix} \|\Gamma\| = \Theta_\Gamma + \delta_3 \quad \text{and} \\ \sigma_{\Delta s} \Gamma \xi_{\Delta z} \begin{pmatrix} \frac{1}{2} \tilde{\mathbf{x}}_\Delta \mathbf{P} \mathbf{S}_{\Delta 2} \mathbf{S}_{\Delta 2}^T \mathbf{P} \tilde{\mathbf{x}}_\Delta T + \\ + \mathbf{R}_\Delta^{-1} \mathbf{B}_\Delta^T \mathbf{P}_\Delta sign(e_\Delta) \end{pmatrix} \geq 0 \end{pmatrix} \\ -\mu_3 \sigma_{\Delta s} \Gamma \xi_{\Delta z} \begin{pmatrix} \frac{1}{2} \tilde{\mathbf{x}}_\Delta \mathbf{P} \mathbf{S}_{\Delta 2} \mathbf{S}_{\Delta 2}^T \mathbf{P} \tilde{\mathbf{x}}_\Delta T + \\ + \mathbf{R}_\Delta^{-1} \mathbf{B}_\Delta^T \mathbf{P}_\Delta sign(e_\Delta) \end{pmatrix} + & \\ +\mu_3 \dfrac{\left(\|\Gamma\|^2 - \Theta_\Gamma \right) \sigma_{\Delta s} \Gamma \xi_{\Delta z} \begin{pmatrix} \frac{1}{2} \tilde{\mathbf{x}}_\Delta \mathbf{P} \mathbf{S}_{\Delta 2} \mathbf{S}_{\Delta 2}^T \mathbf{P} \tilde{\mathbf{x}}_\Delta^T + \\ + \mathbf{R}_\Delta^{-1} \mathbf{B}_\Delta^T \mathbf{P}_\Delta sign(e_\Delta) \end{pmatrix}}{\delta_3 \|\Gamma\|^2} & \begin{matrix} \text{if} \|\Gamma\| = \Theta_\Gamma \quad \text{and} \\ \sigma_{\Delta s} \Gamma \xi_{\Delta z} \begin{pmatrix} \frac{1}{2} \tilde{\mathbf{x}}_\Delta \mathbf{P} \mathbf{S}_{\Delta 2} \mathbf{S}_{\Delta 2}^T \mathbf{P} \tilde{\mathbf{x}}_\Delta T + \\ + \mathbf{R}_\Delta^{-1} \mathbf{B}_\Delta^T \mathbf{P}_\Delta sign(e_\Delta) \end{pmatrix} < 0 \end{matrix} \end{cases} \tag{3.65}$$

where δ_1, δ_2, and δ_3 are the choosing parameter-related boundaries of $f_\Delta(x_\Delta)$, $g_\Delta(x_\Delta)$, Γ, respectively.

3.4 Application to vibration control

3.4.1 Vehicle seat suspension system

To validate superior control performance of the proposed hybrid optimal controller, the vehicle suspension seat system, featuring the MR fluid-based damper (i.e., MR damper) is adopted [32,33]. The mechanical model associated with the hardware is shown in Figure 3.5. The governing equation of the seat suspension is depicted as follows:

$$\begin{aligned} \dot{x}_{\Delta 11} &= \dot{x}_s = x_{\Delta 22} \\ \dot{x}_{\Delta 22} &= f_{\Delta 11}(x_{\Delta 11}, x_{\Delta 22}, x_{\Delta 33}, x_{\Delta 44}) + g_{\Delta 11}(x_{\Delta 11}, x_{\Delta 22}, x_{\Delta 33}, x_{\Delta 44}) u \\ \dot{x}_{\Delta 33} &= \dot{x}_1 = x_{\Delta 44} \\ \dot{x}_{\Delta 44} &= f_{\Delta 22}(x_{\Delta 11}, x_{\Delta 22}, x_{\Delta 33}, x_{\Delta 44}) \end{aligned} \tag{3.66}$$

where

$$\begin{aligned} f_{\Delta 11}(x_{\Delta 11}, x_{\Delta 22}, x_{\Delta 33}, x_{\Delta 44}) &= -\frac{k_s}{m_s}(x_{\Delta 11} - x_0) - \frac{c_s}{m_s}(x_{\Delta 22} - \dot{x}_0) + \frac{k_1}{m_s}(x_{\Delta 33} - x_{11}) \\ &\quad + \frac{c_1}{m_s}(x_{\Delta 44} - x_{22}), \\ g_{\Delta 11}(x_{\Delta 11}, x_{\Delta 22}, x_{\Delta 33}, x_{\Delta 44}) &= \frac{1}{m_s}, \quad u = F_{MR}, \\ f_{\Delta 22}(x_{\Delta 11}, x_{\Delta 22}, x_{\Delta 33}, x_{\Delta 44}) &= -\frac{k_1}{m_1}(x_{\Delta 33} - x_{\Delta 11}) - \frac{c_1}{m_1}(x_{\Delta 44} - x_{\Delta 22}). \end{aligned}$$

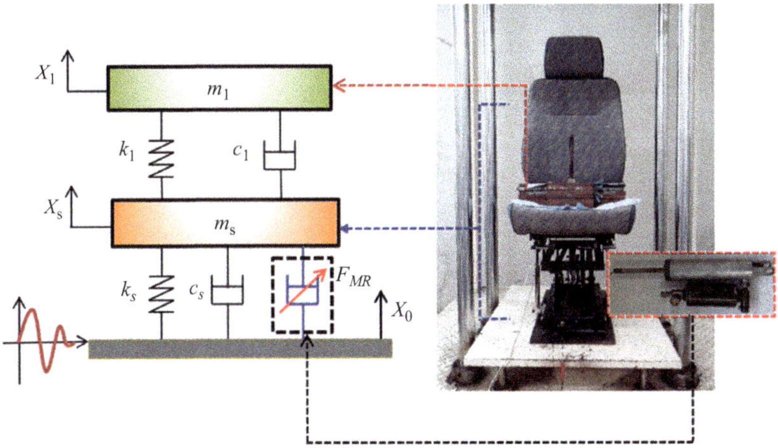

Figure 3.5 Mechanical model of the seat suspension associated with the hardware

The state variables of the system include $x_{\Delta 11}, x_{\Delta 22}, x_{\Delta 33}, x_{\Delta 44}$ and related x_s and x_1. In the proposed control system, the damping force produced from the MR damper is used to control unwanted vibration. Thus, its modeling is required. In this work, the following damping force [32] model of MR damper is used:

$$F_{MR} = (c_a + c_b v)(x_{\Delta 44} - x_{\Delta 22}) + k_0(x_{\Delta 33} - x_{\Delta 11}) + (\alpha_a + \alpha_b v)\phi \tag{3.67}$$

where $\dot{\phi} = -\kappa|x_{\Delta 44} - x_{\Delta 22}| + \lambda_d(x_{\Delta 44} - x_{\Delta 22})|\phi| + \varphi_d(x_{\Delta 44} - x_{\Delta 22})$.

The damping force of the MR damper depends on the magnitude of the input voltage, v, which is directly convertible to the input current. Therefore, the input voltage to be applied to the MR damper for achieving sufficient damping force can be determined as follows:

$$v = \frac{F_{MR} - [c_a(x_{\Delta 44} - x_{\Delta 22}) + k_0(x_{\Delta 33} - x_{\Delta 11}) + \alpha_a\phi]}{c_b(x_{\Delta 44} - x_{\Delta 22}) + \alpha_b\phi} \tag{3.68}$$

This voltage is changed to current, producing a magnetic field around the electrical coil of the damper to control the damping force. The system parameters given by Tables 3.1–3.3 are used for simulations and experiments.

3.4.2 Simulation results and discussions

In simulations, the excitation of random step wave road is used, as shown in Figure 3.6, and two existing controllers are adopted to compare with the proposed hybrid controller. Excitation data are collected from the real road, containing uncertainty and disturbances. The hybrid controller developed in [32] is adopted as *comparative control 1*, and the hybrid controller formulated in [33] is adopted as *comparative control 2*. The simulation is carried out with the control parameters listed in Table 3.4. The simulation includes two parts: displacement tracking control and vibration control. The prescribed parameters listed in Table 3.5 and the

Table 3.1 Parameters of the seat-suspension system

Parameter	**Value**
Mass of seat, m_s	27 kg
Mass of driver, m_1	77 kg
Stiffness coefficient of seat, k_s	17,830 N/m
Stiffness coefficient of torso, k_1	49,340 N/m
Damping coefficient of seat, c_s	1,500 N s/m
Damping coefficient of torso, c_1	2,475 N s/m

Table 3.2 Parameters of MR damper

Parameter	**Value**
Stiffness of gas chamber, k_e	94.6166 N/m
Damping coefficient, c_e	794.4681 N s/m
Piston area, A_p	0.00113354 m^2
Piston rod area, A_r	0.0000785 m^2
Gap of magnetic pole, h	0.01 m
Length of magnetic pole, L	0.05 m
MR inherent coefficient, α_1	26.77185
MR inherent coefficient, α_2	0.60444

Table 3.3 Parameters to calculate the applied current of MR damper

Parameter	**Value**
Linear spring stiffness, k_0	0 N/m
Viscous damping coefficient, c_a	990 N s/m
Viscous damping coefficient influenced by v, c_b	3,095 N s/m V
Stiffness of φ_d, α_a	545 N/m
Stiffness of φ_d influenced by v, α_b	620 N/mV
Positive parameter of hysteresis loop, φ	4
Positive parameter of hysteresis loop, κ	48
Positive parameter of hysteresis loop, λ	48

centroid of the fuzzy model shown in Table 3.6 are used. The damping force of the MR damper is 1,000 N ($\pm$5%) at the applied current, 2 A (4 V).

The simulation results are shown in Figures 3.7–3.10. Results include the displacement of seat, displacement of driver, prescribed sliding surface of the proposed control, comparative control 1, and the acceleration of the driver. The displacements of seat and driver are shown in Figures 3.7 and 3.8, respectively. Three controllers show good performance with vibration. However, from the amplification shown in Figures 3.7(b) and 3.8(b), the control response of the proposed controller is better than the others. In Figure 3.9, the prescribed sliding surface of the proposed

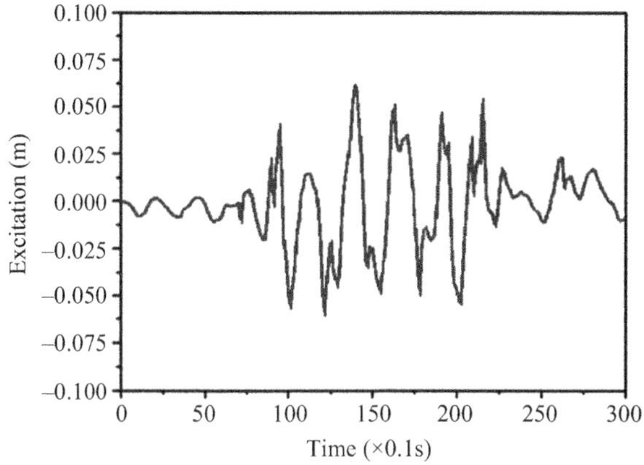

Figure 3.6 Excitation signal of random step-wave road

Table 3.4 Control parameters for the simulation

Simulated parameter	Value
Sigma value for Gaussian function of the fuzzy model	0.4
Values of the classical sliding surface $[k_1, k_2]$	$[1, 20]$
Constant value Γ	10
Constant $c_{\Delta s}$	500
Matrix $\mathbf{L}$	$[-20; 0-2]$
Constants μ_1, μ_2, μ_3 of adaptation laws	10, 10, 10
Values of $\aleph_{f\Delta}$, $\aleph_{g\Delta}$, Θ_Γ	0.1, 0.1, 0.1
Values of δ_1, δ_2, δ_3	0.1, 0.1, 0.1
Initial states for the dynamic states	$[0.035 \quad 2.5]$
Initial states for the observer	$[0.035 \quad 0]$

Table 3.5 Parameters of desired prescribed performance and applied prescribed performance

Parameter	Desired prescribed performance	Applied prescribed performance
Initial value $\lambda(0)$	0.5	0.5
Infinity value λ_∞	0.001	0.001
Exponential value l	1	0.00047

Table 3.6 Centroid vectors of fuzzy rules using Gaussian model

Data	Centroid vector	
	Displacement ($\times 10^{-2}$ m)	Acceleration (m/s^2)
Random step wave	[−8.926 −0.018 −0.007 0.004 0.026 6.017]	[−11.250 −1.355 −0.520 0.085 0.906 7.661]

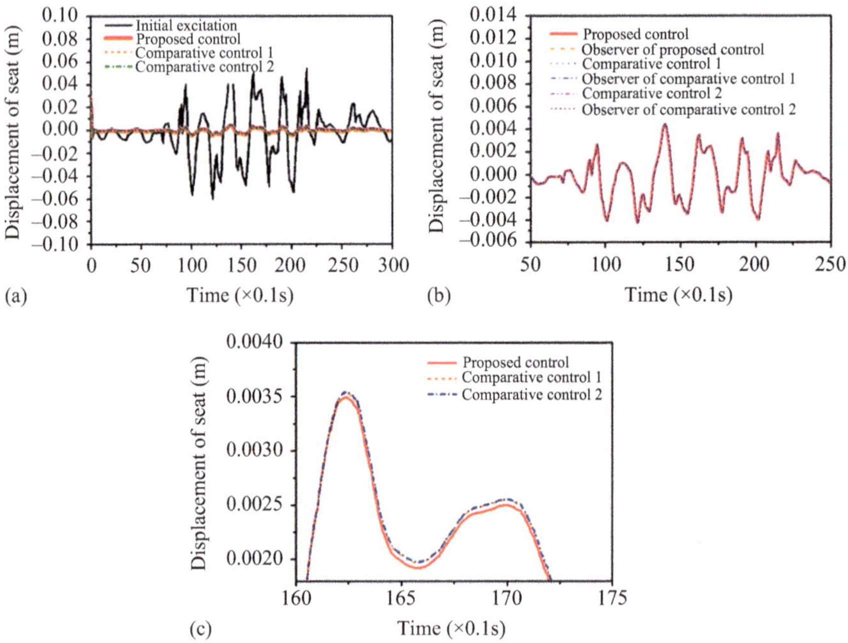

Figure 3.7 Displacement at the seat: (a) full view of simulation, (b) controls and observers, and (c) enlarged view of displacement of seat

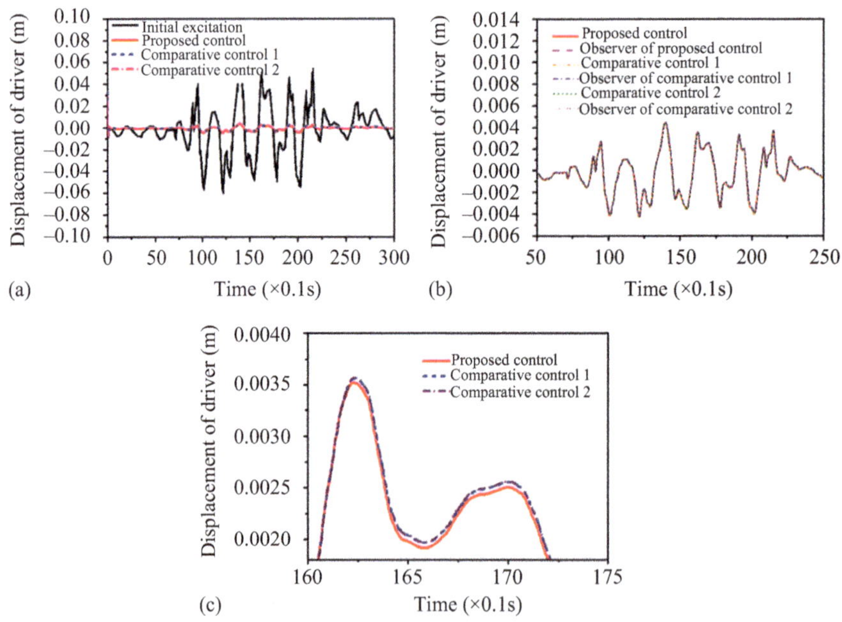

Figure 3.8 Displacement of the driver: (a) full view of simulation, (b) controls and observers, and (c) enlarged view of displacement of driver

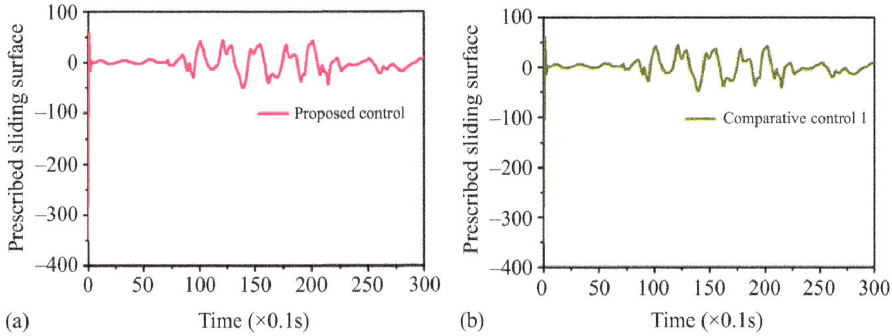

Figure 3.9 Prescribed sliding surface: (a) proposed control and (b) comparative control 1

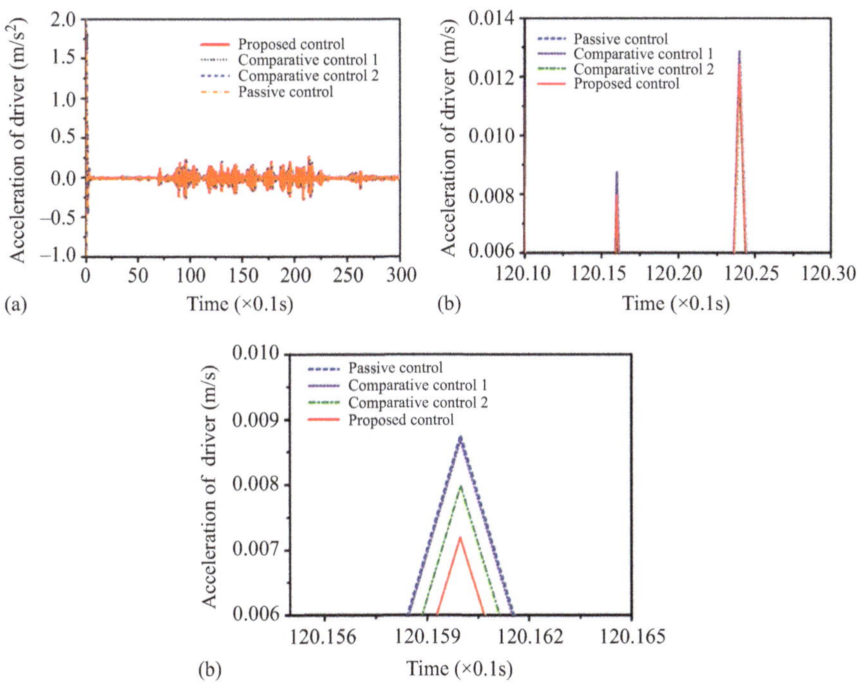

Figure 3.10 Acceleration of driver: (a) full view of simulation, (b) enlarged view of acceleration of driver, and (c) enlarged view of the first peak acceleration of driver

controller is similar to that of comparative controller 1, showing the variation according to the excitation to find the best value for control system. In Figure 3.10, the acceleration of the driver with the proposed controller is better than the others and passive control (no control force). From the above results, the proposed controller provides better control performances than comparative controllers, because more accurate control input based on the optimal controller is applied to the MR damper, satisfying the imposed objective and constraint functions. The higher control performance also indicates better stability of the system subjected to severe disturbances.

3.5 Conclusion

This study presented a new optimal control and a new hybrid optimal controller to enhance the robustness and the stability of dynamic systems subjected to uncertainties and external disturbances.

In the first proposed control, the optimal control law was proposed by adopting adjustable gain of the sliding mode controller based on the objective function of the BM criterion. The eminent property of the proposed optimal control law is to have the value of the gain which is larger than 1. This point is a breakthrough to overcome the disadvantage of the conventional optimal controller showing the values of the definite symmetry matrices less than 1. By adopting the adjustable gain value, the nonnegative definite symmetric matrix function of the Meyer term is reduced, and hence the energy consumption for control action is also reduced. It has been validated from simulation that four crucial values of the proposed optimal control law are less than those obtained from comparative optimal controllers where the gain value is constant. The reduction of these values directly indicates the outstanding performances of the proposed optimal control law showing less energy consumption, higher stability and higher robustness with the faster convergence.

In the second proposed controller, a new composite controller is presented. As a first step, the BM criterion of the optimal controller was modified to meet the system variation. The modified sliding mode control of the proposed controller included two steps: classical control and prescribed performance. These steps were used for two goals: robust control of severe disturbances and guaranteed tracking control. The proposed controller also used the *H*-infinity technique as a bridge to connect the sub-controls and improve system performance. The fuzzy neural networks model was used as a big-filter system to choose the optimal values for the next calculation. Hence, using the proposed controller, advantages of both classical controls (i.e., optimal control, sliding mode control, prescribed performance, and *H*-infinity technique) and intelligent controls (i.e., fuzzy neural networks model) were acquired. This characteristic posits a new super-controller for various dynamic systems to be operated under severe disturbance and uncertainty. After formulating the proposed hybrid controller, it was applied to the vehicle seat suspension for evaluation of vibration control performances. To demonstrate the superiority of the proposed controller, two existing hybrid controllers having very similar structures to the proposed one were adopted. We showed through simulation that the proposed hybrid optimal controller provides better vibration control performances than the two comparative controllers.

Acknowledgment

This research is funded by the Ministry of Education and Training (Vietnam) under grant number B2018-VGU-07 (KYTH-36). This financial support is gratefully acknowledged.

Appendix A

Consider the optimal control problem for a linear system (3.2) with respect to the BM criterion without a non-integral term:

$$J = \frac{1}{2}\int_{to}^{T} \left(u^T(s)R_\Sigma(s)u(s) + x^T(s)L_\Sigma(s)x(s)\right)ds \tag{A.1}$$

As follows from the optimal polynomial-quadratic regulator theory [12], the optimal control law can be represented as

$$u^*(t) = (R_\Sigma(t))^{-1}B_\Sigma^T(t)Q_\Sigma(t)x(t) \tag{A.2}$$

where $Q_\Sigma(t)$ satisfies the Riccati equation:

$$\dot{Q}_\Sigma(t) = -A_\Sigma^T(t)Q_\Sigma(t) - Q_\Sigma(t)A_\Sigma(t) + L_\Sigma(t) - Q_\Sigma(t)B_\Sigma(t)R_\Sigma^{-1}(t)B_\Sigma^T(t)Q_\Sigma(t) \tag{A.3}$$

With the terminal condition $Q_\Sigma(t) = 0$. The optimally controlled system takes the form

$$\dot{x}(t) = A_\Sigma(t)x(t) + B_\Sigma(t)(R_\Sigma(t))^{-1}B_\Sigma^T(t)Q_\Sigma(t)x(t), \quad x(t_0) = x_0 \tag{A.4}$$

In this study, new gain matrix, $Q_\Sigma^{**} = Q_\Sigma(t)\alpha|x(t)|$ which satisfies the equation

$$\begin{aligned}
Q_\Sigma^{**}(t) &= \frac{d(Q_\Sigma(t)\alpha|x(t)|)}{dt} \\
&= \frac{dQ_\Sigma(t)}{dt}\alpha|x(t)| + Q_\Sigma(t)\alpha\frac{d(|x(t)|)}{dt} \\
&= (-A_\Sigma^T(t)Q_\Sigma(t) - Q_\Sigma(t)A_\Sigma(t) + L_\Sigma(t) - Q_\Sigma(t)B_\Sigma(t)R_\Sigma^{-1}(t)B_\Sigma^T(t)Q_\Sigma(t))\alpha|x(t)| \\
&\quad + Q_\Sigma(t)\alpha(A_\Sigma(t)|x(t)| + B_\Sigma(t)R_\Sigma^{-1}(t)B_\Sigma^T(t)Q_\Sigma(t)|x(t)|) \\
&= L_\Sigma(t)\alpha|x(t)| - A_\Sigma^T(t)Q_\Sigma(t)\alpha|x(t)| \\
&= L_\Sigma(t)\alpha|x(t)| - A_\Sigma^T Q_\Sigma^{**}
\end{aligned} \tag{A.5}$$

References

[1] Hamidreza Modares, Frank L. Lewis, and Mohammad Bagher Naghibi Sistani. 'Adaptive optimal control constrained input systems using policy iteration and neural networks'. *IEEE Transactions on Neural Networks and Learning Systems*. 2013; **24**(10): 1513–1525.

[2] Kyriakos G. Vamvoudakis, Marcio Fantini Miranda, and João P. Hespanha. 'Asymptotically stable adaptive–optimal control algorithm with saturating actuators and relaxed persistence of excitation'. *IEEE Transactions on Neural Networks and Learning Systems*. 2016; **27**(11): 2386–2398.

[3] Ding Wang, Derong Liu, Qichao Zhang, and Dongbin Zhao. 'Data-based adaptive critic designs for nonlinear robust optimal control with uncertain dynamics'. *IEEE Transactions on Systems, Man, and Cybernetics Systems*. 2016; **46**(11): 1544–1555.

[4] Manuel Soler, Maryam Kamgarpour, Javier Lloret, and John Lygeros. 'A hybrid optimal control approach to fuel-efficient aircraft conflict avoidance'. *IEEE Transactions on Intelligent Transportation Systems*. 2016; **17**(7): 1826–1838.

[5] Hamidreza Modares, Frank L. Lewis, and Zhong-Ping Jiang. 'Optimal output-feedback control of unknown continuous-time linear systems using off-policy reinforcement learning'. *IEEE Transactions on Cybernetics*. 2016; **46** (11): 2401–2410.

[6] Wencheng Luo, Yun-Chung Chu, and Keck-Voon Ling. 'Inverse optimal adaptive control for attitude tracking of spacecraft'. *IEEE Transactions on Automatic Control*. 2005; **50**(11): 1639–1654.

[7] Tao Bian, Yu Jiang, and Zhong-Ping Jiang. 'Decentralized adaptive optimal control of large scale systems with application to power systems'. *IEEE Transactions on Industrial Electronics*. 2015; **62**(4): 2439–2447.

[8] Ruizhuo Song, Frank L. Lewis, Qinglai Wei, and Huaguang Zhang. 'Off-policy actor-critic structure for optimal control of unknown systems with disturbances'. *IEEE Transactions on Cybernetics*. 2016; **46**(5): 1041–1050.

[9] Michael Basin, Dario Calderon-Alvarez, and Antonella Ferrara. 'Sliding mode regulator as solution to optimal control problem'. *Proceedings of the 47th IEEE Conference on Decision and Control*; Cancun, Mexico, 2008. IEEE; pp. 2184–2189.

[10] Michael Basin and Dario Calderon-Alvarez. 'Optimal controller for uncertain stochastic polynomial systems'. *Journal of the Franklin Institute*. 2009; **346**: 206–222.

[11] Michael Basin and Dario Calderon-Alvarez. 'Sliding mode regulator as solution to optimal control problem for non-linear polynomial systems'. *Journal of the Franklin Institute*. 2010; **347**: 910–922.

[12] Michael Basin, Pablo Rodriguez-Ramirez, Antonella Ferrara, and Dario Calderon-Alvarez. 'Sliding mode optimal control for linear systems'. *Journal of the Franklin Institute*. 2012; **349**: 1350–1363.

[13] Mehdi Nikkhah, Hashem Ashrafiuon, and Kenneth R. Muske. 'Optimal sliding mode control for underactuated systems'. *Proceedings of the 2006 American Control Conference;* Minneapolis, Minnesota, USA, 2006. IEEE; pp. 4688–4693.

[14] Noriaki Sakamoto. 'Optimal control problem via self-adaptation sliding mode controller with neural network'. *Electronics and Communications.* 2011; **94** (11): 1–8. (Translated from Denki Gakkai Ronbunshi, **131-C**(5): 1043–1049)

[15] Yinyan Zhang, Shuai Li, and Xiaoping Liu. 'Adaptive near-optimal control of uncertain systems with application to underactuated surface vessels'. *IEEE Transactions on Control Systems Technology.* 2018; **26**(4): 1204–1218.

[16] Do Xuan Phu and Seung-Bok Choi. 'Vibration control of a ship engine system using high-load magnetorheological mounts associated with a new indirect fuzzy sliding mode controller'. *Smart Materials and Structures.* 2015; **24**: 025009.

[17] Do Xuan Phu, Do Kyun Shin, and Seung-Bok Choi. 'Design of a new adaptive fuzzy controller and its application to vibration control of a vehicle seat installed with an MR damper'. *Smart Materials and Structures.* 2015; **24**: 085012.

[18] Do Xuan Phu, Kruti Shah, and Seung-Bok Choi. 'Design of a new adaptive fuzzy controller and its implementation for the damping force control of a magnetorheological damper'. *Smart Materials and Structures.* 2014; **23**: 065012.

[19] Do Xuan Phu, Seung-Bok Choi, Yang-Sub Lee, and Moon-Sihk Han. 'Vibration control of a vehicle's seat suspension featuring a magnetorheological damper based on a new adaptive fuzzy sliding-mode controller'. *Proceedings of the Institution of Mechanical Engineers, Part D: Journal of Automobile Engineering.* 2015; **230**(4): 437–458.

[20] Do Xuan Phu, Nguyen Vien Quoc, Joon-Hee Park, and Seung-Bok Choi. 'Design of a novel adaptive fuzzy sliding mode controller and application for vibration control of magnetorheological mount'. *Proceedings of the Institution of Mechanical Engineers, Part C: J Mechanical Engineering Science.* 2014; **228**(13): 2285–2302.

[21] Xuan Phu Do, Kruti Shah, and Seung-Bok Choi. 'Damping force tracking control of MR damper system using a new direct adaptive fuzzy controller'. *Shock and Vibration.* 2015; 947937.

[22] Nilesh N. Karnik and Jerry M. Mendel. 'Centroid of a type 2 fuzzy set'. *Information Sciences.* 2001; **132**: 195–220.

[23] Haytham Fayek, Irraivan Elamvazuthi, Perumal Nallagownden, and Bala Venkatesh. 'A controller based on optimal type-2 fuzzy logic: systematic design, optimization and real-time implementation'. *ISA Transactions.* 2014; **53**: 1583–1591.

[24] Xianchang Wang, Xiaodong Liu, and Lishi Zhang. 'A rapid fuzzy rule clustering method based on granular computing'. *Applied Soft Computing.* 2014; **24**: 534–542.

[25] Jerry M. Mendel. 'Computing derivatives in interval type-2 fuzzy logic systems'. *IEEE Transactions on Fuzzy Systems.* 2004; **12**(1): 84–98.

[26] Chia-Feng Juang and Yu-Wei Tsao. 'A self-evolving interval type-2 fuzzy neural network with online structure and parameter learning'. *IEEE Transactions on Fuzzy Systems*. 2008; **16**(6): 1411–1424.
[27] Dongrui Wu. 'Approaches for reducing the computational cost of interval type-2 fuzzy logic systems: overview and comparisons'. *IEEE Transactions on Fuzzy Systems*. 2013; **21**(1): 80–99.
[28] Chia-Feng Juang and Chi-You Chen. 'Data-driven interval type-2 neural fuzzy system with high learning accuracy and improved model interpretability'. *IEEE Transactions on Cybernetics*. 2013; **43**(6): 1781–1795.
[29] Jerry M. Mendel and Xinwang Liu. 'Simplified interval type-2 fuzzy logic systems'. *IEEE Transactions on Fuzzy Systems*. 2013; **21**(6): 1056–1069.
[30] Qilian Liang and Jerry M. Mendel. 'Interval type-2 fuzzy logic systems: theory and design'. *IEEE Transactions on Fuzzy Systems*. 2000; **8**(5): 535–550.
[31] Hongwei Wu and Jerry M. Mendel. 'Uncertainty bounds and their use in the design of interval type-2 fuzzy logic systems'. *IEEE Transactions on Fuzzy Systems*. 2002; **10**(5): 622–639.
[32] Do Xuan Phu, Ta Duc Huy, Van Mien, and Seung Bok Choi. 'A new composite adaptive controller featuring the neural network and prescribed sliding surface with application to vibration control'. *Mechanical Systems and Signal Processing*. 2018; **107**: 409–428.
[33] Do Xuan Phu, Jin Hee An, and Seung Bok Choi. 'A novel adaptive PID controller with application to vibration control of a semi-active vehicle seat suspension'. *Applied Sciences*. 2017; **7**: 1055.

Chapter 4

Developments in modeling of magnetorheological actuators

Janusz Gołdasz[1,2]

A mathematical model of an object or a physical process is always a crucial component of the model development process. Given a set of requirements and objectives, the model is usually required to copy the object/process behavior with a realistic accuracy and reasonable computational effort. It is then important that all the necessary features and characteristics of the examined object/process are incorporated in the modeling process. With the multi-physics objects such as magnetorheological (MR) actuators, the process becomes even more complicated as they require coupling the expertise from various and sometimes distant fields of engineering and science. The multi-physics nature of MR actuators (also known as MR dampers) calls for the expertise in solid body mechanics, fluid mechanics, heat transfer, electromagnetics, power electronics, chemistry, control theory, etc. It requires the use of appropriate tools and methodologies for specific tasks. Therefore, a need to outline various methodologies and models applied in R&D studies on MR dampers/actuators has arisen. Briefly, in the chapter the author discusses how the models and methods can be used in solving engineering and research problems.

4.1 Introduction

By a wide margin, the most common application area of MR actuators regardless of their operating mode is vibration suppression. A typical vibration control system featuring MR devices incorporates a plant, sensors, a controller, a power supply and an MR device. As shown in Figure 4.1, the information from the plant is acquired by a signal processing module via sensors, and then processed through a relevant control algorithm of choice. The calculated algorithm's output is the commanded current signal i_{cmd} of appropriate magnitude. The command is then acquired by the control driver module resulting in the input voltage level u at the actuator control circuit (coil) terminals and the circuital current i_c. The operation results in the

[1]Faculty of Electrical and Computer Engineering, Cracow University of Technology, Poland
[2]Technical Center Kraków, BWI Group, Poland

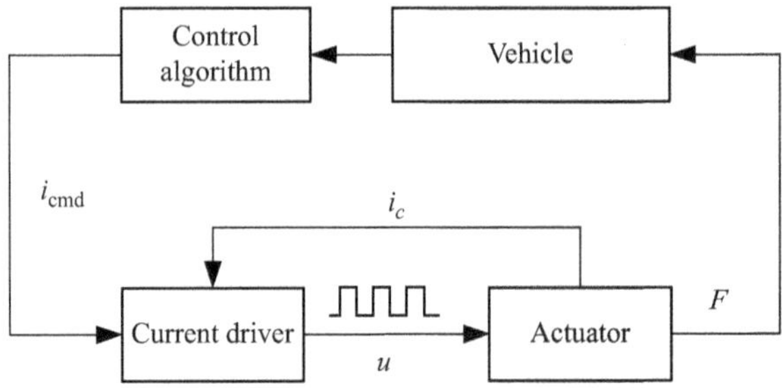

Figure 4.1 Generic layout of a vibration control system with an MR actuator [1]

appropriate output force or torque. The generated force/torque is applied to the plant/object usually in an effort to minimize the plant's vibrations.

By principle, an MR actuator or a damper incorporates housing with the smart fluid and a control valve for the required output generation. The control valve usually integrates a radial and/or annular fluid path(s), an electromagnetic circuit connected to the external power supply. In a typical MR control system, the commanded current is delivered to the coil assembly via a pulse-width modulated type current controller. Next, as shown in Figure 4.2, the current induces the magnetic flux in the structure incorporating the core, the flow path(s) and the outer sleeve surrounding the core. As the flux passes the flow, it energizes the fluid. The magnetized fluid develops a yield stress thus resulting in a resistance-to-flow buildup. The output is force or torque that can be varied according to the external excitation inputs. The valve may contain more flow paths, an arbitrary number of coils connected in series or parallel. The flow paths may be of constant or variable height or move relative to each other.

Modeling the behavior of MR actuators is performed for different reasons by the industry and academia. Researchers usually develop a model to provide an insight into physics governing a particular process of their interest. In the industry, a model is most often employed to aid engineers in the design of the actuator or a complete system. One fairly standard task that an engineer faces is to design a device which meets specific steady-state and dynamic performance goals subject to prescribed geometric constraints. Typical goals when designing MR devices are minimum (maximum) off-state (on-state) pressure drop (force) across the valve at a reference volumetric flow rate (piston velocity), dynamic range, optimum response time, high bandwidth, minimum control volume, maximum operating temperature, etc. [3,4]. To achieve them, manipulating the geometry of the valve as well as fluid's properties and/or characteristics of the electromagnetic circuit needs to be done. Many of these goals are contradictory; e.g. high yielding (threshold) pressure in the on-state condition can be achieved through either decreasing the flow channel height or extending the working length—however, that operation has a negative impact on the off-state pressure drop. The dynamic range goal can be met, e.g. through increasing the coil's ampere turns; however, the resulting higher resistance of the control circuit may exceed the power supply's range. Next, the

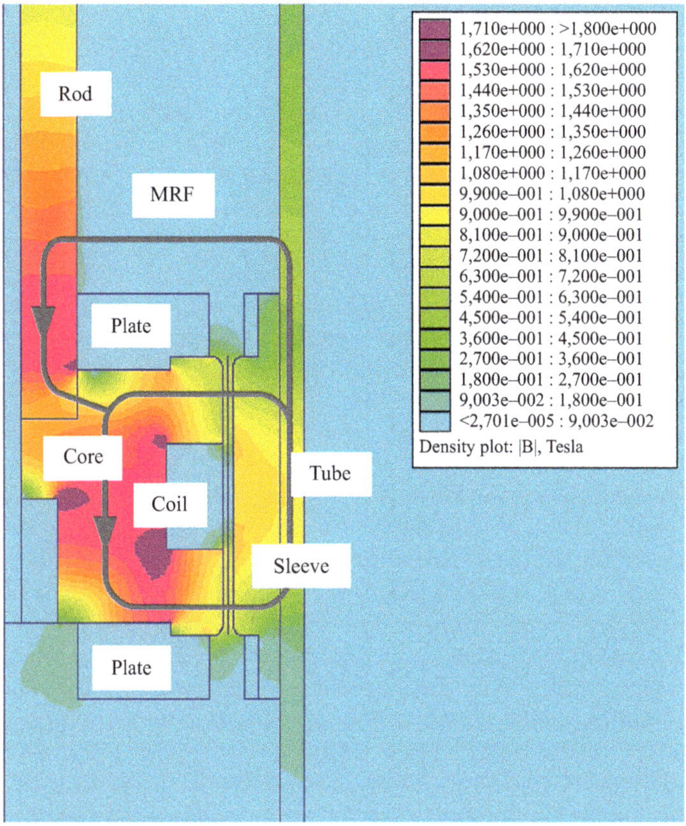

Figure 4.2 MR flow-mode valve—operating principle [2]; primary and secondary (leakage) flux paths

valve's dynamics can be manipulated through the selection of appropriate ferromagnetic high-resistivity materials [5], however, often at an increased cost and lower magnetic saturation flux density.

From the modeling perspective, there are several key factors which affect the magnitude of the generated output of the actuator and should be accounted for. Engineering inputs include the geometry of the flow path, rheology of the fluid, magnetization characteristics of the materials forming the flux path including magnetic saturation, core and magnetic pole dimensions, and coil turns and resistance [6]. In the modeling and/or design process, the inputs are varied until the calculated output and/or the dynamic range (difference between the actuator's output at maximum magnetic field and the actuator's output in the switched-off condition) matches the imposed target. Even more factors influence the actuator's output under unsteady magnetic field and/or transient mechanical excitations. As with any solenoid actuator, any changes in the (unsteady) magnetic flux effectively induce (opposing) eddy currents in the actuator's structure due to Faraday's induction law, and the response of the actuator becomes slower [7]. The phenomenon further delays the output rise/decay in time following the coil circuit (resistor–inductor circuit) response. Next, various inertia and hysteretic effects can be related to the dynamics of a relatively heavy and compressible mass of an MR fluid being pushed through a long annular and narrow flow path of variable resistance (due to the field-induced yield stress) [8]. Magnetic hysteresis is of different nature. First of all, it is an inherent property of ferromagnetic materials forming the solenoid's structure. Second, it does not disappear when the electromagnetic excitation frequency approaches the DC limit [9]. Moreover, there is an overwhelming evidence of mechanical stress, temperature, load history, etc., having a negative influence on the hysteresis and magnetization characteristics of ferromagnetic materials [10,11]. Although studies on the hysteresis of MR fluids have shown that their hysteresis is virtually nonexistent [12], the magnetic hysteresis of materials present in the magnetic circuit of the actuator may impact the performance of both the actuator and the control algorithm [13,14].

Surprisingly, except for MR brakes, heat transfer-related studies on MR dampers are rare [15–17], although such operating conditions may have an impact on the actuator's output. For example, low battery voltage condition and coil high temperature (resulting in a higher resistance of the coil) would degrade the device's dynamic range. Moreover, models which can be directly applied to studies on effects of manufacturing tolerances (eccentricity, straightness, parallelism and cylindricity) are difficult to come across in the available research literature [18].

That brief and limited description of the MR actuator operating principle and factors influencing their output perhaps explains why modeling their behavior is so challenging in various aspects. Again, it requires the consideration of numerous contributors from various domains and the interactions between them, to name only physical and rheological properties of the fluid, sizing and geometry of the flow channel(s), magnetization characteristics of the materials in the actuator's magnetic circuit, sizing of the coil assembly, heat transfer characteristics, etc. The rheological and magnetization characteristics are of prime importance since they determine the magnitude of coupling between the electromagnetic and fluid domains, e.g. through the relationship between the material's yield stress and magnetic field strength or flux density and that is the device's dynamic range. In addition to that, they have an immediate impact on the geometry of the valve and packaging. The others are of secondary importance; however, they need to be considered if particular constraints are to be met and performance goals fulfilled.

The most generic classification of MR actuator models yields two distinct categories: steady state and transient. Steady-state (time-invariant) models vary from simple engineering tools for fast sizing of the actuator's geometry [19] and the control circuit fundamental parameters to complex multi-physics models that are usually developed to analyze specific problems that cannot be easily captured using the fundamental tools [20,21]. For comparison, insight into the dynamics of MR actuators is provided through transient models. In general, they attempt to copy not only the dynamics of the fluid flow through the valve and the flux dynamics but account for the physics outside the control valve as well. Another classification distinguishes between continuum models and lumped parameter models. Lumped parameter models usually utilize a network of element representing physical domains of interest (hydraulic, thermal, electric and magnetic) and connections (interfaces) between them. For example, the electrical circuit of the actuator can be represented in the form of a resistor–nonlinear inductor network model of the second or higher order [6]. Such models allow generating very good numerical solution data or analytical relationships incorporating key variables for fast calculations in prototype studies. Apart from being usable for simple flow geometries, their one disadvantage is the presence of various simplifying assumptions, e.g. uniform yield stress and fully developed flow; continuum models via computational fluid dynamics (CFD) and finite element method (FEM), for comparison, are available for exercising more realistic scenarios, however, at a significant computational expense.

Black-box input–output phenomenological modeling and nonparametric methods (e.g. neural networks) that are more suitable for control studies are considered beyond the scope of this chapter. Such models are based on measurements of the actuating system inputs and outputs, and their usefulness is mostly limited to control prototyping studies.

The problem of MR actuator modeling involves more aspects than just modeling the rheology of the smart fluid. Fluid flow models, in consideration of the inhomogeneous structure of the fluid, can be separated into two additional categories: macroscopic models and microscopic ones [22]. With the macroscopic approach, the fluid is analyzed as a continuum or a generic non-Newtonian fluid. The behavior of such materials is most often quantified using constitutive models, e.g. Bingham, Herschel–Buckley, biviscous, biplastic Bingham and Eyring [23–27]. The approach is capable of good quality results provided the constitutive model parameters are well estimated and assumptions regarding the yield stress distribution made. In the past analytical steady-state solutions for laminar flow have been obtained for the above models for simple flow geometries and uniform yield stress distributions [6]. Scenarios involving complex geometries and unsteady flow can be best addressed with, for example, CFD tools which would allow the spatial variation of yield stress in terms of magnitude and position [21] as well as time. The microscopic models, on the other hand, may provide an interesting insight into the physics governing the MR effect at a particle level [28–30]. The multi-body calculations involved are relatively simple based on Newton's second principle of dynamics but their number requires the application of state-of-the-art computing resources such as GPUs (graphical processing units) or computing multicore clusters. That is a sine-qua-non requirement to accomplish such calculations within a reasonable time frame. That is why the studies have been usually limited to small fluid volumes and diluted fluids with notable exceptions [30]. Here, the most common approach involves the use of discrete micromechanical models for studying the translational and rotational motion of individual particles driven by magnetic and flow forces. With this approach, the base fluid contribution surrounding the particles is usually modeled by means of averaged Stokes' drag forces. Another problem is the treatment of long-range forces [31]. Elastic deformations of the particles due to contact, rolling friction, etc., can be included in those studies, too [32]. An improvement over the traditional micromechanical approach utilizes benefits of CFD techniques to accurately model the flow of fluid around individual particles [33]. Combining CFD and FEM would also allow for the treatment of noncircular geometries of the ferromagnetic particles [34]. The micromechanical models may be useful in studying the behavior of an MR suspension incorporating particles of various size distributions, inclusion of paramagnetic particles, etc. [30]. They are less useful in actuator studies; however, they are invaluable in understanding the physics behind the so-called MR effect.

In general, the purpose of this chapter is to outline various methodologies and models used in the development process of MR actuators. In the chapter, the author discusses various models (and their features) from simple analytical tools to complex finite element (FE) or CFD models, and then highlights their application in various physical domains under steady-state/transient excitation inputs for solving engineering or research problems.

Briefly, the content of the chapter is organized as follows: Section 4.2 discusses modeling efforts in various physical domains. Specifically, electromagnetic modeling is highlighted in Section. 4.2.1, whereas flow modeling efforts are covered in Section 4.2.2. Finally, the chapter is summarized in section 4.3.

4.2 Modeling

Modeling the behavior of MR actuators is multidisciplinary, and it requires considerations of three separate physical domains and the defining links between them—see Figure 4.3.

The current driver (not shown) generates the input voltage at the coil terminals based on the commanded current input (as calculated by a control algorithm) and the feedback current signal i_c. The current driver is often modeled using pulse width modulation components and Proportional-Integral-Derivative (PID) controllers [6,35]. The coil current induces the magnetic flux ϕ in the actuator's structure. Finally, the output force (or torque) F is generated for the received flux and the displacement (velocity) x (v). A thorough knowledge of steady-state and transient performance of the actuator in electric/magnetic/flow domains is then essential. The discussion can be found as follows.

4.2.1 Electromagnetic domain

Electromagnetic modeling is an integral component of the actuator's design process. As already mentioned, electromagnetic models are often grouped into two

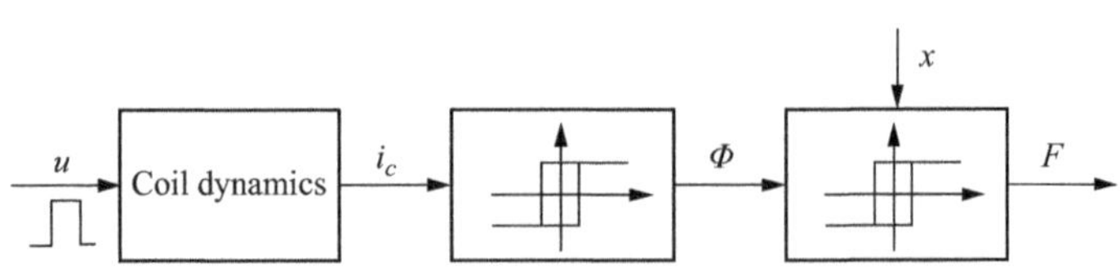

Figure 4.3 Generic layout of a vibration control system with an MR actuator [1]

categories: steady state and transient. Here, typical engineering workflow first involves the application of steady-state lumped parameter models for fast sizing of the geometry followed by the employment of field models, whereas transient calculations are performed to study, e.g. the impact of coil dynamic or eddy current on the time-variant response of the actuator.

Since MR actuators operate under dynamic conditions, their transient performance is even more important than their steady-state characteristics. It is relatively easy to develop actuators with similar steady-state performance and yet having a significantly different transient behavior. Therefore, modeling the electromagnetic behavior of MR actuators should be performed in parallel to similar math-based efforts in flow domain. Zhang *et al.* [36], Farjoud and Bagherpour [7] and Farjoud *et al.* [37] provide a number of steady-state guidelines and criteria for the electromagnetic circuit design, namely, low-reluctance magnetic circuit, maximized magnetic field energy in the air gap (and minimized energy loss in the remaining sections of the electromagnet), no saturation, minimum weight and cost, and optimum coil design (to ensure the required ampere turns at a minimum resistance). The guidelines accompany those of Carlson and Chrzan [38]. Transient analyses, for comparison, are relatively rare, although the steady-state guidelines only ensure the actuator is capable of delivering the required magnetic field energy to the flow gap. Many prior research studies virtually ignored the dynamics of MR dampers and its relationship with the plant's dynamics, and only a brief analysis of prior art in the field shows that it is often a practice nowadays. As a rule of thumb, the response time of MR actuators should not exceed $1/(8f_m)$, where f_m is the upper frequency range [5]. According to this simple guideline, the cut-off frequency of 25 Hz translates to the response time of 5 ms. In general, as already identified, there are several factors contributing to the actuator's response time, namely, response time of the MR fluid, control coil inductance, eddy currents, driver dynamics and exciting current amplitude [5,6,39–42]. The response of the fluid does not exceed 0.5 ms and is often assumed negligible [39,42–44]. Similar observation can be applied to modern current drivers whose response time can be measured within tens of microseconds. In turn, eddy currents create a field that opposes the main flux which again has a detrimental effect on the response time. Moreover, due to the nonlinear magnetization characteristics of the ferromagnetic materials forming the electromagnetic circuit, the response time varies with the exciting current amplitude (ampere turns) [5]. The above factors should be accounted for in the modeling and design process. Next, as already mentioned, other contributors to the output of the solenoid actuators include magnetic hysteresis and internal stresses [9,45–47]. Operating temperature below 100 °C does not affect the materials' magnetization characteristics [48] and can be neglected in electromagnetic models as a result other than the increased resistance of the control coil.

4.2.1.1 Steady-state lumped parameter modeling

The most fundamental approach in the steady-state category simply involves calculations of the required number of ampere turns [3,4,6,7,36,37]. The simple model may also provide an indication of the first-order response time. As an illustration, using the

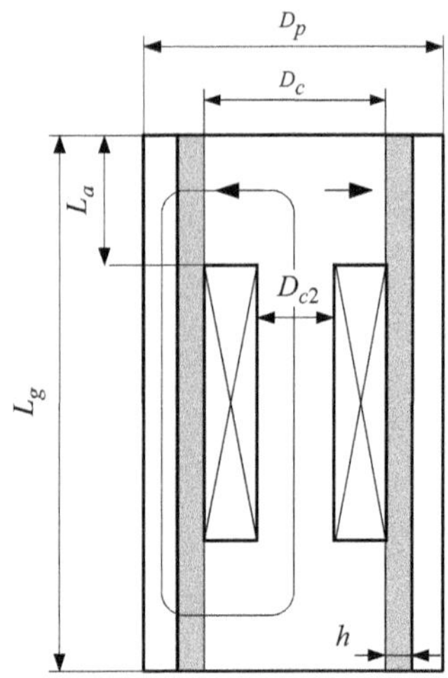

Figure 4.4 Geometry of the MR valve [6]

simple example shown in Figure 4.4 and applying Kirchoff law for magnetic circuit, the following relationships can be derived [6]:

$$\sum_i H_i l_i = N_c I_c \tag{4.1}$$

where N_c is the coil turns, I_c is the coil current, H_i is the magnetic field in the ith segment and l_i is the ith segment effective length. Due to the gap's high reluctance, (4.1) reduces to

$$2hH_g \approx N_c I_c \tag{4.2}$$

assuming the annular gap is the element of the largest reluctance, h is the gap height and H_g is the magnetic field strength (gap). Next, the gap flux density B_g is

$$B_g = \mu_{MR} H_g \tag{4.3}$$

where μ_{MR} is the MR fluid's magnetic permeability. Furthermore, by applying Gauss's law and assuming no flux leakage paths, the magnetic flux ϕ in the electromagnet can be obtained as follows:

$$\phi = B_g A_m = \frac{1}{2} B_g \pi (D_c + h) L_a \tag{4.4}$$

where A_m is the active pole area, D_c is the core diameter and L_a is the active (pole) length. Moreover, the following relationships are valid:

$$\begin{cases} \phi = B_c \dfrac{\pi}{4} D_{c2}^2 \\ \phi = B_s \dfrac{\pi}{4} \left(D_p^2 - D_s^2 \right) \end{cases} \tag{4.5}$$

and $D_s = D_c + 2h$, where B_c is the core flux density, B_s is the outer sleeve flux density, D_{c2} is the core diameter below the coil window and D_s is the sleeve inner diameter. To ensure that no component is magnetically saturated, the following constraints should be met: $B_c \leq B_{sat}, B_s \leq B_{sat}$. The calculations using (4.1)–(4.5) allow obtaining, for example, the coil resistance R_c, the inductance L_c and the response time t_0 [3] as well as the cut-off frequency f_0.

$$\begin{cases} L_c = \dfrac{N_c \phi}{I_c} & R_c = \dfrac{r_c N_c \pi (D_c + D_{c2})}{4} \\ t_0 = \dfrac{L_c}{R_c} & f_0 = \dfrac{1}{2\pi t_0} \end{cases} \tag{4.6}$$

where r_c is the resistance per unit length.

The relationships (4.1)–(4.6) allow for sizing the geometry of the core and the coil window provided the magnetic saturation constraints are met. Engineering constraints may be due to the supply voltage U_{bat}, e.g. $R_c \leq U_{bat}/I_c$. One additional benefit is the raw indication of the coil dynamics through the metrics t_0 and f_0. The basic lumped parameter model can be basis for various optimization schemes and more advanced and detailed models [49]. If used together with flow models, they would provide a ballpark evaluation of the actuator's dynamic range.

The above modeling procedure may be used for MR actuators in all operation modes. In the case of squeeze-flow electromagnets, the modeling procedure is further complicated by their dependency on position (gap height). With such actuators, the steady-state calculations could be performed at the piston smallest gap and the maximum coil current condition, e.g. to satisfy the saturation constraint.

4.2.1.2 Magnetostatic field modeling

Magnetostatic FE models are by far the most common models of MR actuators. Their application is most often limited to the electromagnet design; however, coupling such models with advanced flow models is possible through the yield stress–flux density link.

Modern magnetostatic tools such as Ansys Maxwell address the DC limit of Maxwell equations. Then, the flux density B and the magnetic field strength H should comply with

$$\begin{aligned} \nabla \times H &= J \\ \nabla \cdot B &= 0 \end{aligned} \tag{4.7}$$

where J is the current density, subject to the generic constraint

$$B = \mu H(B) \tag{4.8}$$

Some FE tools such as FEMM 4.2 utilize the magnetic vector potential A approach to solve the above relationship. Then,

$$B = \nabla \times A \tag{4.9}$$

With this formulation, the first relationship of the equation set (4.7) can be simply rewritten for a generic nonlinear material as

$$\nabla \times \left(\frac{1}{\mu(B)} \nabla \times A \right) = J \tag{4.10}$$

Apparently, one benefit of using the vector potential approach is combining all conditions into a single equation (4.10) from which B and H can be developed upon differentiating A. The equations are then solved over the entire problem domain. The approach is the basis for many FE tools used commercially or within the academia.

The FE model is usually developed as a second stage of the modeling (and design) process. If the steady-state lumped parameter approach is used first for providing the first initial (engineering) sizing of the MR valve's geometry, then FE modeling is employed to yield a more accurate description of the actuator.

The approach requires inputs in the form of magnetization characteristics of the ferromagnetic materials forming the electromagnet's circuit and allows for a very good image of the steady-state behavior of the actuators regardless of their operating mode (see Figure 4.5) provided correct boundary conditions are set and excitations defined. FE models also ensure that certain geometric features, namely, contractions, corners and sharp edges, do not affect the magnetic field [7].

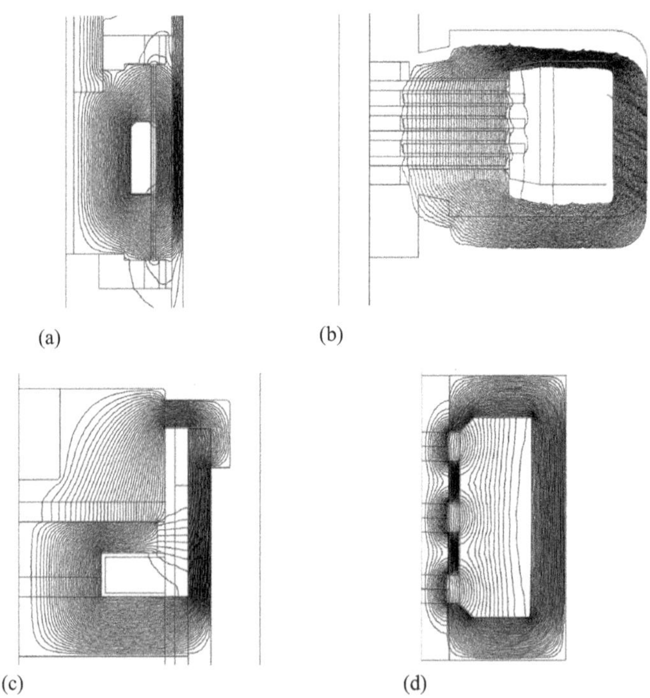

Figure 4.5 Steady-state FE actuator models—flux lines; own simulations. (a) Flow-mode valve; (b) shear-mode brake; (c) squeeze-mode actuators; (d) pinch-mode valve

Inclusion of various types of ferromagnetic materials and permanent magnets can be easily performed with this approach, too. If coupled with flow models, the field approach ensures that the electromagnet is capable of delivering the required dynamic range and no features are incorporated into the flow path which would adversely influence the flow field. The approach is also very effective when performing engineering comparison of alternative designs.

To accomplish that, several metrics can be developed for the assessment of the performance of MR (solenoid) actuators, namely, magnetic energy density, ratio between magnetic energy stored in fluid gap and total magnetic energy stored in the magnet assembly [7], dynamic range [3], packaging [50], response time [3] and temperature [51]. For example, by definition, magnetic energy density in control gap *MED* is obtained via

$$MED(\text{gap}) = \frac{E(gap)}{V(gap)} \tag{4.11}$$

where E is the magnetic energy and V is the volume. Next, the magnetic energy ratio *MFE* is defined as

$$MFE(\text{gap}) = \frac{E(gap)}{E(magnet)} \tag{4.12}$$

which allows for effective comparisons of design proposals as demonstrated in [7].

The modeling approach may be used effectively for MR actuators operating in all fundamental modes. In the case of squeeze-mode actuators, the steady-state calculations are performed at a frozen working gap height [52].

4.2.1.3 Transient lumped parameter modeling

Transient modeling usually follows the steady-state stage and can be integrated into control studies and/or unsteady flow analyses. In control studies, such models are employed for tuning, e.g. the current driver for optimum performance. The models take the form of a resistor–(nonlinear) inductor network [6,7] whose parameters can be extracted from FE analyses or real measurements.

The simplest approach assumes a linear inductor [53] which, however, is valid for coil current small fluctuations only, and does not grasps the transient performance of the actuator in detail. In realistic scenarios, the nonlinear magnetizing inductance should be taken into account. For example, Figure 4.6 illustrates equivalent circuit diagrams of several models used in the past for modeling the transient response of MR actuators. The model shown in Figure 4.6(a) includes the coil resistance R_c and the inductance L_c as well as the eddy current inductance R_{ce} and the inductance L_{ce}. Next, in the model illustrated in Figure 4.6(b) the secondary circuit parameters R_2, L_{22} and L_{c2} refer to the eddy current loop effective resistance, eddy current inductance, mutual inductance between the primary inductance L_c and the eddy current path inductance L_{22}, respectively. The inductances L_c and L_{22} are coupled through the coupling coefficient k_c. Finally, in the model revealed in Figure 4.6(c), the inductance L_1 refers to the leakage path, the resistance R_e is the eddy current resistance and the voltage u_c is the voltage at the inductor L_c.

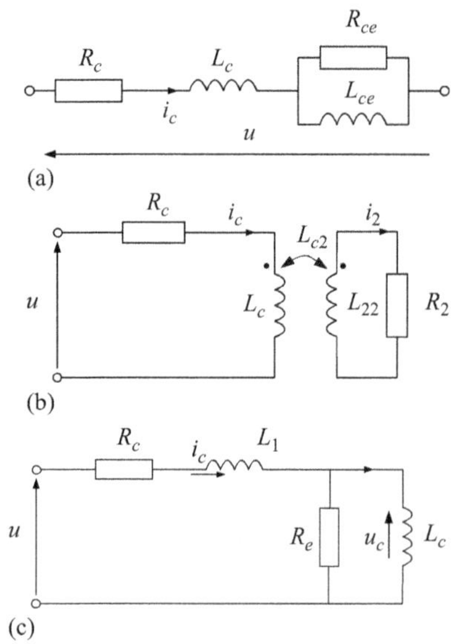

Figure 4.6 Equivalent electric circuits of MR actuators. (a) second-order model [53]; (b) mutual inductor model [6]; (c) Farjoud and Bagherpour [7]

The model reduces to the simple (non)linear model if the (usually small) leakage inductance and the eddy current resistance are neglected. For details on obtaining the parameters of each model, the reader should refer to [6,7].

The voltage at the nonlinear inductor L_c (main coil) can be obtained as

$$\begin{aligned} u_c &= \frac{d\lambda(i_c)}{dt} \quad L_c = \lambda/I_c \\ \lambda &= N_c\phi \quad\;\; L_c = N_c\phi/I_c \end{aligned} \tag{4.13}$$

where λ is the flux linkage. The above models were used on various occasions in lumped parameter studies of flow-mode actuators. Due to the similar electromagnet design, the approach can be extended to shear-mode devices. In the case of squeeze-mode actuators, the magnetizing inductance L_c on the piston position x needs to be accounted for [54,55]. Then, the coil inductance becomes $L_c = L_c(i_c,x)$. Similarly, the inductance map versus position and coil current can be developed through FE studies. The governing equation of the serial resistor–inductor circuit of a squeeze-mode actuator is then

$$R_c i_c + \frac{d\lambda(i_c,x)}{dt} = u \tag{4.14}$$

which finally yields upon differentiation

$$R_c i_c + \frac{d\lambda}{di_c}\frac{di_c}{dt} + \frac{d\lambda}{dx}\frac{dx}{dt} = u \tag{4.15}$$

and introduces an additional component due to motion in (4.15). Next, losses due to magnetic hysteresis have been regularly neglected in MR studies with exceptions; see e.g. [13,14,20,45,56–58]. Some insight can then be gained by inspection of solenoid actuator models. Here, one obvious approach relies on extending the λ–i_c relationship with a hysteretic operator, e.g. Jiles–Atherton, Preisach or Duhem [9,13,56–61].

As an example, let us consider the application of the Duhem hysteresis model [60]. With this approach, the governing equation of the nonlinear Resistor-Inductor (RL) circuit model including hysteresis can be given as follows:

$$\frac{d\lambda}{dt} = a\left|\frac{di_c}{dt}\right|[f(i_c) - \lambda] + g(i_c)\frac{di_c}{dt} \tag{4.16}$$

where $f(i_c)$ and $g(i_c)$ are the shaping functions. The author would propose an exemplary shaping function utilizing the hyperbolic tangent function

$$\begin{cases} f(i_c) = b_1 \tanh(b_2 i_c) \\ g(i_c) = c \end{cases} \tag{4.17}$$

The model is phenomenological with just four parameters (a, b_1, b_2, c) which can be identified from experimental data, although coil inductance calculations may provide initial estimates of the parameter c, for example. It is more suitable for control studies rather than design iterations. Note that for $a = 0$, the hysteresis is effectively eliminated from the model, and the model takes the form of a nonlinear RL circuit. In this context, the hysteretic model is a nonlinear system in which the input is the coil current i_c and the output flux linkage λ. Except for the parameter c, the model parameters are not directly related to the actuator's characteristics or material properties (e.g. remanence and coercivity). One extension of this model would be based on the well-known Bouc–Wen approach to copy the λ–i_c loop although with a considerable increase of the parameter set [61]. Also, combining it with the mutual inductor model [6] would allow for the inclusion of eddy currents.

4.2.1.4 Transient magnetic field modeling

Transient models constitute the key (and obligatory) component of the modeling and development process. This type of modeling usually forms the final stage of the actuator modeling process if coupled with flow models in particular. In the electromagnet design/modeling stage, it is also the most difficult and complex task with even more material data inputs and finer control of the transient solver required. As MR actuators are intended for use in dynamic load conditions, the modeling process should include the effects of fluctuating (transient) magnetic field, namely, losses due to magnetic hysteresis, eddy currents, etc. [58]. Apart from the coil circuit dynamics, controller's dynamics as well as eddy currents are the principal phenomena that adversely affect the flux (and then the force) dynamics [5]. Capturing the hysteretic effect is also a major challenge (as in the field of electrical drives), and full-scale studies on predicting such behavior of MR actuators have been limited [14,58] due to a lack of available tools. Moreover, there is enough experimental

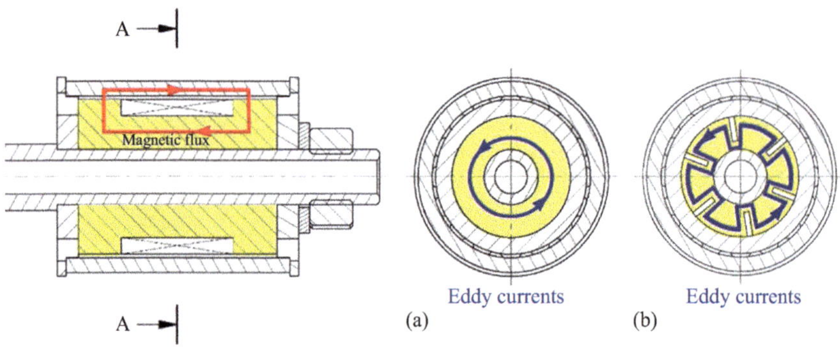

Figure 4.7 Slotted MR valve: (a) regular core and (b) slotted core cross-section [63]

evidence of the influence of magnetic hysteresis on the output of MR actuators [14,45,46,58] and this effect should not be excluded from the investigation. With current computing hardware capabilities, such problems can be addressed within a reasonable amount of time. Problems involving motion and time-variant gap heights as in squeeze-mode MR actuators are possible to be solved with this approach, too.

The generic transient magnetics problem is usually attacked by solving the field problem as summarized in [6]. Only recently, Maxwell's equations have been extended with vector hysteresis [62] and integrated into commercial FE tools (Ansys Maxwell R17).

As way of illustrating the solution of an exemplary transient problem, the author briefly highlights the results of a study to introduce slits into the MR valve's geometry—see Figure 4.7. Narrow deep slits are one technique for improving the response time of solenoid actuators [63]. Their purpose is to increase the effective resistivity of the conducting path [64]; as the resistivity of the conductor increases, the eddy current magnitude decreases. Similar effect may be achieved simply by using low-conductivity materials; however, due to different magnetization characteristics and structural issues, they are not interchangeable.

In the analyzed case, the author developed an FE model of an MR valve having up to four slits in the piston core, and then studied the transient flux output of the MR solenoid subjected to voltage step inputs. Figure 4.8(a) shows the current response, whereas in Figure 4.8(b) it is revealed that the slits improve the flux secondary response time; the primary response is dictated mainly by the coil circuit dynamics. In addition to this, further enhancement of the slit core performance is possible via voltage overdrive as revealed in Figure 4.8(b). The FE model results are compared against a regular geometry without any slits in the core ("Base" curve in Figure 4.8(b)).

4.2.2 Flow domain

Flow modeling is necessary in the modeling and design process. Developing such models should follow electromagnet modeling or be performed concurrently to the electromagnetic domain. Here, the same approach for classifying the models applies as for the electromagnetic domain.

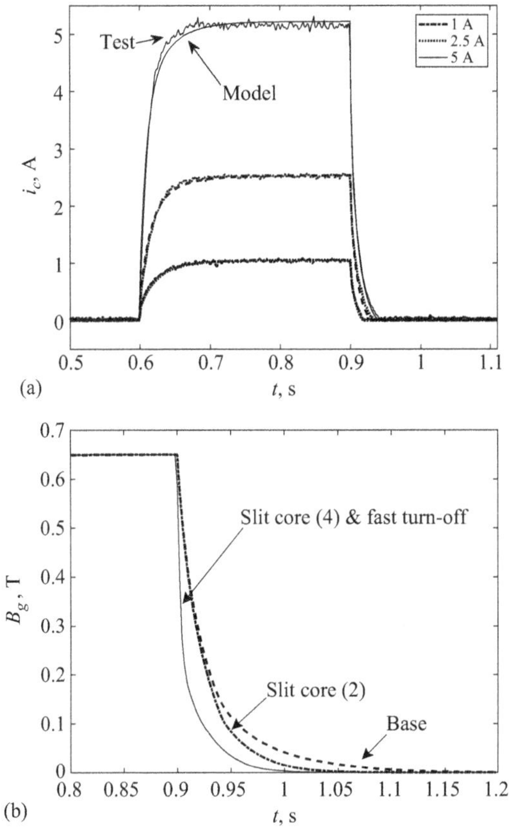

Figure 4.8 Slotted MR valve: (a) current response and (b) flux response; own data

Again, steady-state models are applied in the valve's sizing process. Their primary purpose is to provide means for calculating the dynamic force/torque range of the actuator. For comparison, the employment of transient models allows for studying flow development effects and various phenomena occurring in the flow path [21].

Next, lumped parameter models of MR actuators are based on idealized closed-form solutions of MR fluid flow involving simple flow geometries and significant simplifications in the field-induced yield stress distribution. Unsteady CFD models consume fewer assumptions and are capable of solving problems in which the fluid's yield stress varies both in position and time.

4.2.2.1 Steady-state flow modeling: lumped parameter approach

Steady-state lumped parameter models rely on the description of the MR effect based on constitutive models of the behavior of MR fluids. The application of these models has been proved fairly successful in the process of modeling such actuators. Various authors handled them in parallel plate or annular geometries in various operating modes to develop the relationship between the flow rate through the valve and the resulting pressure drop across it [23–27,65–72]. The Bingham plastic model has been usually the approach of choice, although capturing, for example, post-yield shear-rate thinning/thickening effects, is not possible with it. Although proven effective in handling various scenarios, the models have numerous drawbacks of which some were mentioned earlier in the chapter. Additionally, the assumptions of fully developed flow, omission of entrance/exit losses, exclusion of pressure losses across nonactivated sections of the flow path, etc., render the models invalid at high shear rates in particular. Here, recent advances have resulted in the high shear-rate extension of the Bingham model [21].

One typical manner in which the Bingham plastic model (or others) is used in describing the actuator's output is the flow-mode valve revealed in Figure 4.9(a). In the valve, the fluid is sandwiched between annular surfaces. The parallel plate approach is assumed instead. It is valid for small ratios of the flow path height to the piston diameter. The yield stress is constant in the activated regions and zero elsewhere. The fluid flow in the nonactivated state is assumed fully developed and

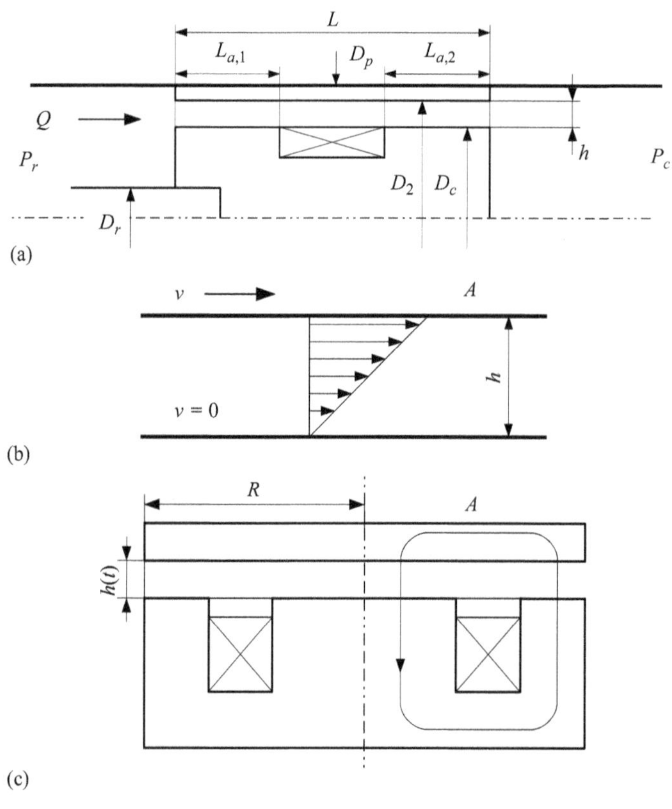

Figure 4.9 Configurations of MR actuators. (a) flow [6]; (b) shear mode; (c) squeeze mode

Newtonian. The piston moves at the velocity v. With the Bingham model of the following form

$$\tau = \tau_0 + \mu \frac{du}{dz} \tag{4.18}$$

where τ is the shear stress, τ_0 is the yield stress, μ is the viscosity and du/dz is the fluid velocity gradient—and applying boundary conditions at the walls, the relationship between the on-state pressure drop Δp_{on} and the flow rate $Q = vA$—where $A = A_p - A_r$, A_p is the piston head cross-section area, A_r is the rod cross-section area—can be obtained as in [3]:

$$-2\left(h\frac{\Delta p_{on}}{2\tau_0}\right)^3 - \left(h\frac{\Delta p_{on}}{2\tau_0}\right)^2\left[3 + 12\left(\frac{\mu Q}{bh^2\tau_0}\right)\right] + 1 = 0 \tag{4.19}$$

where h and b are flow gap height and width, respectively. A further simplification of the cubic equation is the so-called straight-line approximation with which the output pressure drop is simply the sum of the yielding pressure drop across the active sections and the pressure drop due to viscous losses

$$\Delta p_{on} = 2\frac{\tau_0}{h}L_a + 12\frac{\mu L}{bh^3}Q \tag{4.20}$$

In the deactivated condition, the output of the actuator can be described simply by means of the Newtonian model

$$\Delta p_{Off} = 12\frac{\mu L}{bh^3}Q \tag{4.21}$$

from which the force gain K can be deduced as

$$K = \frac{\Delta p_{on}}{\Delta p_{off}} = 1 + \frac{L_a}{L}\frac{\tau_0 bh^2}{6\mu Q} \tag{4.22}$$

Apparently, the model's application scope is fairly limited to simple geometries. If flux bypass or hydraulic bypass features are present in the flow path, the model cannot be applied directly and requires modifications [6].

Next, consider the shear-mode damper shown in Figure 4.9(b) with two parallel plates of which one is stationary and the other moves at the velocity v. The force in the deactivated condition acting on the moving plate is then

$$F_{\text{off}} = \mu\frac{v}{h}A \tag{4.23}$$

In the activated condition, the force output includes the additional term

$$F_{\text{on}} = \left(\tau_0 + \mu\frac{v}{h}\right)A \tag{4.24}$$

Using (4.23) and (4.24), the shear-mode actuator (force) gain becomes

$$K = 1 + \frac{\tau_0 h}{\mu v} \tag{4.25}$$

Finally, let us consider the squeeze-mode damper in Figure 4.9(c). In the exemplary configuration, the circular plate radius is R, and the gap height varies according to the prescribed displacement input $h = h(t)$. Neglecting inertia, the force in the on-state condition is as follows [70]:

$$F_{on} = \frac{3}{2}\frac{\pi R^4}{h^3}v + \frac{4}{3}\frac{\pi R^3}{h}\tau_0\frac{v}{|v|} \tag{4.26}$$

The second term in the above expression disappears as $\tau_0 = 0$. Thus, the dynamic force range (or force gain) becomes

$$K = 1 + \frac{8}{3}\frac{h^2}{\mu R v}\tau_0 \tag{4.27}$$

Equations (4.22), (4.25) and (4.27) provide simple means for calculating the force gain of MR actuators in the three operating modes. Combining them with the expressions in Section 4.2.1.1 yields a simple scheme for calculating the performance range. It is also apparent that the gain K can be expressed simply in terms of flow channel geometry and fluid properties. More accurate approximations may incorporate shear thinning/thickening effects and other configurations [43,68,69], inertia [70] and the like.

Similar or more complex expressions are often cast into nondimensional form. Such approach is often useful as it provides a unifying outlook on the capabilities of analyzed configurations. Since Phillips [23], numerous nondimensional groups have been widely used in the modeling process, namely, the so-called friction factor C_f, Reynolds number Re, Hedstrom number He, Bingham number Bi, Mason number Mn and several others [3,65–73]. Most of them are widely used for studying flow dynamics. Regardless of the approach, the groups are mostly interchangeable; they are based on the same or similar set of variables (geometric dimensions, material properties, etc.).

4.2.2.2 Steady-state CFDs

The scope of lumped parameter models is limited to simplified geometries and uniform yield stress distributions. More complex geometries and realistic scenarios with a spatially varying yield stress require the use of numerical methods for solving such flow problems. Nowadays, CFD tools can be used rather effectively with computing clusters. Nevertheless, the use of CFD in solving MR fluid flow problems is rather rare compared to FE electromagnetics. Some discussion on the topic can be found in [21], and the reader should refer there for further details.

With the modern computing hardware large size, flow problems can be attacked and more design iterations exercised within a given time limit. Their accuracy is somewhat superior to the less sophisticated models especially if flow models can be coupled with electromagnet models; however, non-Newtonian flow problems as in the case of MR fluids present various numerical problems to be dealt with. The problems are twofold. First, due to the singularity of the commonly used Bingham model at zero shear rate, flow field(s) cannot be computed [6,73,74].

Second, accuracy issues can be encountered with poor convergence of some numerical solvers [21].

By principle, steady-state incompressible Navier–Stokes' equation is

$$\rho u_j \frac{\partial u_i}{x_j} - \frac{\partial \tau_{i,j}}{\partial x_j} = -\frac{\partial p}{\partial x_i} \tag{4.28}$$

where u_i is the flow velocity and x_i is the independent variable ($i = 1, 2, 3$). Assuming that

$$\begin{cases} \tau_{i,j} = \mu_{app} \dfrac{\partial u_i}{\partial x_j} \\ \mu_{app} = \mu + \dfrac{\tau_0}{\left|\dfrac{\partial u_i}{\partial x_j}\right|} \end{cases} \tag{4.29}$$

where μ_{app} is the apparent viscosity. The zero shear-rate singularity is handled by limiting the apparent viscosity as shear rate approaches zero. Alternatively, the smoothed approximation of Papanastasiou [75] may be more convenient for convergence and performance. As a result, the Bingham model is handled effectively as a biviscous model with the preyield viscosity being several thousand times the MR fluid's base viscosity. The approach is not rid of faults, however. The technique was developed for handling numerical problems and not fidelity. First, the apparent viscosity model may be often inefficient in predicting the flow field at high yield stress levels and low flow velocities. Effectively, the biviscous approach allows for a fluid "leak" whenever the solidified plug occupies the entire flow channel width. Gołdasz and Sapiński [3] commented, for example, that with this scenario CFD solvers are unable to predict the velocity profile for coarser meshes in particular. This leads to numerically challenging analyses at low Reynolds numbers and high Hedstrom numbers. In practical applications, for example, predicting the performance of an MR valve with flux bypasses [76] in the primary flow path is rather impossible to achieve. That applies to low-speed flow modeling in regular piston geometries, too.

One area where computational flow dynamics tools can provide a valuable insight into the non-Newtonian fluid's behavior is fluid dynamics problems at high speeds and shear rates. Adaptive MR energy absorbers are devices in which fluid is subjected to extremely high flow speeds translating to $Re >> 1{,}000$. In those devices, the reduction in on-state forces was experimentally observed. At such condition, the fluid flow is not fully developed in the flow channel. That makes accurate predictions of the output force of these devices impossible to obtain while using other modeling techniques. It is particularly evident in short flow channels or pistons with short active lengths [43,77]. Using CFD, Sherman in [21] developed a correction for the Bingham model which is capable of capturing the high-speed force falloff.

Another class of problems which can really benefit from the application of fluid dynamics tools is novel MR valve geometries beyond the scope of existing tools [78,79]. As an illustration, let us consider the concept of a pseudo-valve in which the contraction is formed by semi-solidified Bingham fluid regions (of nonzero yield stress) in the direction of flow in a manner resembling the Ventouri pipe. The concept is presented in Figure 4.10. In the flow path, only the portion of the material near the channel walls is solidified. The distance between the solidified regions regulates the size of the pseudo-orifice. In the presented illustration, the channel height is h and its length L. The fluid is incompressible and characterized by the density ρ, viscosity μ and the yield stress τ_0 ($\tau_0 > 0$) in the activated zones and $\tau_0 = 0$ elsewhere. The distance between the activated regions is h_0. The left boundary is the velocity inlet, and the right one is the pressure outlet. The contraction is assumed to be (a)symmetric and may comprise several activated zones in series along the flow path.

In the analysis, the steady-state model was set up with input velocities at the inlet from 0.5 to 15 m/s. The density of the fluid was 3 g/cc, and the base viscosity was equal to 50 cP. The gap height was 1 mm, and the size of the contraction h_0 varied from 0.2 mm to 0.6 mm. Given the geometry and material properties, the flow conditions corresponded to Reynolds numbers from 30 to 900 and Hedstrom numbers up to 36,000. The setup allowed for the acquisition of the pressure drop Δp across the contraction, velocity profiles and the assessment of the valve's concept. Exemplary velocity contours are highlighted in Figure 4.11.

Next, Figure 4.12 reveals the graph of calculated pressure drop versus the average flow velocity for a single symmetric contraction and the highest yield stress of 30 kPa. The results are presented against laminar flow data. The obtained

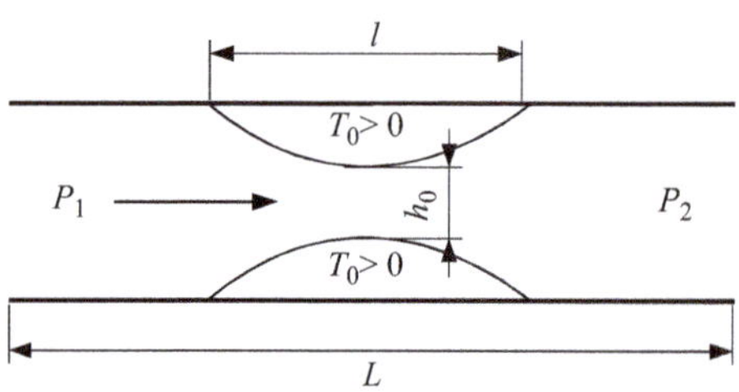

Figure 4.10 Symmetric contraction: flow path geometry; P_1, P_2—pressures [78]

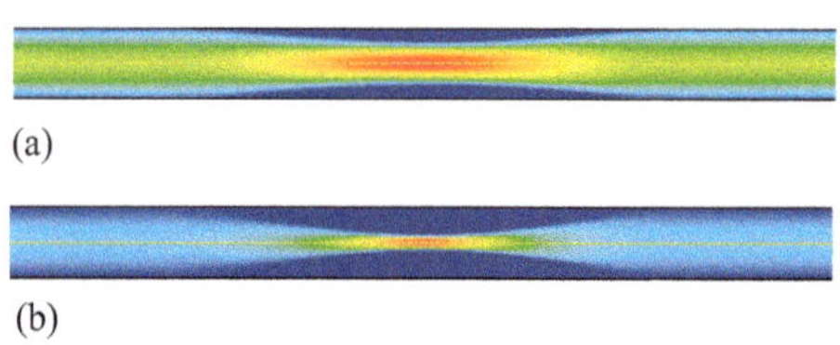

Figure 4.11 Velocity contours: Re = 30 [78]. (a) $h_0 = 0.6$ mm; (b) $h_0 = 0.2$ mm

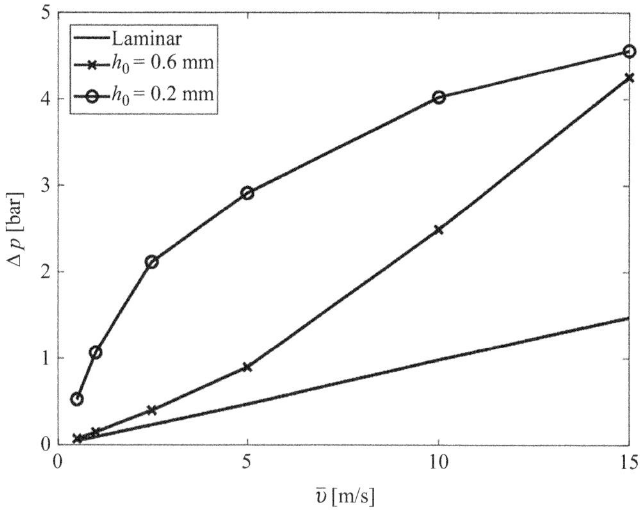

Figure 4.12 CFD results: single symmetric contraction [78]

data show a significant increase of the pressure drop at low and medium velocities and a gradual degradation of the effect due to yielding the material in the nonzero yield stress zones [78].

4.2.2.3 Unsteady fluid dynamics

This category involves lumped parameter models which are described by a set of ordinary differential equations (ODEs) or continuum models represented by partial differential equations (PDEs), namely, Navier–Stokes' equations. The lumped parameter models are often presented in the form of interconnected inertia components, hydraulic resistances and compressible elements [67,80]. The resulting ODE set is then solved numerically with fixed-step or variable-step size solvers. The approach can be fairly productive especially if the phenomena occurring outside the MR valve can be modeled due to fluid compressibility and elastic deformations of the damper cylinder. The approach, however, has drawbacks. In general, it assumes that steady-state resistances still hold for unsteady flows, and the flow driving the inertia element in the working gap is averaged. As such, PDE models will be always wanted for higher accuracy and decent computing cost.

Computationally, this category of modeling is most challenging as it often involves unsteady flow and deforming meshes to simulate piston motion inside the cylinder according to a prescribed displacement input. Unsurprisingly, such modeling studies are rare. Parlak and Engin [81] ran a transient CFD analysis of the MR damper with a single annular flow path in flow mode and using the deformed mesh approach. Case *et al.* [82] executed a fully coupled multi-physics (electromagnetic and CFD) model and validated it experimentally. With a similar CFD model, Zheng *et al.* [20]

included the Jiles–Atherton model of magnetic hysteresis and a simple heat transfer model. Their model was further extended for high-velocity flows in the succeeding analysis [83]. The application of unsteady CFD in flow mode is by far the most common approach; however, the other modes were analyzed with the methodology, too. Gołdasz and Sapiński [84] completed an unsteady analysis of a fundamental squeeze-mode damper model and favorably compared their results against theory using the Laplacian smoothing technique for mesh deformation. Finally, several studies of Hegger and Mass on rotary MR power train clutches possibly involved remeshing techniques [85].

Again, as way of illustration, let us consider the two-dimensional axisymmetric model of the MR valve revealed in Figure 4.13(a). The geometry is that of a standard MR piston valve with one annular flow path and two coils (shown as the three activated areas in the apparent viscosity plot in Figure 4.13(b)). Next, the fluid is assumed non-Newtonian and slightly compressible; bulk modulus $\beta =$ 2,400 MPa. The density of the fluid was 2.7 g/cc, and the base viscosity 63 cP. As indicated in the simple geometry plot, the piston portion was subjected to the prescribed sinusoidal displacement input $x(t) = X_o \sin(2\pi f t)$, X_o—amplitude, f—frequency. The maximum peak velocity was 0.52 m/s and the input frequency range was from 1 to 25 Hz. The exemplary model reflects the actual testing conditions of a real damper in a laboratory.

The unsteady analysis was executed with Ansys Fluent R17 using the time step of 2 μs and the layering scheme for modeling mesh deformations. Figure 4.14 shows exemplary results as (normalized) time histories of damping force, plots of force versus displacement and force versus velocity, respectively.

Specifically, the calculated data reveal the force output phase shift with frequency as well as frequency (for the same velocity) due to the inertia. In the on-state condition, force oscillations can be observed at piston motion reversal points. This could be related to the dynamics of the fluid mass bouncing against two compliant columns of MR fluid—see Figure 4.15.

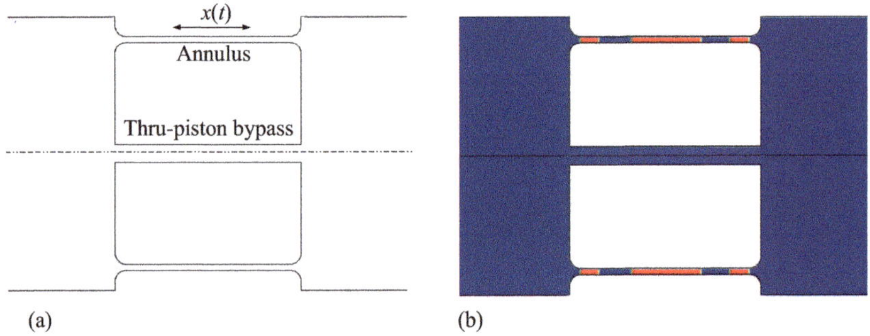

Figure 4.13 Unsteady CFD model; own data. (a) Geometry; (b) apparent viscosity

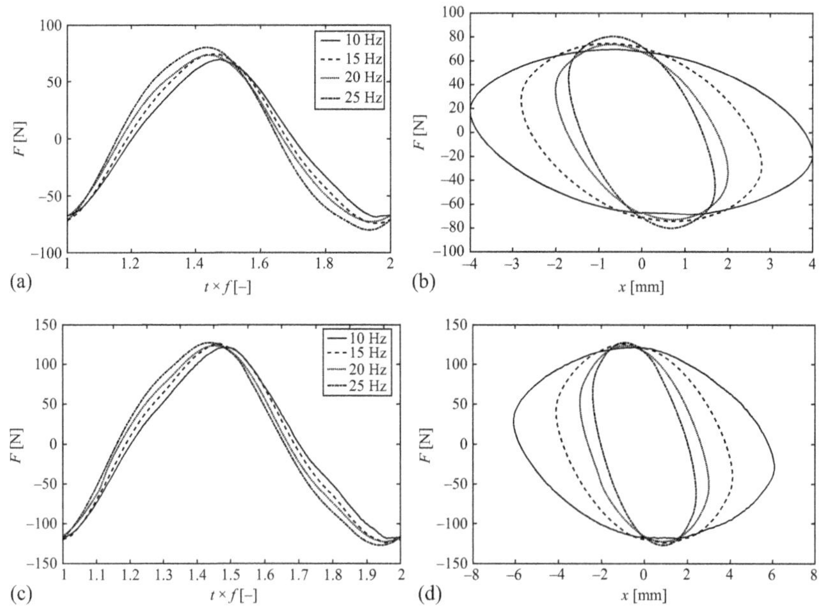

Figure 4.14 *CFD results: off-state plots. (a) F-t: $V_0 = 0.26$ m/s, $\tau_0 = 0$ kPa; (b) F-x: $V_0 = 0.26$ m/s, $\tau_0 = 0$ kPa; (c) F-t: $V_0 = 0.38$ m/s, $\tau_0 = 0$ kPa; (d) F-x: $V_0 = 0.38$ m/s, $\tau_0 = 0$ kPa*

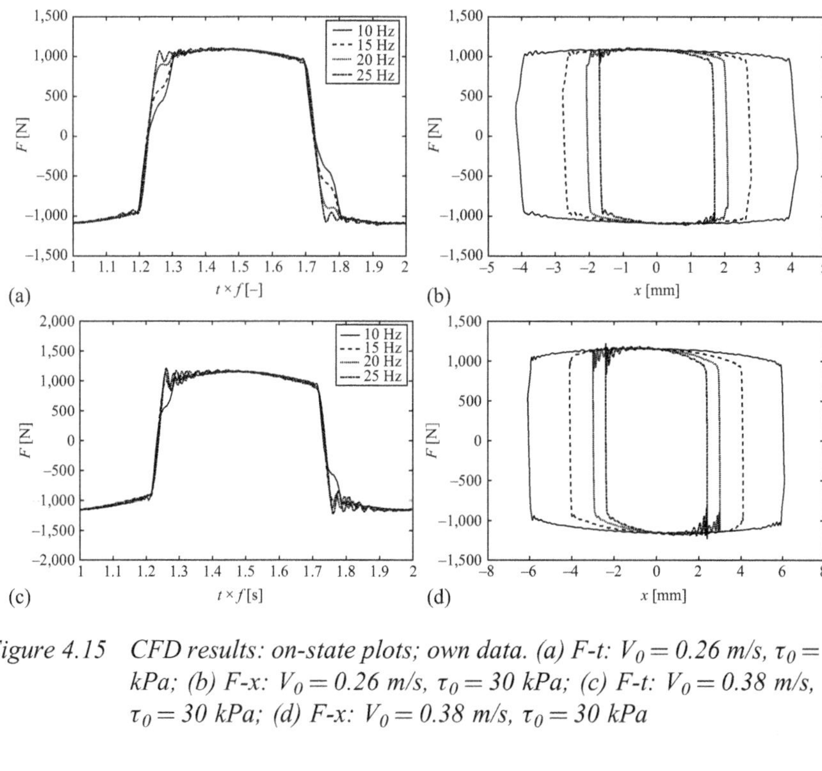

Figure 4.15 *CFD results: on-state plots; own data. (a) F-t: $V_0 = 0.26$ m/s, $\tau_0 = 30$ kPa; (b) F-x: $V_0 = 0.26$ m/s, $\tau_0 = 30$ kPa; (c) F-t: $V_0 = 0.38$ m/s, $\tau_0 = 30$ kPa; (d) F-x: $V_0 = 0.38$ m/s, $\tau_0 = 30$ kPa*

4.3 Summary

Modeling the behavior of MR semi-active actuators is often a challenging task. It entails simultaneous handling of magnetic fields, flow and structural mechanics. Solving various nonlinear problems involved at the same time does not help. This chapter focuses on electromagnetic modeling and flow.

The majority of the developed models which were briefly described in the chapter were developed using the continuum approach with no insight into particle dynamics. Microscale simulations are computationally expensive, which reduces their demand in practical engineering studies. However, effective engineering and research studies are still possible with the continuum models at various levels of complexity.

Most likely, the most accurate results in this research field can be achieved through low-level particle chain simulations coupled to an accurate flow field and magnetic field solutions; however, the computing cost may be overwhelming. Nowadays, first-order continuum tools seem the most sensible solution for fast prototyping damper studies followed by the application of far more accurate FE/CFD tools.

References

[1] Gołdasz J., Sapiński B., Jastrzębski Ł. "Assessment of the magnetic hysteretic behaviour of MR dampers through sensorless measurements." *Shock and Vibration*. 2018;3740208.

[2] Gołdasz J. "Electromechanical analysis of a magnetorheological damper with electrical steel laminations in the piston core." *Przegląd Elektrotechniczny (Electrical Reviews)*. 2013;**2a**:8–12.

[3] Nguyen Q. H., Choi S. B., Wereley N. M. "Optimal design of magnetorheological valves via a finite element method considering control energy and a time constant." *Smart Materials and Structures*. 2008;**17**(**2**): 025024.

[4] Nguyen Q. H., Choi S. B. "Optimal design of a vehicle magnetorheological damper considering the damping force and dynamic range." *Smart Materials and Structures*. 2009;**18**:015013.

[5] Strecker Z., Roupec J., Mazurek I., Machacek O., Kubik M., Klapka M. "Design of magnetorheological damper with short response time." *Journal of Intelligent Material Systems and Structures*. 2015;**26**(**14**):1951–1958.

[6] Gołdasz J., Sapiński B. Insight into magnetorheological shock absorbers. Cham: Springer International Publishing; 2015.

[7] Farjoud A., Bagherpour E. A. "Electromagnet design for magnetorheological devices." *Journal of Intelligent Material Systems and Structures*. 2016;**27**(**1**):51–70.

[8] Gołdasz J., Alexandridis A. A. "Medium- and high-frequency analysis of magnetorheological fluid dampers." *Journal of Vibration and Control*. 2012;**18**(**14**):2140–2148.

[9] Jiles D. C., Atherton D. L. "Theory of ferromagnetic hysteresis." *Journal of Magnetism and Magnetic Materials*. 1986;**61**:48–60.
[10] Pitman K. C "The influence of stress on ferromagnetic hysteresis." *IEEE Transactions on Magnetics*. 1990, **26**(**5**);1978–1980.
[11] Schneider C. S. "Effect of stress on the shape of ferromagnetic hysteresis loops." *Journal of Applied Physics*. 2005;**97**:10E503.
[12] de Vicente J., Durán J. D. G., Delgado A. V., González-Caballero F., Bossis G. "Effect of magnetic hysteresis of the solid phase on the rheological properties of MR fluids." *International Journal of Modern Physics B*. 2002;**17–18**:2576–2582.
[13] Han Y.-M., Choi S. B., Wereley N. M. "Hysteretic behavior of magnetorheological fluid and identification using Preisach model." *Journal of Intelligent Material Systems and Structures*. 2007;**18**(**9**): 973–981.
[14] Jędryczka C., Sujka P., Szeląg W. "The influence of magnetic hysteresis on magnetorheological fluid clutch operation." *COMPEL–The International Journal for Computation and Mathematics in Electrical and Electronic Engineering*, 2009;**28**(**3**):711–721.
[15] Gordaninejad F., Breese D. G. "Heating of magnetorheological fluid dampers." *Journal of Intelligent Material Systems and Structures*. 1999;**10**(**8**):634–645.
[16] Dogruoz M. B., Wang E. L., Gordaninejad F., Stipanovic A. J. "Augmenting heat transfer from fail-safe magneto-rheological fluid dampers using fins." *Journal of Intelligent Material Systems and Structures*. 2003;**14**(**2**): 79–86.
[17] Zalewski R., Nachman J., Shillor M., Bajkowski J. "Dynamic model for a magnetorheological damper." *Applied Mathematical Modelling*. 2014;**38**: 2366–2376.
[18] Choi Y.-T., Wereley N. M., "Flow mode magnetorheological dampers with an eccentric gap." *Advances in Mechanical Engineering*. Retrieved from https://doi.org/10.1155/2014/931683
[19] Duclos T. D. "Design of devices using electrorheological fluids." *SAE Transactions: Journal of Materials*. 1988;**97**:532–536.
[20] Zheng J., Li Y., Li Z., Wang, J. "Transient multi-physics analysis of a magnetorheological shock absorber with the inverse Jiles–Atherton hysteresis model." *Smart Materials and Structures*. 2015;**24**(**10**):105024.
[21] Sherman S. *Magnetorheological fluid dynamics with application to high speed energy absorbers*. Ph.D. Thesis, University of Maryland, MD, 2017.
[22] Ghaffari A., Hashemabadi S. H., Ashtiani M. "A review on the simulation and modeling of magnetorheological fluids." *Journal of Intelligent Material Systems and Structures*. 2015;**26**(**8**):881–904.
[23] Phillips R. W. *Engineering applications of fluids with a variable yield stress*. Ph.D. Thesis, University of Berkeley, CA, 1969.
[24] Wereley N. M. "Non-dimensional Herschel—Bulkley analysis of magnetorheological and electrorheological dampers." *Journal of Intelligent Material Systems and Structures*. 2008;**19**(**3**):257–268.
[25] Wereley N. M., Pang L., Kamath G. M. "Idealized hysteresis modeling of electrorheological and magnetorheological dampers." *Journal of Intelligent Material Systems and Structures*. 1998;**9**:642–649.

[26] Gołdasz J., Sapiński B. "Non-dimensional characterization of flow-mode magnetorheological/electrorheological fluid dampers." *Journal of Intelligent Material Systems and Structures*. 2012;**23(14)**:1545–1562.

[27] Choi Y.-T., Bitman L., Wereley N. M. "Non-dimensional analysis of electrorheological dampers using an Eyring constitutive relationship." *Journal of Intelligent Material Systems and Structures*. 2005;**16**:383–393.

[28] Ly H. V., Reitich F., Jolly M. R., Banks H. T., Ito K. "Simulations of particle dynamics in magnetorheological fluids." *Journal of Computational Physics*. 1999;**155**:160–177.

[29] De Vicente J., Klingenberg D. J., Hidalgo-Alvarez R. "Magnetorheological fluids: a review." *Soft Matter*. 2011;**7(8)**:3701–3710.

[30] Sherman S. G., Paley D. A., Wereley N. M. "Parallel simulation of transient magnetorheological direct shear flows using millions of particles." *IEEE Transactions on Magnetics*. 2012;**48(11)**:3517–3520.

[31] Barski M., Pająk P. "Simulation and numerical homogenization of magnetorheological fluids using the molecular dynamic technique." *Advanced Materials Research*. 2014;**849**:234–241.

[32] Peng Y.-Bo, Ghanem R., Li J. "Investigations of micro-structured behaviors of magnetorheological suspensions." *Journal of Intelligent Material Systems and Structures*. 2012;**23(12)**:1351–1370.

[33] Lagger H. G. *Particle-based simulation of sheared magnetorheological fluids*. Ph.D. Thesis, Fraunhofer Institute, Stuttgart, DE, 2014.

[34] Favier J. F., Abbaspour-Fard M. F., Kremmer M., Raji A. O., "Shape representation of axi-symmetrical, non-spherical particles in discrete element simulation using multi-element model particles." *Engineering Computations*. 1999;**16**:467–480.

[35] Yang G., Spencer Jr B. F., Jung H. J., Carlson J. D. "Dynamic modeling of large-scale magnetorheological damper systems for civil engineering applications." *Journal of Engineering Mechanics*. **130(9)**:1107–1114.

[36] Zhang H. H., *et al*. "A magnetic design method of MR fluid dampers and FEM analysis on magnetic saturation." *Journal of Intelligent Material Systems and Structures*. 2006, **17**:813–818.

[37] Farjoud A., Taylor R., Schumann E., Schlangen T. "Advanced semi-active engine and transmission mounts: tools for modelling, analysis, design, and tuning." *Vehicle System Dynamics*. 2014, **52(2)**:218–243.

[38] Carlson J. D., Chrzan M. J. "Magnetorheological fluid dampers." US Patent No. 5277281, 1994.

[39] Strecker Z., Roupec J., Mazurek I., Klapka M. "Limiting factors of the response time of the magnetorheological damper." *International Journal of Applied Electromagnetics and Mechanics*. 2015, **47(2)**:541–550.

[40] Choi Y.-T., Wereley N. M. "Comparative analysis of the time response of electrorheological and magnetorheological dampers using nondimensional parameters." *Journal of Intelligent Material Systems and Structures*. 2002, **13(7–8)**:443–452.

[41] Koo J. H., Goncalves F. D., Ahmadian M. "A comprehensive analysis of the time response of MR dampers." *Smart Materials and Structures*. 2006, **15**(**2**):351.

[42] Jiang Z., Christenson R. E. "A fully dynamic magnetorheological fluid damper model." *Smart Materials and Structures*. 2012, **21**:065002.

[43] Goncalves F. D., Ahmadian M., Carlson J. "Investigating the magnetorheological effect at high flow velocities." *Smart Materials and Structures*. **15**(**1**):75.

[44] Sedlacik M., Pavlinek V., Saha P., *et al.* "The role of particles annealing temperature on magnetorheological effect." *Modern Physics Letters B*. 2012, **26**(**3**):363–368.

[45] An J., Kwon, D. S. "Modeling of a magnetorheological actuator including magnetic hysteresis." *Journal of Intelligent Material Systems and Structures*. **14**(**9**):541–550.

[46] Erol O., Gonenc B., Senkal D., Alkan S., Gurocak H. "Magnetic induction control with embedded sensor for elimination of hysteresis in magnetorheological brakes." *Journal of Intelligent Material Systems and Structures*. **23**(**4**):427–440.

[47] von Pfingsten G., Hameyer K., Paul D. "Influence of axial mechanical stress on the magnetic properties of non-oriented electrical steel." *6th IEEE Electric Drives Production Conference (EDPC)*. 2016, Nuremberg, DE.

[48] Noh M., Gi M. J., Kim D., Park Y. W., Lee J., Kim J. "Measurements of magnetic properties of electromagnetic actuator in high-temperature environment", *Journal of Magnetics*. 2015, **20**(**1**):86–90.

[49] Zhang X., Li Z., Guo K., Zheng F., Wang Z. "A novel pumping magnetorheological damper: design, optimization and evaluation." *Journal of Intelligent Material Systems and Structures*. 2017, **28**(**17**):2339–2348.

[50] Nguyen Q. H., Han Y. M., Choi S. B., Wereley N. M. "Geometry optimization of MR valves constrained in a specific volume using the finite element method." *Smart Materials and Structures*. 2007, **16**(**6**):2242.

[51] Nguyen Q. H., Choi S. B. "Optimal design of an automotive magnetorheological brake considering geometric dimensions and zero-field friction heat." *Smart Materials and Structures*. 2010, **19**(**11**):115024.

[52] Gołdasz J., Sapiński B. "Modeling of magnetorheological mounts in various operation modes." *Acta Mechanica et Automatica*. 2011, **5**:29–40.

[53] Sapinski B. *Linear magnetorheological fluid dampers for vibration mitigation: modelling, control and experimental testing*. Uczelniane Wydawnictwa Naukowo-Dydaktyczne AGH. 2004.

[54] Sapiński B., Rosół M., Jastrzębski Ł., Gołdasz J. "Development and performance evaluation of an MR squeeze-mode damper." *Smart Materials and Structures*. 2015, **24**:115007.

[55] Sapiński B., Rosół M., Jastrzębski Ł., Gołdasz J. "Outlook on the dynamic behavior of an magnetorheological squeeze-mode damper prototype." *Journal of Intelligent Material Systems and Structures*. 2017, **28**(**20**):3025–3038.

[56] Erol O., Gonenc B., Senkal D., Alkan S., Gurocak H. "Magnetic induction control with embedded sensor for elimination of hysteresis in magnetorheological brakes." *Journal of Intelligent Material Systems and Structures*. 2012, **23**(**4**):427–440.

[57] Yadmellat P., Kermani M. R. "Output torque modeling of a magnetorheological based actuator." *IFAC World Congress, Milano (Italy)*. 2011, 1052–1057.

[58] Guo P., Xie J., Guan X. "Dynamic model of MR dampers based on a hysteretic magnetic circuit." *Shock and Vibration*. 2018, 13. Retrieved from https://doi.org/10.1155/2018/2784950.

[59] Bai X.-X., Cai. F.-L., Chen P. "Resistor-capacitor (RC) operator-based hysteresis model for magnetorheological (MR) dampers." *Mechanical Systems and Signal Processing*. 2019, **117**:157–169.

[60] Hassani V., Tjahjowidodo T., Do T. N. "A survey on hysteresis modeling, identification and control." *Mechanical Systems and Signal Processing*. 2014, **49**:209–233.

[61] Gołdasz J., Sapiński B., Jastrzębski Ł. "On the application of Bouc-Wen approach for modeling of MR actuators." *ACTUATOR 2018, 16th International Conference on New Actuators, Bremen, DE*. 2018.

[62] Visintin A. "Maxwell"s equations with vector hysteresis." *Archive for Rational Mechanics and Analysis*. 2005, **175**(**1**):1–37.

[63] Kubik M. Magnetorheological suspension damper for space application. *Ph.D. Thesis, Brno University of Technology*, 2018.

[64] Kubík M., Koutny D., Palousek D., *et al*. "Skeleton for magnetic core and method for manufacturing the same." *Patent Application*. 2018, **EP3373311** A1.

[65] Hong S. R., Choi S. B., Choi Y.-T., Wereley N. M. "Non-dimensional analysis for effective design of semi-active electrorheological damping control systems." *Proceedings of the Institution of Mechanical Engineers, Part D: Journal of Automobile Engineering*. 2003, **7**:1095–1106.

[66] Hong S. R., Choi S.-B., Choi Y.-T., Wereley N. M. "Non-dimensional analysis and design of a magnetorheological damper." *Journal of Sound and Vibration*. 2005, **288**(**4–5**):847–863.

[67] Gołdasz J., Sapiński B. "Verification of magnetorheological shock absorber models with various piston configurations." *Journal of Intelligent Material Systems and Structures*. 2013, **24**(**15**):1846–1864.

[68] Dimock G. A., Lindler J. E., Wereley N. M. "Bingham biplastic analysis of shear thinning and thickening in magnetorheological damper." *Proc. SPIE 3985, Smart Structures and Materials 2000: Smart Structures and Integrated Systems*. 22 June 2000. Retrieved from https://doi.org/10.1117/12.388847

[69] Farjoud A., Vahdati N., Fah, Y. F. "Mathematical model of drum-type MR brakes using Herschel-Bulkley shear model." *Journal of Intelligent Material Systems and Structures*. 2008, **19**(**5**):565–572.

[70] Zhang X. J., Farjoud A., Ahmadian M., Guo K. H., Craft M. "Dynamic testing and modeling of an MR squeeze mount." *Journal of Intelligent Material Systems and Structures*, 2011, **22**(**15**):1717–1728.

[71] Wereley N. M., Cho J. U., Choi Y.-T., Choi S. B. "Magnetorheological dampers in shear mode." *Smart Materials and Structures*. **17**(**1**):015022.

[72] Hong S. R., Shaju J., Wereley N. M., Choi Y.-T., Choi S.-B. "A unifying perspective on the quasi-steady analysis of magnetorheological dampers." *Journal of Intelligent Material Systems and Structures*. **19**:959–976.

[73] Sherman S. G., Becnel A. C., Wereley N. M. "Relating Mason number to Bingham number in magnetorheological fluids." *Journal of Magnetism and Magnetic Materials*. 2015, **380**:98–104.

[74] Wong A. P., Tozer R. C., Bullough W. A., Ellam D. J. "Computational fluid dynamics in the flow of ERF/MRF in control devices and of oil flow through piezo-hydraulic valves." *Computers and Structures*. 2008, **86**:266–280.

[75] Papanastasiou T. C. "Flows of materials with yield." *Journal of Rheology*. 1987, **31**:385.

[76] Foister R., Nehl T. W., Kruckemeyer W., Raynauld O. "Magnetorheological (MR) piston assembly with primary and secondary channels to improve MR damper force." US Patent 832798 B2. 2009.

[77] Mao M. *Adaptive magnetorheological sliding seat systems for ground vehicles*. Ph.D. Thesis. University of Maryland, 2011.

[78] Gołdasz J., Kluszczyński K., Sapiński B. "Fluid flow pinched by near-wall Bingham plastic zones: simulation study." *19th International Conference on Research and Education in Mechatronics (REM)*. 2018, doi:10.1109/REM.2018.8421800.

[79] Potnuru M. R., Wang X., Mantripragada S., Gordaninejad F. "A compressible magnetorheological damper-liquid system." *International Journal of Vehicle Design*. 2013, **63**:256–274.

[80] Nguyen Q.-H., Choi S. B. "A new approach for dynamic modeling of an electrorheological damper using a lumped parameter method." *Smart Materials and Structures*. 2009, **18**:115020.

[81] Parlak Z., Engin T. "Time-dependent CFD and quasi-static analysis of magnetorheological fluid dampers with experimental validation." *International Journal of Mechanical Sciences*. 2012, **64**:22–31.

[82] Case D., Taheri B., Richer E. "Multiphysics modeling of magnetorheological dampers," *International Journal of Multiphysics*. 2013, **7**(**1**):61–76.

[83] Zheng J., Li Z., Koo J., Wang J. "Magnetic circuit design and multiphysics analysis of a novel MR damper for applications under high velocity." *Advances in Mechanical Engineering*. 2014, 402501.

[84] Gołdasz J., Sapiński B. "Application of CFD to modeling of squeeze mode magnetorheological dampers." *Acta Mechanica et Automatica*, 2015, **9**(**3**):129–134.

[85] Hegger C., Maas J. "Torque enhancement possibilities for energy-efficient MRF-based coupling elements." *SPIE Conference on Active and Passive Smart Structures and Integrated Systems XII*. 2018, **10595**:1059531.

Chapter 5

Use of magnetorheological shock absorber for impact loading mitigation with individually controllable coils

Wang Jiong[1], Zhang Guang[1], Wang Huixing[1], Ouyang Qing[2], Zheng Jiajia[3], Li Yancheng[4] and Li Zhaochun[5]

This book chapter presents comprehensive investigations on the use of magnetorheological (MR) shock absorber to mitigate the effect of impact loading, with emphasis on the recoil system. A physical model of the field gun is established and a long-stroke MR recoil absorber with four-stage parallel electromagnetic coils is designed to apply separate current respectively and generate variable magnetic field distributions in the annular flow channel. The response time and the compensation method are investigated to facilitate the application. Based on the dynamic analysis and firing stability condition during the process of shock, the ideal recoil force-stroke profiles of MR absorber at different limiting firing angles are obtained. The experimental studies are carried out on the impact test rig under different combinations of current loading: conventional unified control mode, separate control mode and timing control mode.

5.1 Introduction

For recoil impact system, small recoil force and high firing accuracy are the basic requirements. The recoil forces impart substantial vibrations and affect the fatigue life of the structure and firing accuracy, although the firing event has a short duration [1]. To eliminate the negative effects caused by excessive recoil forces, hydraulic recoil absorbers utilizing variable orifice area to minimize recoil impact forces are most widely used in the recoil system [2]. In addition, the other passive

[1]School of Mechanical Engineering, Nanjing University of Science and Technology, China
[2]School of Mechanical and Electrical Engineering, Jiaxing University, China
[3]School of Technology, Zhejiang Normal University, China
[4]Faculty of Engineering and Information Technology, University of Technology Sydney, Australia
[5]College of Mechanical and Electronic Engineering, Nanjing Forestry University, China

components can be used independently or in combination in the recoil impact system, such as gas shock absorber, rubber shock absorber, stiff spring absorber or muzzle brakes. Generally, the hydraulic absorber plays a leading role in term of vibration absorption and mitigated the recoil load transmissions to the gun cradle in the large caliber gun recoil system. As for the aircraft gun, rifle and those guns with relatively small caliber, they are usually adopted gas, rubber or spring shock absorbers, respectively.

These traditional passive absorbers have been designed and applied in various recoil systems [3,4]. However, they have limitations in controlling recoil performance accurately due to the involved passive elements. Thus, they were typically used for coping with a fixed load condition rather than an irregularly varying impact load. Once the structural parameters of these traditional absorbers (i.e., orifice profile, material property, size of absorber) are determined, the performances of the absorbers are set. In other words, the passive absorbers cannot make adaptive intelligent response for the real-time external excitation and obtain optimal shock absorption effect.

Considering the inherent capabilities of MR fluid that can execute specific functions autonomously in response to changing external magnetic field [5,6], the adaptive MR absorbers have also been applied in recoil systems for absorbing vibration and controlling recoil performance [7–11] and other shock engineering applications, such as seat suspension system [12,13], military vehicles [14,15] and seismic isolation system [16]. MR fluids are suspensions of magnetically responsive particles in liquid carrier. When the external magnetic field is applied to the MR fluid, the fluid behavior would change from a freely flowing liquid into a semisolid reversibly. Thus, the load-stroke profile of the MR absorber can be reversibly and rapidly adjusted, making it desirable for varying impact conditions [17]. Many present scientific studies have been focused on minimizing the recoil load transmissions to the gun cradle and improving the firing accuracy by installing the semi-passive MR absorbers. However, these previous studies have rarely paid attention to the shock mitigation of MR absorber, which have special requirements of recoil force at different limiting firing angles. Moreover, the structure configuration of those MR recoil absorbers in previous studies were usually designed with single stage electromagnetic coil or multistage coils connected in series, which can only generate uniform magnetic field distribution in the annular flow channel. Although these conventional MR absorbers have successfully demonstrated a high-performance in shock mitigation, in practice, it has still not fully development their performance potential due to the inflexible magnetic field distribution. Especially, when the absorber piston moves at a high speed under impact loading and drives the fluid into the annular flow channel in a split second, it is prone to generate the "fluid locking" effect [18], which explains the fluid is not flowing through the gap at a rate that would allow the piston to enter the housing during the initial stage of the impact.

In this chapter, a new concept of parallel coil multistage absorber with controllable damping force is proposed to resolve the inherent shortcomings of traditional series multistage coil-based absorber. Following this, the details of new adaptive base absorber are proposed and introduced. In particular, in order to exploit the potential of MR absorber and ease the fluid-locking effect during impact process, we propose a new design of the MR absorber with four-stage parallel electromagnetic coils, which can apply separate current respectively. Since the electromagnetic coils configuration has a modular concept, the variable magnetic field distribution in the annular flow channel can be easily obtained by rearranging the number of working coils and the current magnitude of each coil to meet various requirements of the output recoil impact force.

5.2 Gun-recoil system

The mathematical model of the gun recoil system is to be established in this part. As shown in Figure 5.1, when the gun is launched with angle ϕ, the force of the recoil part contains the chamber force F_{pt} and recoil part gravity $m_z g$, which act on the axis of the gun bore and the centroid of the recoil. The restraining forces include the damping force of the MR damper F_d, the force of the recuperating machine F_h, the friction of the sealing device F_f, the normal force of the rocking guide F_{N1}, and the corresponding friction F_{T1}, F_{T2}. When the internal ballistic parameters, structural dimensions and gravity are determined, the recoil motion and the force regulation of gun are controlled by regulating the variation law of damping force.

The physical model of MR recoil device can be described by a single degree of freedom system as shown in Figure 5.2 for the single stroke recoil without reentry. The MR damper is fixed to the gun by the guide rail. The recoil mass moves along the x direction when the chamber acts on the recoil. The process can be described with the following differential equation:

$$(m_b + M_r)\ddot{x} = F_{pt}(t) + M_r g\sin\phi - \varsigma M_r g\cos\phi - F_f - F_d(x, \dot{x}, B) \qquad (5.1)$$

where x, $\dot{x}$ and $\ddot{x}$ are displacement, velocity and acceleration of recoil system, respectively. $F_d(x, \dot{x}, B)$ is the damping force of the MR damper, and the amplitude depends on the velocity and displacement of the recoil and magnetic field. m_b and M_r are the mass of projectiles and recoil, respectively. Σ is the friction coefficient

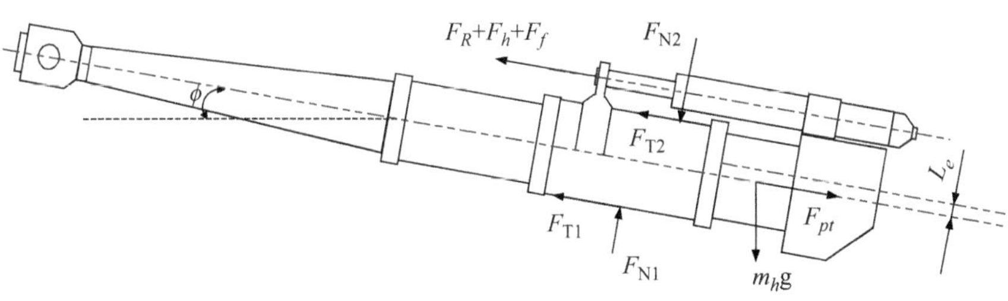

Figure 5.1 Force analysis of the gun recoil

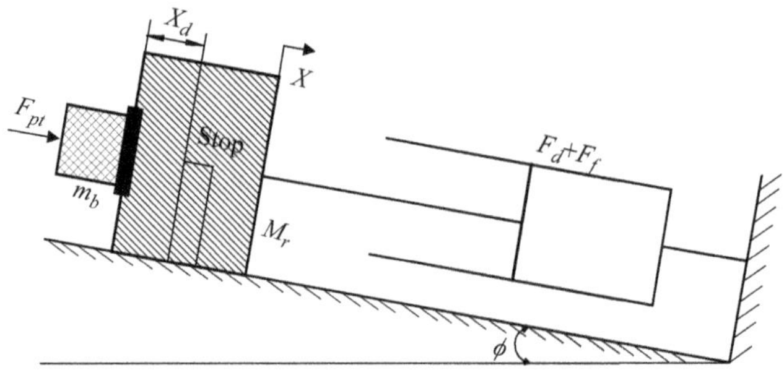

Figure 5.2 Physical model of MR damper system

between the recoil part and the guide surface. g is gravitational acceleration. F_{pt} is impact force caused by gunpowder explosion.

When the gun is lunched horizontally ($\phi = 0°$), (5.1) can be simplified as

$$(m_b + M_r)\ddot{x} = F_{pt}(t) - \overline{F}_d(x, \dot{x}, B) \tag{5.2}$$

where $\overline{F}_d(x, \dot{x}, B)$ is the damping force generated by the recoil part, including the output damping force and contact friction force of the MR damper. The initial state is $(0) = 0$ and $\dot{x}(0) = 0$.

The impact force caused by gunpowder explosion can be divided into two stages. In the first stage, the instantaneous pressure inside the chamber explodes, and the pressure in the chamber increases sharply. In the second stage, the propellant gas rapidly expels from the closed chamber, the pressure and density of the cavity rapidly decreases to zero. The descending trend is similar to the exponential distribution, which has been proved by experiments. Therefore, the impact force of the second stage can be expressed as

$$F_{pt}(t) = F_g e^{-((t-t_g)/\overline{b})} \tag{5.3}$$

$$F_g = \frac{1}{\tilde{\varphi}}\left(1 + \frac{1}{2}\frac{\varpi}{m_b}\right)Ap_c(t_g) \tag{5.4}$$

where ϖ is the gunpowder gravity, A is the cross section area of the bore part, $\tilde{\varphi}$ is the coefficient of secondary work calculation, p_c is the average pressure of the cavity, t_g is the beginning time of the second stage, time constant $\overline{b}$ reflects the size of the total impulse in the second stage.

$$\overline{b} = \int_{t_g}^{t_g+t_e} F_{pt}(t)/F_g dt \tag{5.5}$$

In general, the amplitude of impact depends on the amount of the powder in the chamber. In the simulation process, the impact force of gunpowder explosion can be expressed as a combination of sinusoidal and negative exponential function as

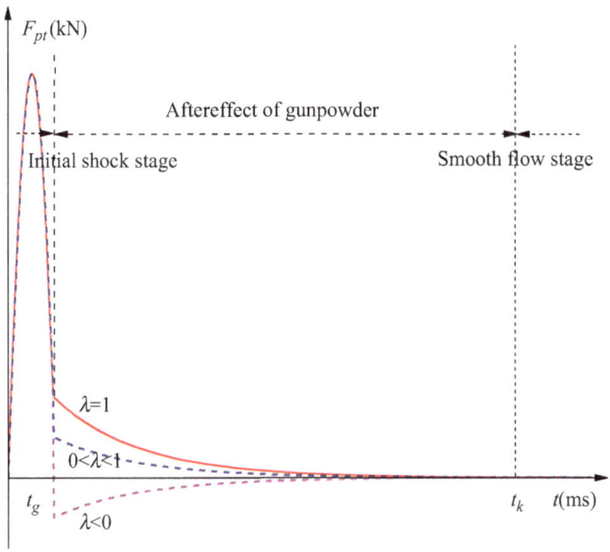

Figure 5.3 Resultant force of the bore with time

stated below. Therefore, the curve of the resultant force of the bore with time is shown in Figure 5.3.

$$\overline{F}_{pt}(t) = \begin{cases} A\sin\left(\dfrac{\pi}{t_e}t\right) & 0 \le t \le t_g \\ \lambda A\sin\left(\dfrac{\pi}{t_e}t_g\right)e^{-((t-t_g)/\overline{b})} & t_g < t \le t_k \\ 0 & t > t_k \end{cases} \tag{5.6}$$

By replacing (5.6) into (5.2), the three stage of the MR damper system, namely, the initial shock period, the post-impact period and the steady flow period, can be obtained and expressed as

$$\begin{cases} (m_b + M_r)\ddot{x} = A\sin\left(\dfrac{\pi}{t_e}t\right) - \overline{F}_d(x, \dot{x}, B) & 0 < t < t_g \\ (m_b + M_r)\ddot{x} = \lambda A\sin\left(\dfrac{\pi}{t_e}t_g\right)e^{-((t-t_g)/\overline{b})} - \overline{F}_d(x, \dot{x}, B) & t_g \le t \le t_k \\ M_r\ddot{x} = -\overline{F}_d(x, \dot{x}, B) & t_k < t \le t_\lambda \end{cases} \tag{5.7}$$

5.3 Design of multi-coil MR absorber

Figure 5.4 displays the schematic and a photograph of the novel MR absorber. The number of turns of each coil is 500 with 0.69 mm diameter copper wire.

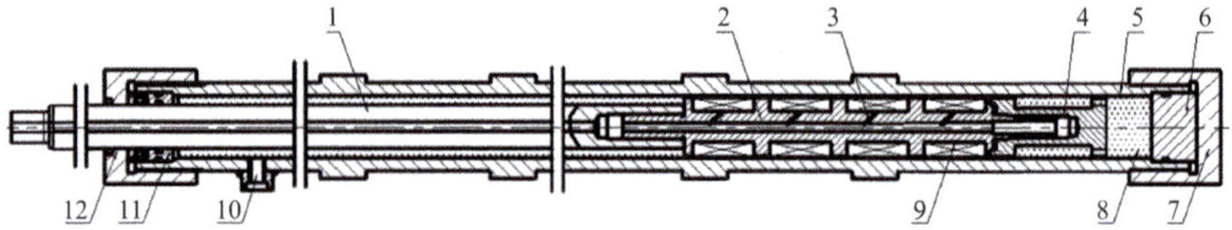

(a)

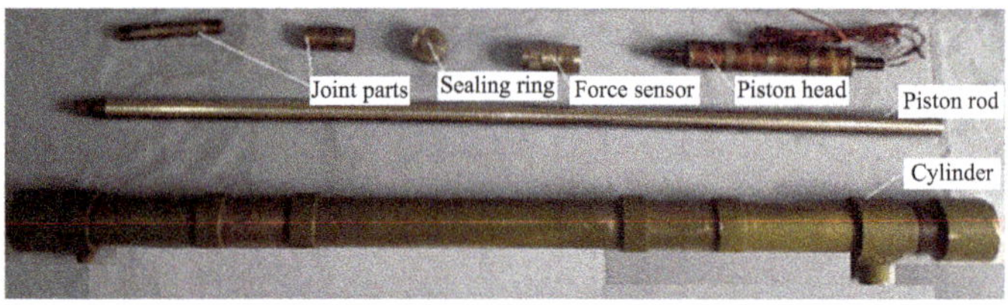

(b)

Figure 5.4 Novel MR absorber: (a) schematic of the designed absorber and (b) photograph of the designed absorber

Table 5.1 Main parameters of the novel MR absorber

Maximum stroke (mm)	Number of coils	Turns of coil	Gap width (mm)	Inner diameter of cylinder (mm)	Effective length of piston head (mm)
±500	4	500	1.5	40	40

The enameled wires come from the wire-leaded hole 3 and wind around the groove to form four parallel magnetic circuits. There are four wire-leaded holes inside the piston, so each coil can be applied with input current separately. The magnetic field path is composed of the cylinder, working gap and piston head. To make sure the magnetic field lines are going along the expected route, the piston rod and guide head connecting the piston head are made of aluminum and copper, respectively. The basic specifications of the novel MR absorber are shown in Table 5.1.

When currents are applied to the coils of the MR absorber, the induced electromagnetic will control damping force to regulate the reciprocating movement of piston. By adjusting the number of working coils as well as value of input current, the MR absorber can produce controllable damping force. For the situation with no current applied, the MR absorber acts as a conventional viscous absorber.

5.4 Time response of MR buffer system under impact loading

5.4.1 Impact test platform system

Figure 5.5 shows an impact buffer system with an MR absorber, which is intended to control the damping characteristics continuously by varying the input power of the electromagnet. The experimental setup consists of an industrial control computer, a real-time simulation system (dSPACE), a current controller and an experimental platform [19]. The absorber is installed in shaft sleeves with a recoil mass, and the piston rod is connected to the ground mount.

During the experiment, the explosion of gunpowder in the close bomb produces a large impulsive force, which pushes the recoil mass along with the MR absorber cylinder back on the guide rail relative to the ground mount. The elastomeric bumper is installed at the rear of the test setup to prevent the recoil from bumping into the test rig in case that the recoil exceeds the maximum stroke length. The current controller is used to control the damping force according to the control input prescribed by the dSPACE control system, which is also used as the data-acquisition system.

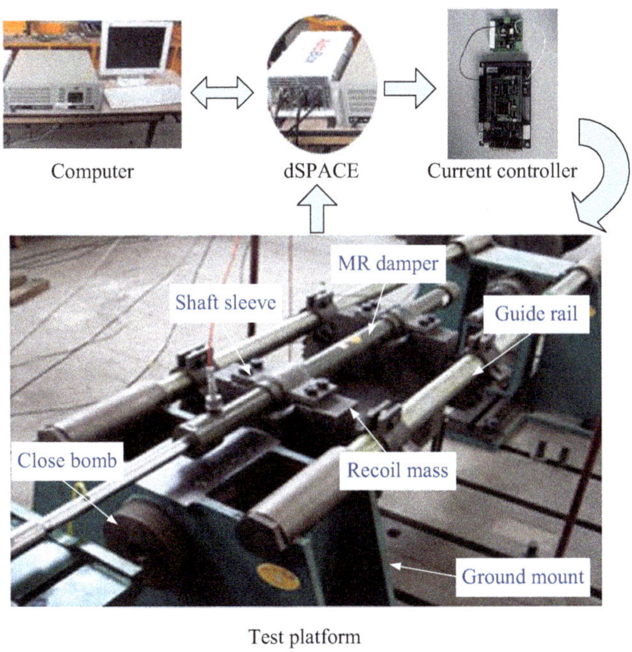

Figure 5.5 Impact test platform system

The MR absorber in the buffer system functions as a shock absorber and is used to dissipate energy of the impact loading. The impact loading within an extremely short period of time is transformed into small damping force acting on a long displacement. Such dynamics can be expressed as the following differential equation:

$$m_h \frac{dv_r}{dt} = F_{pt} - F_{cv} - F_f \tag{5.8}$$

where F_{pt} is the impact load, m_h is the recoil mass, v_r is the recoil velocity, F_{cv} is the sum of Coulomb force (F_τ) and viscous force (F_η) and F_f is the friction force between the contact surfaces.

5.4.2 Current controller

In order to illustrate the principle of the electromagnetic circuit for an MR absorber, an equivalent model was first investigated. The electromagnetic circuit of an MR absorber is generally simplified into an equivalent resistance and an equivalent inductance connected in series. The control current cannot make a sudden change because the inductance needs a time to store and release energy, which delays the "activation" of electromagnetic coil. Figure 5.6 illustrates an electromagnetic circuit of an MR absorber by assuming R_m to be an equivalent resistance and L_m an equivalent inductance. According to the Kirchhoff laws, the differential equation of a control electric circuit in the zero state, which can be described as follows:

$$L_m \frac{dI_m(t)}{dt} + R_m \cdot I_m(t) = V(t) \tag{5.9}$$

If the loading voltage is a constant, the solution of (5.9) can be given as

$$I_m(t) = \frac{V_0}{R_m}\left(1 - e^{(1-(R_m/L_m)t)}\right) \tag{5.10}$$

where V_o is the initial value of the loading voltage.

Figure 5.7 shows the simulated results of the relationship between the equivalent resistance R_m and the response time under the designated coil current. Figure 5.8 shows the relationship between the equivalent inductance L_m and the response time under the designated coil current. To be consistent with the literature, the authors

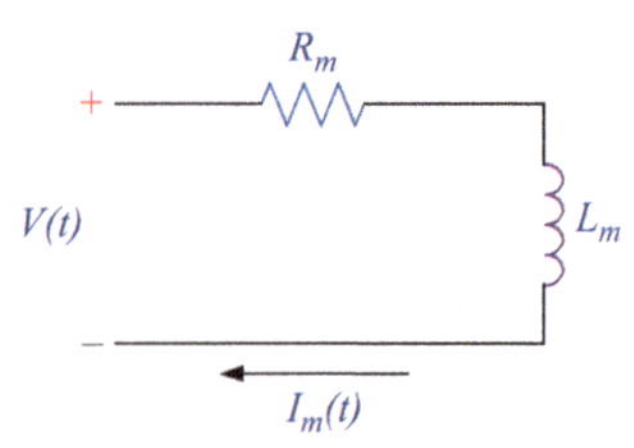

Figure 5.6 Electromagnetic circuit of MR absorber

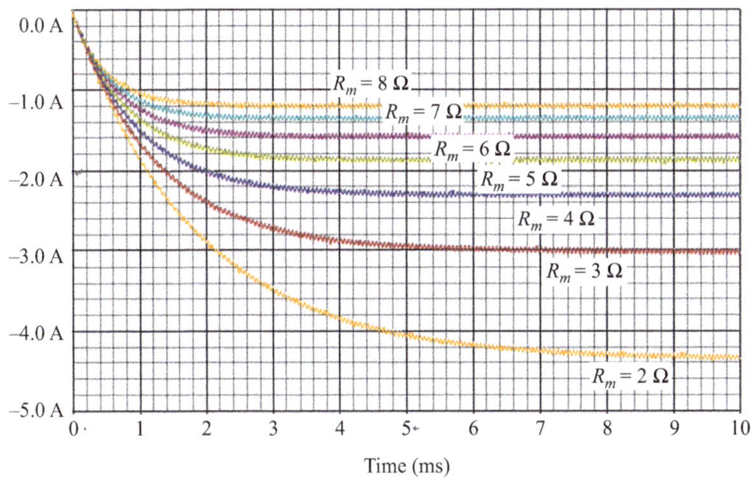

Figure 5.7 Resistance vs time

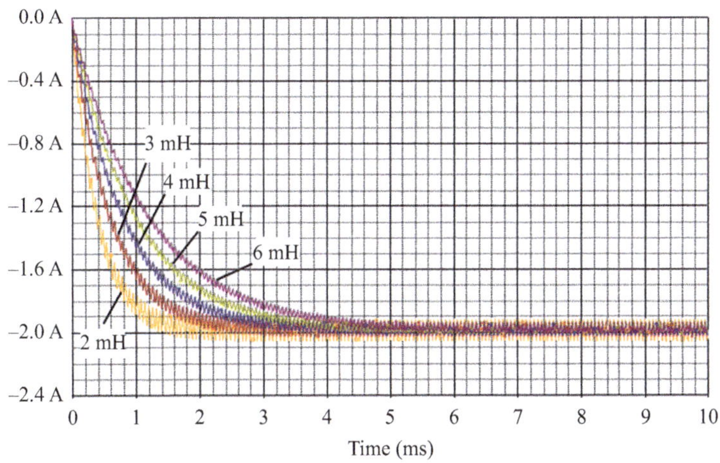

Figure 5.8 Inductance vs time

define the response time of the coil current as the time required to reach 95% of the final state from the initial state [20]. As shown in Figure 5.7, the response time of current triples when the resistance reduces from 8 to 2 Ω. However, it decreases 4 ms when the inductance reduces from 6 to 2 mH. The response time of the magnetic field and the response time of the current should be the same in the ideal case. However, due to the presence of eddy currents, the response of the magnetic field always lags the changing current. Besides, the magnetic loop of MR absorber passes through steel and ferroalloy which have ability to keep magnetisms. In other words, the magnetic flux strength B is always lagging behind the magnetic field intensity H, which makes the time delay last longer.

In order to control the output force of the MR absorber, the input current must be continuously adjustable by a current controller. To this regard, the PWM (pulse width modulation) current drive technology is adopted to control the current on the electromagnetic coil of the MR absorber, which generates magnetic fields and controls the output damping forces. Figure 5.9 shows a current controller which includes a current driver and a digital signal processor (DSP) controller. The PWM signal V_c is generated by the DSP controller, and the current driver is connected to the electromagnetic coil to produce the controllable current. Figure 5.10 shows a

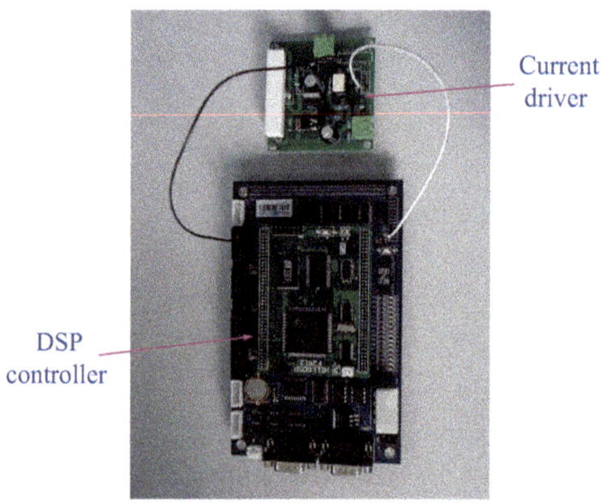

Figure 5.9 Current controller

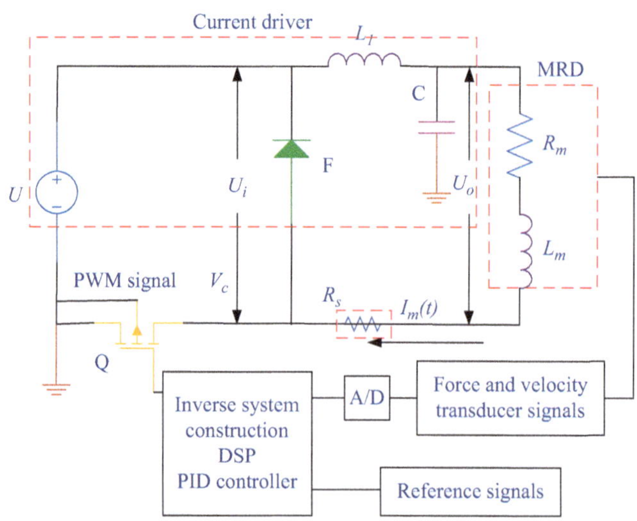

Figure 5.10 Schematic of MR control system

schematic of the control system for MR absorber, in which the control algorithm is carried out by the DSP controller when the A/D-converted force and velocity transducer signals together with the reference signal are transmitted as the inputs. The electromagnetic drive circuit is activated by the PWM signal with a certain duty cycle. Finally, the desired operating current is applied to the electromagnetic coil to achieve the desired damping force.

The PWM signal V_c with a certain duty cycle, which is produced by the DSP controller, switches the MOSFET Q to on or off state, transforming the input voltage U_i to pulse signals. The coil current I_m shows an upward trend when the tube voltage U_Q is at low level, setting the MOSFET at off state, as shown in Figure 5.11. The LC filter circuit is designed to decrease the current fluctuation. The loop circuit is composed of the fast recovery diode (F) and the inductance (L_1), thus the current is continuously supplied when the input voltage (U_i) becomes a zero. Based on this current driver circuit, the steady current flowing through the MR absorber coil can be achieved, and its amount can be controlled by the duty cycle of the input PWM signal (V_c).

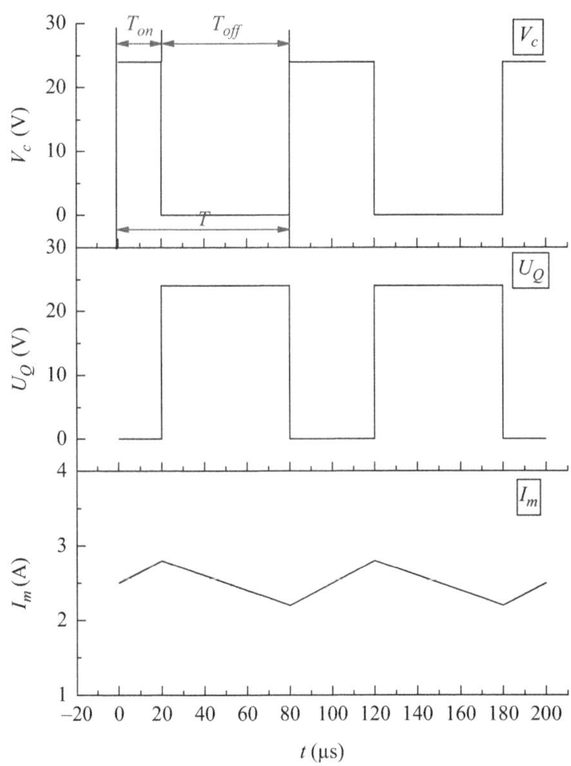

Figure 5.11 Waveform diagram for the switch regulator circuit

As shown in Figure 5.11, the duty cycle is defined as

$$D = \frac{T_{on}}{T} \tag{5.11}$$

where T is the period of a pulse, and T_{on} is the working time for one cycle.

The response of the drive circuit can be described by the following transfer function:

$$\frac{U_o}{U_i} = \frac{L_m s + R_m}{L_1 L_m C s^3 + R_m C L_1 s^2 + (L_1 + L_m)s + R_m} \tag{5.12}$$

$$\frac{U_o}{I_m} = L_m s + R_m \tag{5.13}$$

where U_i is the input voltage and U_o is the output voltage working on the electromagnetic coil. Dividing (5.12) by (5.13), the circuit transfer function can be achieved as

$$G_I = \frac{I_m}{U_i} = \frac{1}{(L_1 L_m C/R_m)s^3 + L_1 C s^2 + ((L_1 + L_m)/R_m)s + 1} \times \frac{1}{R_m} \tag{5.14}$$

The input voltage U_i is a pulse signal and its Laplace transform can be described as

$$U_i(s) = \frac{U(1 - e^{-T_{on}})}{s(1 - e^{-sT})} \tag{5.15}$$

where U is the constant voltage (12 V) supplied by the DC-regulated power supply.

Since the coefficient of s^3 is small enough to be neglected, it can be simplified as

$$G_I = \frac{I_m}{U_i} \approx \frac{1}{L_1 C s^2 + ((L_1 + L_m)/R_m)s + 1} \times \frac{1}{R_m} \tag{5.16}$$

Then the Laplace transformation of the output current in the electromagnetic coil of MR absorber can be expressed as

$$I_m(s) = U_i(s) \cdot G_I(s) = \frac{1}{L_1 C s^2 + ((L_1 + L_m)/R_m)s + 1} \cdot \frac{U(1 - e^{-sT_{on}})}{s(1 - e^{-sT})} \times \frac{1}{R_m} \tag{5.17}$$

According to the final value theorem of Laplace, the stable value of I_m is achieved as

$$\lim_{t\to\infty} I_m(t) = \lim_{s\to 0} sI_m(s) = \lim_{s\to 0} \frac{\tau e^{-sT_{on}}}{Te^{-sT}} \times \frac{U}{R_m} = \frac{T_{on}}{T} \times \frac{U}{R_m} = D \times \frac{U}{R_m} \tag{5.18}$$

The above analysis indicates that when the current drive circuit approaches a stable state, the current I_m in the electromagnetic coil of the MR absorber is proportional to the duty cycle of input PWM signal (V_c). Besides, if the pulse period (T) is much smaller than the time constant τ (L_m/R_m), that is, $T \ll \tau$, the current fluctuation can be neglected.

When a current controller is added in series with electromagnetic coils, the corresponding electromagnetic circuit can be simplified as a second-order linear system with the PWM signal (V_c) as the input and the coil current (I_m) as the output. This can be expressed as follows:

$$G_{Ic} = \frac{I_m}{V_c} \approx \frac{1}{L_1 C s^2 + ((L_1 + L_m)/R_m)s + 1} \times K \tag{5.19}$$

where K is a proportional coefficient, $K = (D \times U)/(V_c \times R_m)$.

The model of (5.19) is used to design a simulation model, and the simulation results are compared with the experimental data in the subsequent sections. The transfer functions are obtained by a parameter identification technique to the second-order model of current responses.

5.4.3 Time delay compensation

5.4.3.1 Response analysis of electromagnetic drive circuit

This part of the study evaluates the time delay of an electromagnetic drive circuit with four different PWM signals. The experiment was carried out by measuring the voltage U_s upon the sampling resistance R_s(0.5 Ω), when the electromagnetic drive circuit was applied with various PWM signals. The coil current I_m was then obtained by the division of U_s/R_s. In order to compare the resulting responses distinctively, the PWM signal V_c was normalized by scaling down to the corresponding final values of current I_m. Figure 5.12 shows the measured current responses of the electromagnetic coil by applying a normalized signal $\overline{V}_c$ with four different values from 0.5 to 2.0 with a step of 0.5.

The current response time exhibits an approximated increase when the PWM signal V_c rises. Furthermore, there is a large amount of overshoot which increases as the input voltage rises, and its dynamic characteristics perform undesirably. The response time varies approximately 40, 30, 40, 65 ms when the final output currents

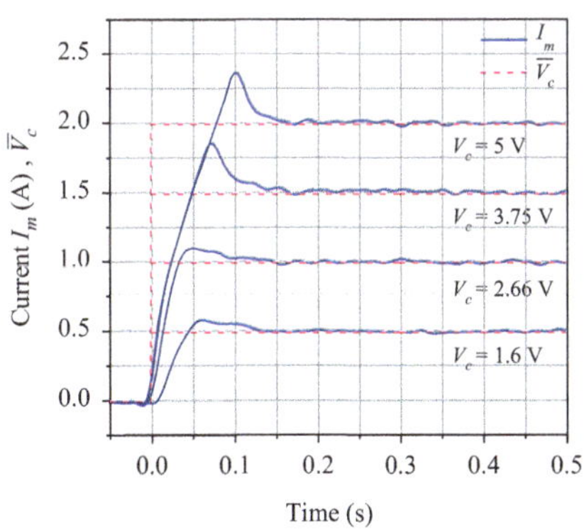

Figure 5.12 Response of coil current

I_m changes 0.5, 1.0, 1.5, 2 A, respectively. However, the magnitudes of the PWM signal V_c and the output current I_m are proportional and the output current I_m tends to be saturated when the PWM signal is applied greater than 5 V (see Figure 5.13).

5.4.3.2 Response analysis of MR impact system

To further evaluate the time delay of the MR buffer system, a series of tests were conducted under different PWM signals V_c. The output damping force of the MR absorber consists of viscous damping force (F_η) and Coulomb damping force (F_τ). The symbol F_η is only relevant to the recoil velocity v_r and reflects the characteristics of fluid viscous damping. The symbol F_τ is determined by the magnetic field intensity in the working gap and controlled by the input current, which reflects the characteristics of adjustable damping force. In order to determine the response time of Coulomb damping force (F_τ) with respect to output current I_m response, the dynamic force model can be obtained as the following form which neglects the friction force F_T [21]:

$$F_\tau \tanh(a_1 v_r + a_2) = F_R - \alpha v_r^n \tag{5.20}$$

where $\tanh(a_1 v_r + a_2)$ represents the Coulomb force (F_τ) when the recoil velocity v_r approaches to a zero, $a_1 = 10$, $a_2 = 0.5$, $\alpha = 1,100$ and $n = 1.6$. The coefficients were obtained by the nonlinear least squares fitting technique based on the experimental data, and the total damping force F_R and the recoil velocity v_r were measured real-time utilizing the force and velocity sensors.

Table 5.2 shows the relationship between the Coulomb force (F_τ) and the current (I_m). The Coulomb force increases when the coil current rises. However, it

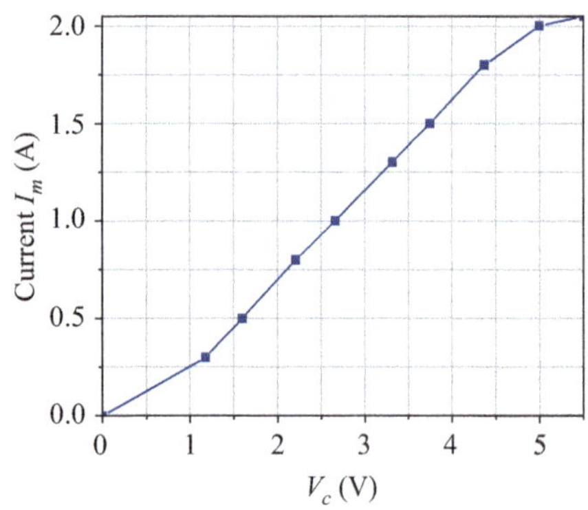

Figure 5.13 Current vs input voltage

Table 5.2 Relationship between I_m and F_τ

PWM signal V_c	1.18	1.60	2.21	2.66	3.32	3.75	4.37	5.0
Current I_m (A)	0.3	0.5	0.8	1.0	1.3	1.5	1.8	2.0
Coulomb force F_τ(N)	1,550	2,058	3,220	3,800	4,550	5,100	6,150	6,150

reaches the saturated point when the current I_m exceeds 1.8 A as shown in Table 5.2.

Figure 5.14 was shown the comparison of the response between the Coulomb damping force (F_τ) and the electromagnetic coil current I_m intuitively. Similarly, the value of F_τ is normalized through divining by an integer M which depended on the corresponding final value of the current I_m. After calculation, the values of the integer M are 1,600, 3,400, 4,300, 6,500, when the current I_m is applied with 0.5, 1.0, 1.5, 2.0 A, respectively.

As shown in Figure 5.14, the response of Coulomb force (F_τ) is slower than that of the coil current I_m. The total time delay τ seems to be affected by (1) time delay τ_α, due to the electronics parts of the current controller, and (2) time delay τ_β, the response time of mechanical parts of the absorber and the MR fluid. Thus, the total time delay can be expressed as

$$\tau = \tau_\alpha + \tau_\beta \tag{5.21}$$

When the PWM signal V_c (currents I_m) increases, the time delay τ_α slightly increases; the time delay τ_β decreases; and the total time delay stabilizes at about 90 ms. As shown in Figure 5.15, the time delay τ_β is about 15 ms which contains the hysteresis of magnetic materials and the response time of MR fluid when the

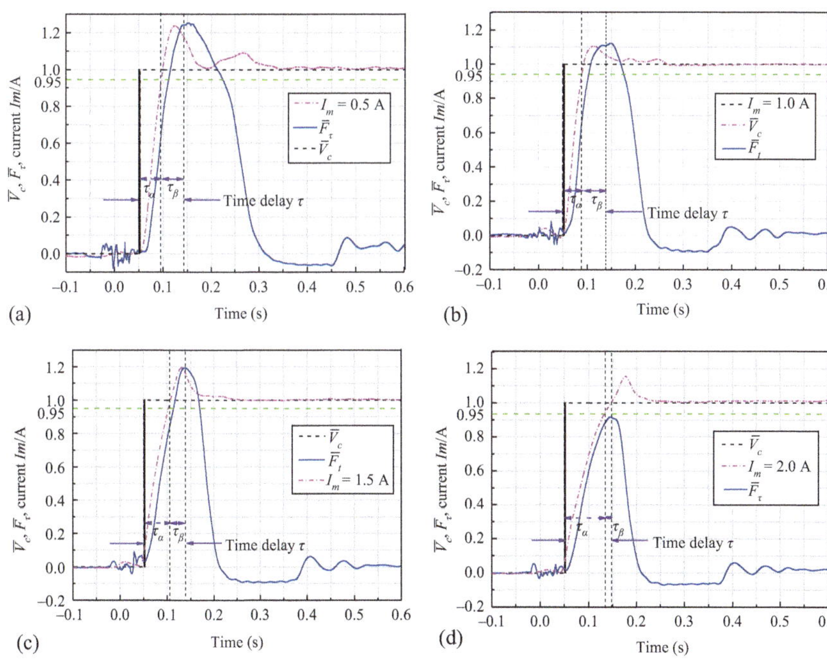

Figure 5.14 Response of MR buffer system: (a) $V_c = 1.6$ V, (b) $V_c = 2.66$ V, (c) $V_c = 3.75$ V and (d) $V_c = 5$ V

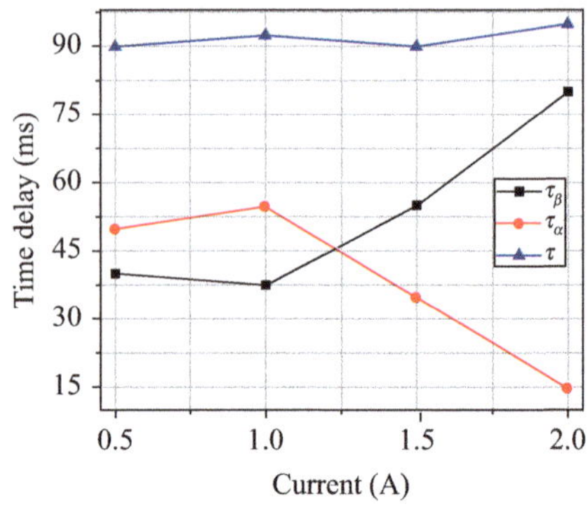

Figure 5.15 Time delay of MR buffer system

input current I_m is 2 A. Hence, the response of the electromagnetic circuit has an increasing effect on the time delay of the impact MR buffer system along with the increment of PWM signal V_c (currents I_m).

The above results indicated that the response time of the MR buffer system to step PWM control signals is the summation of those coming from electromagnet circuit, magnetic hysteresis, MR effect, etc. The electromagnetic circuit can be simplified to a linear second-order system response and the hysteresis and MR effect can be regarded as simple delays. Besides, compared with other factors, the response of electromagnetic circuit has the most significant influence on the response time of the entire MR buffer system.

5.4.3.3 Time delay compensation methods

In order to reduce the time delay in the electromagnetic circuit, this study proposes an advanced circuit for the MR absorber. A new equation can be obtained by Laplace transforming of (5.10),

$$I(s) = \frac{1}{s} \cdot \frac{V_0}{L_M s + R_m} \tag{5.22}$$

It is noted that the electromagnetic model of an MR absorber can be simplified as the transfer function of a first-order inertial system and the current regarded as step signal. Therefore, it is possible to modify the transfer function of electromagnetic circuit model by incorporating an advanced correcting transfer function.

Figure 5.16 shows the advanced correcting circuit with input $U_a(s)$ and output $U_o(s)$, which is supposed to connect the electromagnetic circuit of an MR absorber shown in Figure 5.10 in series. Generally, a simple transfer function of the advanced correcting circuit can be written as

$$G(s) = \alpha \frac{Ts + 1}{\alpha Ts + 1} \tag{5.23}$$

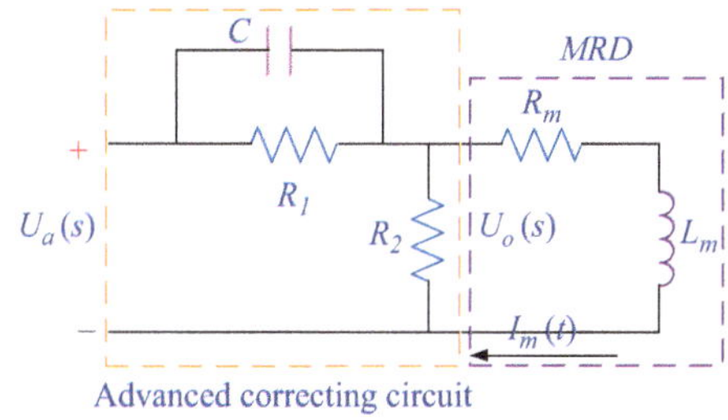

Figure 5.16 Advanced correcting circuit diagram

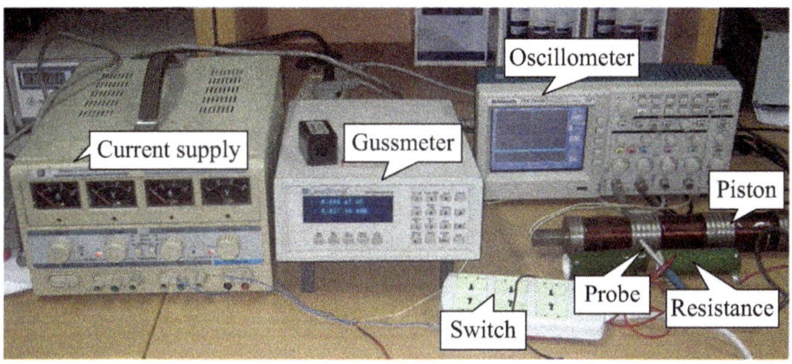

Figure 5.17 Experimental test

where the coefficients α and T should be determined by the parameters of capacitance and resistance in the advanced correcting circuit, $T = R_1C, \alpha = R_2/(R_1 + R_2)$.

To determine the parameters involved in the transfer function and validate the response model of an MR absorber under a constant voltage, a testing experiment was set up as shown in Figure 5.17. During testing, an external resistance was connected in series in order to monitor the changes of the current by measuring the voltage signal at the ends for the wire. A hall probe was positioned at the magnetic pole to measure the magnetic field intensity by a Gussmeter (450 Lake Shore), which was connected to an oscilloscope.

Since the time delay is unrelated with the values of the input voltage, the response of the magnetic field and the current are compared by normalizing their magnitudes to one, respectively. As shown in Figure 5.18, the time delay of the magnetic field is much longer than that of the current due to the eddy current effects. Based on the experimental data, the responses of the current and the magnetic field were fitted into two first-order inertial transfer functions, $G_I(s)$ and $G_M(s)$. Therefore, the numerical values of parameters $R_1 3\Omega$, $R_2 1\Omega$, $C2.17$mF were obtained by assuming $T0.0065$ s and $\alpha 0.25$ in (5.12). By applying the obtained parameters into advanced correcting circuit, it takes 5 ms to achieve 95% of the final state, which increases 75% of the response time as compared with the original response time as shown in Figure 5.19.

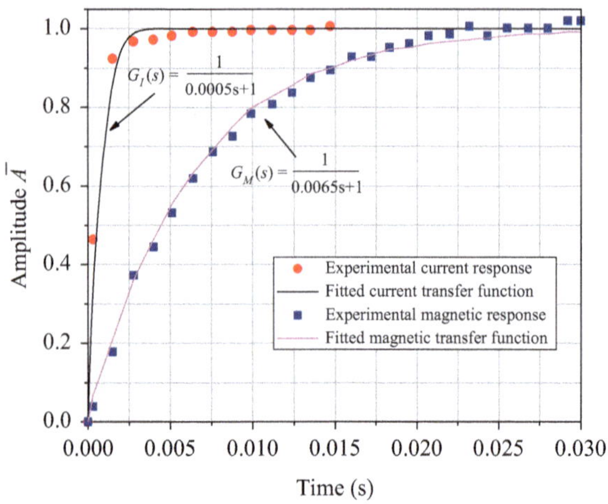

Figure 5.18 Current and magnetic field response

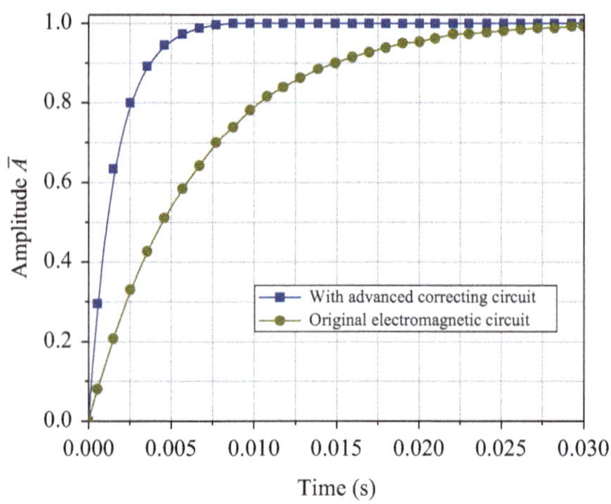

Figure 5.19 Magnetic field response time

5.4.3.4 DSP PID controller

Figure 5.20 shows the comparison of parameterized second-order models $G_{Ic}(s)$ and the experimental results. Due to nonlinear effect of inductance by the electromagnetic coil with the iron core, the natural frequency and the damping ratio of the electromagnetic loop control system keep changing within a certain range.

A PID controller was designed to adjust the delayed response using the DSP controller. The typical PID equation can be written in Laplace transform form:

$$G_{pid}(s) = K_p + \frac{K_i}{s} + K_d s \tag{5.24}$$

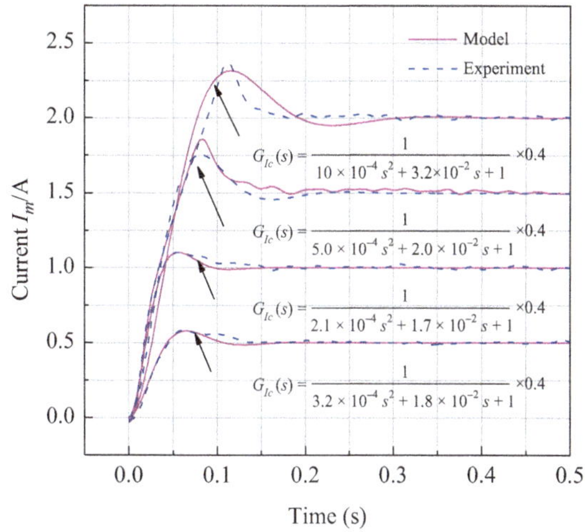

Figure 5.20 Comparison of second-order model and experimental results

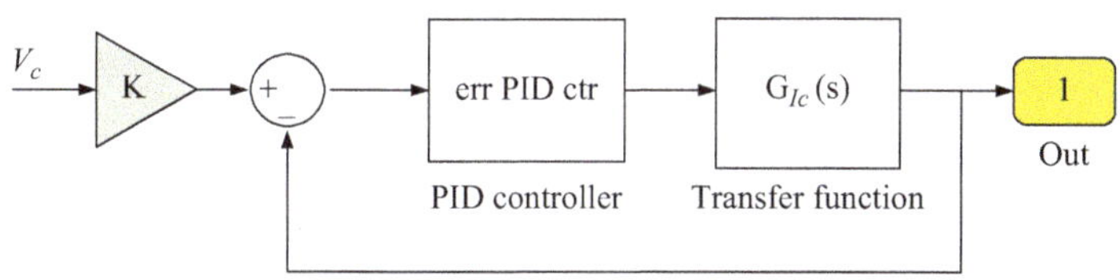

Figure 5.21 Block diagram of PID controller

In order to reduce the overall time delay, the transfer function model $G_{Ic}(s)$ which had the longest response time of coil current ($I_m = 2$ A) was chosen for the PID controller. Figure 5.21 shows a block diagram of the PID controller with a unit feedback loop, in which PWM signal V_c was treated as the input control. An error value, which is defined as the difference between the PWM signal V_c and the output current I_m, was minimized to obtain the optimal three constant parameters (K_p, K_i, K_d). In order to balance the performance and the robustness, the PID gains were obtained as 10, 0.5, 0.08, respectively, by setting a rise time (the time required for the response to rise from 10% to 90% of its final value) $t_r = 0.005$ s, the phase margin >45° and the bandwidth $w_c \geq 628$ rad/s.

The results show that the step response of the coil current after the PID regulation is greatly improved as compared with original system (Figure 5.22). It is obvious that the current response (I_m) decreases its overshoot and improves its performance under different PWM signals V_c. The response time with the PID control is reduced to 5 and 15 ms as compared with that of original cases for 0.5 and 1 A. However, the effect of PID regulation becomes less significant due to the saturation characteristic of current controller for higher current considered in this study (i.e., 1.5 and 2.0 A). Overall, such designs (advanced correction circuits, PID

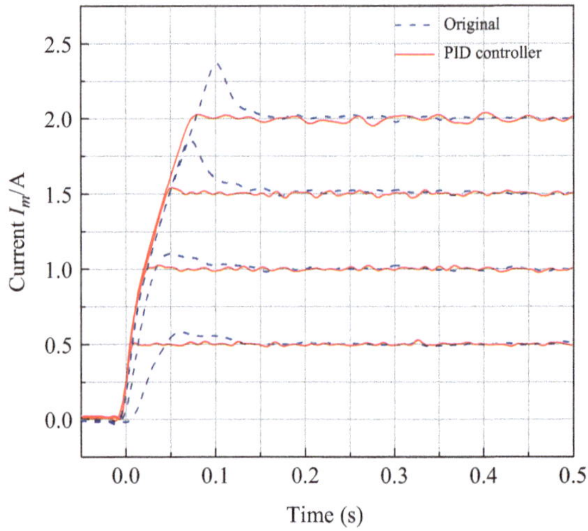

Figure 5.22 Effect of PID controller

regulations, etc.) effectively reduce the time delay of MR systems subjected to impact loadings.

5.5 Dynamic characteristics of MR buffer system

To validate the effectiveness of novel MR recoil absorber, the impact test is carried out on the test platform, as shown in Figure 5.23. The test platform consists of industry control computer, dSPACE system, current controllers, impact testing rig, four power supplies for each coil and several sensors. The required data for experimental analysis including damping force, impact velocity and displacement of MR absorber are collected by Kistler 9351B force sensor and MTS Tempo Sonics RPV750 sensor, respectively.

The control signal that powers the MR absorber originates within dSPACE, which was used to perform data acquisition and control. The output signal from dSPACE is limited in its power, so a current driver circuit and amplifier must be used in order to generate a control current that is proportional to the control voltage coming from dSPACE [22]. The Wonder Box RD-3002_03 current controllers (LORD Corporation) were chosen for controlling the coil current. This current controller operates in 12 V/1.5 A power supply and can output 2.0 A maximum current.

The MR absorber is connected to the recoil mass part, and the piston rod end is fixed with the base. During the impact test, the explosion of powder in the closed bomb generated a large impulsive force on the recoil mass part, and then the MR absorber moved along the guide rail with the recoil mass part. The elastomeric bumper is used at the end of guide rail to absorb the residual impact energy in case the recoil exceeds the maximum length of 600 mm on this rig. The current

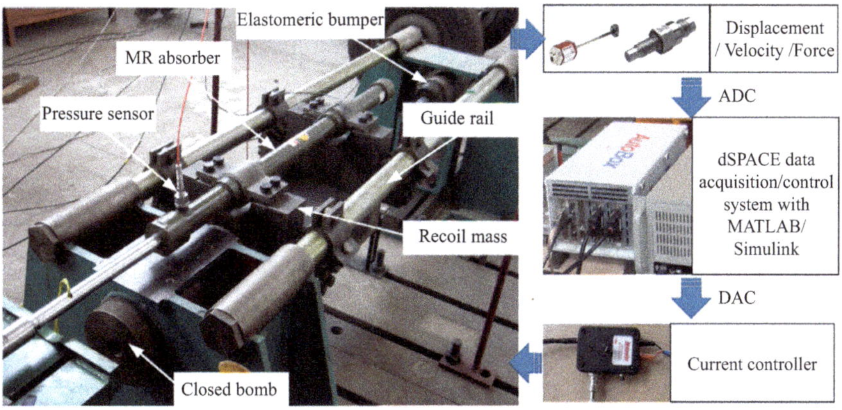

Figure 5.23 Impact test platform

controller is used to control the damping force according to the applied control algorithms. It should be noted that although the impact test platform here is horizontal, it does not affect the performance evaluation of novel MR anti-impact absorber under various firing angles.

The amount and type of gunpowder should keep consistent during the multi-group experiment. Hereon, we used 6 g gunpowder in this study. In order to exploit the flexibility of current loading in the novel MR absorber, which can input current separately, three experimental groups were conducted under the conventional unified control mode, separate control mode and timing control mode, respectively.

1. In the conventional unified control mode, the current on each coil kept same. In this mode, four cases of current loading were conducted, respectively: 0, 0.5, 1.0 and 2.0 A.
2. In the separate control mode testing, four coils are able to make combined work: single coil working mode (SWM), double coil working mode (DWM), triple coil working mode (TWM) and quadruple coil working mode (QWM), respectively, as shown in Table 5.3, which lists the 16 cases of experimental testing with different input currents. The total input current in four coils is kept as 2 A for comparison. There are various combination ways that can be obtained by principle of permutation and combination in the SWM, DWM and TWM, respectively. In these control modes, the total current was averaged on the working coils. To be specific, each working coil was applied 2, 1 and 0.67 A in the SWM, DWM and TWM, respectively. And these various combinations of different control modes reflect the spatial change of the magnetic distribution in the annular gap of MR absorber. However, in the case of QWM, there is only one combination way, that is, all the electromagnetic coils have been energized. In order to analyze the influence of spatial magnetic field distribution change on damping force under the QWM, we applied the various currents on the coils rather than just averaged currents.

Table 5.3 Experimental case with different input current

Coil	SWM				DWM			
I	2	0	0	0	1	0	0	0
II	0	2	0	0	1	1	1	0
III	0	0	2	0	0	1	0	1
IV	0	0	0	2	0	0	1	1
Coil	**TWM**				**QWM**			
I	0.67	0.67	0.67	0	0.3	0.5	0.7	0.7
II	0.67	0.67	0	0.67	0.7	0.5	0.6	0.3
III	0.67	0	0.67	0.67	0.3	0.5	0.4	0.7
IV	0	0.67	0.67	0.67	0.7	0.5	0.3	0.3

3. As for the novel MR absorber with parallel coils configuration, the combined work of coils should not be confined to the separate control mode that only focuses on the space dimension combination. The timing control mode focuses on the influence of energized coils in time dimension on the recoil performance. Each coil was energized from Coil IV to Coil I in sequence, and the time interval is 10 ms.

5.5.1 Conventional unified control mode

Figure 5.24(a) shows the relationship between damping force and impact velocity. It can be found that the damping force is the function of the velocity as well as the input current, which is in accordance with general conclusions [23,24]. The total damping force includes viscous damping force and field-dependent damping force, which change with the velocity and current, respectively. The damping force increases with the input current due to the strengthened apparent viscosity of MR fluid. The peak damping force increases from 3,584 N at zero current to 4,800 N at 2.0 A current when the maximum velocity values keep consistently at 3.6 m/s. The field-dependent force under different input currents can be obtained when the velocity dropped to zero. The maximum field-dependent force almost reaches to 539, 921 and 1,086 N when the input current is applied 0.5, 1.0 and 2 A, respectively, as shown in Figure 5.24(a).

As seen in Figure 5.24(a), there are different dynamic behaviors in the acceleration and deceleration phases during recoil process: the *F*–*v* profile is strong nonlinear in the acceleration stage, the force does not grow significantly with the increase of velocity of MR absorber immediately, while a near linear *F*–*v* relationship is appeared in the deceleration stage due to the compression of air bubbles inside the MR absorber. Figure 5.24(b) shows the relationship of force and displacement of MR absorber. The air compression stroke is about $S_0 = 65$ mm, and the volume of air bubbles is equal to the press-in volume of piston rod in the compression stroke S_0. The volume of air bubbles is roughly 40 ml by estimating. Although the presence of air bubbles within the fluid chamber will influence the MR absorber performance, especially on the *F*–*v* dynamic characteristics,

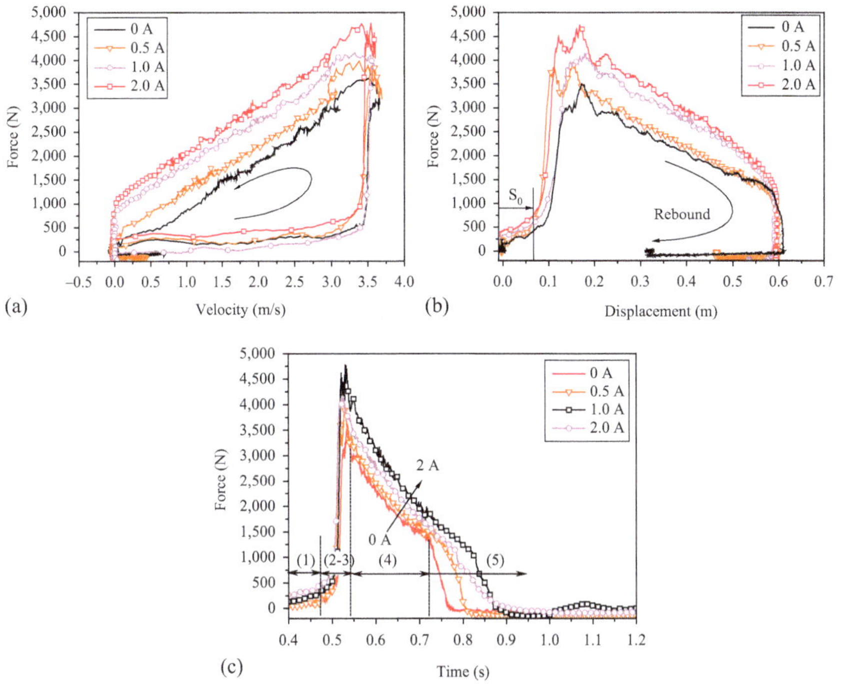

Figure 5.24 Dynamic characteristics of MR absorber in the conventional unified control mode: (a) force–velocity profile, (b) force–displacement profile, (c) force–time profile

compared with the maximum stroke ($S_m = 600$ mm) of absorber, the air compression stroke S_0 is relatively small and it has less influence on the *F–S* characteristics. It means that a little amount of air bubbles can be neglected in the controllability analysis of MR absorber. The areas under *F–S* curve indicate the dissipated energy of MR absorber. As seen in Figure 5.24(b), the larger the input current, the larger the area under *F–S* curve will be. It means that more impact energy is dissipated under the larger input current. The maximum stroke of this MR absorber is 600 mm and then stops by the rubber bumper with a return stroke motion to consume residual impact energy. In the case of input current of 0 A, the springback displacement can nearly reach to 300 mm, while with the increase in input current, the springback displacement reduces to 50 mm.

Many studies indicated that the entire dynamic response of electrorheological (ER) absorber or MR absorber was characterized by several distinct stages. Zhang *et al.* [25] divided the impact process of ER absorber into three stages, which is, an initial shock stage, a transition stage and a stable flow stage. And Zheng *et al.* [26] studied the dynamic performance of single piston rod MR absorber under impact load and took the air compression into account the impact stages. However, in the recoil system application, the MR absorber is usually subjected to springback process after recoil shock process. In the impact stopping and springback process, the inertia effect of MR fluid and the emergence of the vacuum during the pulling out of piston rod will affect the dynamic behavior of MR absorber in the return stroke process. Thus, as for the MR absorber in recoil system, the impact stopping and springback process should be taken into account. In other words, the entire impact response of long-stroke MR absorber can be featured with five stages, namely, (1) air compression stage, (2) instant shock stage, (3) transition stage, (4) flow stage, (5) stop-springback stage, as shown in Figure 5.24(c). To a large extent, the air compression stage is caused by structure feature of MR absorber, which is the single-ended structure feature and without rod-volume compensator.

5.5.2 Separate control mode

The overall change trend of dynamic curve of MR absorber in the separate control mode is same as that in the conventional unified control mode. So, in this section, the analysis is focused on the *F–v* dynamic curve of MR absorber, and the main purpose is to analyze the diversity of controllable dynamic damping force due to the high magnetic energy density in the separate control mode. According to Table 5.3, various experimental cases were conducted and the *F–v* curve of novel MR absorber in the separate control mode are shown in Figure 5.25.

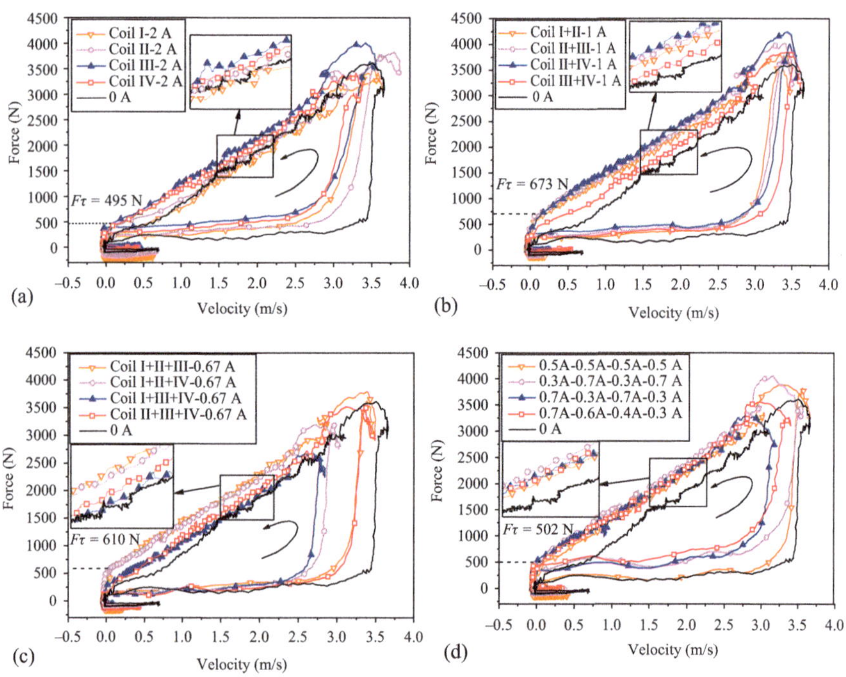

Figure 5.25 Dynamic characteristics of MR absorber in separate control mode: (a) SWM, (b) DWM, (c) TWM and (d) QWM

In the case of SWM, as shown in Figure 5.25(a), the maximum field-dependent damping force F_τ is near to 495 N when the Coil III is energized. While the Coil I is energized, the MR effect will be weak as the zero-current case. Figure 5.25(b) shows the *F–v* characteristics of MR absorber with double working coil. Each coil applied 1 A current, respectively. The maximum and minimum controllable damping force corresponds to 673 N in the Coil II–Coil III combined working case and 500 N in the Coil III–Coil IV case, respectively. And as shown in Figure 5.25(c), when three coils work jointly at the same current 0.67 A, the peak field-dependent damping force is less than that at the case of DWM. As for the QWM with 0.5 A current in Figure 5.25(d), the field-dependent damping force is almost 500 N, and the differences of field-dependent damping force among these *F–v* profiles are pretty tiny. However, compared with zero-current case, the MR effect in the QWM is more obvious.

Overall, in the single working coil mode, the controllable field-dependent damping force is smaller than other cases with 2 A total current. This is because the magnetic field distribution of single working coil mode is so concentrated that the activated area of MR fluid is limited. While two coils were energized, the wider range of magnetic field distribution and the larger activated area of MR fluid were obtained; thus, the larger controllable field-dependent force was output. However, with the current distributed on the multistage coil, such as in the QWM, the field-dependent damping forces were nearly equal. The experiment results of separate control model show that the magnetic field distribution in space dimension could affect the recoil performance, and the different combinations of working coil can lead to different dynamic characteristics.

5.5.3 Timing control mode

To investigate the influence of current loading in time dimension on the damping performance, the impact experiments were performed in the timing control mode, which is applied current on the Coil IV to Coil I in sequence and the time interval is 10 ms between two coils loading current. The current magnitude is 0.5, 1.0 and 2 A, respectively. It should be noted that the remarkable difference between timing control mode and conventional unified control mode or separate control mode is the time difference of current loading on each coil rather than the magnitude of current or magnetic field in annular gap. As shown in Figure 5.26, take the case of 1 A applied current as an example, the current of each coil was measured. It can be found that the response time of electromagnetic circuit is obvious. Furthermore, there is a little amount of overshoot which can influence the dynamic characteristics of MR absorber. These problems should be considered in the control strategy study.

The results of the differences between conventional unified control mode and timing control mode are shown in Figure 5.27. It can be found that the peak damping forces in the timing control mode are less than those in the conventional unified control mode. To be more specific, in the case of 0.5 A input current, the peak damping force reaches to 4,002 N in the conventional unified control mode,

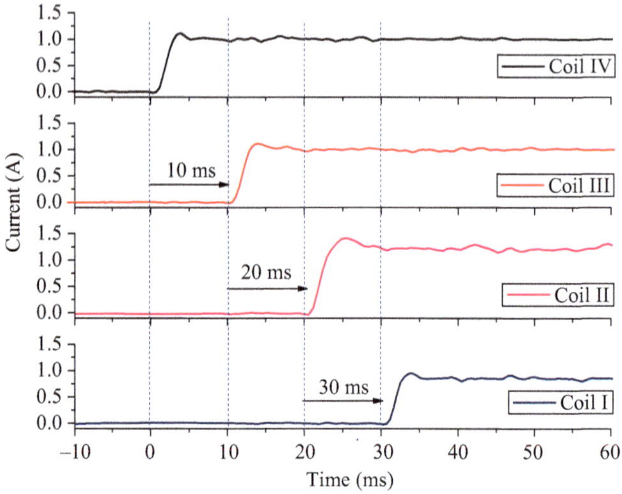

Figure 5.26 Measured current of each coil in the timing control (1.0 A)

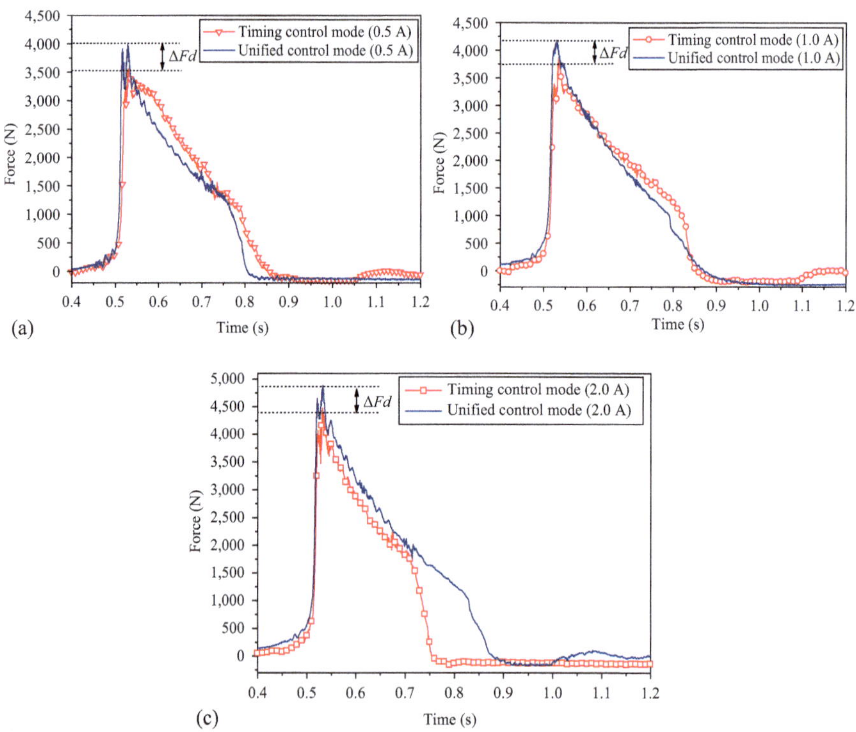

Figure 5.27 Comparison of peak damping force in timing control mode and unified control mode: (a) 0.5 A, (b) 1.0 A and (c) 2.0 A

while reduces by 12.3% to 3,510 N in the timing control mode, the difference peak force is $\Delta F_d = 492$N. Moreover, the decrease trend of peak damping force in the timing control mode is also happened in the case of 1.0 and 2.0 A input current, where peak damping force reduces by 10.47% from 4,189 to 3,750 N and by 8.3% from 4,800 to 4,398 N, respectively. This can be attributed to the improvement of "fluid locking" effect during recoil process by the timing control mode. The essence of timing control mode is that the variation of magnetic field distribution in time dimension.

Figure 5.28 presents the force-stroke profiles by experimental measurement vs the ideal damping force-stroke profile by theoretical derivation at various limiting firing angles. At the beginning of the impact process, the change of damping force has a strong nonlinear phenomenon due to some factors such as the air compression and fluid inertia effect. The phase of impact acceleration process can be viewed as uncontrolled stage, where the influence of input current on the damping force is not obvious. Only when the relative velocity of MR absorber drops to a threshold value, will the fluid became controllable [27]. In general, the peak point of damping force is the cutoff point for the uncontrollable and controllable stage, as shown in Figure 5.28. Thus, this section mainly focuses on the controllability analysis of novel MR absorbers in the controllable stage.

In the case of the zero input current, the *F–S* curve is below the ideal *F–S* curve at 0° limiting firing angle. It means that this case failure to dissipate the impact energy in the most effective way although it is possible to ensure firing stability. When the coils are energized in this way: Coil I = 0.3 A, Coil II = 0.7 A, Coil III = 0.3 A, Coil IV = 0.7 A, the experimental *F–S* profile is more closely to the target curve at $\phi_j = 0°$ than the zero input current case. Furthermore, with the increase of input current, it can obtain the required damping force for a larger limiting firing angle. For example, when the input current is 1 A, the results by experimental

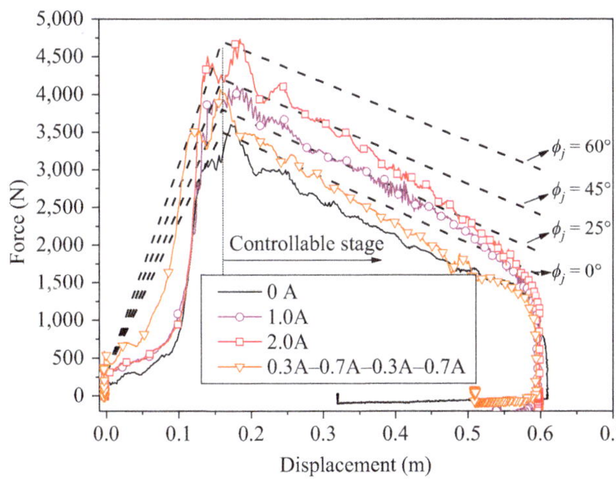

Figure 5.28 Comparison of measured and theoretical F–S curve

measurement are in good agreement with ideal *F–S* profile at $\phi_j = 25°$ deduced by theoretical calculation. However, in the case of 2 A input current, the measured damping force declines sharply from peak force to approximately 1,500 N in the controllable stage. An extreme descending slope makes it unable to meet the requirements of ideal *F–S* curve, both for $\phi_j = 45°$ and $\phi_j = 60°$.

In order to quantitatively evaluate the controllability of MR absorber during the whole recoil motion, the fullness degree index (FDI) α is defined as the ratio of the motion displacement S_d when the measured damping force deviate 10% from the ideal damping force to the total displacement S_d in the controllable stage, and it can be expressed as follows:

$$\alpha = \frac{S_d}{S_t} \times 100\% = \frac{S_d}{S_m - S_a} \times 100\% \tag{5.25}$$

Figure 5.29 shows the deviation of experimental curve under various input currents and the ideal curve under different limiting firing angle in the controllable stage, where 0.16 m $< S \leq$ 0.6 m. As shown in Figure 5.29(a), in the case of zero input current, the S_d is about 0.104 m and the FDI parameter $\alpha = 23.64\%$ at $\phi_j = 0°$. When applying 0.3, 0.7, 0.3, 0.7 A input current on Coil I–Coil IV respectively, the effective displacement S_d reaches to 0.282 m and the FDI parameter α increase to

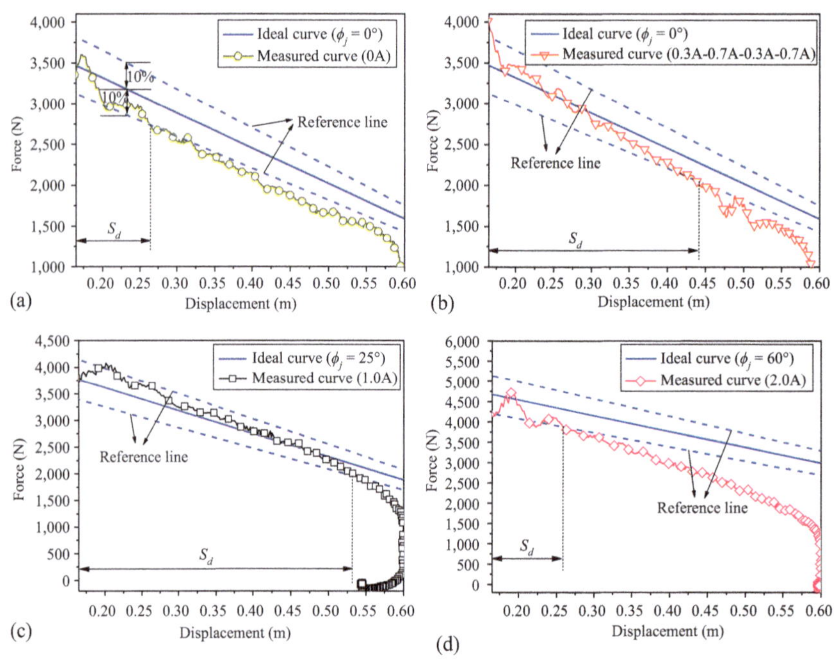

Figure 5.29 Damping force vs displacement for measured and theoretical results in the controllable stage: (a) off-field, (b) input current (A): 0.3-0.7-0.3-0.7, (c) input current (A): 1.0 and (d) input current (A): 2.0

64.09% at $\phi_j = 0^\circ$ accordingly, as indicated in Figure 5.29(b). Furthermore, it is apparent from Figure 5.29(c) that the data coincidence between experimental curve and theoretical curve is higher than other case. The FDI parameter α is almost 84.49%. It means that when the limiting firing angle ϕ_j increases to 25°, the MR absorber can output the quite reasonable damping force by applying 1 A current. As for the case of 2 A input current, although the peak force of measured curve approaches to the ideal curve at $\phi_j = 60^\circ$, the parameter S_d and α is only about 0.099 m and 22.43%, respectively. From Figure 5.29(b) and (c), it can be found that the force-stroke profile of MR absorber in the recoil system can approach to the ideal curve by the adjustable rheological properties of MR fluid at different limiting firing angles (from $\phi_j = 0^\circ$ to $\phi_j = 25^\circ$). However, when the limiting firing angle ϕ_j exceeds 45°, the controllability of MR absorber will deteriorate, as shown in Figure 5.29(d).

The damping force of MR absorber can be mainly divided into two independent viscous components: one is the pure rheological component F_η, and the other is the MR component F_τ that depends on the magnetic field [28]. The deterioration of controllability of MR absorber at the higher limiting firing angle (i.e., $\phi_j = 45^\circ$, $\phi_j = 60^\circ$) can be attributed to the insufficient MR component. In the recoil impact process, the velocity of MR absorber increases dramatically to peak value and then declines to zero gradually. The field-dependent force F_τ should increase gradually to fill the shortfall of the F_η due to the velocity decrease and then make the output damping force follow with the ideal *F–S* curve in the recoil period. At the higher limiting firing angle, for example, in the case of $\phi_j = 60^\circ$, the required ideal damping force at the final stroke exceeds 3,000 N, as shown in Figure 5.29(d), and it is greater than the maximum controllable field-dependent damping force $F_\tau = 100$N that in the 2 A current case. The insufficiency of MR component F_τ would be solved to satisfy the higher limiting firing angle application in the future work.

5.6 MR shock buffer device control strategy and experimental verification

5.6.1 Methods of MR damper control

Under the action of the artillery force F_{pt}, the recoil speed quickly rises to the peak and then begins to decline. From the time of receding until the speed drops to about 2.5 m/s, i.e., in the high acceleration region ($0 \leq t \leq t_s$), the speed change trend does not change with the applied current. And the peak speed is the same, indicating that the current size has no effect on the recoil speed at this stage. The recoil movement is mainly caused by the momentum applied by the artillery force F_{pt}. The speed in this area is not controllable [29]. Therefore, the relevant part of the damping force related to velocity is also uncontrollable, indicating that the adjustable part of the damping force $F_\tau(B)$ and the additional resistance $F_a(v, a, B)$ related to the magnetic field have a great influence.

The selected control area should be followed in the following principles: when the maximum recoil displacement of the passive control satisfies the constraint conditions, the area after the peak damping force should be selected as control area to ensure minimize the maximum recoil displacement. When the maximum recoil displacement of the passive control does not meet the constraint conditions, the selected control area is the entire recoil process in order to increase the peak value of the recoil resistance appropriately so as to ensure minimize the maximum recoil displacement.

Under a certain recoil stroke condition, the optimal shock buffer control effect should make $F_R(t) = const$. In the gun project, the control objective of maintaining the recoil resistance constant throughout the entire recoil stroke is called the "platform effect." Any practical impact buffer control cannot achieve the theoretical platform effect. Therefore, in the gun project, the extent of the recoil resistance curve approaching the platform effect is defined as "fullness," which is an important indicator for evaluating the performance of the anti-recoiling device of the gun.

However, in the gun engineering test, the recoil fullness index was only described qualitatively and was not quantified. In order to more accurately evaluate the performance of the MR anti-recoiling device, the recoil resistance fullness index as shown in Figure 5.30 was quantified. In Figure 5.30, the ordinate represents the recoil force F_R, and the abscissa represents the recoil time t, A is the peak of recoil force and defines two quantitative indices to evaluate the fullness, which are

1. Fullness $\alpha\%$ and $\beta\%$: The ratio of the time elapsed from the initial moment of recoil to the time when the recoil force drops to 85% of the peak value T_d to the total time of recoil T_a, ie:

$$\alpha\% = \frac{T_d}{T_a} \times 100\% \tag{5.26}$$

Alternatively, a similar definition can also be applied to the recoil resistance–displacement curve: the ratio of the displacement elapsed from the initial displacement of recoil to the displacement when the recoil force drops to 85% of the peak value S_d to the total displacement of recoil S_a, i.e.,:

$$\beta\% = \frac{S_d}{S_a} \times 100\% \tag{5.27}$$

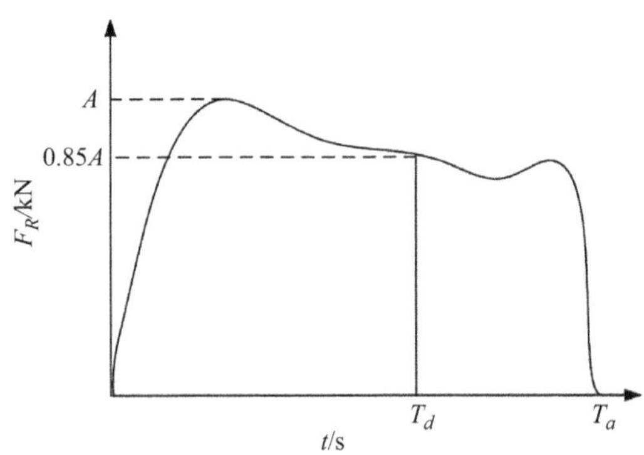

Figure 5.30 Recoil fullness curve

It should be noted that if the recoil force does not fall to 85% of the peak value during the whole recoil process, when calculating T_d or S_d, the time or displacement after the end of the recoil should be taken as the ending value. Obviously, the fullness $\alpha\%$ and $\beta\%$ are the percentage from 0% to 100%, and the ideal value of $\alpha\%$ and $\beta\%$ is 100%.

2. The relative change in the recoil resistance $\delta\%$: The ratio of the maximum change in the recoil resistance $\Delta F_{\max}$ to the peak value of the recoil resistance F_{peak} during the period from the peak of the recoil resistance to the end of the recoil, i.e.,

$$\delta\% = \frac{\Delta F_{\max}}{F_{peak}} \times 100\% \tag{5.28}$$

Obviously, the smaller the value of the relative change in recoil resistance $\delta\%$, the better the corresponding plateau effect, the ideal $\delta\%$ is zero.

5.6.2 MR shock buffer control system

MR shock buffer control system is mainly composed of dSPACE real-time system, LORD current driver, displacement velocity sensor, force sensor, charge amplifier, connection box and industrial personal computer. The connection diagram and physical composition diagram of the system are shown in Figure 5.31. Among them, the dSPACE real-time system is the core component of the control system

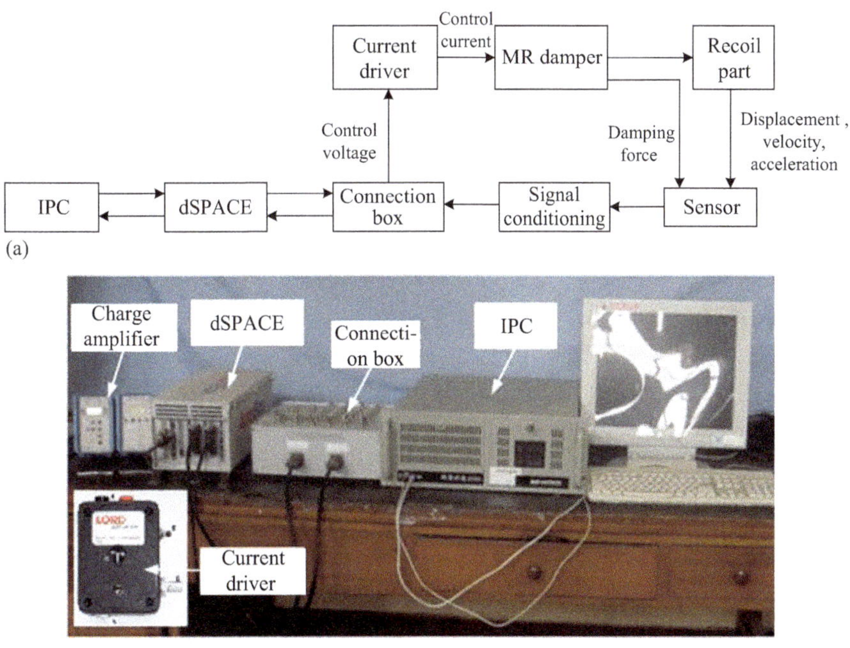

Figure 5.31 Composition of MR shock buffer control system: (a) connection schematic diagram and (b) physical composition diagram

developed by the North American branch of dSPACE GmbH in Germany. It is a semi-physical simulation platform developed based on the MATLAB®/Simulink™ control system and can fully realize seamless connection with MATLAB, Simulink and Stateflow. It is mainly used in system design, rapid control prototype, automatic code generation and hardware-in-the-loop testing.

The dSPACE real-time system is composed of two parts: hardware module and software platform. The hardware module is mainly composed of auto control box, DS1005 main processing board, DS2002A/D module and DS2103D/A module. The software platform is mainly composed of MATLAB/Simulink, RTW, RTI (real-time interface) and dSPACE ControlDesk.

The development process of dSPACE real-time simulation system is as follows:

1. Build an object mathematical model by MATLAB/Simulink, design a control scheme and perform off-line simulation.
2. Retain the modules that need to be downloaded to dSPACE, replace the original logical connection relationship with the hardware interface relationship, configure the IO and set the hardware and software interrupt priority.
3. Use RTW and RTI provided by dSPACE to automatically generate code and download it to hardware.
4. Create an experiment interface in the control development package and associate the target file (*.ppc) with the hardware system.

The data acquisition of the system is realized by the DS2002A/D module. This module contains two independent A/D converters and provides 32 channels of analog sampling. The input range is ± 5 or ± 10 V, programmable selection, the maximum resolution is 16 bits, the fastest sampling time of two channels is 3.9 μs. The control output of the system is realized through the DS2103D/A module. This module contains 32 independent D/A converters, provides 32 channels of analog output, the output range is ± 5 or ± 10 V, programmable selection, resolution is 14 bits, refresh time is 10 μs.

ControlDesk is a software platform of dSPACE real-time system, mainly used for hardware management, data acquisition and display, variables and parameter management. The control experiment interface developed using ControlDesk is shown in Figure 5.32.

5.6.3 Fuzzy logic control strategy

Due to the nonlinear mechanical characteristics of MR damper in impact buffer process, it is feasible to choose the fuzzy control strategy in the impact buffer system. The fuzzy control system is composed of controlled object, fuzzy controller and sensor test link. The system composition is shown in Figure 5.33.

5.6.3.1 One-dimensional fuzzy control strategy

According to the control target, the error value e between the ideal recoil resistance signal F_{Ri} and the recoil signal F_R measured in real time were used as the input

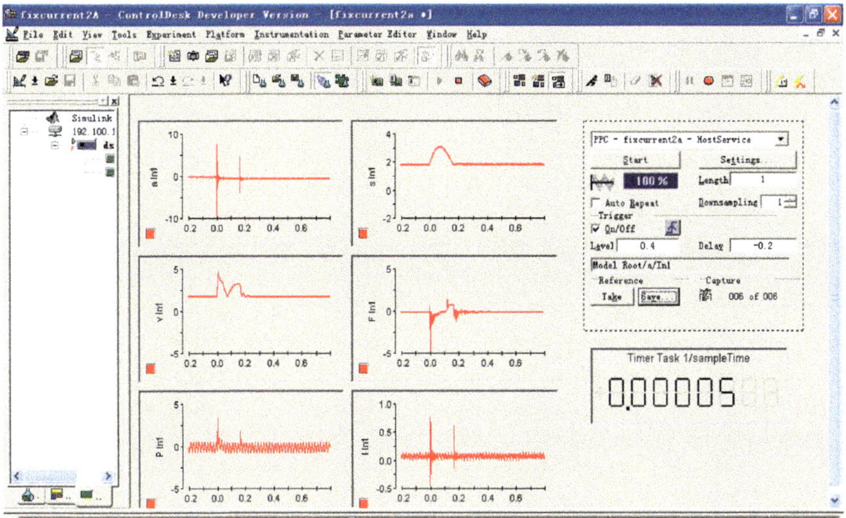

Figure 5.32 The control experiment interface developed by ControlDesk

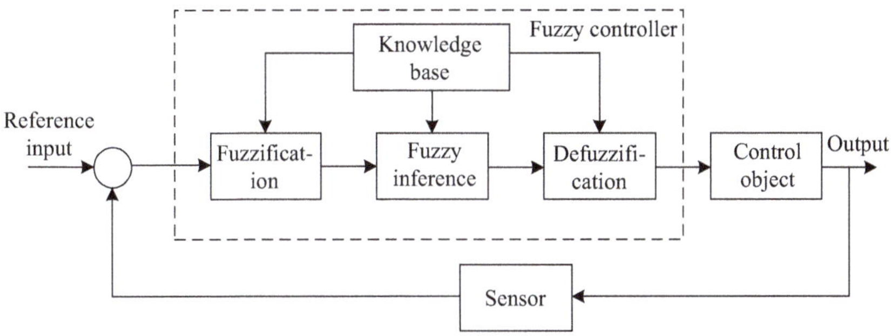

Figure 5.33 Control principle of fuzzy control system

signal of the fuzzy controller, and the ideal damping force is determined by the fuzzy control rule. The error of recoil resistance signal is blurred, and the input signal is normalized to the range of [−1,1]. The basic domain of the output damping force is [0,10], indicating that the output damping force is 0–10 kN. The fuzzy input signal is defined as five fuzzy sets, namely, represents the input fuzzy state, represented by letters: {NB NS ZO PS PB}. The membership functions of fuzzy control input and output signals are all defined as normal distribution. The fuzzy criterion of the output information of the fuzzy controller adopts the MIN–MAX barycentric method. The setting range of one dimension fuzzy control is the whole recoil stroke.

5.6.3.2 Two-dimensional delay fuzzy control strategy

The area where the two-dimensional fuzzy control function is set as the area after the damping force reaches the peak value. Therefore, the principle of time delay is that the control current is not applied before the peak of the control area when the damper reaches its peak value, and the fuzzy control strategy is adopted in the range of control area after the peak reaches the peak until the end of the recoil.

The basic domain of the recoil speed is [0,4], which indicates that the input velocity range is 0–4 m/s; the basic domain of the recoil displacement is [0,0.2], indicating that the input displacement range is 0–0.2 m, and their fuzzy domain is quantified as 7 level {ZS ZM ZB M PS PM}, the quantification factors k_e and k_{ec} are both 1. The basic discourse field of the output damping force is [0,10], which means that the output damping force is 0–10 kN. It is quantized to 9 levels {ZS ZM ZL ZB M PS PM PL PG}, the proportional factor k_u is 1. The membership function of the input is a trigonometric function that is easy to implement and has good control performance. The membership function of the output is Gaussian function with good anti-interference ability. According to the goal of "the minimum recoil peak and the best filling degree," the fuzzy control strategy is summed up. The general experience is as follows: while the delay effect is added, the nonadjustable hindrance is bigger in the initial stage of the recoil because of the high speed. In order to prevent the heavy recoil resistance, the smaller damping force should be generated in the later stage of the recoil process since the resistance force cannot be adjusted due to the rapidly decreasing velocity of the shock absorber. The detailed fuzzy control rules are summarized as shown in Table 5.4. It should be noted that due to the selection of the control area, some of the rules in the rules table will not appear, i.e., blank areas in the table. The fuzzy criterion of the output information of the two-dimensional fuzzy controller is the same as that of the one-dimensional fuzzy control, and the MIN–MAX barycentric method is adopted.

5.6.4 *Fuzzy control experiment*

The MR shock buffer control system introduced by Section 5.6.2 is used to test the effect of passive control, 1.5 A fixed current control, one-dimensional fuzzy control and two-dimensional delay fuzzy control, respectively. The test conditions are all

Table 5.4 Two-dimensional fuzzy control rule table

F_d		S						
		ZS	**ZM**	**ZB**	**M**	**PS**	**PM**	**PL**
v	ZS						PS	PG
	ZM						M	PL
	ZB					ZL	ZB	PM
	M			ZL	ZB	ZB	ZL	PS
	PS			ZS	ZM	ZS	ZM	
	PM	ZM	ZS	ZS	ZM	ZM		
	PL	ZS	ZM	ZM	ZM			

positive charge and 0° angle, and the recoil quality remains unchanged under various control strategies. Each control strategy was repeated three times and the mean of three times was taken as the final result.

The data acquisition module DS2002 is used to collect the six channel signals, which correspond to the recoil acceleration, the recoil displacement, the recoil speed, the damping force, the working chamber pressure and the working current of the damper excitation coil. The sampling frequency of each channel is 20 kHz, and the control voltage is transferred from the DS2103 module to the current driver. The Simulink flow chart of one-dimensional fuzzy control and two-dimensional time-delay fuzzy control is shown in Figures 5.34 and 5.35.

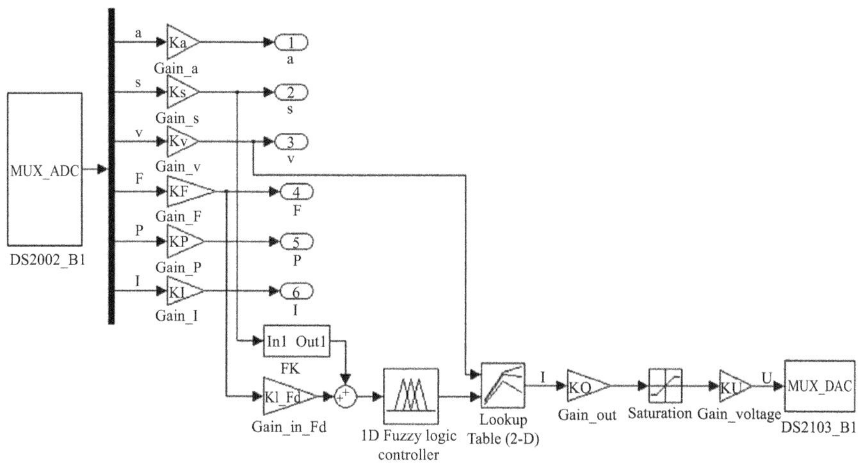

Figure 5.34 The block diagram of one-dimensional fuzzy control Simulink experimental

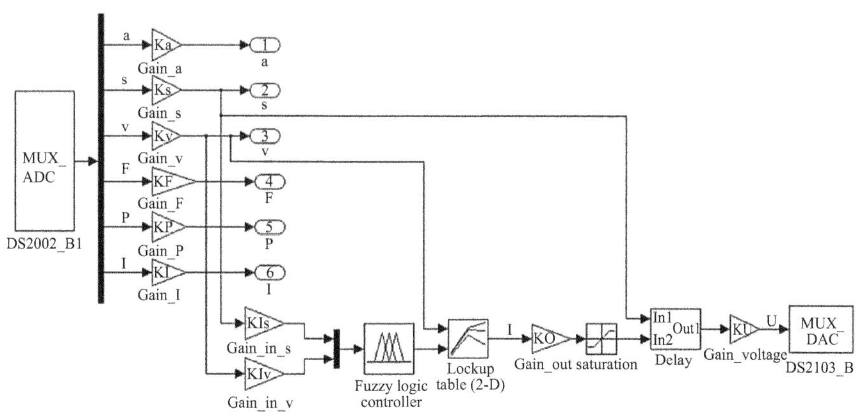

Figure 5.35 The block diagram of two-dimensional time delay fuzzy control Simulink

The “Lookup Table” in Figures 5.34 and 5.35 corresponds to the determined inverse model of the MR damper dynamics; the “Saturation” module is the voltage control range that prevents the output from exceeding the 0–5 V. Using the RTI provided by dSPACE to automatically generate code and download the function to the hardware, the Simulink experiment block diagram is converted and downloaded to the dSPACE hardware, and the data collection results are displayed on the test interface designed by the ControlDesk. It should be noted that different from the simulation model used to determine the differential speed signal, the two-dimensional fuzzy control delay function in the experiment is achieved through the detection of the displacement signal. When the displacement signal is less than a threshold, the output signal is zero; otherwise, the output current is controlled. This is because in the actual squat process, the signal will be affected by the more severe high-frequency interference. The differential zero of the speed signal does not mean that the speed at this moment reaches the peak.

5.6.5 Experimental results and analysis of fuzzy control

The damping force, squat displacement and squat velocity provided by the MR damper during the squat movement under two kinds of fuzzy control actions are shown in Figure 5.36. As a contrast, the experimental results under two conditions of the passive control (the original recoil) and the fixed current control are given. The recoil resistance of each control strategy and its relationship with displacement are compared, and the comparison results are shown in Figure 5.37.

As shown in Figure 5.37, the experimental results under various control strategies are in agreement with the simulation results. In passive control, the recoil resistance peak value is smaller because of no control current, but the peak value of the peak recoil is larger. Compared to passive control results, the control of 1.5 A fixed current reduces the squat displacement and increases the peak of recoil force, which leads to reduced fullness. The recoil resistance is relatively flat under one-dimensional fuzzy and two-dimensional time-delay fuzzy control. Among them, the whole seat back resistance relatively large and the peak displacement is small because the delay effect has not been added to one-dimensional fuzzy control. The peak value of the receding resistance of the two-dimensional time-delayed fuzzy control is almost the same as that of the passive control, but the peak value of the recoil displacement decreases, and the fullness and the relative change of the recoil resistance are all superior.

The specific performance indexes calculated from the experimental results are shown in Table 5.5. From Table 5.5, it can be seen that the performance indicators of the experiments and simulations all achieved consistent evaluation under various control conditions. It is worth noting that under the two-dimensional fuzzy control, the relative variation of recoil resistance of the experimental results is poorer than that of the corresponding simulation results. The reason is that there is a certain measurement error in the input signal of the experiment; on the other hand, the delay time of the delay fuzzy control is different in the experiment and simulation because of the time delay factors in the test system. The fullness curve of the

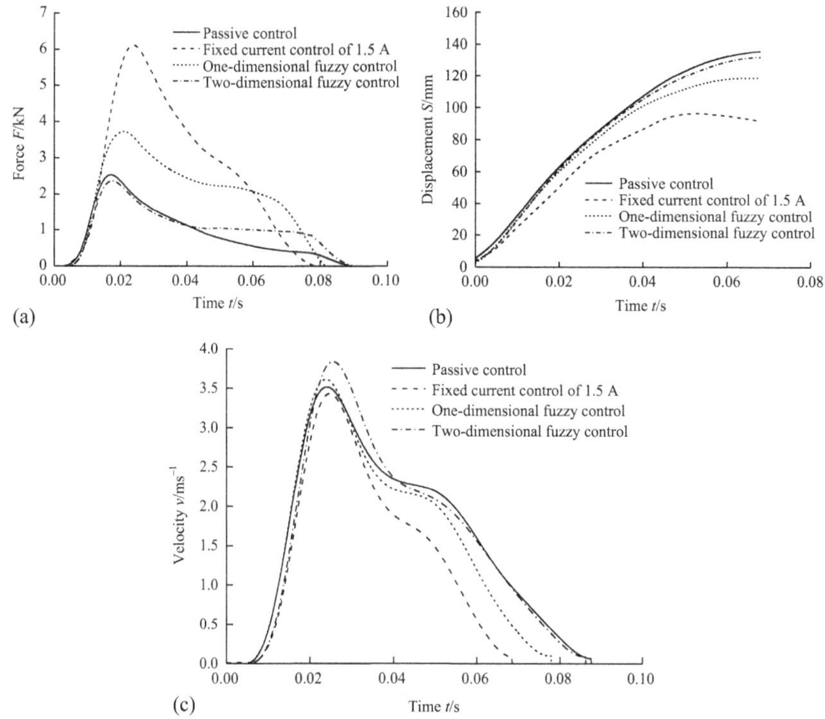

Figure 5.36 Recoil variables in different control strategies: (a) damping force–time, (b) backseat displacement–time and (c) recoil speed–time

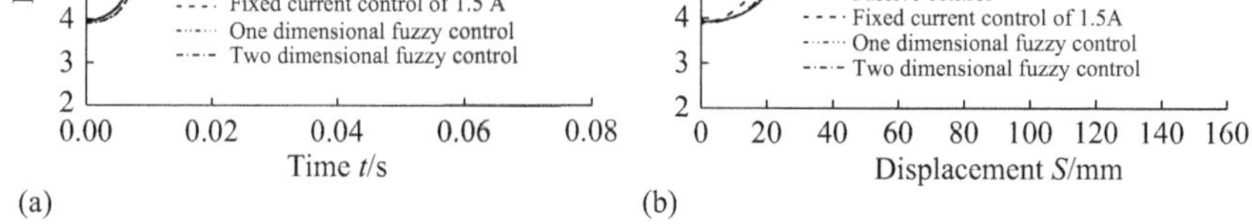

Figure 5.37 Experimental comparison of buffer effect under different control strategies: (a) recoil resistance-time and (b) backseat resistance-displacement

Table 5.5 Comparison of shock buffering performance indexes under different control algorithms (experiment)

Control strategy	Passive control	Fixed current control of 1.5 A	One-dimensional fuzzy control	Two-dimensional fuzzy control
Full degree α(%)	55.4	48.4	82.7	100
Full degree β(%)	75.6	80.4	98.3	100
Relative change of recoil resistance δ(%)	22.4	29.8	16.26	11.69
Maximum displacement S_{max} (mm)	135.6	96.5	118.6	131.2
Peak recoil F_{Rmax} (N)	6,782	10,413	8,041	6,607

1.5 A fixed current control indicates that the MR damper's tunable coefficient is large enough, so the further increase in fullness depends on further accurate measurement of the input signal, the accurate selection and real-time execution of the delay time.

It is necessary to explain that the recoil resistance of the experimental results is slightly larger than the simulation results, and the peak seat shift peak is slightly smaller. This is mainly due to the difference between the theoretical model and the actual situation, but the difference of the two is less than 10%. Therefore, it is considered that the experiment is in good agreement with the simulation, and the theoretical model can describe the experimental system well. From the experimental results, the rising time of the recoil resistance time curve or the recoil resistance displacement curve is longer than the simulation curve, which is due to the high-frequency interference in the signals in the dynamic measurement process. In the process of data processing, low pass filtering is needed to remove the high-frequency noise signal, and some useful signals in the fast-rising process are also filtered, but after the recoil peak, the signal change is more stable and the low pass filter does not make the signal distortion. Therefore, the data processing does not affect the overall evaluation of the results.

5.7 Summary

In this chapter, a long-stroke MR recoil absorber with four-stage parallel electromagnetic coils which can apply separate current respectively was designed. Based on the experimental analysis, the effects of combined working coils on the performance of the MR absorber under impact loading are investigated. Furthermore, the controllability of MR absorber in the field gun is discussed.

MR damper's electromagnetic circuit can be regarded as a first-order inertial system, whose transfer function can be changed to reduce the time delay. A comprehensive research on the time delay of an impact buffer system based on an MR damper was investigated. It is found that the total time delay of the impact buffer system is contributed by the electromagnetic circuit delay, the hysteresis delay (*B* vs *H*) and the MR fluid's response time. It is also found that the time delay is highly dependent on the effectiveness of the electric part of the control hardware. Rheology characteristics of MR fluids seem to play a negligible role on the time delay of impact buffer systems; thus the focus should be placed on the analysis of the electronic parts for such systems.

Three experimental groups were conducted under the conventional unified control mode, separate control mode, and timing control mode, respectively. The strong nonlinear phenomena in the acceleration period appear due to the air compression. From the separate control mode and timing control mode, it can be found that the influences of current loading in space and time dimension on the damping performance are obvious. In the separate control mode, when two coils were energized, the larger controllable field-dependent force was obtained due to the wider range of magnetic-field distribution and the larger activated area of MR fluid. While in the timing control mode, the peak force is less than the conventional unified control mode.

The insufficiency of controllability of MR absorber at higher limiting firing angle would be solved by increasing the controllable field-dependent force in the future work.

Acknowledgments

This book is a compilation of the authors' recent work on the application of MR fluids and other smart materials to use in anti-impact absorber system. In particular, this book attempts to thread together the concepts that have been separately introduced through the papers published by the authors internationally. The authors owe a debt of gratitude to many individuals; foremost are Yancheng Li at University of Technology Sydney, Zhaochun Li at Nanjing Forestry University, Hongsheng Hu and Qing Ouyang at Jiaxing University, Jiajia Zheng at Zhejiang Normal University, Zhaochun Li at Nanjing Forestry University, Lijie Zhang, Na Sheng at Nanjing University of Science and Technology, and others and acknowledge the contribution of many graduate and doctoral students at Nanjing University of Science and Technology. In particular, the authors wish to acknowledge the financial support provided by National Natural Science Foundation of China (Fund number: 51675280; 51175265; 51705467 and 50675106).

References

[1] Bhatnagar R M 2005 Recoil motion theorem. *Proceedings of the Institution of Mechanical Engineers, Part K: Journal of Multi-body Dynamics*, 219(2): 173–176.

[2] Jumper Jr G Y and Floroff S G 1985 Feasibility of a microprocessor controlled recoil mechanism for large caliber artillery weapons. *Army Armament Research and Development Center, Dover, NJ, Large Caliber Weapon System Lab.*

[3] Hajihosseinloo M A, Hooke C J and Walton D 1989 Gun recoil system performance measurement and prediction. *Proceedings of the Institution of Mechanical Engineers, Part C: Journal of Mechanical Engineering Science*, 203(2): 85–92.

[4] Bond D and Ebener W 2004 Shock absorber for a rifle *U.S. Patent Application 10/962,136*.

[5] Carlson J D and Jolly M R 2000 MR fluid, foam and elastomer devices. *Mechatronics*, 10(4): 555–569.

[6] Peer P, Stenicka M, Sedlacik M, *et al.* 2015 Magnetorheological behaviour and electrospinning of poly (ethylene oxide) suspensions with magnetic nanoparticles. *Journal of Intelligent Material Systems and Structures*, 27(7): 898–903.

[7] Bajkowski M and Bajkowski J M 2012 Design of the magnetorheological damper for the recoil damping of the special object 7.62 mm calibre. *Machine Dynamics Research*, 36(1): 15–23.

[8] Ahmadian M and Poynor J C 2001 An evaluation of magnetorheological dampers for controlling gun recoil dynamics. *Shock and Vibration*, 8: 147–155.

[9] Singh H J and Wereley N M 2014 Optimal control of gun recoil in direct fire using magnetorheological absorbers. *Smart Materials and Structures*, 23(5): 055009.

[10] Hu H, Jiang X, Wang J, *et al.* 2012 Design, modeling, and controlling of a large-scale magnetorheological shock absorber under high impact load. *Journal of Intelligent Material Systems and Structures*, 23(6): 635–645.

[11] Bajkowski M, Makuch A and Lindemann Z 2015 Determining parameters of recoil reduction system with spring and magnetorheological damper intended for special object. *Machine Dynamics Research*, 38(3): 87–96.

[12] Bai X X, Hu W and Wereley N M 2013 Magnetorheological damper utilizing an inner bypass for ground vehicle suspensions. *IEEE Transactions on Magnetics*, 49(7): 3422–3425.

[13] Cho C H and Choi S B 2015 Designing requirement of spring and MR damper for new type baby car seat. *Applied Mechanics and Materials*, 741: 28–31.

[14] Ha S H, Seong M S and Choi S B 2013 Design and vibration control of military vehicle suspension system using magnetorheological damper and disc spring. *Smart Materials and Structures*, 22(6): 065006.

[15] Hosseinloo A H, Vahdati N and Yap F F 2012 Performance of spade-less wheeled military vehicles with passive and semi-active suspensions during mortar firing. *Vehicle System Dynamics*, 50(10): 1515–1537.

[16] Gordaninejad F, Wang X, Hitchcock G, *et al.* 2010 Modular high-force seismic magneto-rheological fluid damper. *Journal of structural engineering*, 136(2): 135–143.

[17] Choi Y T and Wereley N M 2005 Mitigation of biodynamic response to vibratory and blast-induced shock loads using magnetorheological seat suspensions. *Proceedings of the Institution of Mechanical Engineers, Part D: Journal of Automobile Engineering*, 219(6): 741–753.

[18] Ahmadian M and Norris J A 2008 Experimental analysis of magnetorheological dampers when subjected to impact and shock loading. *Communications in Nonlinear Science and Numerical Simulation*, 13(9): 1978–1985.
[19] Li Z C, Wang J and Zhang L J (2009) Investigation of the response time of magnetorheological recoil dampers. *International Conference on Mechanical Engineering and Mechanics (ICMEM).*
[20] Yang G (2001) Large-scale magnetorheological fluid damper for vibration mitigation: modeling, testing and control. *PhD Dissertation University of Notre Dame, Notre Dame, IN.*
[21] Zhang L J, Hu H S, Wang J (2008) Control system simulation of magnetorheological dampers under impact load. *15th International Congress on Sound and Vibration, Daejeon, Korea.*
[22] Koo J H, Goncalves F D, Ahmadian M 2006 A comprehensive analysis of the response time of MR dampers. *Smart Materials and Structures*, 15(2): 351.
[23] Mao M, Hu W, Wereley N M, *et al.* 2007 Shock load mitigation using magnetorheological energy absorber with bifold valves. *The 14th International Symposium on: Smart Structures and Materials & Nondestructive Evaluation and Health Monitoring. International Society for Optics and Photonics*: *652710-652710-12.*
[24] Wang D H and Liao W H 2011. Magnetorheological fluid dampers: a review of parametric modelling. *Smart Materials and Structures*, 20(2): 023001.
[25] Zhang X W, Yu T X and Wen W J 2010 Electro-rheological cylinders used as impact energy absorbers. *Journal of Intelligent Material Systems and Structures*, 21(7): 729–745.
[26] Zheng J, Ouyang Q, Li Z, *et al.* 2015 Experimental analysis of separately controlled multi-coils on the performance of magnetorheological absorber under impact loading. *Journal of Intelligent Material Systems and Structures. 1045389X15600902.*
[27] Ahmadian M and Norris J A 2004 Rheological controllability of double-ended MR dampers subjected to impact loading smart structures and materials. *International Society for Optics and Photonics*, 5386: 185–195.
[28] Alghamdi A A, Lostado R and Olabi A G 2014 Magneto-rheological fluid technology. *Modern Mechanical Engineering. Springer Berlin Heidelberg*: 43–62.
[29] Li Z 2012 Characteristics and control methods of magnetorheological shock buffering device. Nanjing University of Science and Technology.

Chapter 6

Influence of magnetorheological stabilizer bar on vehicle roll stability

Xian-Xu Bai[1], Shi-Xu Xu[1,2], Chao Tang[1] and Sheng Shen[1]

In order to ensure the driving safety (the anti-roll performance) of vehicles when steering and the ride comfort when passing through uneven roads, this chapter proposes the principle and configuration of a novel reciprocating rotary damper based on a smart material—magnetorheological (MR) fluid with a large controllable torque range and a semi-active stabilizer bar system based on the MR damper. The dynamic model of the MR semi-active stabilizer bar system is established. The mathematical model of the proposed MR damper is derived, and the corresponding controllable torque characteristics are obtained and evaluated using the finite element analysis results via software ANSYS. A piecewise control strategy with respect to vehicle body roll angular velocity is proposed, and a preliminary control of MR stabilizer bar is conducted and compared with the passive-on stabilizer bars on the vehicle anti-roll performance.

6.1 Introduction

The number of deaths caused by vehicle rollover accounts for more than 60% of all traffic accident deaths [1,2]. At present, vehicle body-control systems have been already applied [3–7], such as semi-active or active suspension systems, differential braking systems, active front wheel steering systems, electronic stability program (ESP) and chassis integrated control systems. They will implement vehicle anti-roll control to different extents. In addition to the effective improvement of the vertical vibration performance of vehicles, the anti-roll performance of vehicles will also be improved by the active suspension systems. However, the configurations of the active suspension systems and control system are more complex. The energy consumption and the cost are much higher, and the anti-roll control is not the objective for the active suspension systems. Differential braking and active steering control systems, mainly considering the semi-active or active control of the braking

[1]Laboratory for Adaptive Structures and Intelligent Systems (LASIS), Department of Vehicle Engineering, Hefei University of Technology, People's Republic of China
[2]SAIC Motor Corporation Limited, People's Republic of China

systems and the steering systems, will improve the performances of braking, steering and also the anti-roll performance of the vehicles. But the active-front-wheel steering systems are complicated and tend to generate large lateral forces of tires, which will influence the ride comfort of the vehicle at the beginning of the system's functioning.

Stabilizer bar, as an elastic component of the vehicle suspension systems, can effectively reduce the body roll of vehicles. For the vehicle anti-roll control, as compared to other vehicle body-control technologies that can only partially take into account the vehicle anti-roll control performance, the stabilizer bar systems show the advantages of strong pertinence, simple configuration and low cost and is a decent approach to enhance the anti-roll performance of vehicles. In recent years, more and more attention has been paid to the active and semi-active stabilizer bar systems [8–13]. Kim *et al.* [14,15] investigated the electric anti-roll control systems equipped with linear motor. An $H\infty$ control algorithm based on the three-degree-of-freedom linear model of the vehicle with the new anti-roll system was proposed and experimentally studied via the hardware-in-the-loop tests. Yim [16] studied the anti-roll performance of the joint system of active anti-roll bar and ESP. Active braking can be realized by the longitudinal speed control of ESP, which will reduce the lateral acceleration of the vehicles and prevent vehicle rollover. Cebon *et al.* [17–19] used two simple actuators with high and low damping (equivalent lateral stiffness) for semitrailer trucks to realize a semi-active anti-roll system. Low stiffness will be provided from the system for low-speed cornering and high stiffness for high-speed cornering. Since cornering is often a transient event, such a damping system can also improve the stability of heavy vehicles. At the same time, the anti-roll system requires little power. The anti-roll system they proposed and developed increased the upper limit of the vehicle lateral acceleration by up to 12.5%. BMW [8] developed an active lateral stability control system based on a rotary actuators providing anti-roll torques for vehicle anti-roll control. Toyota Motor Corporation [10,20–22] proposed an electric active stabilizer suspension system with a direct current (DC) motor. The "normal mode" and "sport mode" of the DC motor realizing different working states of the stabilizer bar are set according to the expected angle set with respect to the real lateral accelerations. Aisin Group [23] investigated a two-degree-of-freedom robust control on the DC motor in the electric active stabilizer suspension system proposed by Suzuki [20]. BWI Co., Ltd. proposed an active stabilizer bar system based on a ball screw [24] and a hydraulic pump [25]. The stabilizer bar torque can provide a torque range of 0.9–1.5 kN m when the hydraulic pump is under 180 bar working pressure, which improves the ride comfort and driving safety of the vehicles, and the off-road performance of the vehicles is enhanced.

MR fluids are smart material that changes their rheological properties under applied magnetic fields reversibly. Figure 6.1(a) and (b) shows the distribution of ferromagnetic particles in MR fluids without and with applied magnetic field, respectively. As shown in Figure 6.1(a), when no magnetic field is applied, the ferromagnetic particles in MR fluids are freely dispersed in the carrier liquid, and MR fluids present the characteristics of the Newtonian fluid. At this state, the shear

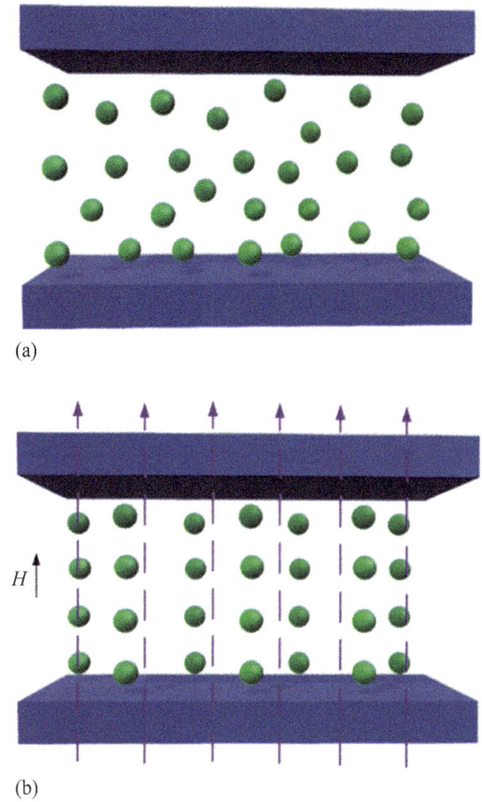

Figure 6.1 Rheological properties of MR fluids: (a) no magnetic field applied and (b) magnetic field applied

stress of MR fluids is small, which is equal to the product of viscosity and shear rate. As shown in Figure 6.1(b), when under an external magnetic field, the ferromagnetic particles are arranged in a chain. MR fluids exhibit Bingham fluid properties, and its yield shear stress is composed of the viscous force and the yield stress of the MR fluids. The rheological property changes as the yield stress monotonously increases with the increase of the magnetic field strength in the range of the magnetic field saturation, while the viscosity of MR fluids does not change [26–29]. Because of the short response time and large controllable damping range of MR fluids, the actuators based on MR fluids and MR semi-active control applications have been extensively studied. MR technology is widely applied on vehicle systems such as vehicle suspensions [30–33], seats suspensions [34,35], engine mounting and vibration isolation systems [36–38] and military equipment areas [39].

The main technical contribution of this chapter is aiming at improving the anti-roll performance of vehicles and keeping ride comfort to investigate a semi-active vehicular stabilizer bar based on a newly proposed reciprocating rotary MR damper.

The mathematical model of the MR semi-active stabilizer bar system and the entity model for software ADAMS are established. The influences of MR stabilizer bar on vehicle roll stability under passive-on and piecewise control strategies are compared and analyzed.

6.2 MR semi-active stabilizer bar system

6.2.1 Vehicle roll dynamic model

Lateral acceleration is a basic index of vehicle anti-roll capability. Assuming that the road and vehicle conditions are fair, the vehicle roll is only generated by steady steering, and the generated lateral acceleration depends on the driving speed and the turning radius. The vehicle front suspension with MR stabilizer bar is shown in Figure 6.2. According to Figure 6.2, the expression of roll torque balance of the sprung mass relative to the roll center is given by

$$\begin{cases} (k_f + k_r)\varphi_0 + T_{MRf} + T_{MRr} = m_s a_y h_s + m_s g h_s \varphi_0 \pi/180 \\ k_f \varphi_0 = 2(k_{cf} + k_{tf})\varphi_0 \\ T_{MRr} = \dfrac{1-\alpha}{\alpha} T_{MRf} \end{cases} \tag{6.1}$$

where φ_0 is the roll angle between the installation of stabilizer bar and vehicle body or frame; k_f and k_r are the roll angle stiffnesses of front and rear suspensions, respectively; a_y is the lateral acceleration of vehicle body; h_s is the distance from the roll center to the center of mass; k_{cf} and k_{tf} are the roll angle stiffnesses of the coil spring; T_{MRf} and T_{MRr} are the torques of the front and rear MR stabilizer bars; and α is the ratio of the front roll angle stiffness to full-vehicle roll angle stiffness.

According to (6.1), the lateral acceleration of vehicle body can be expressed as

$$a_y = \frac{T_{MRf} + \alpha\varphi_0(k_f + k_r - m_s h_s g\pi/180)}{\alpha m_s h_s} \tag{6.2}$$

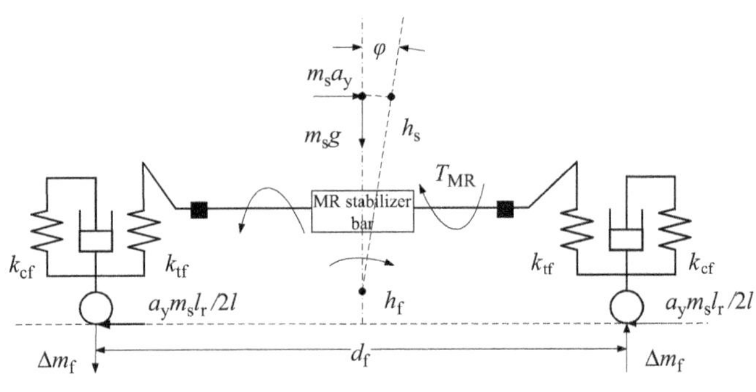

Figure 6.2 Vehicle roll model when equipped with MR stabilizer bar

Vehicle's unsprung mass equilibrium equation relative to the center of mass is

$$\begin{cases} k_f\varphi_0 + T_{MRf} = \Delta m_f d_f - a_y m_s l_r h_f/l \\ k_r\varphi_0 + T_{MRr} = \Delta m_r d_r - a_y m_s l_f h_r/l \end{cases} \tag{6.3}$$

where l_f and l_r are the distance from full load center of mass to front and rear axles respectively; l is the wheelbase; h_f and h_r are the heights of the front and rear roll centers, respectively; d_f and d_r are the front and rear wheel tracks, respectively; Δm_f and Δm_r are the difference between the vertical forces of the front and rear tires, respectively, and Δm_f is given by

$$\Delta m_f = \frac{\left(k_f\varphi_0 + T_{MRf} + a_y m_s l_r h_f/l\right)}{d_f} \tag{6.4}$$

Under a vertical force, pneumatic tires will vertically deform. The stiffness characteristics of the tires are usually expressed by the curve of the load and deformation of the tires. It is approximately a linear relationship between the applied load and the deformation. The body roll angle generated by the wheels under the vertical force is expressed by

$$\Delta\varphi = \arctan\frac{2\Delta m_f}{K_z d_f} \tag{6.5}$$

where K_z is the vertical stiffness of the tires.

According to (6.2) and (6.5), within the working range of the stabilizer bar, the relationship between the body roll angle φ and lateral acceleration a_y can be expressed as

$$\varphi = \frac{a_y \alpha m_s h_s - T_{MRf}}{\alpha(k_f + k_r - m_s h_s g\pi/180)} + \arctan\frac{2\Delta m_f}{K_z d_f} \tag{6.6}$$

6.2.2 MR stabilizer bar

6.2.2.1 Mathematical model

The proposed configuration of the MR stabilizer bar is presented in Figure 6.3. As shown in Figure 6.3, its overall appearance is U-shaped. The MR stabilizer bar is mainly composed of an MR rotary damper, right and left bars rigidly connected with MR damper, flange nuts and bushings. The outer casing and central shaft of the MR damper are fixedly connected by the flange nut and the left and right bars, respectively. The left and right bars are, respectively, installed by bushing onto the vehicle body or frame and the longitudinal part of the U-shaped structure of stabilizer bar is connected with the suspension guide arm (not shown in the figure).

The structural dimensions of the stabilizer bar are shown in Figure 6.4. As seen in Figure 6.4, one side bar of stabilizer bar is fixed onto the suspension guide arm, and the views on the *xoy* and *yoz* planes are also provided. The bushing restrains the middle part of stabilizer bar to rotate around the *y*-axis and produces a displacement

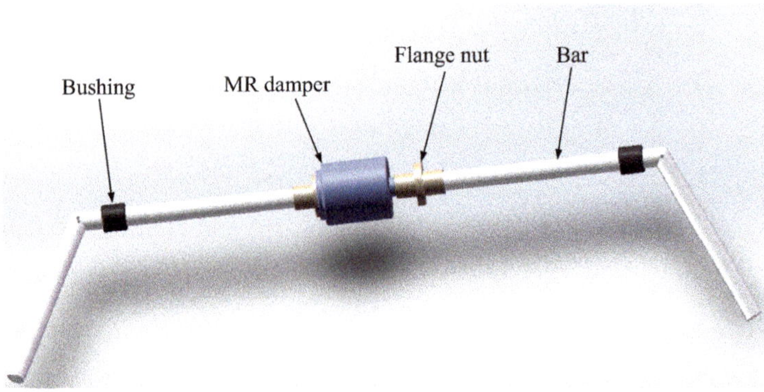

Figure 6.3 The configuration of the MR stabilizer bar

along the z-direction at the mounting point of stabilizer bar and the suspension guide arm.

The deflection or rotation angle of the stabilizer bar under the joint action of several loads is equal to the algebraic sum of deflection or the rotation angle under the separate action of individual load. The displacement h_{bar} along the z-direction is obtained

$$h_{bar} = \frac{F_{bar}}{3EI_{bar}}\left[l_1^3 - a^3 + \frac{l_b}{2}(a+b)^2 + 4l_2^2(b+c)\right] \tag{6.7}$$

$$I_{bar} = \frac{\pi d^4}{64} \tag{6.8}$$

where a, b, c, d, l_1, l_2 and l_b are all the dimensional parameters of the stabilizer bar, as shown in Figure 6.4; F_{bar} is the z-direction force at the mounting point of the stabilizer bar and the suspension guide arm; E is the elastic modulus; I_{bar} is the inertia torque of the stabilizer bar.

The force that can produce a unit displacement along the z-direction of the end of stabilizer bar is defined as the linear stiffness k_{bar}. The relationship between k_{bar} and h_{bar} and F_{bar} is

$$F_{bar} = 2h_{bar}k_{bar} \tag{6.9}$$

According to (6.7) and (6.8), k_{bar} can be rewritten as

$$k_{bar} = \frac{3EI_{bar}}{2\left[l_1^3 - a^3 + (l_b/2)(a+b)^2 + 4l_2^2(b+c)\right]} \tag{6.10}$$

The roll stiffness of the stabilizer bar acting on the suspension is approximately equal to the equivalent roll stiffness K_{bar} at the wheel. The relationships between

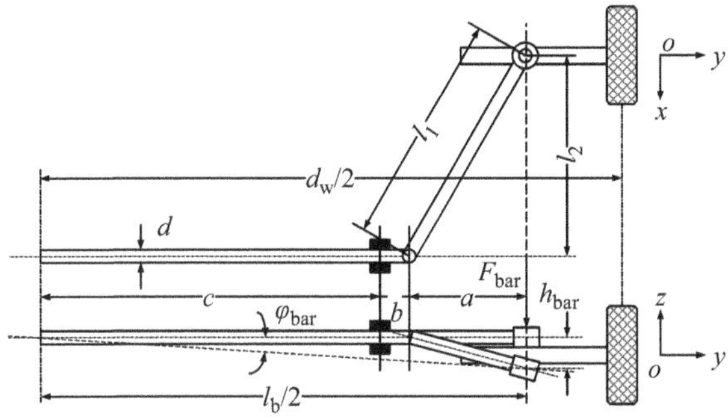

Figure 6.4 The structural dimensions of the MR stabilizer bar

the linear stiffnesses k_{bar} and K_{bar}, between the linear dampings c_{bar} and C_{bar}, are expressed by

$$K_{bar} = \frac{k_{bar} l_b^2}{2} \tag{6.11a}$$

$$C_{bar} = \frac{c_{bar} l_b^2}{2} \tag{6.11b}$$

The torques that the stabilizer bar is twisted by 1° and 1°/s around the y-axis are the torsional stiffness k_{rb} and the torsional damping coefficient c_{rb}. k_{rb} and c_{rb} can be, respectively, expressed as

$$k_{rb} = \frac{k_{bar} l_2^2}{2 \times 57.3} \tag{6.12a}$$

$$c_{rb} = \frac{c_{bar} l_2^2}{2 \times 57.3} \tag{6.12b}$$

Since the stabilizer bar is hingedly connected to the vehicle body through the bushing, only the rotation about the y-axis direction is retained without relative frame movement. When with no consideration of the unsprung mass roll and bar bending along x-direction, the generated angle φ_{bar} between h_{bar} and the horizontal plane is the vehicle body roll angle. The rotation angle around the y-axis generated by the z-direction displacement h_{bar} on both sides of the stabilizer bar is the angle φ_{arb} of the MR damper and the bar. φ_{bar} and φ_{arb} can be expressed as

$$\varphi_{bar} = \arctan \frac{2h_{bar}}{l_b} \tag{6.13a}$$

$$\varphi_{arb} = 2\arctan \frac{h_{bar}}{l_2} \tag{6.13b}$$

Table 6.1 The dimensional parameters of the stabilizer bar model

Parameters	Value	Unit
a	113.17	mm
b	80	mm
c	317.5	mm
d	24.145	mm
l_1	353.6	mm
l_2	353	mm
l_b	1,021.6	mm

In the case of small corners, φ_{bar} and φ_{arb} can be approximated by the radian system, and (6.13a) and (6.13b) are rewritten as

$$\varphi_{bar} = \frac{2h_{bar}}{l_b} \tag{6.14a}$$

$$\varphi_{arb} = \frac{2h_{bar}}{l_2} \tag{6.14b}$$

The total torque T_{bar} of the stabilizer bar can be expressed by k_{rb} and c_{rb} as

$$T_{bar} = k_{rb}\varphi_{bar} + c_{rb}\dot{\varphi}_{arb} \tag{6.15}$$

In addition, the anti-roll torque M_{arb} of the stabilizer bar to the vehicle body can be expressed according to the external body roll state and the roll stiffness and damping

$$M_{arb} = C_{bar}\dot{\varphi} + K_{bar}\varphi \tag{6.16}$$

The simplified model dimensions of the stabilizer bar designed in this chapter are listed in Table 6.1.

6.2.2.2 Force analysis of the MR stabilizer bar

The MATLAB®/Simscape Multibody model with the MR stabilizer bar and the MATLAB/Simulink™ model are shown in Figures 6.5 and 6.6, respectively. It should be noted that the left and right bars of the stabilizer bar and the outer casing and the central shaft of the reciprocating rotary MR damper are rigidly connected by the flange without displacement in the middle. Therefore, when the multi-body dynamics model is established in the MATLAB/Simscape Multibody, the flanges of installation and the installation relationship between the end and the suspension are ignored.

The MR stabilizer bar model is motion simulated according to the simplified model dimensions of the designed stabilizer bar as listed in Table 6.1. One end of the stabilizer bar is fixedly constrained, and the other end is excited by sinusoidal

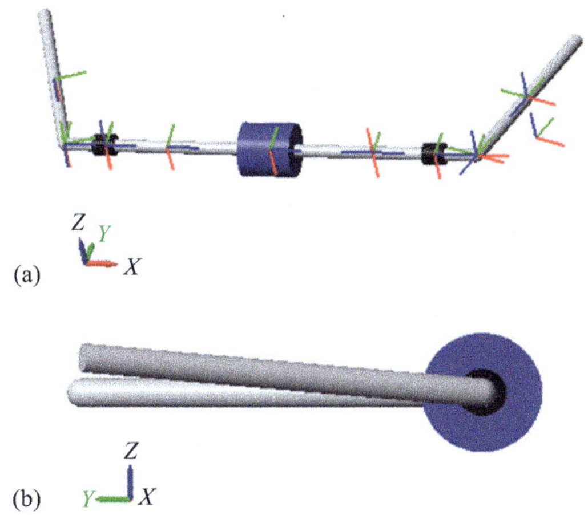

Figure 6.5 MATLAB/Simscape Multibody model of the MR stabilizer bar: (a) model entity and (b) side view

velocity excitation. The velocity amplitudes (*Fr*) are 0.2, 0.6 and 1 rad/s, respectively. The vertical forces acting on the end of the stabilizer bar at different sinusoidal velocities and the torques in the middle of the bar are shown in Figure 6.7(a) and (b), respectively. As the rotational velocity increases, both of the amplitude of the force and the set roll torque applied to the stabilizer bar on one end increase. Since the set damping coefficient has a quite small relative stiffness coefficient and the rotational velocity is not large, the end force increases from 1.13 to 1.35 kN, and the corresponding torque increases from 547 to 682 N m.

6.2.3 The MR damper

6.2.3.1 MR fluids properties

The rheological properties of MR fluids are one of the important factors affecting the output performance of MR dampers. The employed MR fluid (type: MRF-J01T) has a maximal yield stress of 60 kPa and a working temperature in the range of −40 °C to 130 °C. Figure 6.8 shows the relationship between the yield stress of the MR fluid and the applied magnetic flux density.

6.2.3.2 Structural principle

The three-dimensional structure and the MR fluid channel of the MR damper are shown in Figure 6.9(a) and (b), respectively. As seen from Figure 6.9(a), the MR damper consists of the body casing, the rotor assembly, the stator assembly and the seal. The rotor assembly includes a rotor axle and a pair of blades. The stator assembly includes a fixed part, a stator cover, a stator core, a positioning ring and an electromagnetic coil winding. As shown in Figure 6.9(b), the MR fluid channel is formed by the upper and lower stator covers, fixed parts, positioning rings, stator

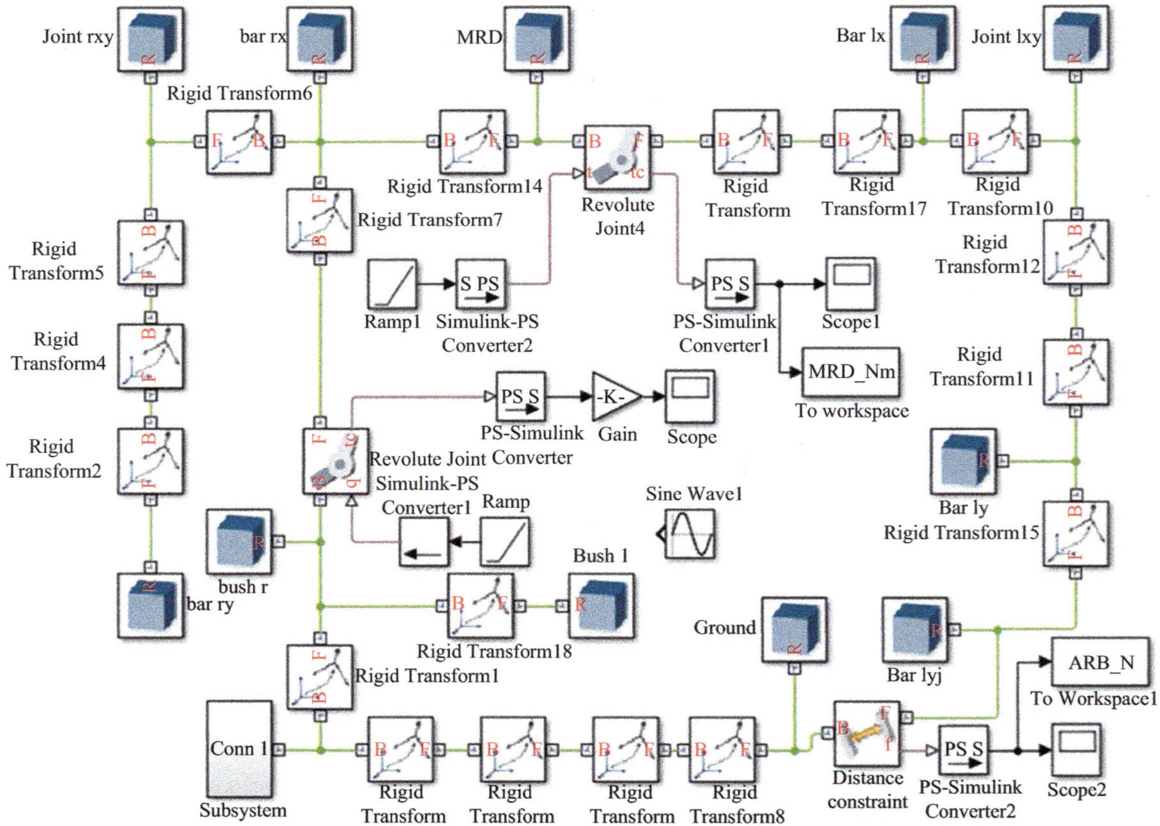

Figure 6.6 MATLAB/Simulink model of the MR stabilizer bar

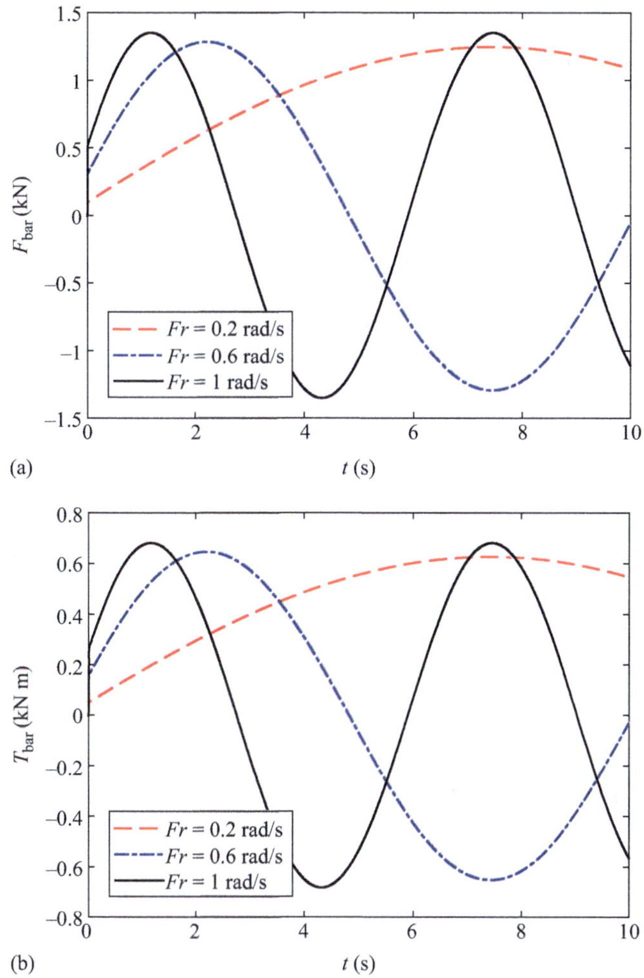

Figure 6.7 Force analysis results of the MR stabilizer bar model: (a) vertical force at the end point and (b) generated torque in the middle of the bar

cores and coils. Each side of the stator has two fan-shaped gap and a rectangular gap, and the MR fluid chambers separated by the blades on the same side can be connected via the circulation holes of the upper and lower stator covers. During the rotation, the MR fluid is pushed to flow through the MR fluid channel in the stator assembly into the chamber at the other side. The electromagnetic field, as presented in Figure 6.9(a), is generated by the coil winding in the stator with applied current. The generated magnetic field passes vertically through the three gaps in the MR fluid channel, the damping force (torque) is controllable by changing the viscosity of the MR fluid in the channel.

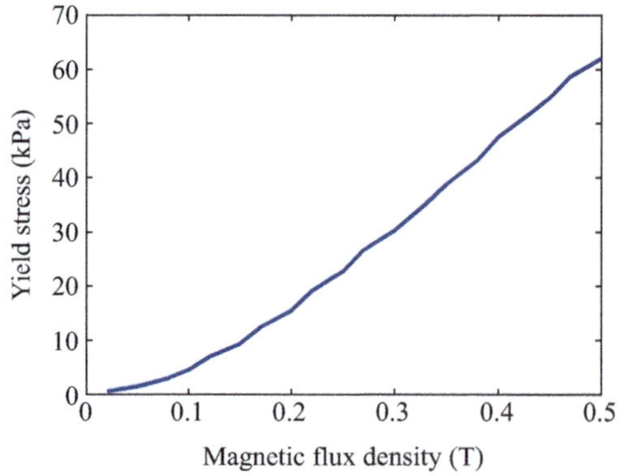

Figure 6.8 The relationship between the yield stress and magnetic flux density of the MR fluid (type: MRF-J01T)

According to the proposed structural principle of the MR damper, there are at least three particular advantages:

1. The proposed MR damper is compact and is suitable for the applications where the stabilizer bar is small. The diameter of the proposed MR damper casing is 110 mm and the height of the casing is 85 mm.
2. The MR damper provides very large controllable damping force range. There are two stator assemblies inside the MR damper, consisting total of six magnetic-field activated gaps. In other words, the MR damper has very large effective area for MR damping force or torque.
3. The MR damper utilizes only 81.8 mL MR fluids, which will effectively increase the activate proportion of MR fluids (i.e., 13.4%) [40–42] and decrease the actuator costs.

6.2.3.3 Controllable torque

The structural dimensions and flow model of the MR fluid channel in one stator assembly are shown in Figure 6.10. Ignoring the torque generated by the internal mechanical friction, the output torque of the MR damper consists of the torque due to the viscosity of the MR fluids and the torque due to the MR effect, which can be expressed as

$$T_{\mathrm{MR}} = T_\eta + T_\tau \tag{6.17}$$

where T_η is the viscous torque and T_τ is the torque due to the MR effect.

T_η and T_τ can be expressed, respectively, as

$$T_\eta = 2(2\Delta P_{\eta \mathrm{r}} + \Delta P_{\eta \mathrm{v}})h_{\mathrm{d}} \int_{r_{\mathrm{s}}}^{r_{\mathrm{c}}} r\,dr \tag{6.18a}$$

$$T_\tau = 2(2\Delta P_{\tau \mathrm{r}} + \Delta P_{\tau \mathrm{v}})h_{\mathrm{d}} \int_{r_{\mathrm{s}}}^{r_{\mathrm{c}}} r\,dr \tag{6.18b}$$

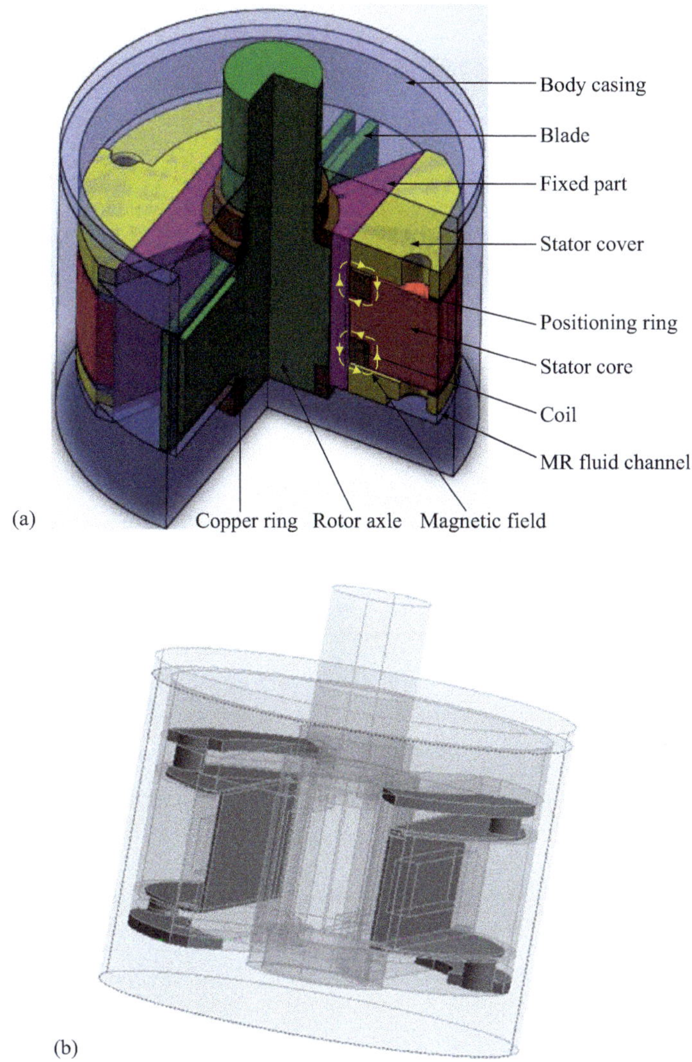

Figure 6.9 The MR damper: (a) three-dimensional structure and (b) MR fluid channel (in dark gray)

where $\Delta P_{\eta r}$ is the pressure drop due to the viscosity of the MR fluids in the fan-shaped gap; $\Delta P_{\eta v}$ is the pressure drop due to the viscosity of the MR liquid in the rectangular gap; $\Delta P_{\tau r}$ is the pressure drop due to the MR effect in the fan-shaped gap; $\Delta P_{\tau v}$ is a rectangle the pressure drop due to the MR effect on the rectangular gap surface; h_d is the height of the stator; r_s is the distance between the axis and the rectangular surface of the fixed part and r_c is the inner radius of the body casing.

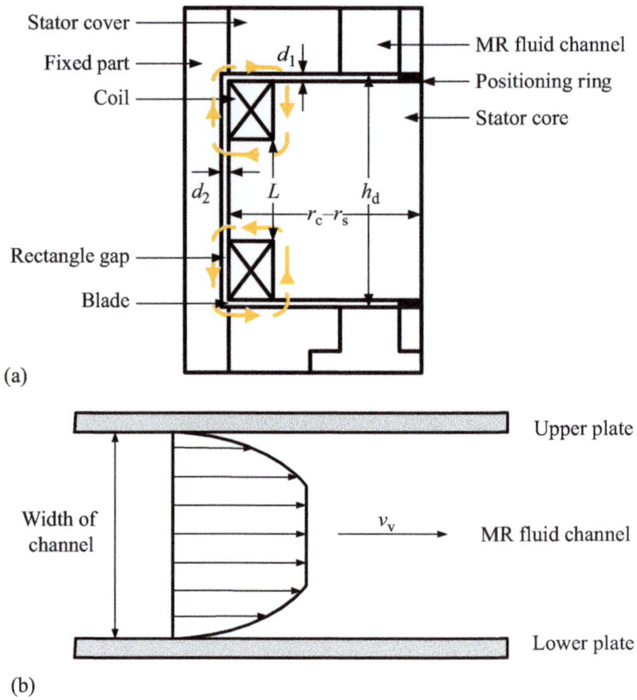

Figure 6.10 MR fluid channel in the stator assembly: (a) channel structure and (b) parallel plate model of MR fluids

$\Delta P_{\eta r}$, $\Delta P_{\eta v}$, $\Delta P_{\tau r}$, and $\Delta P_{\tau v}$ are, respectively, expressed as

$$\Delta P_{\eta r} = \frac{n_r}{2\pi}\frac{6\eta Q}{\pi d_1^3}\ln\frac{r_c}{r_0} \tag{6.19a}$$

$$\Delta P_{\eta v} = \rho v_v^2 \frac{f_d N h_v}{2d_2} \tag{6.19b}$$

$$\Delta P_{\tau r} = \frac{n_r}{2\pi}\frac{c_t \tau_y(B)}{d_1}(r_c - r_0) \tag{6.19c}$$

$$\Delta P_{\tau v} = \frac{2Nh_v \tau_y(B)}{d_2} \tag{6.19d}$$

where n_r is the angle of the fan-shaped gap; η is the viscosity of the MR fluids; Q is the flow volume of the MR fluids; d_1 and d_2 are the gap thickness of fan-shaped gap and rectangular gap, respectively; r_0 is the equivalent radius of effective area of the fan-shaped gap; ρ is the density of the MR fluids; v_v is the average flow rate of MR fluids in the rectangular gap; N is the number of the coil windings; h_v is the height of rectangular gap; f_d is Darcy factor; c_t is a correction factor and $\tau_y(B)$ is the shear yield stress related to the magnetic field.

Q, v_v, c_t and $\tau_y(B)$ are given by

$$Q = \frac{1}{2}\omega\left(r_c^{\,2} - r_s^{\,2}\right)h_v \tag{6.20a}$$

$$v_v = \frac{Q}{h_w d_2} \tag{6.20b}$$

$$c_t = 2.07 + \frac{12Q\eta}{12Q\eta + 0.8\pi r_c d_1^2 \tau_y(B)} \tag{6.20c}$$

$$\tau_y(B) = k_1 B^{\alpha_1} \tag{6.20d}$$

where ω is the angular velocity of the MR damper, h_w is the width of the rectangular gap, k_1 and α_1 are the coefficients and B is magnetic flux density.

f_d is determined by Reynolds number (Re). The Reynolds number can be used to distinguish fluids flow states: laminar or turbulent or the flow state between the two. Re is expressed as

$$Re = \frac{\rho v_v 2d_2}{\eta} \tag{6.21}$$

f_d is expressed according to the value of Re by

$$\begin{cases} f_d = \dfrac{96}{Re}(Re \le 2{,}000) \\ f_d = (1-\alpha_e)\dfrac{96}{Re} + \dfrac{\alpha_e}{\left\{1.8\lg\left[(\varepsilon/(3.7\times 2d_2))^{1.11} + (6.9/4{,}000)\right]\right\}}(2{,}000 < Re < 4{,}000) \\ f_d \approx \dfrac{1}{\left\{-1.8\lg\left[(\varepsilon/(3.7\times 2d_2))^{1.11} + (6.9/Re)\right]\right\}^2}(Re \ge 4{,}000) \end{cases} \tag{6.22}$$

where ε is the relative surface roughness of the gaps and α_e is expressed as

$$\alpha_e = \frac{Re - 2{,}000}{2{,}000} \tag{6.23}$$

It is worth noting that the maximal angular speed of the stabilizer bar (i.e., MR damper) caused by the suspension is generally below 1 rad/s. Based on the MR fluid properties and structural parameters of the MR damper listed in Table 6.2, the value of Re is much lower than 2,000. In other words, the MR fluids in the flow channel of the MR damper is at laminar state.

6.2.3.4 Electromagnetic simulation

The output torque of the MR damper is mainly determined by the torque due to the MR effect, which depends on electromagnetic fields of the MR fluids. In order to analyze the electromagnetic fields generated by the electromagnetic coil winding at

Table 6.2 Parameters of the MR fluid and the MR damper

Parameters	Specification	Value	Unit
ρ	Density of MR fluid	2,650	kg/m^3
η	Viscosity of MR fluid	0.11	Pa s
τ_y	Maximal yield stress	60	kPa
r_c	Inner radius of the body casing	50	mm
r_s	Distance between the axle and the rectangular gap surface of fixed part	18	mm
h_d	Stator height	64	mm
N	Number of coil winding	1	–
d_1	Thickness of fan-shaped gap	0.6	mm
d_2	Thickness of rectangular gap	0.6	mm
h_v	Height of rectangular gap	56	mm
h_w	Width of rectangular gap	71.9	mm
n_r	Angle of fan-shaped gap	110	°
r_0	Equivalent radius of effective area of fan-shaped gap	28.5	mm
ε	Relative roughness of the gap surface	1.6	μm

specific applied currents, the finite element analysis software ANSYS is used to conduct electromagnetic simulation of the static magnetic field of the designed MR damper.

When 1 A current is applied to the electromagnetic coil winding, the magnetic flux lines and the contour of the magnetic flux density are shown in Figure 6.11(a) and (b), respectively. As seen from Figure 6.11(a), one closed magnetic circuit can be seen around electromagnetic coil winding. The magnetic flux density at the rectangular gap surface is around 0.97 T and around 0.33–0.41 T at the upper and lower fan-shaped gap surfaces. The strong magnetic field is distributed on the fixed part near the upper and lower end of the rectangular gap surface, and the maximal magnetic flux density on the fixed part reaches as high as 2.13 T, as shown in Figure 6.11(b).

6.2.3.5 Output characteristics

According to the parameters of the employed MR fluid and the MR damper as listed in Table 6.2 and the electromagnetic simulation results of the reciprocating rotary MR damper, the output torque of the MR damper is numerically simulated. The controllable torque range of the reciprocating rotary MR damper is shown in Figure 6.12. As seen from Figure 6.12, when no applied current, the torque of the MR damper is caused by the viscosity of the MR fluids and increases as the rotating speed increases. When the current increases to maximal, the output torque reaches as high as 1.65 kN m. Before the saturation of the MR fluids, the larger the applied current in the coil (i.e., the stronger magnetic flux density), the greater the output torque. At the same time, the output torque increases with the rotating speed. As seen, from the figure, the MR damper provides a very large controllable torque range.

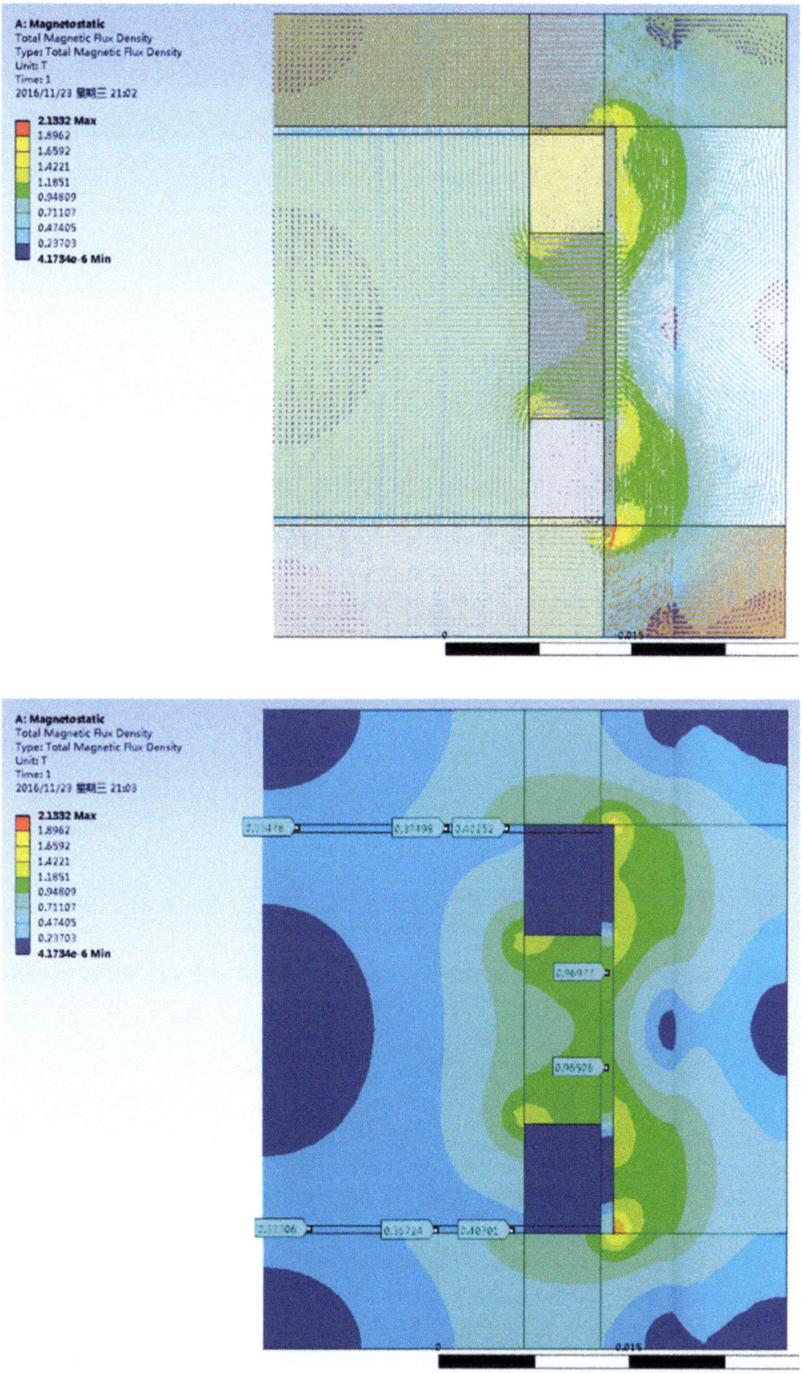

(a)

(b)

Figure 6.11 Electromagnetic simulation results: (a) magnetic flux lines and (b) contours of magnetic flux density

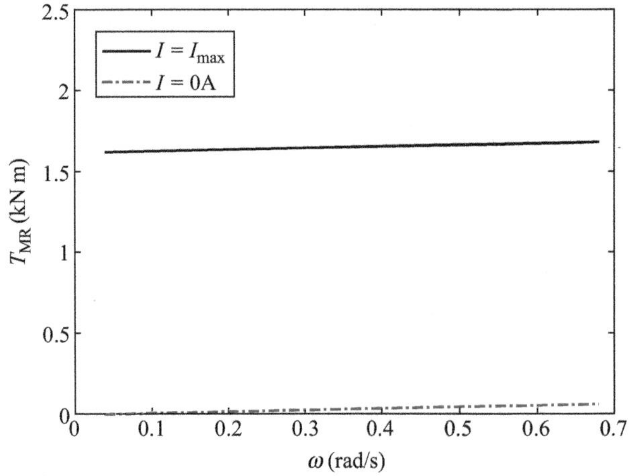

Figure 6.12 Controllable torque range of the MR damper

Table 6.3 The proposed piecewise control algorithm

a_y (m/s)	f (mm)	T_{MR} (N m)
<0.5	<4	0.7M
0.5–1.5	4–12	1M
1.5–3	12–24	1.5M
≥ 3	≥ 24	2M

6.2.4 Control strategy for MR stabilizer bar

In order to study the influence of MR stabilizer bar on the roll performance of the vehicle when providing different torques, a piecewise control based on lateral acceleration of the vehicle body is proposed, and the semi-active control is performed on MR stabilizer bar, as shown in Table 6.3 and Figure 6.13. The maximum torque T_{MR} of the MR damper is set equal to 0.7M, 1M, 1.5M and 2M (M is the torque of the passive stabilizer bar at the limit working position), and are proportional to the relative deformation displacement f of the tip of the longitudinal part of both sides of the U-shaped stabilizer bar. The relationship between the different torques T_{MR} and the roll angle and the lateral acceleration under piecewise control is shown in Figure 6.13. In the passive control mode, as the T_{MR} increases, the maximum lateral acceleration that the vehicle body can withstand at the same roll angle is significantly increased by 2.80 m/s^2 compared to the passive stabilizer bar. The body roll angle is only increased by 0.02° due to the tire deformation, indicating that a large T_{MR} can improve the roll safety of the vehicle.

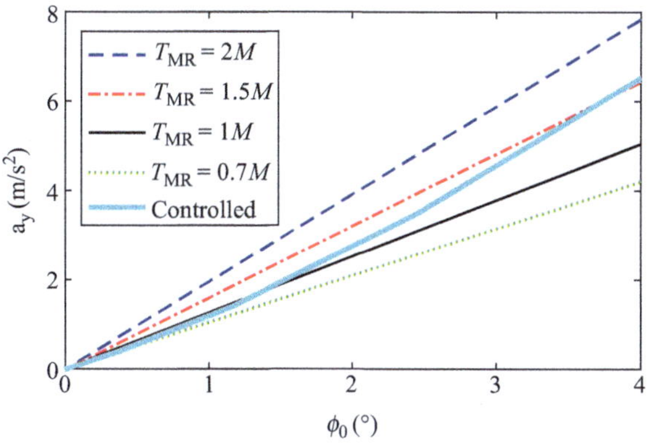

Figure 6.13 Relationship between the roll angle and lateral acceleration when under different T_{MR}

6.3 ADAMS/Car modeling and simulation

6.3.1 Comparison between ADAMS/Car model and mathematical model

The ADAMS/Car vehicle model of a certain vehicle is shown in Figure 6.14. As shown in Figure 6.14, the established vehicle includes the powertrain system, brake system, steering system, wheels, suspension and passive stabilizer bar. In order to verify the consistency of the established full-vehicle ADAMS/Car model and the mathematical model, full-vehicle-simulation analysis is conducted to investigate the relationship between the roll angle and lateral acceleration and is compared with the curve of the mathematical model, as shown in Figure 6.15. In the scope of the stabilizer bar when vehicles being in steady state, the simulation model and mathematical model agree well with each other. When out of the scope of the stabilizer bar, vehicles will have a tendency to instability, and the roll angle will increase rapidly with the increase of lateral acceleration.

6.3.2 Full-vehicle simulation

6.3.2.1 Fish hook maneuver

The fish hook maneuver test, also known as the rollover test, is used to evaluate the vehicles' dynamic rollover performance and road maintenance. According to the new vehicle evaluation regulations issued by the US Highway Safety Administration, the rollover test requires that the vehicle travels at a speed of 56 km/h in a straight line, and the steering wheel first turns 270° within 0.5 s and then quickly reverses to turn 540° within 1 s [43]. In the fish hook maneuver simulation, the displacement of the center of mass of vehicle, the vehicle body lateral acceleration and vehicle body roll angle are shown in Figure 6.16(a), (b) and (c), respectively. D_x and D_y are the displacements of the vehicle in the x and y directions, respectively.

Figure 6.14 ADAMS/Car model of a vehicle

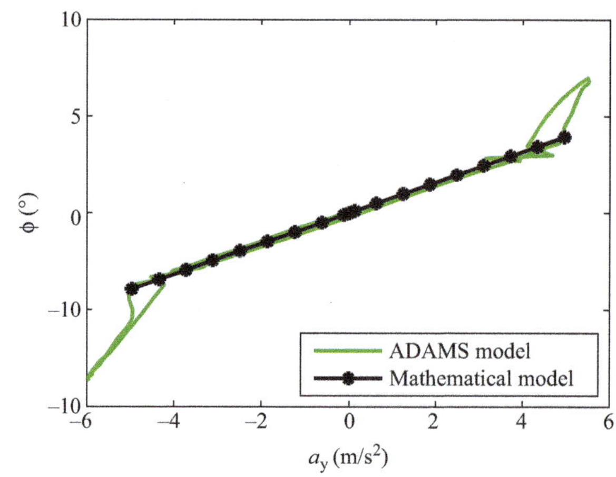

Figure 6.15 Comparison of ADAMS/Car and mathematical models

When under passive-on control of the MR damper, the turning radius of the vehicle increases with the increase of T_{MR}, and the maximal lateral acceleration of center of mass of increases too. The maximal roll angle decreases as T_{MR} increases. When T_{MR} is $0.7M$, $1M$, $1.5M$ and $2M$, the maximal roll angles are 9.20°, 8.95°, 8.61° and 8.39°, respectively. The maximal roll angle under piecewise control is 8.47°, which is next to the case when T_{MR} is $2M$.

6.3.2.2 Single lane change maneuver

Single lane change simulation refers to driving the car in required time and through an S-shape curve road to simulate change lanes. Under the single lane change maneuver simulation, the steering wheel input angle, the vehicle body lateral acceleration and vehicle body roll angle are shown in Figure 6.17(a) and (b), respectively. The steering wheel angle in this simulation is provided as the requirement of the roll tests.

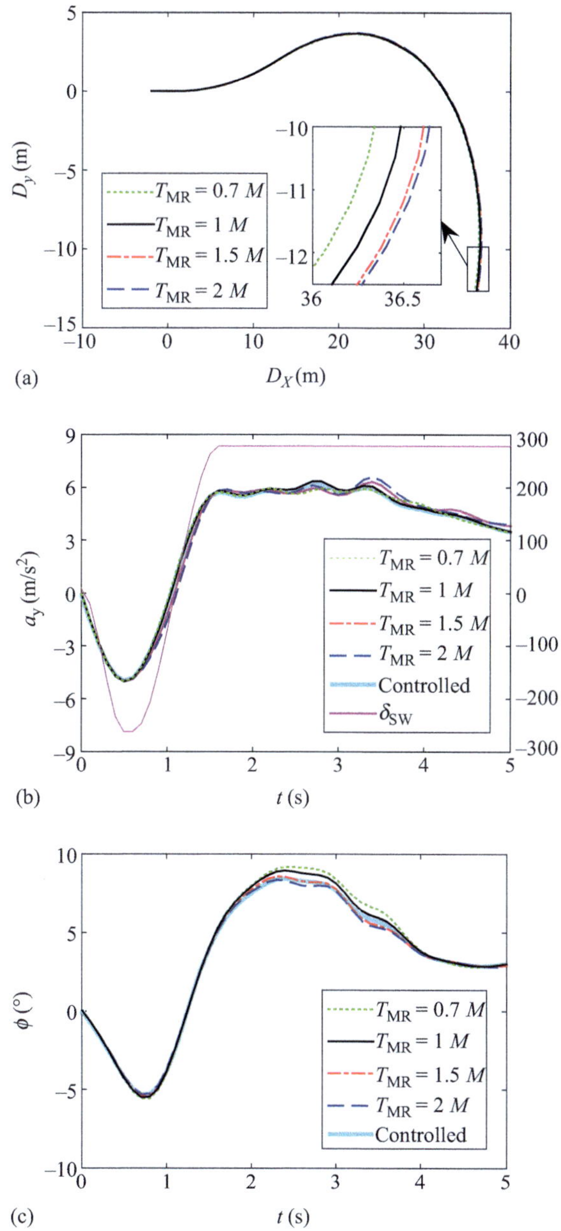

Figure 6.16 Simulation results in the fishhook maneuver: (a) motion curve of the center of mass, (b) steering wheel angle and lateral acceleration, and (c) roll angle

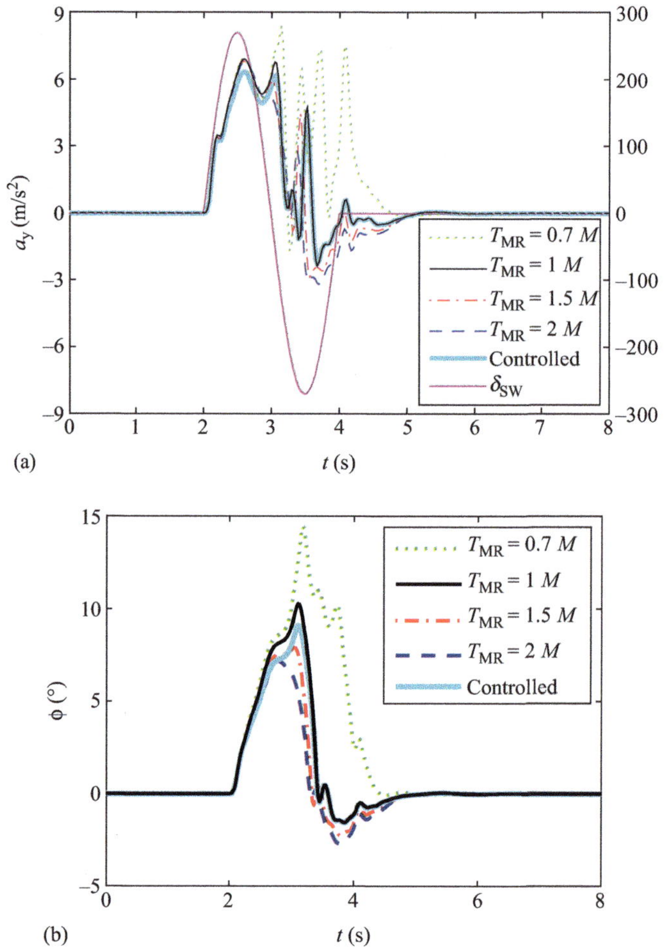

Figure 6.17 Simulation results in the single lane change maneuver: (a) steering wheel angle and lateral acceleration and (b) roll angle

As shown in Figure 6.17, the maximal vehicle body roll angle decreases with the increase of T_{MR}, and when T_{MR} is 0.7M, 1M, 1.5M and 2M, the maximal roll angles are 14.5°, 10.28°, 7.93° and 7.17°, respectively, all of which are beyond the working scope of the stabilizer bar. When T_{MR} is equal to 0.7M, the lateral acceleration and its fluctuation are much larger than the others, and the vehicle is in a critical state of risk of rollover. Less lateral acceleration and roll angle during the first half course under the piecewise control than those when T_{MR} is 0.7M and 1M.

6.3.2.3 Pylon course maneuver

Pylon course simulation is mainly used to evaluate the transient response and roll stability. Steering wheel angle and lateral acceleration and roll angle of pylon course simulation are shown in Figure 6.18(a) and (b), respectively. In the simulation, at a speed of 50 km/h, the steering wheel first turns with 200° within 0.2 s and remains stable for 4.7 s, then it turns reversely with 400° within 0.4 s and ends until the second steering hold time.

As shown in Figure 6.18(b), the maximal vehicle body roll angles decrease with the increase of T_{MR}. When T_{MR} is 0.7*M*, 1*M*, 1.5*M* and 2*M*, the maximal roll angles are 6.89°, 6.47°, 5.91° and 5.54°. When T_{MR} is 1.5*M* and 2*M*, after the second turn, their overshot are small and tend to be stable soon since the roll angle stiffness is large. Under the piecewise control, a smaller roll angle (5.92°) is realized than when T_{MR} is 0.7*M* and 1*M*.

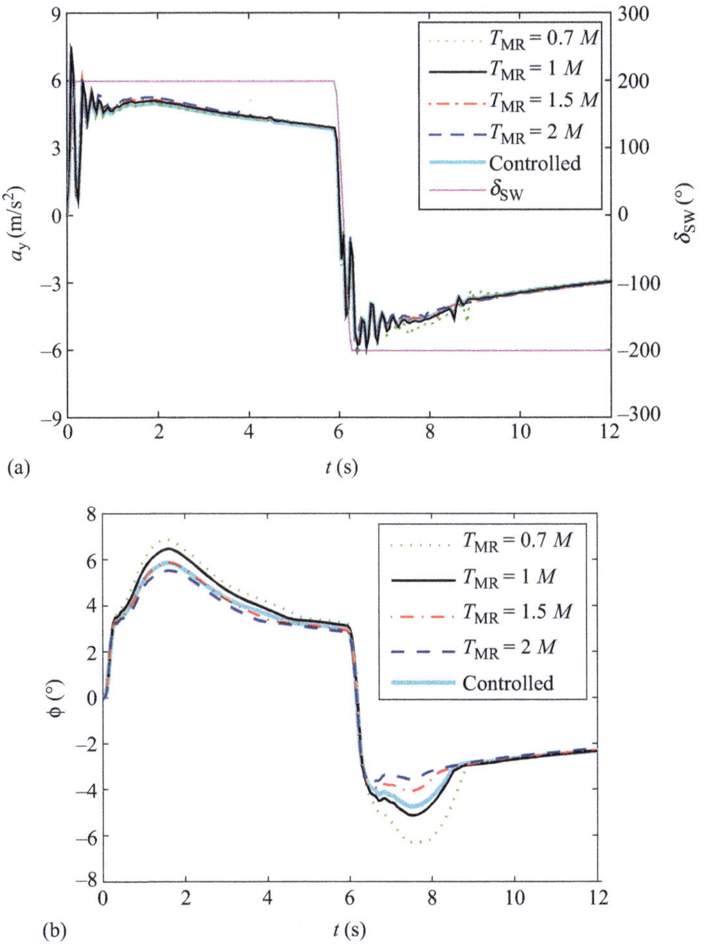

Figure 6.18 Simulation results in the pylon course maneuver: (a) steering wheel angle and lateral acceleration and (b) roll angle

6.4 Conclusion

The smart material MR fluids have the properties of short response time and large controllable damping range, and the application-based MR semi-active actuators are more and more widely applied. In order to ensure both the driving safety (the anti-roll performance) of the vehicle while steering and the ride comfort when passing the uneven road, an MR semi-active stabilizer bar system is designed and studied in this chapter. A stabilizer bar based on a reciprocating rotary MR damper is proposed. The motion force of the stabilizer bar is simulated and analyzed. Multi-body dynamics modeling and simulation analysis are conducted in MATLAB/Simscape Multibody and MATLAB/Simulink. The forces and torques of the MR stabilizer bar are obtained. The structural principle of reciprocating rotary MR damper with large controllable torque range is proposed, and the detailed mathematical model is established. The electromagnetic simulation of the MR damper is carried out via ANSYS, and the corresponding controllable torque is obtained. For the mathematical model with the MR stabilizer bar and ADAMS/Car model, under the passive-on and piecewise control strategies, the simulations under roll conditions, such as fishhook maneuver, single lane change maneuver, and pylon course maneuver, are carried out. The research results show that (i) with the increase of the torque of the MR stabilizer bar, the roll angle of the vehicle decreases; (ii) the vehicle equipped with the MR stabilizer bar increases the bearable lateral acceleration significantly at the same roll angle within the working range of stabilizer bar, i.e., better safety; and (iii) the MR stabilizer bar under piecewise control provides better vehicle roll performance than the traditional stabilizer bars.

References

[1] NHTSA, 1995, Traffic Safety Facts 1994: A compilation of motor vehicle crash data from the fatality analysis reporting system and the general estimates system, Washington, DC: US Department of Transportation, National Center for Statistics & Analysis, National Highway Traffic Safety Administration.

[2] NHTSA, 2016, Traffic Safety Facts 2015: A compilation of motor vehicle crash data from the fatality analysis reporting system and the general estimates system, Washington, DC: US Department of Transportation, National Center for Statistics & Analysis, National Highway Traffic Safety Administration.

[3] Gao H J, Sun W C and Shi P, 2010, Robust sampled-data H∞ control for vehicle active suspension systems, *IEEE Transactions on Control Systems Technology*, 18(1): 238–245.

[4] Priyandoko G, Mailah M and Jamaluddin H, 2009, Vehicle active suspension system using skyhook adaptive neuro active force control, *Mechanical Systems and Signal Processing*, 23(3): 855–868.

[5] Anubi O M and Crane C, 2015, A new semiactive variable stiffness suspension system using combined skyhook and nonlinear energy sink-based controllers, *IEEE Transactions on Control Systems Technology*, 23(3): 937–947.

[6] Morales A L, Nieto A J, Chicharro J M and Pintado P A, 2018, A semi-active vehicle suspension based on pneumatic springs and magnetorheological dampers, *Journal of Vibration and Control*, 24(4): 808–821.
[7] Li L, Lu Y, Wang R and Chen J, 2017, A three-dimensional dynamics control framework of vehicle lateral stability and rollover prevention via active braking with MPC, *IEEE Transactions on Industrial Electronics*, 64 (4): 3389–3401.
[8] Konik D, 2002, Development of the Dynamic Drive for the new 7 Series of the BMW Group, *International Journal of Vehicle Design*, 28(1–3): 131–149.
[9] Strassberger M and Guldner J, 2004, BMW's dynamic drive: an active stabilizer bar system, *IEEE Control Systems*, 24(4): 28–29.
[10] Buma S, Ookuma Y, Taneda A, Suzuki K, Cho J S and Kobayashi M, 2010, Design and development of electric active stabilizer suspension system, *Journal of System Design and Dynamics*, 4(1): 61–76.
[11] Yoon J, Yim S, Cho W, Koo B, Yi K, 2010, Design of an unified chassis controller for rollover prevention, manoeuvrability and lateral stability, *Vehicle System Dynamics*, 48(11): 1247–1268.
[12] Kong Z X, Pi D W, Wang X, Wang H and Chen S, 2016, Design and evaluation of a hierarchical control algorithm for an electric active stabilizer bar system, *Strojniški vestnik—Journal of Mechanical Engineering*, 62(10): 565–576.
[13] Sun H, Chen Y H, Zhao H and Zhen S, 2018, Optimal design for robust control parameter for active roll control system: a fuzzy approach, *Journal of Vibration and Control*, 24(9): 4575–4591.
[14] Kim H J, Yang H S and Park Y P, 2002, Robust roll control of a vehicle: Experimental study using a hardware-in-the-loop set-up, *Proceedings of the Institution of Mechanical Engineers, Part D: Journal of Automobile Engineering*, 216(1): 1–9.
[15] Kim H J and Park Y P, 2004, Investigation of robust roll motion control considering varying speed and actuator dynamics, *Mechatronics*, 14(1): 35–54.
[16] Yim S, 2012, Design of a robust controller for rollover prevention with active suspension and differential braking, *Journal of Mechanical Science and Technology*, 26(1): 213–222.
[17] Sampson D J M and Cebon D, 2003, Achievable roll stability of heavy road vehicles, *Proceedings of the Institution of Mechanical Engineers, Part D: Journal of Automobile Engineering*, 217(4): 269–287.
[18] Stone E J and Cebon D, 2008, An experimental semi-active anti-roll system, *Proceedings of the Institution of Mechanical Engineers, Part D: Journal of Automobile Engineering*, 222(12): 2415–2433.
[19] Stone E J and Cebon D, 2010, Control of semi-active anti-roll systems on heavy vehicles, *Vehicle System Dynamics*, 48(10): 1215–1243.
[20] Suzuki S, Buma S, Urababa S, Nishihara A and Taneda A, 2006, Development of electric active stabilizer suspension system, *SAE Technical Paper*, 2006-01-1537.

[21] Ohkuma Y, Buma S, Urababa S, *et al.*, 2006, Development of electric active stabilizer suspension system for Lexus GS, *Toyota Technical Review*, 54(2): 88.

[22] Cho J S, Buma S and Doi S, 2007, The application of CAE tools in the development of an Active Stabilizer Suspension System, *Control, Automation and Systems, 2007, ICCAS'07*, International Conference on, IEEE, 91–95.

[23] Ohta Y, Kato H, Yamada D, *et al.*, 2006, Development of an electric active stabilizer system based on robust design, *SAE Technical Paper*, 2006-01-0758.

[24] Beijing West Industries (BWI) Co, Ltd, 2018, http://www.bwigroup.com/product/asbs-active-stabilizer-bar-systems/.

[25] Hac A, Brown T and Martens J, 2004, Detection of vehicle rollover, *SAE Technical Paper*, 2004-01-1757.

[26] Rabinow J, 1948, The magnetic fluid clutch, *Electrical Engineering*, 67(12): 1167–1167.

[27] Li W H, Chen G and Yeo S H, 1999, Viscoelastic properties of MR fluids, *Smart Materials and Structures*, 8(4): 460.

[28] Claracq J, Sarrazin J and Montfort J P, 2004, Viscoelastic properties of magnetorheological fluids, *Rheologica Acta*, 43(1): 38–49.

[29] Chen P, Qian L J, Bai X X and Choi S B, 2017, Velocity-dependent characteristics of magnetorheological fluids in squeeze mode considering the hydrodynamic and the magnetic field interactions, *Journal of Rheology*, 61 (3): 455–465.

[30] Dong X, Yu M, Li Z, Liao C and Chen W, 2008, Neural network compensation of semi-active control for magneto-rheological suspension with time delay uncertainty, *Smart Materials and Structures*, 18(1):015014.

[31] Yu M, Dong X M, Choi S B and Liao CR, 2009, Human simulated intelligent control of vehicle suspension system with MR dampers, *Journal of Sound and Vibration*, 319(3–5): 753–767.

[32] Bai X X and Wereley N M, 2014, A fail-safe magnetorheological energy absorber for shock and vibration isolation, *Journal of Applied Physics*, 115 (17): 17B535.

[33] Dutta S and Choi S B, 2016, A nonlinear kinematic and dynamic modeling of Macpherson suspension systems with a magneto-rheological damper, *Smart Materials and Structures*, 25(3): 035003.

[34] Jolly M R, Bender J W and Carlson J D, 1998, Properties and applications of commercial magnetorheological fluids, *Smart Structures and Materials 1998: Passive Damping and Isolation*, International Society for Optics and Photonics, 3327: 262–276.

[35] Bai X X, Jiang P and Qian L J, 2017, Integrated semi-active seat suspension for both longitudinal and vertical vibration isolation, *Journal of Intelligent Material Systems and Structures*, 28(8): 1036–1049.

[36] Chen P, Bai X X, Qian L J and Choi S B, 2016, A magneto-rheological fluid mount featuring squeeze mode: analysis and testing, *Smart Materials and Structures*, 25(5): 055002.

[37] Xin F L, Bai X X and Qian L J, 2017, Principle, modeling, and control of a magnetorheological elastomer dynamic vibration absorber for powertrain mount systems of automobiles, *Journal of Intelligent Material Systems and Structures*, 28(16): 2239–2254.

[38] Qian L J, Xin F L, Bai X X and Wereley N M, 2017, State observation–based control algorithm for dynamic vibration absorbing systems featuring magnetorheological elastomers: principle and analysis, *Journal of Intelligent Material Systems and Structures*, 28(18): 2539–2556.

[39] Hiemenz G J, Hu W and Wereley N M, 2008, Semi-active magnetorheological helicopter crew seat suspension for vibration isolation, *Journal of Aircraft*, 45(3): 945–953.

[40] Bai X X, Shen S, Wereley N M and Wang D H, 2018, Controllability of magnetorheological shock absorber. Part I: Insights, modeling and simulation, *Smart Materials and Structures*, 28(1): 015022.

[41] Bai X X, Deng X C and Shen S, 2018, Controllability of magnetorheological shock absorber. Part II: Testing and analysis, *Smart Materials and Structures*, 28(1): 015023.

[42] Bai X X, Zhong W M, Zou Q, Zhu A D and Sun J, 2018, Principle, design and validation of a power-generated magnetorheological energy absorber with velocity self-sensing capability, *Smart Materials and Structures*, 27(7): 075041.

[43] ISO 3888-2:2002, 2003, Passenger cars-test track for a severe lane-change manoeuvre, British Standards Institute.

Chapter 7
Hybrid active and semi-active seat suspension

Donghong Ning[1], *Haiping Du*[1] *and Weihua Li*[2]

The seat suspension plays an important role in improving drivers' ride comfort and protecting their health from the vibration caused by the rough road [1,2]. Especially, heavy-duty vehicles work in severe conditions where the vibration magnitude is always higher than passenger vehicles and their drivers need to be exposed in the vibration for a long working time. Much research has been done on the vibration of heavy-duty vehicles by studying their seat suspensions. Generally, the highest vibration magnitude of the heavy-duty vehicle seat is in the frequency range of 2 to 4 Hz, and heavy-duty vehicle drivers usually experience vibration around 3 Hz which will increase fatigue and drowsiness [3]. The semi-active seat suspension which can change the damping in real time has been widely studied with the electrorheological fluid damper and the magnetorheological fluid (MRF) damper [4–7]; the vibration magnitude at resonance can be easily suppressed with these semi-active dampers. The active seat suspension can reduce the vibration magnitude by outputting an active force. It applied several kinds of active actuators, e.g. a hydraulic absorber and a controlled air spring [8], an electric servomotor with a ball screw mechanism [9], the electromagnetic linear actuators [10] and rotary motors [11,12]. The active and semi-active seat suspensions have their own advantages and disadvantages. The active seat suspension has better performance in vibration control. However, the cost and energy consumption, which are closely related to the active actuator's maximum continuous force output, are much higher than the semi-active one. The semi-active seat suspension has limited performance in vibration control.

In recent years, some research into active and semi-active seat suspension control has been conducted, e.g. H_∞ control [13], linear–quadratic–Gaussian control [14,15] and adaptive control [16]. A new adaptive fuzzy sliding mode controller is proposed for magnetorheological (MR) damper [17]. An active vibration control strategy is proposed based on a primary controller and actuator's reverse dynamics [18]. In practical application, the relative seat suspension displacement

[1]School of Electrical, Computer and Telecommunications Engineering, University of Wollongong, Australia
[2]School of Mechanical, Materials, Mechatronic, and Biomedical Engineering, University of Wollongong, Australia

and acceleration can be easily measured. When the controller of a seat suspension is designed, the measurable variables should be considered as feedbacks.

In this chapter, a hybrid active and semi-active seat suspension for heavy-duty vehicles is experimentally investigated to fill the performance gap between the two kinds of seat suspensions [19]. An active actuator with low continuous force output, which is insufficient for an active seat suspension system, is applied together with a semi-active MRF damper. It has several advantages. First, the energy consumption will be less than a merely active seat suspension. Second, the benefits of active and semi-active seat suspension in vibration control will be integrated. The MRF damper can assist the active actuator to suppress the high resonance vibration with low energy consumption, and at the higher frequency (2–4 Hz) vibration, the semi-active seat suspension cannot further improve the seat suspension performance, but the active actuator further reduces the vibration with small force output. Third, it is a fail-safe system because of the semi-active actuator; the failure of the active actuator will have less influence on the system security. Additionally, the low-power active actuator has a small size and a low current requirement which can benefit the installation and application of the hybrid system in an existing semi-active seat suspension. In this study, the proposed hybrid controller only applies measurable variables in practical application.

Notation: I is used to denote the identity matrix of appropriate dimensions. The symbol * indicates the transposed element in the symmetric position of a matrix.

7.1 Hybrid active and semi-active seat suspension design and prototype

7.1.1 Motivation

This hybrid active and semi-active seat suspension is inspired by the research of an active seat suspension with active motor [11] and a semi-active seat suspension with MRF damper [20]. The resonance frequency of a seat suspension, which is depended on the spring stiffness and the loaded mass, is often set to lower than 2 Hz by the manufacturer. The vibration from the vehicle cabin floor will be amplified at the resonance frequency by the seat suspension, and attenuated at frequencies above the resonance; the vibration amplification and attenuation magnitude can be controlled by the suspension damping. We mainly discuss two categories of vibrations on the seat suspension for heavy-duty vehicles, the resonance period (generally 1–2 Hz vibration) and the isolation period (2–4 Hz vibration). For a passive seat suspension, as its resonance frequency is designed to lower than 2 Hz, the ride comfort can be obtained when seat suspension is capable to isolate vibration at 2–4 Hz which has the highest magnitude in heavy-duty vehicles. The comparison of the active and semi-active suspension performance is shown in Table 7.1. The semi-active seat suspension can vary its damping within a high and a low values; the high damping requires a current to energize the coil in the MRF damper. The energy is consumed when the semi-active seat suspension suppresses

Table 7.1 The comparison of semi-active and active seat suspension performance

	Semi-active seat suspension	**Active seat suspension**
1–2 Hz vibration	Good performance in suppressing the resonance vibration	Effective
	Energy consumed	High energy consumption
2–4 Hz vibration	Keep the lowest damping of the seat suspension	Effective
	Almost no energy consumed	Low energy consumption

the high magnitude resonance vibration where the active seat suspension needs much higher energy consumption. However, with 2–4 Hz vibration, the active seat suspension only requires a small force output to further reduce the vibration and the semi-active one can only keep the performance of when it is with its lowest damping. The active actuator's price and energy consumption are closely related to its highest continuous force output. Therefore, a small power actuator can be applied to reduce the 2–4 Hz vibration, and with the assistance of a semi-active damper, the resonance vibration can be also handled with. Two actuators will certainly increase the installation complexity. However, the small power actuator will be cheaper than the actuator for a barely active system, thus the total cost of the hybrid system maybe not higher than the active one. The energy consumption of the hybrid system will certainly be lower than the active one, and its performance will be much better than the semi-active one.

7.1.2 Prototype

A hybrid active and semi-active seat suspension prototype is designed and built (see Figure 7.1). This prototype is a modification of a normal commercial vehicle seat (GARPEN GSSC7) where the original damper has been removed. An active rotary motor (Panasonic MSMJ042G1U) and a semi-active rotary MR damper are installed on scissors structures centre of the suspension's two sides, respectively. The rated torque output of the motor is 1.3 N m and its power is 400 W. Gear reducers, with ratio 20:1 and 8:1, are applied to amplify torque output of the active motor and semi-active MR damper. The active rotary motor with its gear reducer is the active actuator; similarly, the rotary MR damper with its gear reducer is the semi-active actuator.

Figure 7.2 shows the structural design of the applied rotary MR damper. This rotary MR damper mainly consists of a cylinder, a coil, a shaft, a rotor and the enclosed reservoir that is filled with MRF. Unlike the linear MR damper, there is no spring or gas chamber to provide the initial pressure in the rotary MR damper. For the consideration of magnetic circuit design, the rotor is made of low-carbon steel and the shaft is made of aluminium. When the coil, which is composed by winding the copper wire around the rotor, is energized, the magnetic circuit is formed as shown in Figure 7.2. This design can guarantee that the magnetic flux passes through the MRF.

Figure 7.1 The hybrid active and semi-active seat prototype

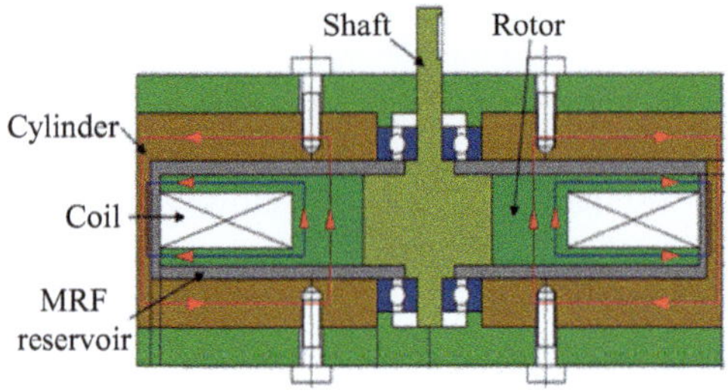

Figure 7.2 Rotary MR damper schematic

7.2 The seat suspension prototype test and model identification

7.2.1 Testing method

The dynamic properties of the seat suspension system were tested and evaluated with an MTS machine (Load Frame Model: 370.02, MTS Systems Corporation) as shown in Figure 7.3. The seat suspension is fixed between the upper and lower gripers of the MTS machine. When the testing started, the upper griper will force the suspension to move in accordance with a preprogrammed sinusoidal routine. The response force of the seat suspension can be measured and saved with the movement routine together.

7.2.2 Test results

The field, amplitude and frequency-dependent performance of the seat suspension are evaluated with the above-described test setup. For the field-dependent test, different currents were chosen to energize the magnetic field with same sinusoidal movement routine (with 10 mm amplitude and 1 Hz frequency). Figure 7.4 shows the force–displacement relationships of the seat suspension with different currents.

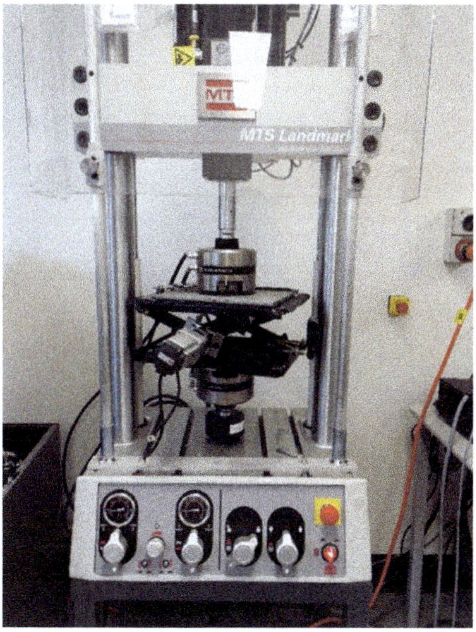

Figure 7.3 Test system

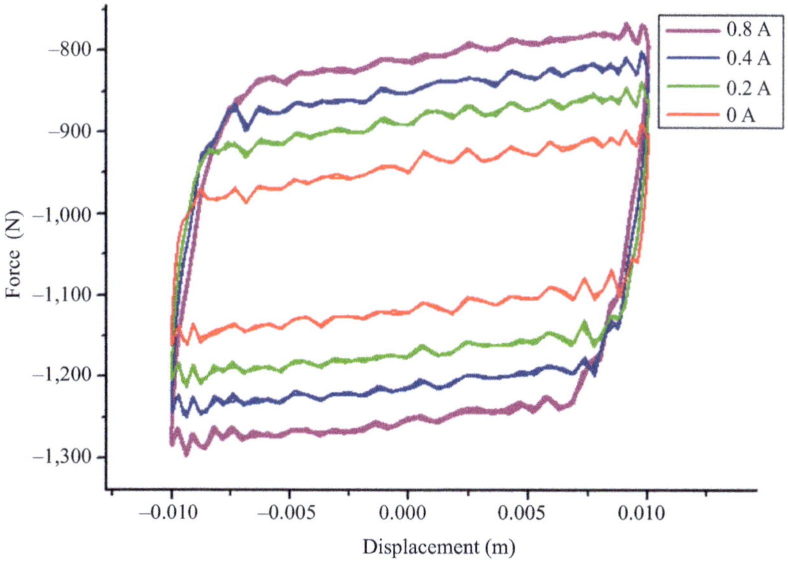

Figure 7.4 Test results with different currents

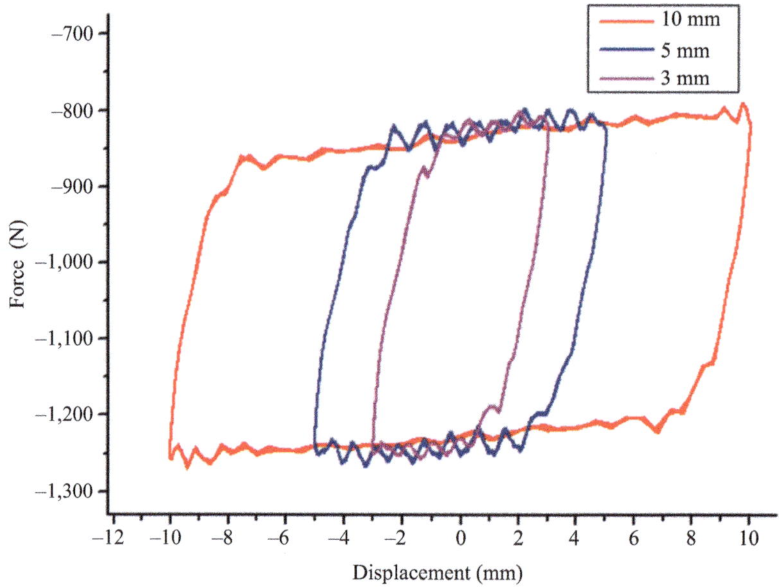

Figure 7.5 Test results with different amplitudes

As the area of the enclosed force–displacement loops can indicate the system damping, it can be seen that the damping of this seat suspension can be controlled by the applied current.

The amplitude-dependant performance of the seat suspension is shown in Figure 7.5, when the seat suspension was loaded with sinusoidal signals with different amplitudes (3, 5 and 10 mm) at a constant frequency (1 Hz) and current (0.5 A). The response force of seat suspension is nearly equal in 0 mm where the spring force of seat suspension is 0 N. It indicates that with the same current, the saturation force of the MR damper is equal with different velocity. Similarly, the effects of changing frequencies are presented in Figure 7.6, where the amplitude and current are constants with 10 mm and 0.5 A, respectively. It can be seen that the maximum force has slight variations with different frequencies.

The above experiments show that the semi-active actuator of the hybrid seat suspension is controllable and has stable properties in different velocities. On the other hand, the torque output of the active actuator can be accurately controlled by the motor drive. As the transformation of active actuator from torque output to vertical force output can be easily implemented [11], the whole seat suspension system is controllable.

7.2.3 Model identification

As two gear reducers are applied in the system, the friction force in the system cannot be ignored. The seat suspension model is shown in Figure 7.7, where z_s and z_v are, respectively, the displacement of upper and lower platform of the seat

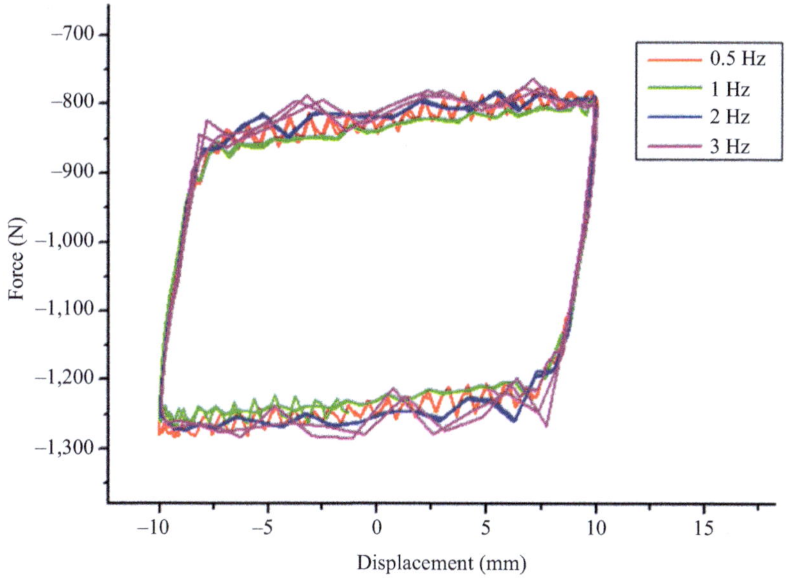

Figure 7.6 Test results with different frequencies

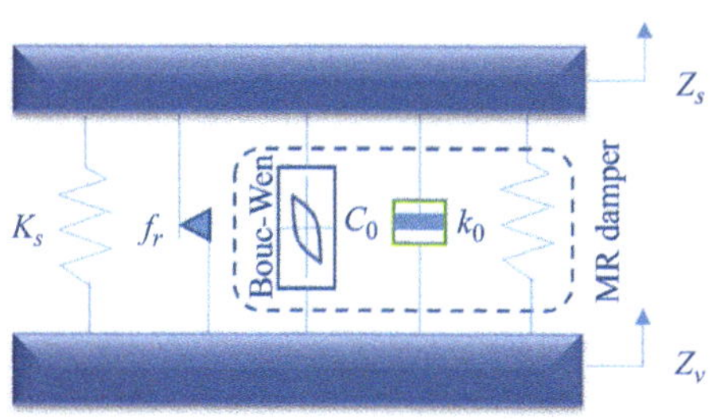

Figure 7.7 Seat suspension model

suspension, k_s is the spring stiffness and f_r is the inner friction. The Bouc–Wen model is applied to describe the MR damper. Considering that the friction has similar hysteresis phenomenon as MR damper, the inner friction is modelled into MR damper force together.

The seat suspension model is defined as follows:

$$x_d = z_s - z_v \tag{7.1}$$

$$F_t = k_s x_d + f_r + F_{mr} \tag{7.2}$$

$$F_d = f_r + F_{mr} = c_1 \dot{y}_d \tag{7.3}$$

$$\dot{y}_d = \frac{1}{c_0 + c_1}\left[\alpha z_d + c_0 \dot{x}_d + k_0 (x_d - y_d)\right] \tag{7.4}$$

$$\dot{z_d} = -\gamma_d|\dot{x_d} - \dot{y_d}|z_d|z_d|^{n-1} - \beta_d(\dot{x_d} - \dot{y_d})|z_d|^n + A_d(\dot{x_d} - \dot{y_d}) \tag{7.5}$$

$$\alpha = \alpha_1 + \alpha_2 I + \alpha_3 I^2 \tag{7.6}$$

$$c_0 = c_{01} + c_{02} I + c_{03} I^2 \tag{7.7}$$

$$c_1 = c_{11} + c_{12} I + c_{13} I^2 \tag{7.8}$$

where I is the current applied to the MR damper. The parameters γ_d, β_d, A_d, n, k_0, α_1, α_2, α_3, c_{01}, c_{02}, c_{03}, c_{11}, c_{12} and c_{13} are used to characterize the MR damper [21,22]. The optimized values of those parameters are determined by fitting the model with experimental data using MATLAB® parameter estimation toolbox. Table 7.2 shows the identified model parameters. The fitting result in Figure 7.8 shows the simulation data matches the experimental data very well.

Table 7.2 The identified model parameters

γ_d	7.263e + 05	α_3	-1.005e + 05
β_d	7.359e + 05	c_{01}	95.873
A_d	1.860	c_{02}	9.428e + 02
n	2	c_{03}	-5.249e + 02
k_s	4,600	c_{11}	1.338e + 04
k_0	3.519e + 02	c_{12}	-5.981e+04
α_1	7.553e + 04	c_{13}	3.078e + 05
α_2	1.972e + 05		

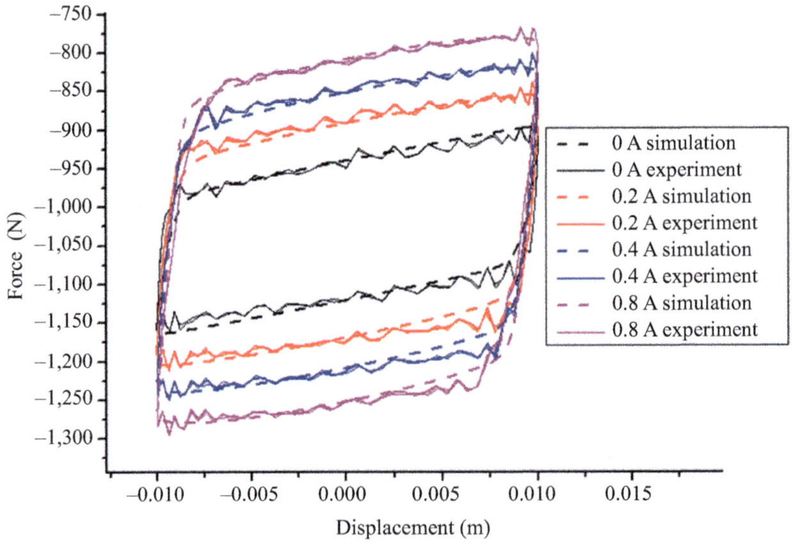

Figure 7.8 The fitting result of model

7.3 Control algorithm

This section proposes a control algorithm for the hybrid active and semi-active seat suspension. This algorithm is based on a modified semi-active on–off control method with the compensation of small active force on the off state of the MR damper. The relative displacement and seat acceleration, which can be easily obtained in practical application, are used as control feedback.

7.3.1 Hybrid seat suspension model

The hybrid seat model is shown in Figure 7.9, where m is the mass loaded on the suspension, F_d includes inner friction and variable damping force and u is the output of active force. The dynamic model of the system can be described as

$$m\ddot{z}_s = -k_s(z_s - z_v) - F_d + u \tag{7.9}$$

In the practical application, the suspension relative displacement and seat acceleration can be measured; the suspension relative velocity can be obtained from the differential of relative displacement. Therefore, the state variables are chosen as $\mathbf{X} = \begin{bmatrix} z_s - z_v & \dot{z}_s - \dot{z}_v \end{bmatrix}^{\mathrm{T}}$, the vibration disturbance is $d = \ddot{z}_v$ and the measurement variables are $\mathbf{Y_1} = \begin{bmatrix} z_s - z_v & \dot{z}_s - \dot{z}_v \end{bmatrix}^{\mathrm{T}}$ and $Y_2 = \ddot{z}_s$. Thus, combing with (7.9) the system model is defined as

$$\dot{\mathbf{X}} = \mathbf{AX} + \mathbf{B}_1(u - F_d) + \mathbf{B}_2 d \tag{7.10}$$

$$\mathbf{Y}_1 = \mathbf{C}_1\mathbf{X} \tag{7.11}$$

$$Y_2 = \mathbf{C}_2\mathbf{X} + D_2(u - F_d) \tag{7.12}$$

where $\mathbf{A} = \begin{bmatrix} 0 & 1 \\ -\frac{k_s}{m} & 0 \end{bmatrix}$, $\mathbf{B}_1 = \begin{bmatrix} 0 \\ \frac{1}{m} \end{bmatrix}$, $\mathbf{B}_2 = \begin{bmatrix} 0 \\ -1 \end{bmatrix}$, $\mathbf{C}_1 = \begin{bmatrix} 1 & 0 \\ 0 & 1 \end{bmatrix}$, $\mathbf{C}_2 = \begin{bmatrix} -\frac{k_s}{m} & 0 \end{bmatrix}$ and $D_2 = \frac{1}{m}$.

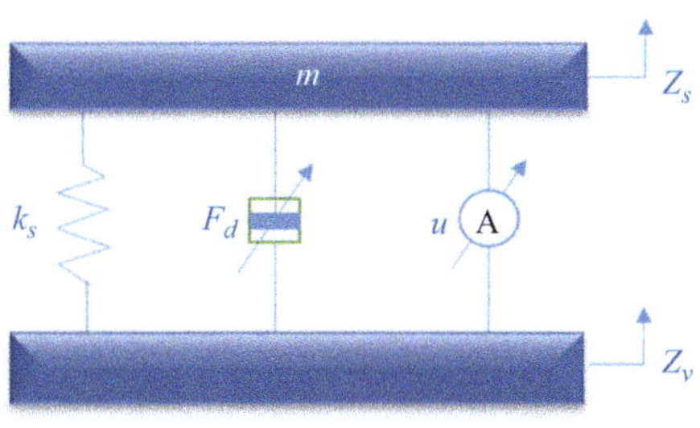

Figure 7.9 Hybrid active and semi-active seat model

The seat acceleration is the main optimization objective in the controller design. Therefore, the controlled output is defined as

$$Z_1 = \mathbf{C_3X} + D_3(u - F_d) \tag{7.13}$$

where $\mathbf{C_3} = \alpha\mathbf{C_2}$, $D_3 = \alpha D_2$ and α is a constant.

7.3.2 Controller design

First, the controllers for the two actuators are designed.

Controller 1

The modified on–off control method is a very efficient way for the semi-active actuator. A modified on–off controller for the MR damper is designed as $u_I = \text{sat}(I)$, where $\text{sat}(I)$ is a saturation function defined as

$$I = k_i\ddot{z}_s(\dot{z}_s - \dot{z}_v) \tag{7.14}$$

$$\text{sat}(I) = \begin{cases} I_{\max}, & I > I_{\max} \\ I & I_{\max} > I > 0 \\ 0, & \text{Others} \end{cases} \tag{7.15}$$

where k_i is positive constant and $I_{\max}$ is the maximum input current for MR damper. This controller is very intuitive. When the direction of the generated force of the MR damper is opposite to the inertia force $M\ddot{z}_s$, it can impede the increase of the acceleration. Thus, a large damping is required by loading current into the MR damper, otherwise it will keep its smallest state (smallest damping). For instant, if the directions of both $\ddot{z}_s$ and $\dot{z}_s - \dot{z}_v$ are positive, then the force of the MR damper is opposite to the inertia force. In this case, the acceleration can be reduced when the force of the MR damper is increased by loading current. On the contrary, when $\ddot{z}_s$ and $\dot{z}_s - \dot{z}_v$ have different directions, there should have no current input applied to the MR damper to keep the MR damper to be soft.

In the practical application, the measurement noise of sensors is inevitable. This modified on–off controller with saturation can attenuate the frequent variation of MR damper between maximum current and no current caused by the measurement noise.

Controller 2

The inner friction cannot be measured; and the force output of MR damper, which varies based on suspension relative velocity and the output of Controller 1, is also unmeasured. A disturbance observer is applied to estimate the unmeasurable force F_d.

$$\dot{\widehat{F_d}} = L\left[Y_2 - \left(\mathbf{C_2X} + D_2\left(u - \widehat{F_d}\right)\right)\right] \tag{7.16}$$

where the observer gain L is a constant to be designed.

An H_∞ controller with disturbance compensation is constructed as

$$u = \mathbf{KX} + \widehat{F_d} \tag{7.17}$$

where **K** is the state feedback gain to be designed.

The disturbance observer gain L and H_∞ controller gain **K** can be obtained by solving following linear matrix inequality (LMI) with MATLAB LMI toolbox:

$$\begin{bmatrix} \mathbf{AQ} + \mathbf{B_1R} + (\mathbf{AQ} + \mathbf{B_1R})^{\mathrm{T}} & * & * & * \\ -\mathbf{B_1}^{\mathbf{T}} & (\mathbf{G}D_2)^{\mathbf{T}} + \mathbf{G}D_2 & * & * \\ \mathbf{B_2}^{\mathbf{T}} & 0 & -\lambda^2 & * \\ \mathbf{C_3Q} + \mathbf{D_3R} & -D_3 & 0 & -\mathbf{I} \end{bmatrix} < 0 \tag{7.18}$$

where $\mathbf{Q} = \mathbf{Q}^{\mathbf{T}}, \mathbf{Q} > 0$, and λ is a given performance index. After solving (7.18) for matrices **Q**, **R** and *G*, the controller gain is obtained as $\mathbf{K} = \mathbf{RQ}^{-1}$, and the observer gain is $L = \mathbf{P_2}^{-1}\mathbf{G}$. The computation procedure is similar as that used in [11] and omitted here for brevity. In this study, L is chosen as −2,000; **K** is chosen as $[3,000 \quad -20]$.

Second, the two controllers are integrated together. Apparently, the MR damper can only do negative work which means the direction of the force generated by the MR damper should be opposite to the direction of suspension relative movement. When the negative work can dissipate the vibration energy and thus suppress the seat vibration, the output of Controller 1 is not equal to zero. On the contrary, when the seat suspension needs positive work to reduce seat vibration, the MR damper should keep its softest state to do least negative work due to its semi-active control characteristic. In the proposed hybrid seat suspension, when the MR damper keeps its softest state, the output of Controller 1 is equal to zero, but the motor will do positive work to reduce seat acceleration. Figure 7.10 shows the control strategy. The Controller 1 is the main controller which controls the current for rotary MR damper. The Controller 2's output will send to the active actuator when Controller 1's output is equal to zero. This control strategy can save the active actuator's energy consumption.

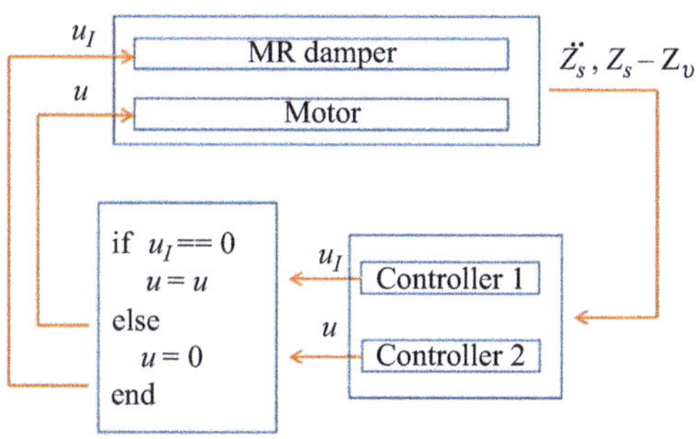

Figure 7.10 Hybrid controller

7.4 Evaluation

7.4.1 Numerical simulation

In this test, the performance of the hybrid active and semi-active controller applied to the seat suspension model is evaluated. The cushion model is also included in the simulation with a stiffness 90,000 N/m and damping 2,000 N s/m. The suspension and driver body mass are 28 and 70 kg, respectively. The saturated active force is chosen as 70 N which is insufficient to control the resonance vibration independently. The saturated semi-active control current is chosen as 0.7 A, and $k_i = 100$ is applied.

The harmonic excitation test, which was a sweep frequency signal from 1 to 3 Hz in 20 s with 30 mm amplitude, was first implemented. Figure 7.11 shows the seat absolute displacements of semi-active control compared with seat suspension with different currents (0, 0.1 and 0.7 A). The soft suspension (0 A current) has resonance around the 5 s and performs best in higher frequency. When 0.1 A current is loaded, the resonance amplitude has been greatly suppressed. Although the hard suspension (0.7 A current) has best performance around resonance frequency, it works worst in high frequency. The semi-active control suspension can successfully suppress the resonance vibration. At the same time, it performs close to a soft suspension (0 A current) at high frequency. This result indicates that the semi-active control can improve the seat suspension performance around the resonance frequency. Figure 7.12 shows the result comparison of the semi-active and hybrid control. The seat displacement of the hybrid control is further reduced in the whole test time range. Especially around the resonance frequency, the MR damper can

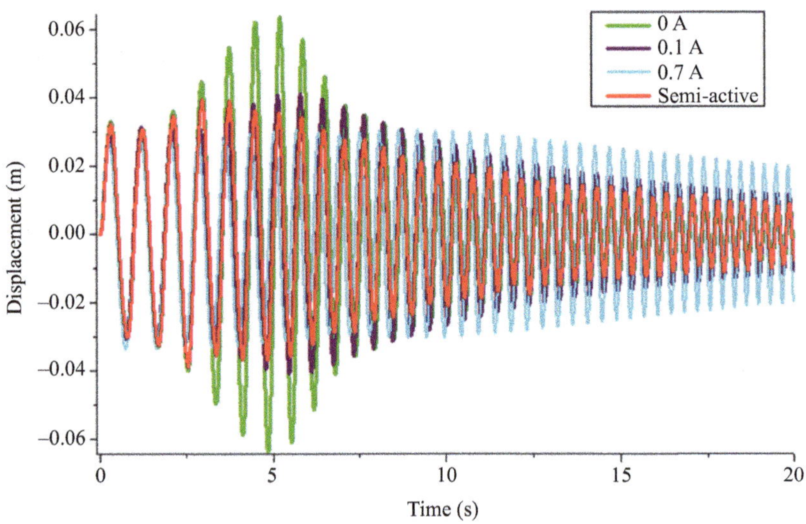

Figure 7.11 Semi-active control result comparison

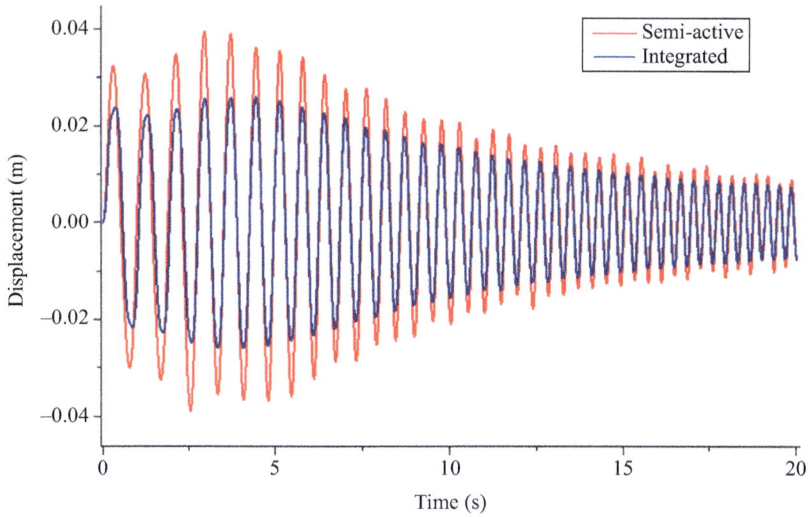

Figure 7.12 Hybrid control result comparison

Table 7.3 RMS acceleration of random vibration (m/s^2)

Hybrid	Semi-active	Power on	Power off
1.09	1.29	1.40	1.46

suppress the resonance vibration, and then a small active force can further isolate the vibration.

The random vibration test was implemented. Table 7.3 shows the root mean square (RMS) acceleration comparison, where the hybrid control has the lowest value, and has a 15.5 per cent reduction compared with semi-active control. The time domain and frequency domain acceleration in Figure 7.13 further validates the control algorithm. The hybrid control has the lowest power spectral density (PSD) value around the resonance frequency and has a low value in higher frequency vibrations.

7.4.2 Experimental setup

The experimental setup is shown in Figure 7.14. One six-degrees-of-freedom (6-DOF) vibration platform was applied to generate the vertical vibration excitation based on preprogrammed routine. The seat suspension was fixed on the upper platform of the 6-DOF vibration platform. An NI CompactRio 9074 was applied to acquire data and implement designed controller; the control frequency is 500 Hz. Two accelerometers (ADXL203EB) were used to measure the vertical seat acceleration and excitation acceleration. The suspension relative displacement was measured by a laser displacement sensor (Micro Epsilon ILD1302-100).

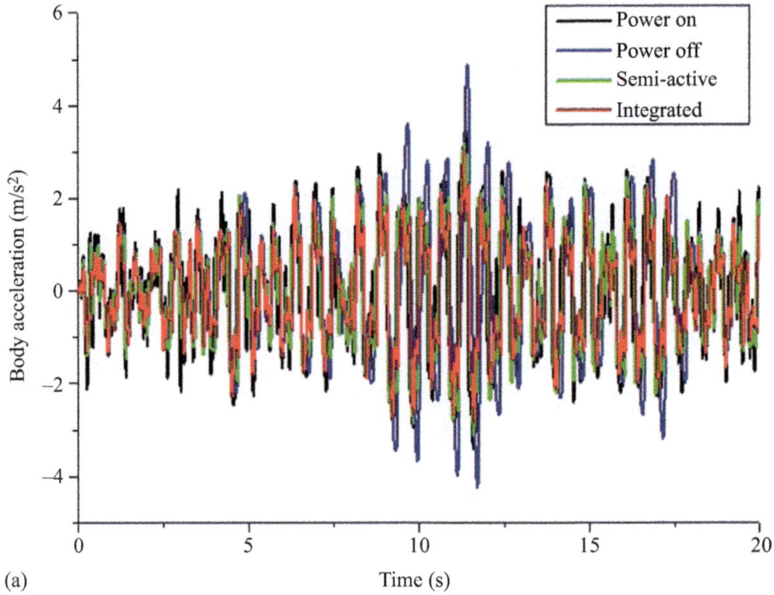

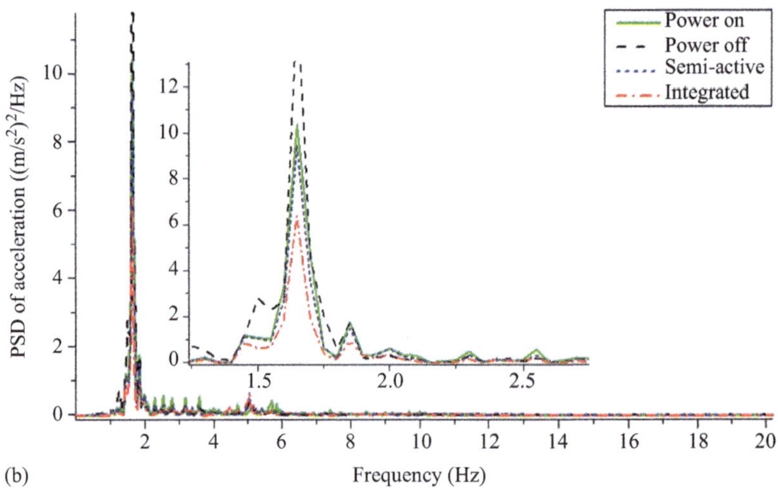

Figure 7.13 Acceleration with random vibration. (a) Time domain. (b) Frequency domain

The desired current signal is amplified by a power amplifier and then used to control the rotary MR damper. The active actuator applied in this prototype can output a maximum force of 150N. For verifying the proposed controller, the active force was limited as ±70N.

Three experiments are implemented. The passive experiment tests the seat suspension's passive response when the MR damper is loaded with different

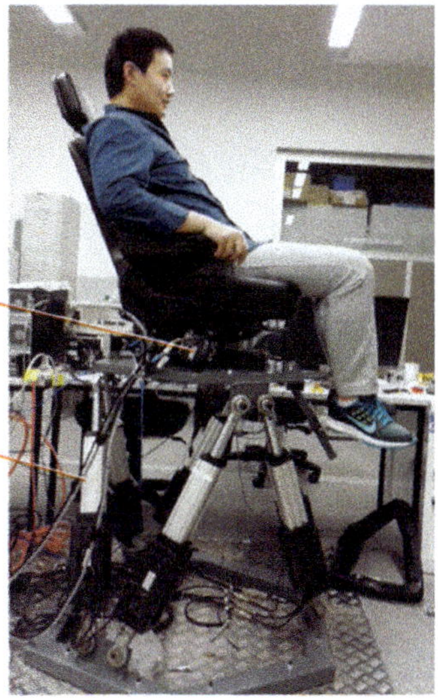

Figure 7.14 Experimental setup

currents (0, 0.1 and 0.7 A). In the semi-active control experiment, the semi-active controller is applied. In the hybrid control experiment, the proposed hybrid controller is implemented. It should be emphasized that the active actuator works only when the MR damper is in off state (0 A current). Disturbance may be introduced because the active actuator is turned on and off suddenly. For example, when MR damper turns off, and a 50 N active force is required, the motor output needs to change from 0 to 50 N. So, a single-order low-pass point-by-point filter module in Libview (software for developing program in NI CompactRio 9074) was applied to the active control output to avoid its sudden change.

7.4.3 Experimental results

The sinusoidal excitations were applied to the seat suspensions with 75 kg load. Figure 7.15 shows the seat acceleration comparison in both time and frequency domain with resonance frequency (1.6 Hz). When the MR damper is power off (0 A), the seat has maximum peak acceleration. The hybrid control has the best performance. When the vibration frequency is higher than resonance frequency range, a soft seat suspension should perform better than a hard suspension. Figure 7.16 shows the seat acceleration with 2.4 Hz vibration when the soft seat suspension (0 A) can successfully isolate vibration. When the MR damper is loaded with 0.1 A current, the peak acceleration will greatly increase. The performance of the semi-actively controlled seat suspension is better than the case of MR damper with 0.1 A current and worse than the case of MR damper with 0 A current. The hybrid control can greatly reduce the acceleration; its acceleration is even lower than the 0 A current case.

The transmissibility of seat suspension is shown in Figure 7.17. When the MR damper is powered with 0.7 A current, the vibration transmissibility is around 1.05

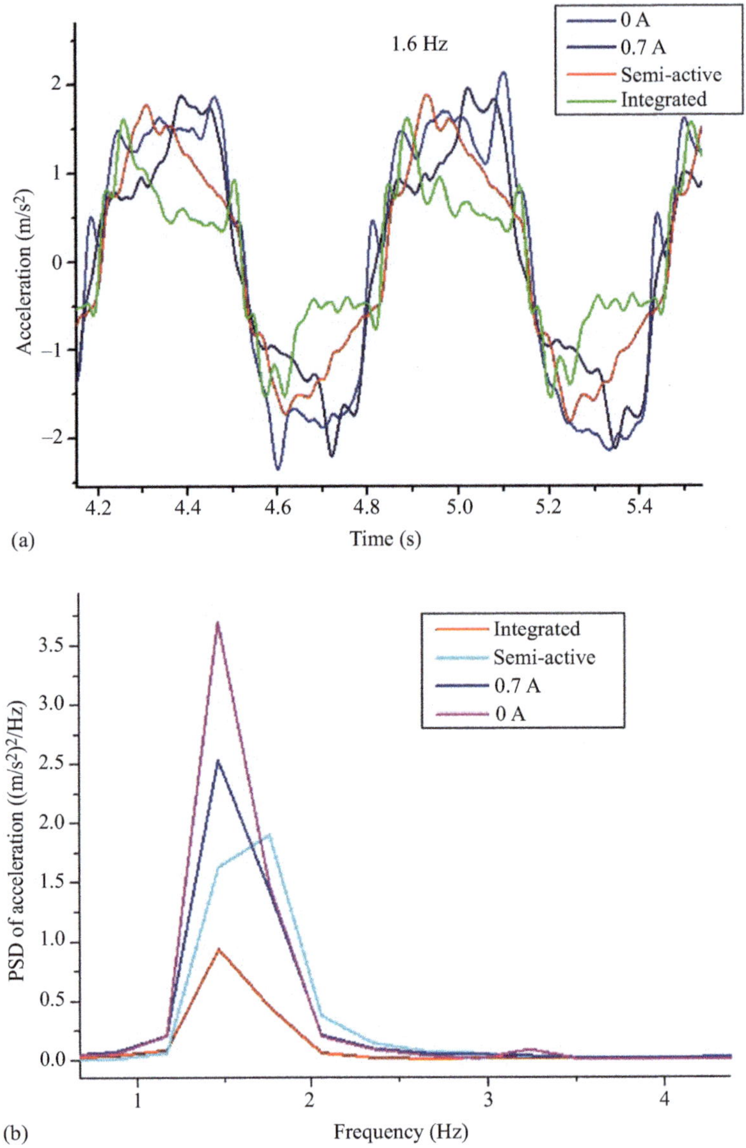

Figure 7.15 Seat acceleration at 1.6 Hz vibration. (a) Time domain; (b) Frequency domain

in the whole testing frequency range. On the contrary, when the MR damper is power off (0 A), the vibration transmissibility is highest in 1.6 Hz, and then going down. In other words, the soft seat suspension has resonance around 1.6 Hz and starts to isolate vibration when the vibration frequency is above resonance

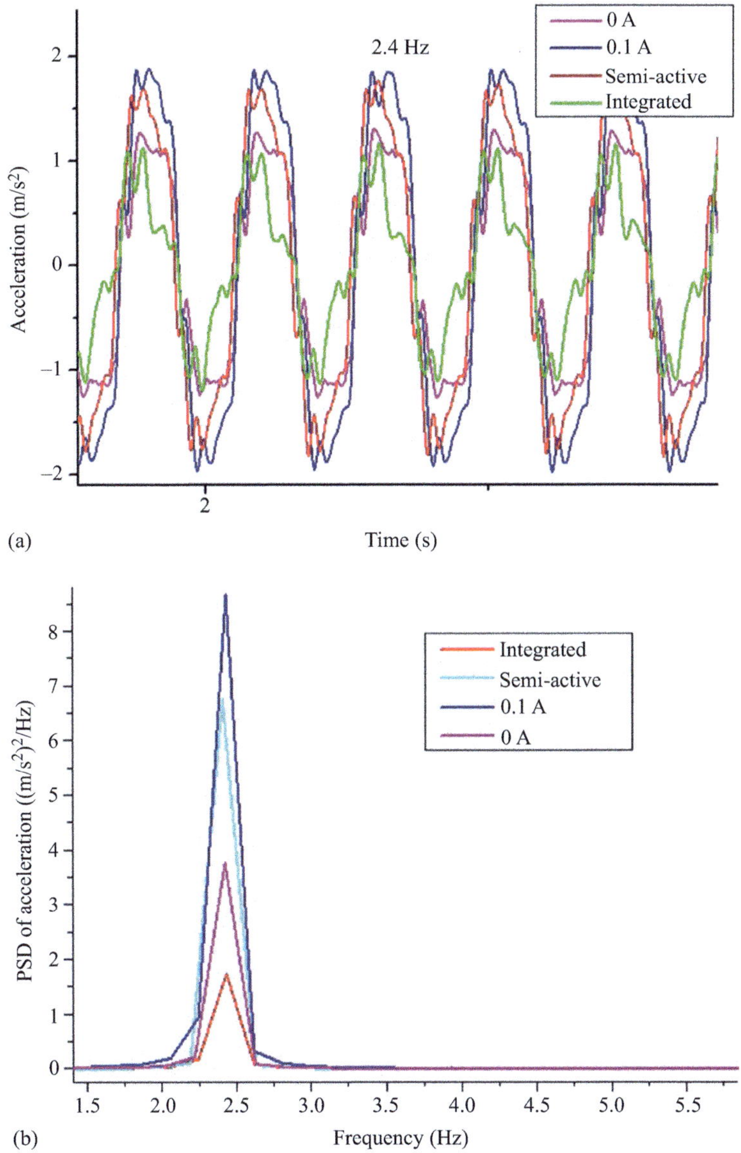

Figure 7.16 Seat acceleration at 2.4 Hz vibration. (a) Time domain; (b) Frequency domain

frequency. The semi-active control seat suspension can suppress the resonance vibration and even performs better than a hard (0.7 A) suspension in 1.6 Hz. The hybrid control has lowest transmissibility in 1.6 Hz and performs better than the soft seat suspension (0 A) in higher frequency. The experiments result verified

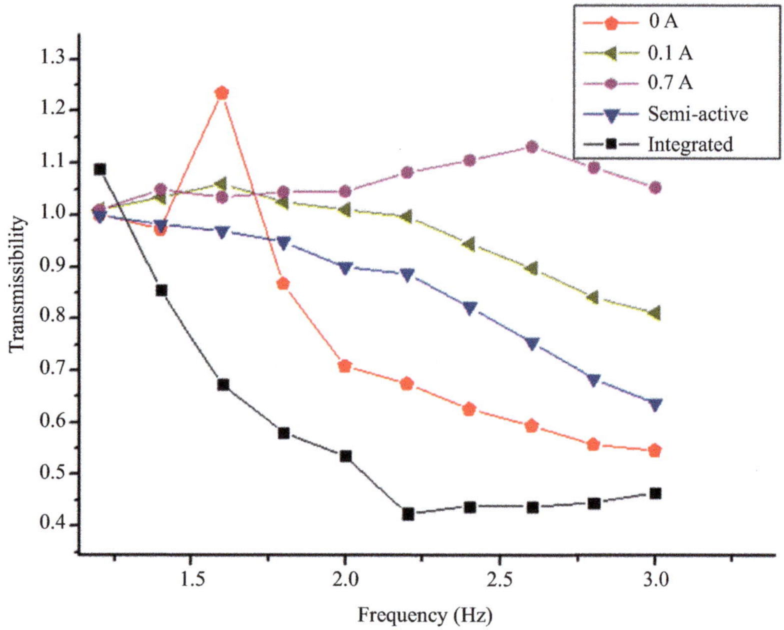

Figure 7.17 Transmissibility

the seat suspension's design idea and validated the proposed control algorithm. When the vibration is around resonance frequency, the semi-active control MR damper can dissipate much vibration energy, and then a small active force, which is insufficient to control vibration independently, can successfully further isolate the vibration to a low magnitude. In the higher frequency vibration, the required active force is relatively small, so the applied active force can reduce the vibration transmissibility to values lower than soft seat suspension (0 A). Generally, the semi-active control is hard to perform as good as a passive soft seat suspension in practical application.

The random vibration is often used to evaluate seat suspension performance in the time domain. Figure 7.18 shows the seat acceleration with random vibration. When the MR damper is power off (0 A), there is resonance around the 8 s. On the other hand, when the MR damper is power on (0.7 A), its acceleration is obviously larger than other cases except in the resonance range. Table 7.4 shows the RMS acceleration comparison. The acceleration of hybrid control seat suspension has 47.7 per cent and 33.1 per cent reductions when compared with power on and power off seat suspension, respectively. Furthermore, there is 26.5 per cent reduction of RMS acceleration when compared with semi-active control seat suspension. Figure 7.19 shows the PSD value of acceleration, which further indicates that the hybrid seat suspension with small active force can improve the performance of a semi-active control seat suspension greatly.

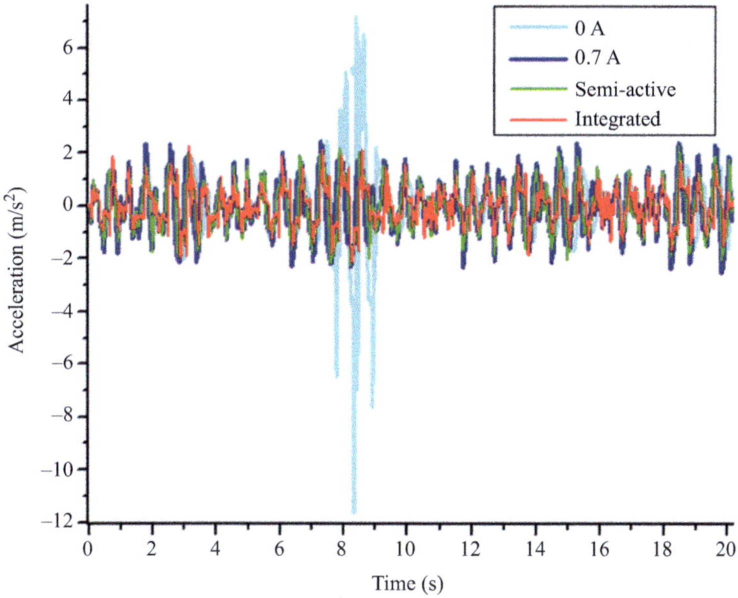

Figure 7.18 Seat acceleration with random vibration

Table 7.4 RMS acceleration of random vibration (m/s^2)

Power off	Power on	Semi-active	Hybrid
1.3054	1.0195	0.9296	0.6825

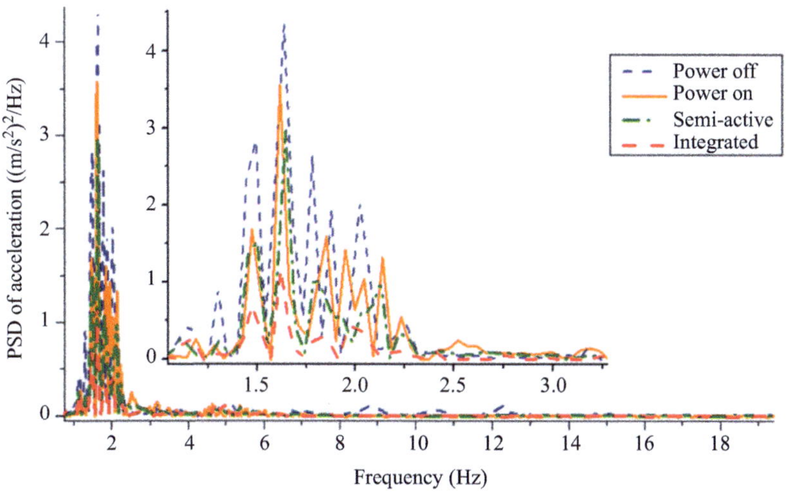

Figure 7.19 PSD of acceleration with random vibration

7.5 Conclusion

In this chapter, a hybrid active and semi-active seat suspension for heavy-duty vehicles has been proposed, built and tested. The proposed control algorithm for the seat suspension prototype applied measurable variables in practical, namely, the relative displacement and seat acceleration. The hybrid seat suspension has an active motor and a semi-active MR damper; gearboxes are applied to amplify the actuators torque output. The field, amplitude and frequency-dependent performance of the seat suspension have been tested with an MTS system. The tests results show that the semi-active actuator of the hybrid seat suspension is controllable. The simulation and experiment were implemented and both validated the proposed hybrid seat suspension and controller. In the random vibration experiment, the RMS acceleration of hybrid control seat suspension has 47.7, 33.1 and 26.5 per cent reductions when compared with power on, power off and semi-active control seat suspension, respectively. The acceleration PSD value further indicates that the hybrid seat suspension has best performance with random vibration. With the proposed seat suspension, the MR damper can suppress resonance vibration with low power consumption, and then a small active force can further isolate the resonance vibration; when vibration frequencies are above resonance frequency, the MR damper is difficult to further improve the ride comfort, and then the small active force can also improve the vibration isolation performance.

References

[1] H. Metered, Z. Šika, Vibration control of a semi-active seat suspension system using magnetorheological damper, in: *Mechatronic and Embedded Systems and Applications (MESA), 2014 IEEE/ASME 10th International Conference on, IEEE*, 2014, pp. 1–7.

[2] G.S. Paddan, M.J. Griffin, Effect of seating on exposures to whole-body vibration in vehicles, *Journal of Sound and Vibration*, 253 (2002) 215–241.

[3] N. Mabbott, G. Foster, B. McPhee, *Heavy Vehicle Seat Vibration and Driver Fatigue*, Department of Transport and Regional Services/Australian Transport Safety Bureau, 2001, Report No. CR 203: 1–35.

[4] S.-B. Choi, Y.-M. Han, Vibration control of electrorheological seat suspension with human-body model using sliding mode control, *Journal of Sound and Vibration*, 303 (2007) 391–404.

[5] S.-B. Choi, M.-H. Nam, B.-K. Lee, Vibration control of a MR seat damper for commercial vehicles, *Journal of Intelligent Material Systems and Structures*, 11 (2000) 936–944.

[6] G.J. Hiemenz, W. Hu, N.M. Wereley, Semi-active magnetorheological helicopter crew seat suspension for vibration isolation, *Journal of Aircraft*, 45 (2008) 945–953.

[7] X. Zhu, X. Jing, L. Cheng, Magnetorheological fluid dampers: a review on structure design and analysis, *Journal of Intelligent Material Systems and Structures*, 23 (2012) 839–873.

[8] I. Maciejewski, L. Meyer, T. Krzyzynski, The vibration damping effectiveness of an active seat suspension system and its robustness to varying mass loading, *Journal of Sound and Vibration*, 329 (2010) 3898–3914.

[9] M. Kawana, T. Shimogo, Active suspension of truck seat, *Shock and Vibration*, 5 (1998) 35–41.

[10] Z. Gan, A.J. Hillis, J. Darling, Adaptive control of an active seat for occupant vibration reduction, *Journal of Sound and Vibration*, 349 (2015) 39–55.

[11] D. Ning, S. Sun, H. Li, H. Du, W. Li, Active control of an innovative seat suspension system with acceleration measurement based friction estimation, *Journal of Sound and Vibration*, 384 (2016) 28–44.

[12] D. Ning, S. Sun, J. Zhang, H. Du, W. Li, X. Wang, An active seat suspension design for vibration control of heavy-duty vehicles, *Journal of Low Frequency Noise, Vibration and Active Control*, 35 (2016) 264–278.

[13] Y. Zhao, W. Sun, H. Gao, Robust control synthesis for seat suspension systems with actuator saturation and time-varying input delay, *Journal of Sound and Vibration*, 329 (2010) 4335–4353.

[14] S. Türkay, H. Akçay, Aspects of achievable performance for quarter-car active suspensions, *Journal of Sound and Vibration*, 311 (2008) 440–460.

[15] Y. He, J. McPhee, Multidisciplinary design optimization of mechatronic vehicles with active suspensions, *Journal of Sound and Vibration*, 283 (2005) 217–241.

[16] J. Darling, A. Hillis, Z. Gan, Adaptive control of an active seat for occupant vibration reduction, *Journal of Sound and Vibration*, (2015).

[17] S.-B. Choi, Y.-S. Lee, M.-S. Han, Vibration control of a vehicle's seat suspension featuring a magnetorheological damper based on a new adaptive fuzzy sliding-mode controller, *Proceedings of the Institution of Mechanical Engineers, Part D: Journal of Automobile Engineering*, (2015) 0954407015586678.

[18] I. Maciejewski, Control system design of active seat suspensions, *Journal of Sound and Vibration*, 331 (2012) 1291–1309.

[19] D. Ning, S. Sun, H. Du, W. Li, Integrated active and semi-active control for seat suspension of a heavy duty vehicle, *Journal of Intelligent Material Systems and Structures*, 29 (2018) 91–100

[20] S. Sun, D. Ning, J. Yang, H. Du, S. Zhang, W. Li, A seat suspension with a rotary magnetorheological damper for heavy duty vehicles, *Smart Materials and Structures*, 25 (2016) 105032.

[21] J. Yang, H. Du, W. Li, *et al.*, Experimental study and modeling of a novel magnetorheological elastomer isolator, *Smart Materials and Structures*, 22 (2013) 117001.

[22] H. Du, J. Lam, K. Cheung, W. Li, N. Zhang, Direct voltage control of magnetorheological damper for vehicle suspensions, *Smart Materials and Structures*, 22 (2013) 105016.

Chapter 8

Development of magnetorheological brake with magnetic coils placed on side housings

Nguyen Quoc Hung[1], *Nguyen Ngoc Diep*[2], *Le Duc Thang*[1] *and Choi Seung Bok*[3]

This chapter focuses on new development of MRB with magnetic coils placed on side housings (side-coil MRB). First, an overview of configurations of MRB is introduced; from this, the reason for development of side-coil MRB is figured out. Then, the chapter focuses on modeling, optimal design and evaluation of different configurations of side-coil MRB such as single side-coil, double side-coil and triple side-coil MR brakes. In addition, experimental works are conducted to validate performance of the proposed multiple side-coil MRB. The chapter is closed by a summary on the advantages and potential development of the side-coil MRBs.

8.1 Introduction

In last two decades, researches on development and application of brake featuring magnetorheological fluid (MRF) have interested many researchers. In the development of MRBs, the size, mass and maximum braking toque of the brake are important factors that should be taken into account. In order to improve these factors, various configurations of MRB have been proposed and reported in the literature. The most popular configuration of MRB is the disc-type MRB which is also the first one described in the seminal work of Rabinow [1]. This type of brake is characterized by simple structure, manufacture convenience and high compactness. Therefore, the disc-type MRBs are used in many studies [2–6]. Figure 8.1 shows a typical configuration of a disc-type MRB, in which the rotor of the brake is very thin that is referred as a rotary disc, made of magnetic material such as silicon steel. The disc is placed inside a cylindrical housing, on which a coil is placed. The housing is also made of magnetic material. When a current is applied to the coil, a magnetic field is generated and the MRF at the end faces of the disc is energized

[1]Faculty of Engineering, Vietnamese German University, Vietnam
[2]Department of Mechanical Engineering, Industrial University of Ho Chi Minh City, Vietnam
[3]Department of Mechanical Engineering, Inha University, South Korea

and becomes solid like. The shear friction between the rotating disc and the solidified MRF provides a braking torque to the shaft. Significant geometric dimensions of the disc-type MRB are also shown in Figure 8.1.

Similar to mechanical or hydraulic brakes, the drum-type MRBs have been considered in some studies. Huang *et al.* [7] give some of the analytical expressions about the drum-type MRBs, and LORD Corporation uses them as an example in a tutorial about the design of devices making use of MRF. Figure 8.2 shows a typical configuration of drum-type MRB. In the drum-type MRB, magnetic flux goes across the cylindrical MRF duct only. Obviously, drum-type MRB is suitable for such applications that the working space of the MRB is a long cylinder.

Nguyen *et al.* [8] have proposed a hybrid configuration of MRBs for a haptic wrist. This configuration is a combination of disc-type and drum-type. In the hybrid MRB, all parts of the housing are made of magnetic materials; therefore, the magnetic flux goes across both the cylindrical duct and end-face ducts of MRF. Figure 8.3 shows the hybrid MRB with one coil placed in the middle, while Figure 8.4 shows the hybrid MRB with two coils placed at two ends of the cylindrical housing.

In order to improve braking performance of MRB, especially braking torque, the MRB with a T-shaped rotor (T-type MRB), as shown in Figure 8.5, was proposed for motorcycle application [9]. With this configuration, the magnetic flux goes across both the MRF ducts at the leg and the flange of the T-shape rotor. Therefore, a high braking torque can be archived. The results showed that the braking torque generated by the T-type MRB can reach the required braking torque for a middle size motorcycle (430 N m), while its size is compact enough to replace conventional drum-type brake of the rear wheel.

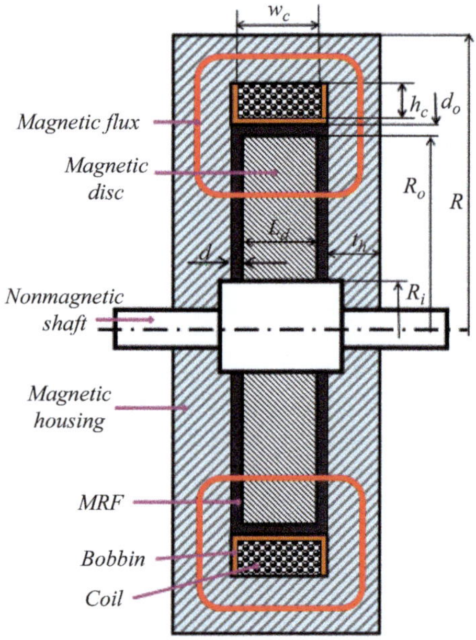

Figure 8.1 Conventional disc-type MRB

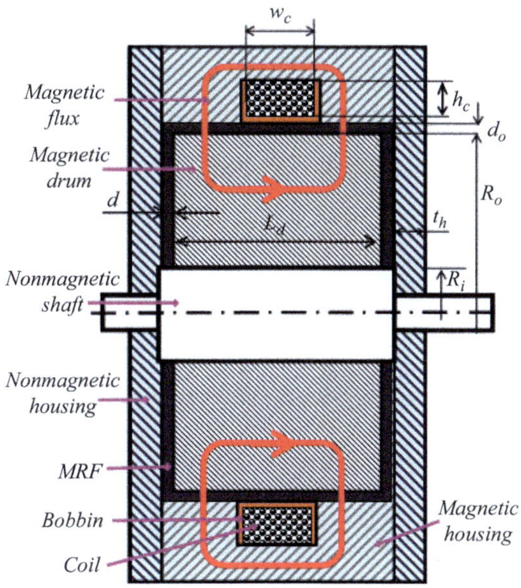

Figure 8.2 Conventional drum-type MRB

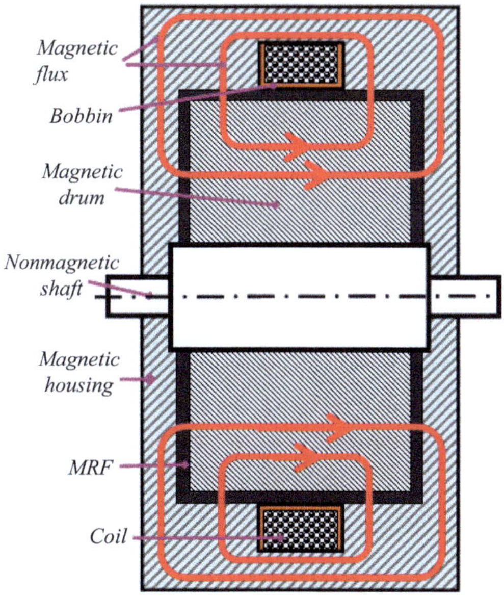

Figure 8.3 Hybrid MRB with one coil

In order to evaluate performance characteristics of the above MRB types, Nguyen *et al.* have performed researches on optimal design of different types of MRB [10]. The objective of the optimization is to maximize braking torque of the MRBs constrained in a specific volume. The results are shown in Figure 8.6. From the

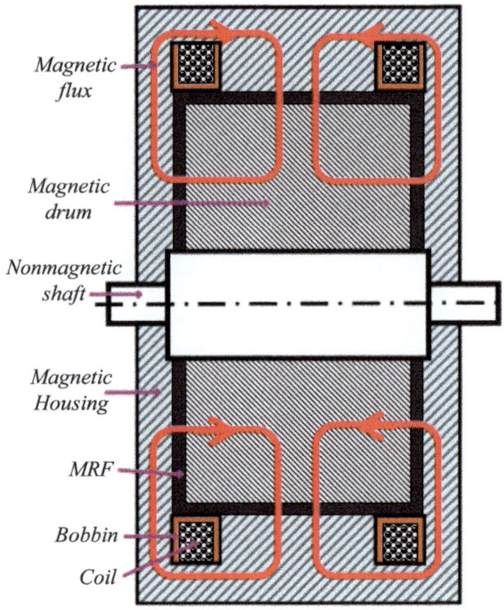

Figure 8.4 Hybrid MRB with two coils

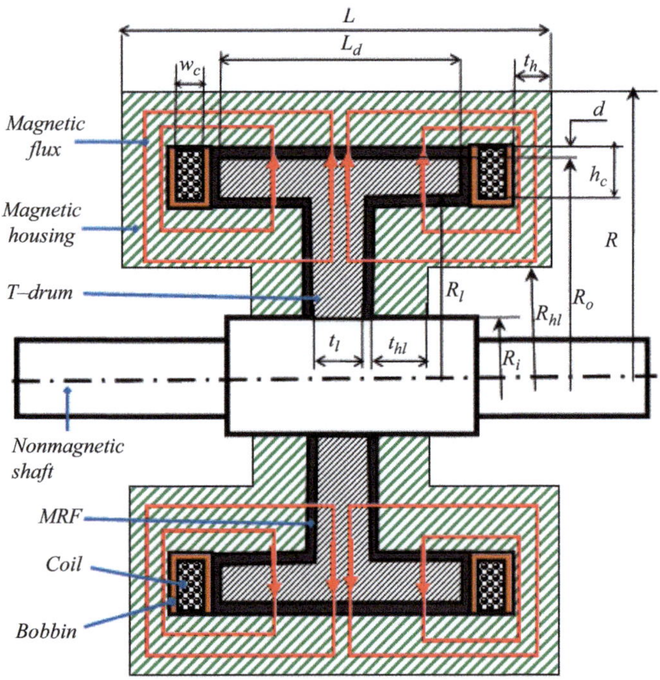

Figure 8.5 Conventional drum-type MRB

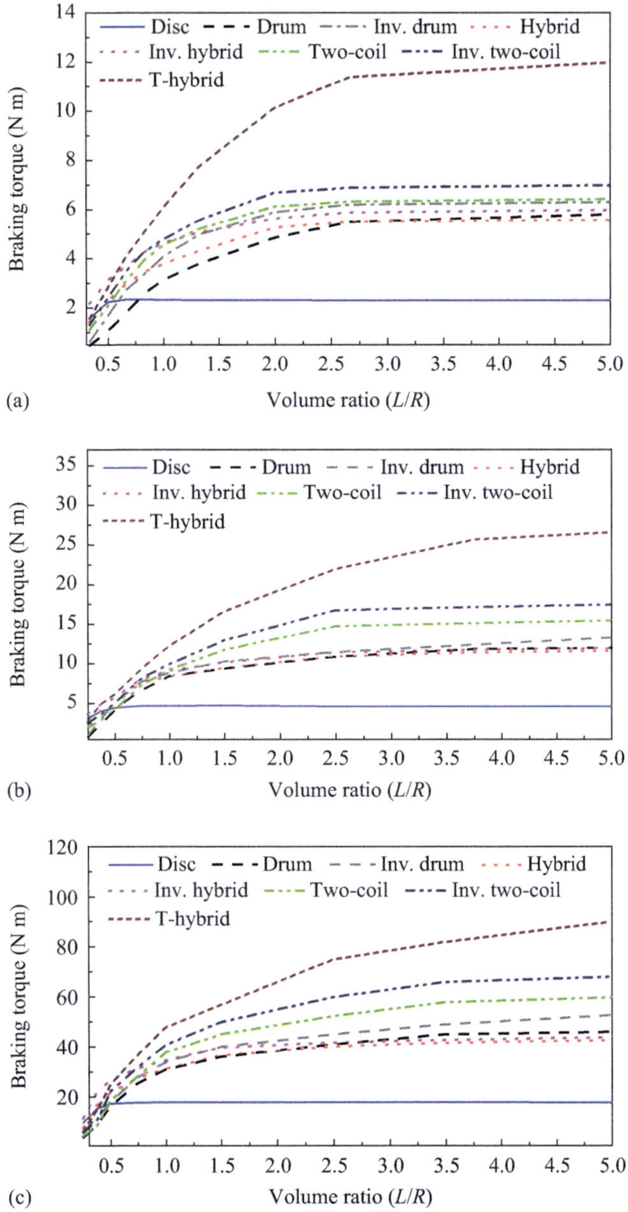

Figure 8.6 Braking torque of MRBs featuring MRF-132-DG as a function of volume ratio (L/R): (a) R = 30 mm, (b) R = 40 mm and (c) R = 60 mm

results, an optimal selection of MRB type, which can produce a maximum braking torque depending on the constrained volume of the brake, was given as following:

- When the geometry aspect *R/L* is smaller than 0.36, a disc-type of MRB should be employed taking simplicity of the brake into consideration.
- When the geometry aspect *R/L* is greater than 8.5, a T-type of MRB should be employed because its maximum braking torque is far greater than that of the others.
- When the geometry aspect *R/L* is from 0.36 to 0.75, a single-coil hybrid configuration of MRB should be used.
- When the geometry aspect *R/L* is from 0.75 to 1.5, the two-coil hybrid configuration and the T-type configuration of MRB should be considered. The T-type MRB can produce a higher braking torque (up to 130% of the two-coil hybrid configuration). However, the complexity and high cost of T-type MRB are also issues that should be taken into account.
- The drum-type MRBs should be considered only in case the geometry aspect *R/L* is greater than 2 and the high value of braking torque is not necessary.

In many applications of MRB-based system, performance of the system is significantly affected by the mass of MRBs such as haptic systems, robotics. Therefore, an optimal selection of MRB type considering braking torque and mass of the brakes is very significant. Recently, Nguyen *et al.* [11] performed a research on optimal design of the abovementioned MRB. The objective of the optimization is to minimize the mass of the MRB that can produce a required braking torque. The results are shown in Figure 8.7. It was showed that the disc-type is the best choice when required braking torque is smaller than around 15 N m. At high value of required braking torque (>15 N m), the T-type is the most compact size. However, the maximum braking torque of the T-type is not more than 110% of that of the disc type while its power consumption is much higher (150% or greater). Therefore, the much more complex structure and high power consumption of the T-type compared to the disc-type are also important issues in selection of the most suitable MRB type.

As aforementioned, conventional drum-type MRB is suitable for the application when the geometry aspect *R/L* is greater than 2 and the high value of braking torque is not necessary. In order to improve braking performance of drum-type MRB, Doruk Senkal *et al.* [12,13] have develop a drum-type MRB with serpentine magnetic flux line as shown in Figure 8.8. With this configuration, the magnetic flux density across the MRF duct distributed uniformly and can reach to high value. This results in a high-compact MRB compared to conventional drum type. Experimental results showed that a prototype of the MRB with diameter of 63.5 mm and the length of 89.7 mm (the mass is approximately 2 kg) can generate a maximum torque of 10.9 N m. This is better than conventional drum-type MRB. However, the serpentine MRB is much more complicated and has high cost for manufacturing.

In all above MRBs, the coils are conventionally placed on the inner cylindrical face of the housing. This may result in larger size of the brake, severe the bottleneck

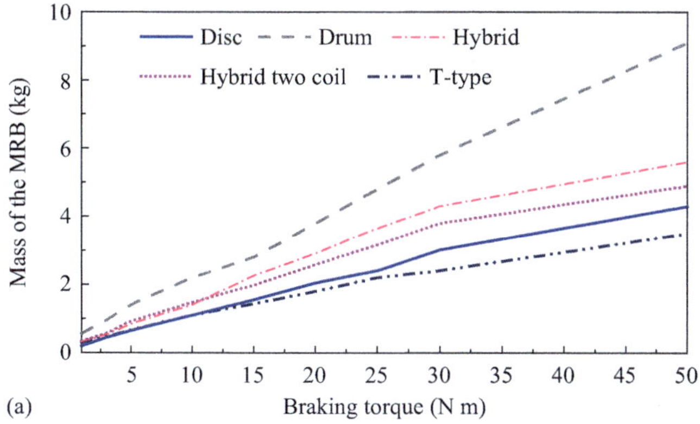

Figure 8.7 Mass and power consumption of the optimized MRBs featuring MRF-132DG as a function of maximum braking torque: (a) mass vs. braking torque and (b) power consumption vs. braking torque

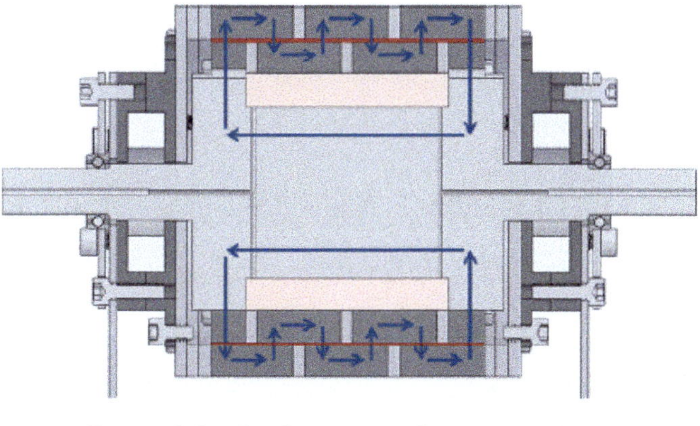

Figure 8.8 Configuration of serpentine MRB

problem of the brake magnetic circuit and difficulty in manufacturing, assembling and maintenance. Recently, Nguyen *et al.* [14–16] have developed a new configuration of MRB called side-coil MRBs, in which the coils are placed on the side housing. With this configuration, the coil can be wound directly on the housing without a bobbin. This may facilitate the manufacturing, reduce bottleneck problem and the size of the brake. The rest of this chapter focuses on development of side-coil MRBs.

8.2 Configuration of the side-coil MR brake

Figure 8.9(a) shows the single side-coil MRB developed by Nguyen *et al.* In this configurations, a disc (rotor) made of magnetic steel is fastened to the flange of the MRB shaft made of nonmagnetic steel. The disc is embedded inside a stationary envelop (housing) made of magnetic steel. A coil is placed on each side housing of the brake. By using the side-coil configuration, it is observed that the coils can be placed directly on the housing, the magnetic circuit can be checked, measured before the MRB is assembled and the MRB can be dissembled without removing the coils out of the housing. In addition, multiple coils can be implemented, especially for high braking torque MRBs. In Figure 8.9(b), the double side-coil MRB with two coils placed on each side of the housing is shown. It is noted that counter currents are applied to the coils to generate a mutual magnetic field as shown in this figure. In the same way, configuration of the triple side-coil MRB with three coils placed on each side housing is shown in Figure 8.9(c). From the above, configuration of side-coil MRBs with more than three coils placed on each side housing can be also obtained.

8.3 Modeling of the side-coil MR brake

From Figure 8.9, it is obviously observed that the braking torque of the proposed MR brake comes from two sources: the friction of MRF acting on the two end-faces and on the outer cylindrical face of the disc. First, the friction between MRF and the end-faces of the disc (MRF in radial duct), as shown in Figure 8.10, is analyzed.

The induced torque from MRF in the radial duct acting on one end-face of a disc can be expressed as follows:

$$T_e = \int_A \tau_e r dA = \int_{R_1}^{R_2} \tau_e r 2\pi r dr = 2\pi \int_{R_1}^{R_2} \tau_e r^2 dr$$

where A is the area of the end-face of the disc, r is the radius of an infinitesimal area of the disc, R_1 and R_2 are the inner and outer radius of the duct. τ_e is the shear stress of the MRF at the interface, which can be predicted from the Bingham-plastic model of the MRF as follows:

$$\tau_e = \tau_{ye} + \mu_e \dot{\gamma}_e$$

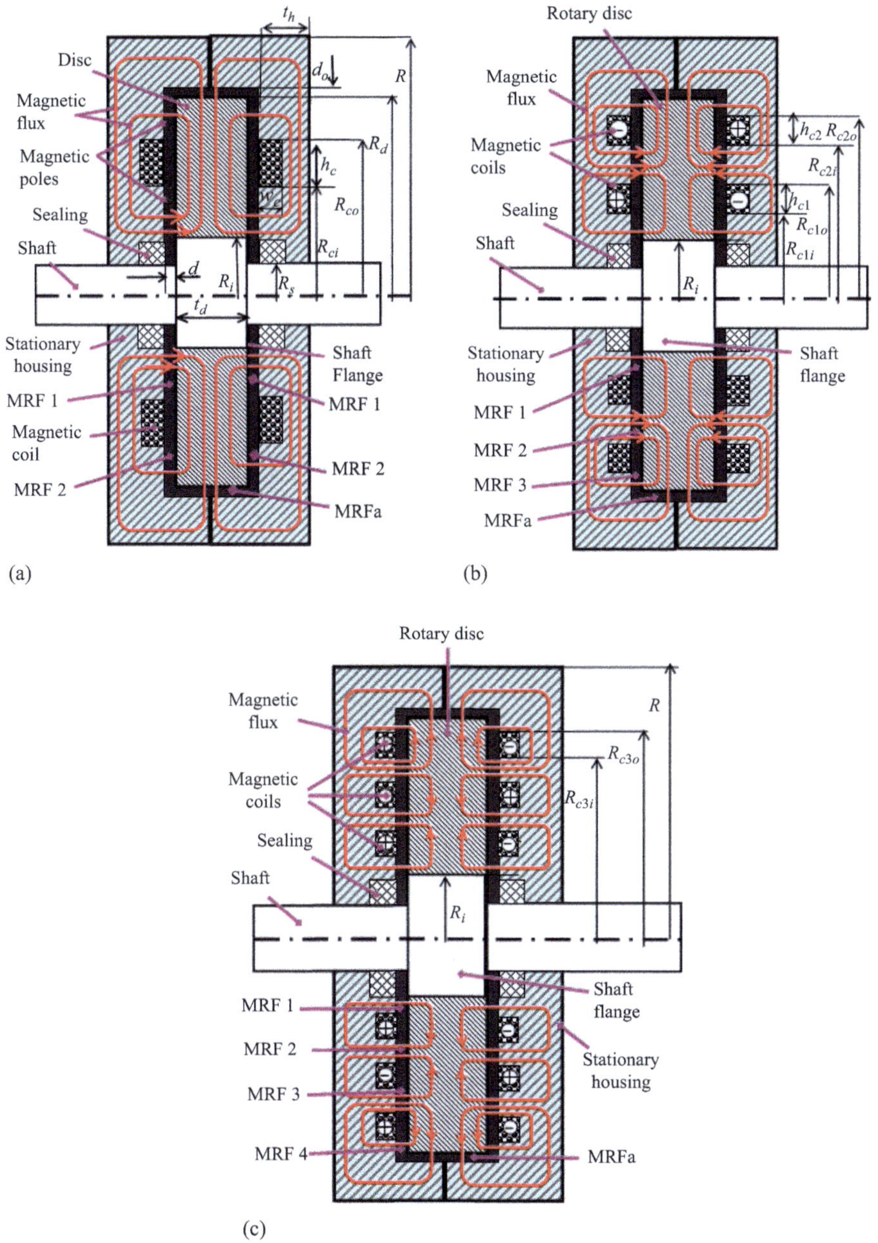

Figure 8.9 Configuration of side-coil MRBs: (a) single side-coil MRB, (b) double side-coil MRB and (c) triple side-coil MRB

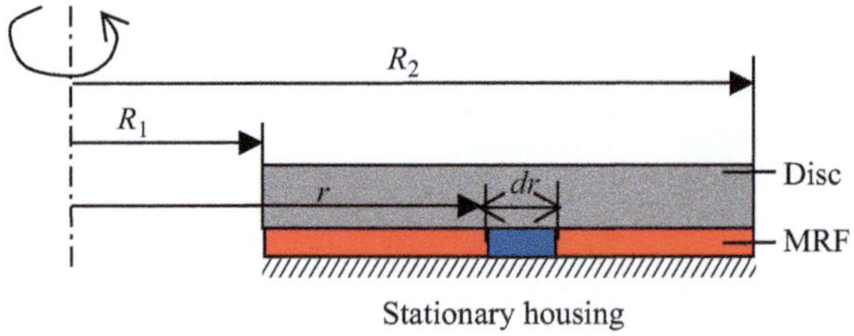

Figure 8.10 Calculation of friction torque induced by MRF in a radial duct

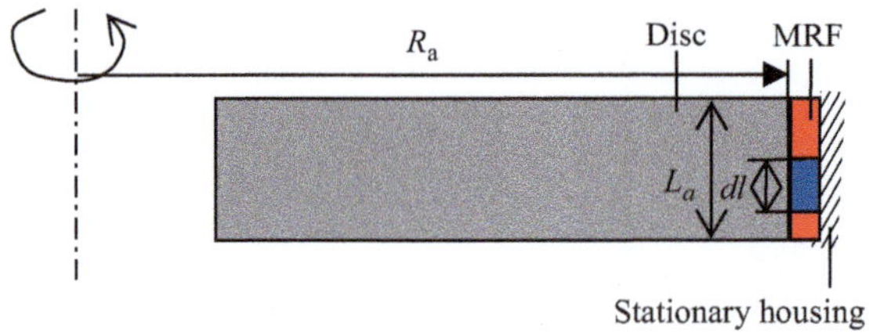

Figure 8.11 Calculation of friction torque induced by MRF in the annular duct

where τ_{ye} and μ_e are, respectively, the yield stress and post yield viscosity of the MRF. $\dot{\gamma}_e$ is the shear rate of MRF at the interface. Because the gap size of the duct is very small, the shear rate of MRF in the gap is assumed to be linearly distributed, which can be approximately determined as follows:

$$\dot{\gamma}_e = \frac{r\Omega}{d}$$

where d is the gap size and Ω is the angular velocity of the drum. From above, the following equation can be obtained:

$$T_e = 2\pi \int_{R_1}^{R_2} r^2 \mu_e \left(\frac{r\Omega}{d} \right) dr + 2\pi \int_{R_1}^{R_2} r^2 \tau_{ye} \, dr$$

Generally, the magnetic density in the MRF gap is a function of radius r, thus the values of τ_{ye} and μ_e of the MRF in the gap are also functions of r. Therefore, generally the induced torque has to be calculated by numerical integration. In order to facilitate the calculation, it is assumed that the magnetic density in the MRF gap is constant and an average value of the magnetic density obtained from numerical integration is used. In this case, the induced yield stress can be calculated by

$$T_e = \frac{\pi \mu_e R_2^4}{2d} \left[1 - \left(\frac{R_1}{R_2} \right)^4 \right] \Omega + \frac{2\pi \tau_{ye}}{3} (R_2^3 - R_1^3)$$

Next, the friction between MRF and the end faces of the disc (annular duct), as shown in Figure 8.11, will be analyzed.

From the figure, the friction torque due to MRF in annular gap acting on the outer circular face of the disc can be determined by

$$T_a = \int_{A_a} R_a \tau_a dA = \int_0^{L_a} R_a \tau_a 2\pi R_a dl = 2\pi R_a^2 \int_0^{L_a} (\tau_{ya} + \mu_a \dot{\gamma}) dl$$

where A_a is the area of the outer annular face of the disc, L_a and R_a are radius and length of annular duct which are also the thickness and the outer radius of the disc, τ_a is the shear stress at the annular face of the disc, τ_{ya} and μ_a are the yield stress and post yield viscosity of the MRF in the annular duct. Similar to MRF in the end-face gap, by using average magnetic density obtained from numerical integration and assuming a linear distribution of shear rate in the MRF gap, the following can be obtained:

$$T_a = 2\pi R_a^2 L_a \left(\tau_{ya} + \mu_a \frac{R_a \Omega}{d} \right)$$

From the above, the induced braking torque of the side-coil MRBs can be correspondingly obtained [16]

$$T_{s1} = 2(T_{e1} + T_{e2}) + T_a + 2T_{sf} = \frac{\pi \mu_1 R_{ci}^4}{d} \left[1 - \left(\frac{R_i}{R_{ci}} \right)^4 \right] \Omega + \frac{4\pi \tau_{y1}}{3} (R_{ci}^3 - R_i^3)$$
$$+ \frac{\pi \mu_2 R_d^4}{d} \left[1 - \left(\frac{R_{co}}{R_d} \right)^4 \right] \Omega + \frac{4\pi \tau_{y2}}{3} (R_d^3 - R_{co}^3) + 2\pi R_d^2 t_d \left(\tau_{ya} + \mu_a \frac{\Omega R_d}{d_o} \right) + 2T_{sf}$$

$$T_{s2} = 2(T_{e1} + T_{e2} + T_{e3}) + T_a + 2T_{sf} = \frac{\pi \mu_1 R_{c1i}^4}{d} \left[1 - \left(\frac{R_i}{R_{c1i}} \right)^4 \right] \Omega + \frac{4\pi \tau_{y1}}{3} (R_{c1i}^3 - R_i^3)$$
$$+ \frac{\pi \mu_2 R_{c2i}^4}{d} \left[1 - \left(\frac{R_{c1o}}{R_{c2i}} \right)^4 \right] \Omega + \frac{4\pi \tau_{y2}}{3} (R_{c2i}^3 - R_{c1o}^3) + \frac{\pi \mu_3 R_d^4}{d} \left[1 - \left(\frac{R_{c2o}}{R_d} \right)^4 \right] \Omega$$
$$+ \frac{4\pi \tau_{y3}}{3} (R_d^3 - R_{c2o}^3) + 2\pi R_d^2 t_d \left(\tau_{ya} + \mu_a \frac{\Omega R_d}{d_o} \right) + 2T_{sf}$$

$$T_{s3} = 2(T_{e1} + T_{e2} + T_{e3} + T_{e4}) + T_a + 2T_{sf} = \frac{\pi \mu_1 R_{c1i}^4}{d} \left[1 - \left(\frac{R_i}{R_{c1i}} \right)^4 \right] \Omega$$
$$+ \frac{4\pi \tau_{y1}}{3} (R_{c1i}^3 - R_i^3) + \frac{\pi \mu_2 R_{c2i}^4}{d} \left[1 - \left(\frac{R_{c1o}}{R_{c2i}} \right)^4 \right] \Omega + \frac{4\pi \tau_{y2}}{3} (R_{c2i}^3 - R_{c1o}^3)$$
$$+ \frac{\pi \mu_3 R_{c3i}^4}{d} \left[1 - \left(\frac{R_{c2o}}{R_{c3i}} \right)^4 \right] \Omega + \frac{4\pi \tau_{y3}}{3} (R_{c3i}^3 - R_{c2o}^3) + \frac{\pi \mu_4 R_d^4}{d} \left[1 - \left(\frac{R_{c3o}}{R_d} \right)^4 \right] \Omega$$
$$+ \frac{4\pi \tau_{y4}}{3} (R_d^3 - R_{c3o}^3) + 2\pi R_d^2 t_d \left(\tau_{ya} + \mu_a \frac{\Omega R_d}{d_o} \right) + 2T_{sf}$$

$$T_{sN} = \frac{\pi\mu_1 R_{c1i}^4}{d}\left[1-\left(\frac{R_i}{R_{c1i}}\right)^4\right]\Omega + \frac{4\pi\tau_{y1}}{3}(R_{c1i}^3 - R_i^3)$$
$$+\sum_{j=2}^{N}\left\{\frac{\pi\mu_j R_{cji}^4}{d}\left[1-\left(\frac{R_{c(j-1)o}}{R_{cji}}\right)^4\right]\Omega + \frac{4\pi\tau_{yj}}{3}\left(R_{cji}^3 - R_{c(j-1)o}^3\right)\right\} + \frac{\pi\mu_{(N+1)}R_d^4}{d}$$
$$\left[1-\left(\frac{R_{c(N+1)o}}{R_d}\right)^4\right]\Omega + \frac{4\pi\tau_{y(N+1)}}{3}(R_d^3 - R_{cNo}^3) + 2\pi R_d^2 t_d\left(\tau_{ya} + \mu_a\frac{\Omega R_d}{d_o}\right) + 2T_{sf}$$

and the off-state force (the torque of the MRB when no magnetic field is applied to MRF) of all considered MRBs can be expressed as

$$T_0 = \frac{\pi\mu_0 R_d^4}{d}\left[1-\left(\frac{R_s}{R_d}\right)^4\right]\Omega + \frac{4\pi\tau_{y0}}{3}(R_d^3 - R_s^3) + 2\pi R_d^2 t_d\left(\tau_{y0} + \mu_0\frac{\Omega R_d}{d_o}\right) + 2T_{sf}$$

where T_{sN} is the braking torque of the side-coil MRB with N coils placed on each side of the housing, T_0 is the off-state torque of MRBs, T_{sf} is the friction torque between the shaft of the brake and the sealing, R_d is the outer radius of the disc, R_i is the inner radius of the active MRF volume in the end-face duct, R_s is the shaft diameter at the sealing, d is the gap size of the radial MRF ducts, d_o is the gap size of the annular MRF duct, t_d is the thickness of the disc, R_{ci} and R_{co} are the inner and outer radius of the coil in case of the single side-coil MRB, R_{cji} and R_{cjo} are the inner and outer radius of the jth coil, Ω is the angular velocity of the rotor, μ_j is the average post yield viscosity of the jth active MRF volume (denoted by MRF$_j$ in Figure 8.9) in the end-face duct of the side-coil MRBs while τ_{yj} is the corresponding yield stress, μ_a and τ_{ya} are respectively the average post yield viscosity and yield stress of MRF in the annular duct of the side-coil MRBs, τ_{y0} and μ_0 are the zero-field yield stress and viscosity of the MRF. The rheological properties of MRF, such as the induced yield stress τ_{ye}, τ_{yj}, τ_{ya} and the corresponding average post yield viscosity μ_e, μ_j, μ_a, depends on the exerted magnetic flux density across the active MRF volumes and can be approximated by [17]

$$Y = Y_\infty + (Y_0 - Y_\infty)(2e^{-B\alpha_{SY}} - e^{-2B\alpha_{SY}})$$

where Y represents for a rheological parameters of MRF such as the yield stress and the post yield viscosity, Y_0 and Y_∞, respectively, represents for the value of Y in the absence of magnetic field and in saturated state of MRF, α_{SY} is the saturation moment index of the Y parameter, B is the applied magnetic density. The sealing used in this study is the lip seal from Parker Corporation. From the manual book of Parker Corporation, the friction torque of the seal T_{sf} can be approximately estimated by [18]

$$T_{sf} = 0.65(2R_s)^2\Omega^{1/3}$$

In the above, T_{sf} is the friction torque of a lip seal in ounce–inches, Ω is the rotation speed of the brake shaft measured in rounds per minute and R_s is the shaft diameter at the sealing measured in inches.

8.4 Optimization of the side-coil MR brake with rectangular housing profile

In this part, the optimal design problem of the conventional disc-type MRB and the side-coil MRBs whose the housing profile has rectangular shape are conducted. The optimization objective is to minimize the mass of the MRBs while the braking is constrained to be grater and a required values. Generally, the MRB mass can be approximately calculated by

$$m_b = V_d\rho_d + V_h\rho_h + V_s\rho_s + V_{MR}\rho_{MR} + V_c\rho_c$$

where V_d, V_h, V_s, V_{MR} and V_c are, respectively, the geometric volume of the disc, the housing, the shaft, the MRF and the coil of the brake, ρ_d, ρ_h, ρ_s, ρ_{MR} and ρ_c are corresponding densities.

In order to determine the braking torque of the MRBs, first, finite element analysis is used to evaluate the magnetic density across the ducts of MRF. In detail, the finite element models generated by 2D-axisymmetric couple element (PLANE 13) of commercial ANSYS software are applied to solve the magnetic circuits of the MRB. After that, the induced MRF rheological properties in the ducts such as the yield stress (τ_{yj}, τ_{ya}) and the post yield viscosity (μ_{ej}, μ_a) are calculated via the average magnetic density and its initial rheological parameters. From the obtained values of all abovementioned parameters, the on-state braking torque of each MRB can be then evaluated. The mass of the MRBs which is also the objective function of MRB optimal design problem is determined from CAD model of the MRBs in the ANSYS software and updated in each optimization loop. In this chapter, the optimal solution is solved by using the first order optimization method with the gradient decent algorithm. A detailed description of this algorithm integrated with ANSYS software is shown in previous research [19,20].

The materials for components of MRBs are selected as followings: the commercial silicon steel is used for magnetic components of the MRB such as the housing and the disc, MRF is the MRF132-DG which is a commercial MRF made by Lord Corporation, the copper coil wires are sized as 24-gage (diameter = 0.511 mm) whose maximum working current is around 3 A and the shaft is made of nonmagnetic stainless steel. From experimental results at different values of magnetic density, the relevant parameters to determine rheological properties of the MRF132-DG such as yield stress and post yield viscosity are determined using least square curve fitting method. The results are [15,16]: $\mu_0 = 0.1$ Pa s, $\mu_\infty = 3.8$ Pa s, $\tau_{y0} = 15$ Pa, $\tau_{y\infty} = 40,000$ Pa, $\alpha_{s\mu} = 4.50\ \mathrm{T}^{-1}$, and $\alpha_{st_y} = 2.9\ \mathrm{T}^{-1}$. The number of turns of a coil can be calculated by

$$n_{turn} = n_w n_h; \quad \text{in which } n_w = \frac{\xi_w w_c}{d_c}; \quad n_h = \frac{\xi_h h_c}{d_c}$$

where w_c and h_c are, respectively, the height and the width of the coil; n_w and n_c are the corresponding number of wire layers, ξ_w and ξ_h are the corresponding filling ratio in the width and the height direction which is set by 0.9 in this chapter, d_c is the diameter of the wire. In the ANSYS software, an excitation current density is applied to the coils which is determined by

$$j = \frac{n_{turn} I}{S_{coil}} = \frac{n_w \cdot n_h I}{w_c \cdot h_c} = \xi_w \xi_h \frac{I}{d_c^2} = 0.81 \frac{I}{d_c^2}$$

where n_{turn} and S_{coil} are correspondingly the numbers of turns in the winding and the cross-sectional area of the coil, I is the applied current to the coils with is 2.0 A.

In the optimization, the design variable are significant geometric dimensions of the MRBs such as the coil height (h_c, h_{cj}), the coil width (w_c, w_{cj}), the inner radius of the disc R_i, the outer radius of the disc R_d, the disc thickness t_d, outer radius of the brake R, the housing thickness t_h and the inner radius of the coils in the case of side-coil MRBs R_{ci} are considered as design variables. It is noted that in design of MRB, the gap size of the MRF ducts should be as small as possible considering manufacturing availability and cost. Therefore, during optimization process, the MRF gap size is not considered as a design variable and fixed at a certain value. In this research, two typical values of the gap size, which are 0.8 and 1.0 mm, are considered. The convergence rate is set by 0.1%. It is also noted that the shaft radius, R_s, is calculated considering the strength of the shaft. In order to solve magnetic circuit of the MRBs, the finite element model using 2D-axisymmetric couple element (PLANE 13) of ANSYS software as shown in Figure 8.12 is used.

Figure 8.13 shows optimal solutions of the conventional disc-type and the side-coil MRBs when the braking torque is constrained to be greater than 10 N m with 2% of accuracy. The shaft radius is set by $R_s = 6$ mm. From Figure 8.13(a), it is seen that the mass of the optimized conventional MRB is 1.69 kg. As shown in Figure 8.13(b), for single side-coil MRB, the minimum mass is 1.67 kg which is a bit smaller than that of the conventional one. Figure 8.13(c) shows the optimal solution of the double side-coil MRB. It is observed that the minimum mass is 1.34 kg which is significantly smaller than that in the case of single side-coil MRB. From Figure 8.13(d), it is found that the minimum mass is 1.26 kg, which is a bit smaller than that of the double side-coil one. Figure 8.14 shows magnetic density distribution of the MRBs at their optimal solution. From Figure 8.14(a) and (b), it is observed that the magnetic distribution in the housing of the side-coil MRBs is more uniformed than that of the conventional one. This results in a more compact size of the side-coil MRB as abovementioned. From Figure 8.14(b)–(d), it is observed that the more coils are used, the more uniformed magnetic distribution density in the housing can be obtained.

The optimal solutions of the MRBs are summarized in Table 8.1. In this table, it is seen that the mass of the side-coil MRBs is significantly smaller than the conventional one; however, its power consumption has significantly increased. It is also noted that the width of the coils in case of double side-coil MRB is almost similar. In the case of triple side-coil MRB, the width of middle and outer coils is almost the same, while the width of the inner coil is smaller. However, the difference is not very much (around 0.3 mm), which is smaller than diameter of the coil wire used (0.51 mm). Therefore, in practical, it is suggested that the width of the coils should be set equal to each other to facilitate the optimal design and manufacturing.

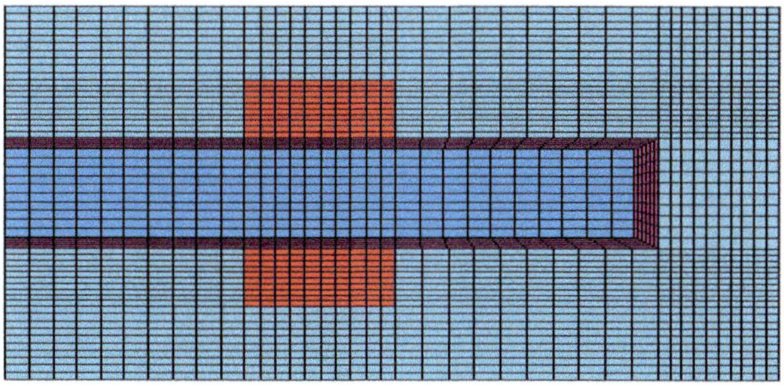

(a)

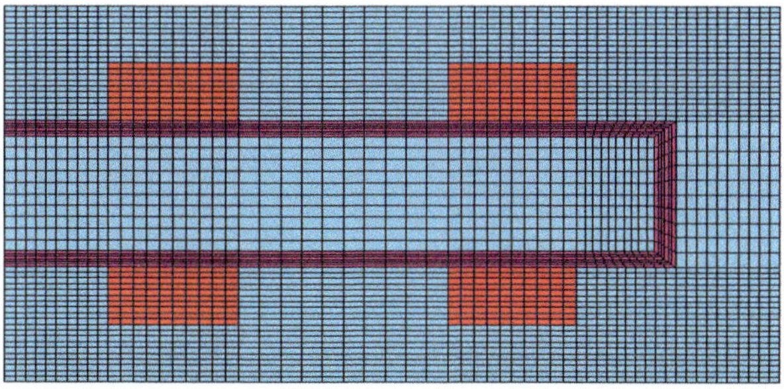

(b)

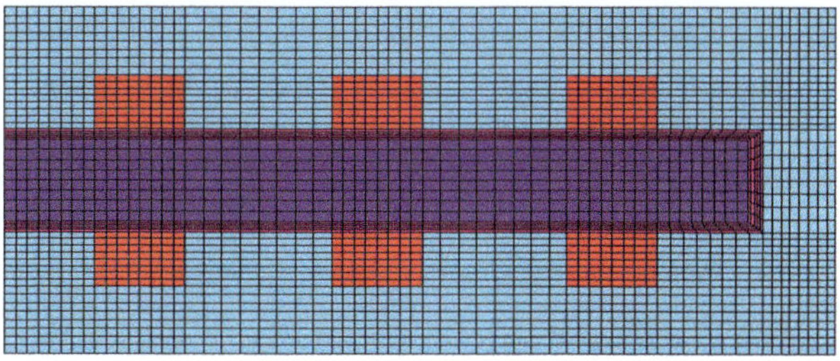

(c)

Figure 8.12 Finite element models to analyze magnetic circuit of the MRBs: (a) single side-coil MRB, (b) double side-coil MRB and (c) triple side-coil MRB

In order to investigate the characteristics of the MRBs as a function of maximum achievable braking torque, optimal solutions of the MRBs at different values of required (constrained) braking torque are obtained in the same manner of that in the case of 10 N m and showed in Figure 8.15. From Figure 8.15(a), it is seen that at small values of required braking torque (≤ 5 N m), the mass of all MRBs are almost similar. In addition, the mass of single side-coil MRB is almost the same as that of the conventional one with the required braking torque being smaller than 50 N m. With some values of the braking torque being higher than 50 N m, the mass of the single side-coil MRB is a bit smaller than that of the conventional one. It is also observed that the mass of double side-coil and triple side-coil MRB are

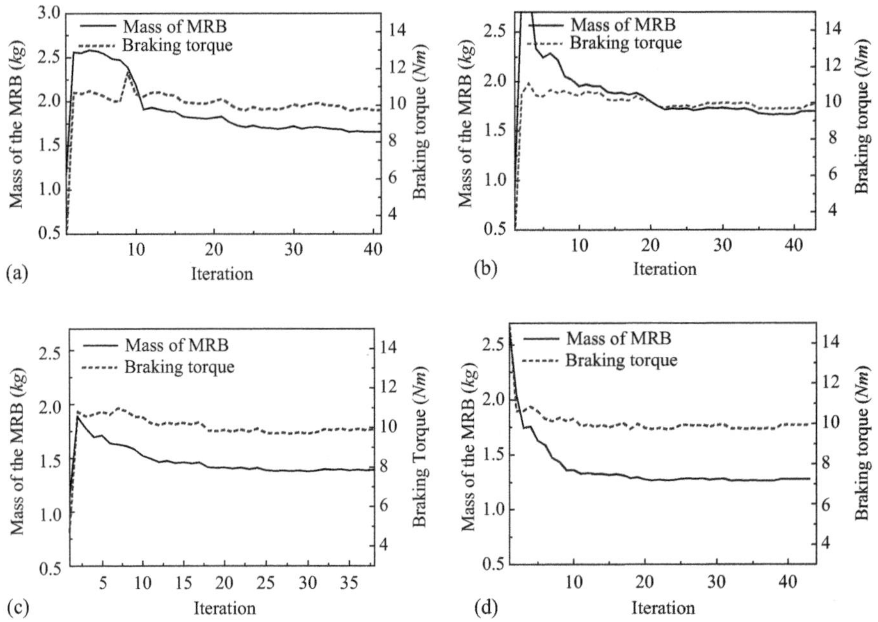

Figure 8.13 Optimization solutions of the MRBs with required braking torque being 10 N m and the MRF duct size being 0.8 mm: (a) conventional MRB, (b) single side-coil MRB, (c) double side-coil MRB and (d) triple side-coil MRB

significantly smaller than that of the single side-coil and the conventional ones in cases where the required braking torque is greater than approximately 5 N m. From Figure 8.15(b), it is observed that the off-state torque of the conventional MRB is generally smaller than that of the side-coil MRBs. In addition, the more coils are used in MRB systems, the greater off-state torques are resulted. This is mainly due to a larger diameter of the disc which is required for multiple side-coil MRBs which is presented in Figure 8.15(c). In Figure 15(d), it can be realized that the width of the double and the triple side-coil MRBs is significantly smaller than that of the conventional and the single side-coil ones. This is the reason that the mass of the multiple side-coil MRBs is smaller than that of the rest. As aforementioned, the power consumption of the side-coils MRBs is significantly greater than that of the conventional one, especially the multiple side-coil MRBs. This is more clearly presented in Figure 8.15(e). Noteworthily, power consumption is also a critical issue. The high power consumption will result in a high temperature of the MRBs, which may degrade the braking performance (maximum working temperature of most of commercial MRFs is limited by 130 °C). Therefore, the temperature issue should be taken into account when the MRBs work continuously, especially in automotive application [9,21]. In addition, in many applications, electric power is limited and should be as small as possible such as in haptic systems.

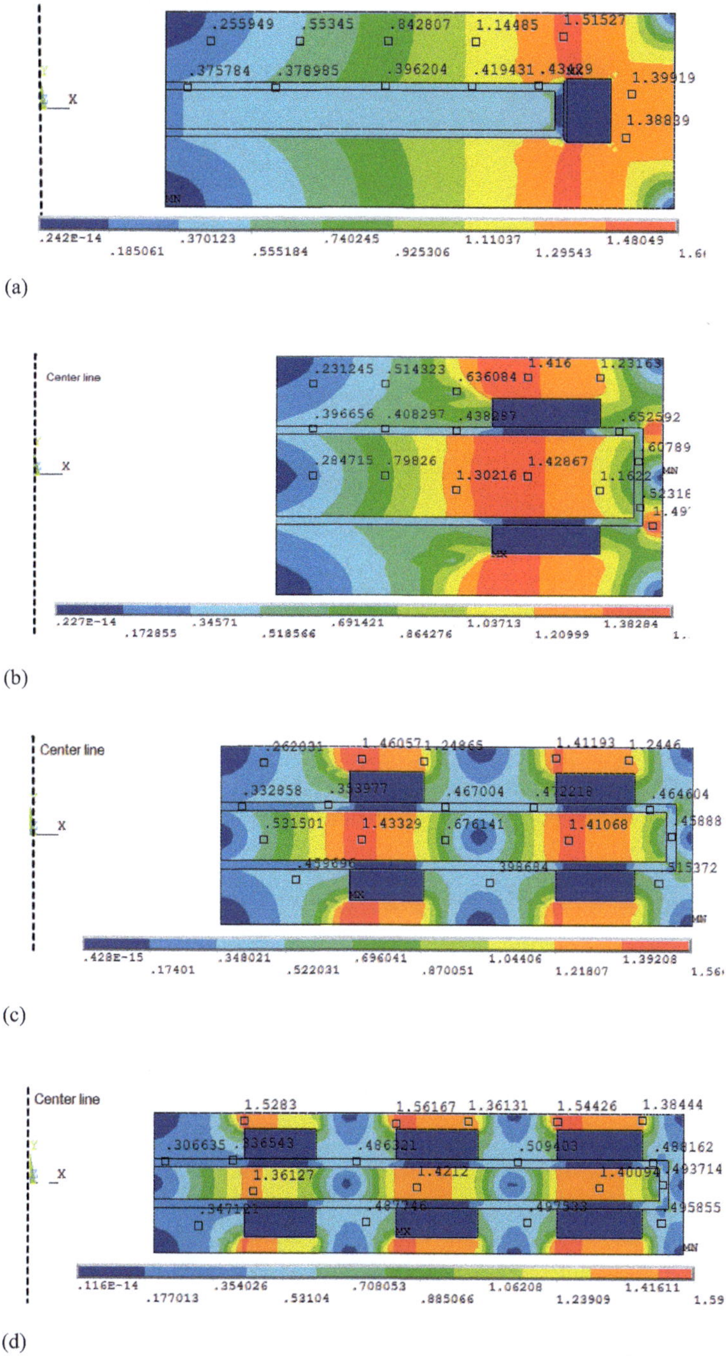

Figure 8.14 Magnetic flux density of the brake at the optimum: (a) conventional MRB, (b) single side-coil MRB, (c) double side-coil MRB (d) triple side-coil MRB

Table 8.1 Optimal solution of the MRBs when the required torque is 10 N m and the MR duct gap is 0.8 mm

Column heads	Design parameters (mm)	Mass (kg)	Off-state torque (N m)	Fed power (W)/n_{turn}
Conventional	$w_c = 6.6$; $h_c = 4.7$; $R_{ci} = 54.2$; $R = 65.2$; $L = 20$; $t_d = 4$; $R_d = 53$; $R_i = 13$	1.69	0.08	15.1/96
Single side-coil	$w_c = 2.83$; $h_c = 9.67$; $R_{ci} = 40.57$; $R = 56.3$; $L = 22.7$; $t_d = 7.95$; $R_d = 54.0$; $R_i = 25.5$	1.67	0.09	22.2/168
Double side-coil	$w_{c1} = 2.8$; $h_{c1} = 6.7$; $w_{c2} = 2.7$; $h_{c2} = 7.3$; $R_{c1i} = 28.8$; $R_{c2i} = 47.5$; $R = 59.8$; $L = 15.9$; $t_d = 4.3$; $R_d = 57.5$; $R_i = 17.3$	1.34	0.1	27.8/238
Triple side-coil	$w_{c1} = 2.8$; $h_{c1} = 6.6$; $w_{c2} = 2.5$; $h_{c2} = 7.2$; $w_{c3} = 2.4$; $h_{c3} = 7.4$; $R_{c1i} = 15.6$; $R_{c2i} = 31.7$; $R_{c3i} = 48.2$; $R = 59.6$; $L = 15.5$; $t_d = 4.0$; $R_d = 58.1$; $R_i = 8.6$	1.28	0.08	40.7/338

In order to take power consumption of the MRBs into account, optimal solutions of the side-coil MRBs with power consumption, constrained to be equal or smaller than that of the conventional one, are obtained and presented in Figure 8.16. It is observed that in the case of power consumption constraint, the mass of the double and triple MRBs is significantly increased, while that of the single side-coil MRB is slightly increased compared to that in the case of no power consumption constraint shown in Figure 8.15. In this case, the mass of the double MRB is almost equal to that of conventional one and a bit smaller than that of the single and triple MRB. This correspondingly leads to a significant increase of the overall length. Besides, the overall radius of the side-coil MRBs is also increased slightly.

In order to evaluate the effects of the MRF duct size on performance characteristics of all considered MRBs, the optimal design procedure is generated with the MRF duct size being 1.0 mm. The optimal solutions in this case are presented in Figure 8.17. From this figure, almost the entire observation is similar to that in the case where MRF duct is 0.8 mm. Obviously, the mass, the outer radius, the width and the power consumption of all optimal MRBs in this case are greater than those in the previous case. However, the off-state torque is somewhat smaller, which can reduce the heat friction generating into MRF duct. As abovementioned, it is realized that the duct size of MRF has an important role with the performance of MRBs. Therefore, choosing suitable values for MRF duct size is a critical issue which should be determined carefully when considering the mass and the thermal effect simultaneously in MRB optimal design problem. Similar to the case of 0.8 mm MRF duct, in this case, optimal solutions of the side-coil MRBs with power consumption, constrained to be equal to that of the conventional one, are obtained and presented in Figure 8.18. From the results, similar conclusions can be archived.

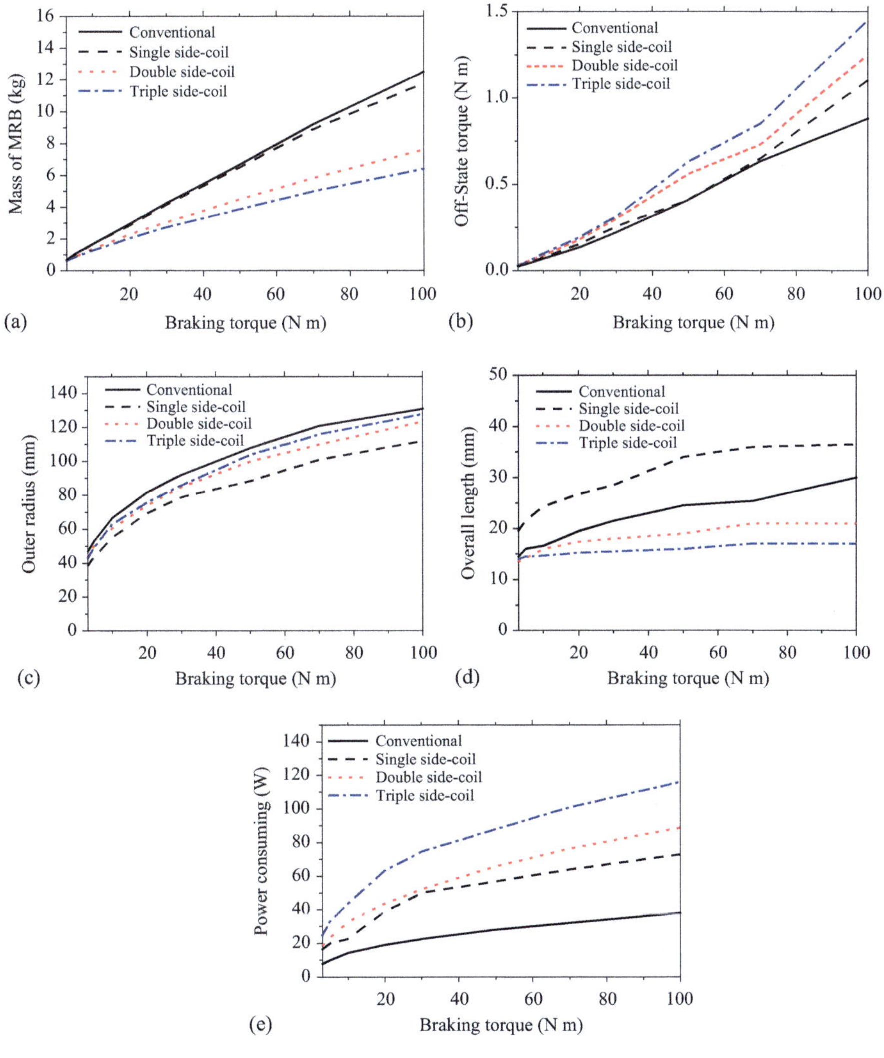

Figure 8.15 Optimal results of the MRBs as a function of brake torque, d = 0.8 mm: (a) MRB mass, (b) off-state torque, (c) overall radius, (d) overall width and (e) power consumption

In order to have a better comparison between the single side-coil MRB and the multiple ones when power consumption is taken into account, optimal solutions of the multiple side-coil MRBs with power consumption, constrained to be equal to that of the single side-coil one are obtained and presented in Figure 8.19. In this case, it is observed that the mass of the double and triple MRB is significantly increased compared to that in the case of no constraint on power consumption. However, the mass of the multiple side-coil MRBs is still smaller than that of the

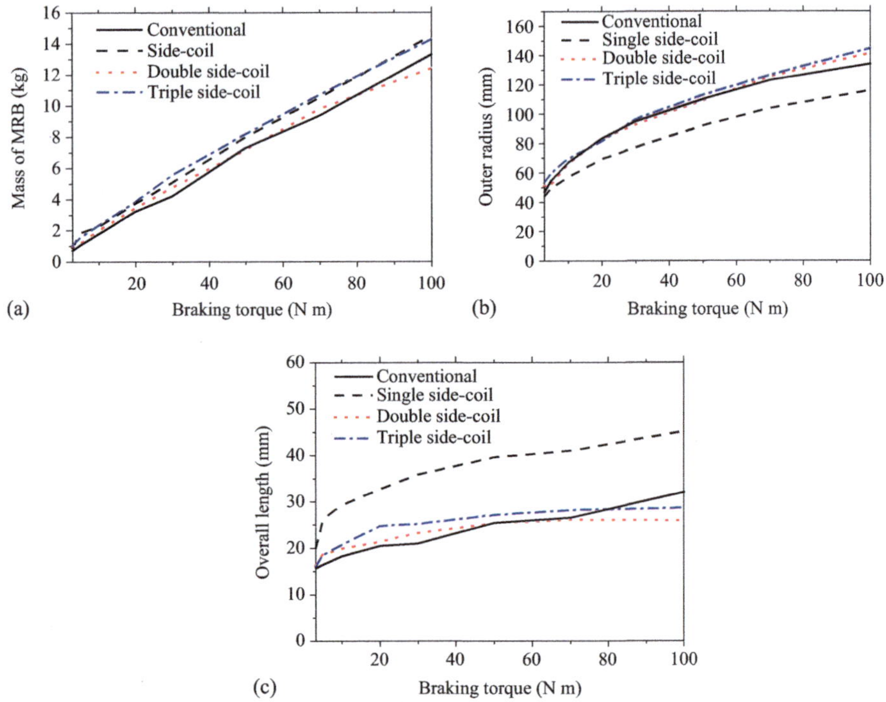

Figure 8.16 Optimal results of the MRBs when the power consumption is constrained to be smaller than that of conventional MRB, d = 0.8 mm: (a) mass vs. braking torque, (b) outer radius vs. braking torque and (c) overall length vs. braking torque

single one, especially when the braking torque is higher than 20 N m. It is also observed that the mass of the double side-coil and the triple side-coil MRB is almost similar. Therefore, in this case, it is better to use double side-coil because of its convenience in manufacturing and operation compared to triple side-coil ones.

In order to validate the above optimal results, the prototypes of the optimized MRB are manufactured and some relative results are also presented here. In this study, the experiments are performed for a prototype of single side-coil MRB, a prototype of double side-coil MRB and a prototype of triple side-coil MRB. For the single and double side-coil MRBs, optimal results of the MRBs with a required braking torque of 10 N m shown in Table 8.1 are implemented. It is noted that, for convenience in manufacturing, actual parameters of the prototypes are a bit different from those in Table 8.1. The actual parameters of the manufactured prototypes of single and double MRBs are shown in Table 8.2. In order to investigate actual performance of MRBs at higher braking torque, the prototype of triple side-coil MRB with a required braking torque of 70 N m is considered, and its optimal design parameters are shown in Table 8.3. Figure 8.20 shows some major parts of the prototype MRBs. It is reminded that, in the prototypes, first, the coils are wound

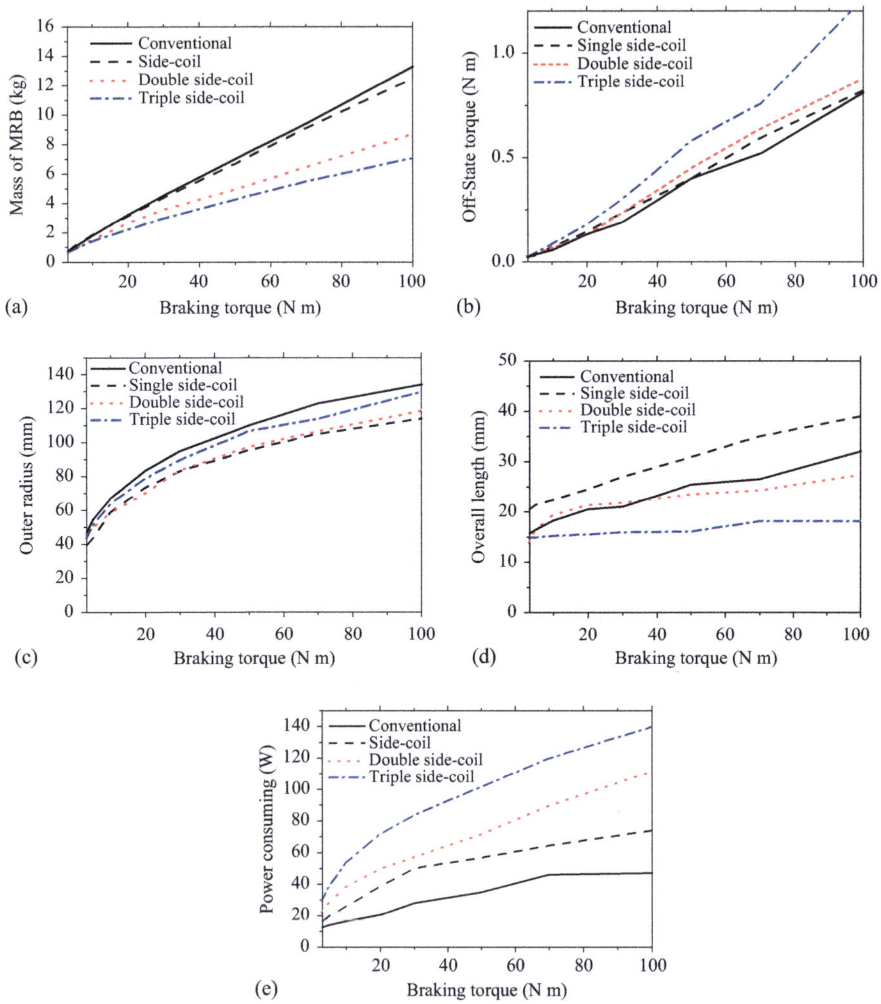

Figure 8.17 Optimal results of the MRBs as a function of braking torque, $d = 1$ mm: (a) MRB mass, (b) off-state torque, (c) overall radius, (d) overall width and (e) power consumption

using outside bobbins, and then they are taken off the bobbins and placed directly on the side housings of the MRBs.

Figure 8.21 shows the experimental setup for testing the performance of the abovementioned prototypes. In this experimental system, a gear-box DC motor controlled by a computer revolves the shaft of the MRB rotor at a constant angular speed of 4π rad/s. A torque sensor is used to measure the torque generated by the MRB. Via the A/D converter, analog output signal from the torque sensor is sent to the computer for evaluation. When the experiment process is started, a step current

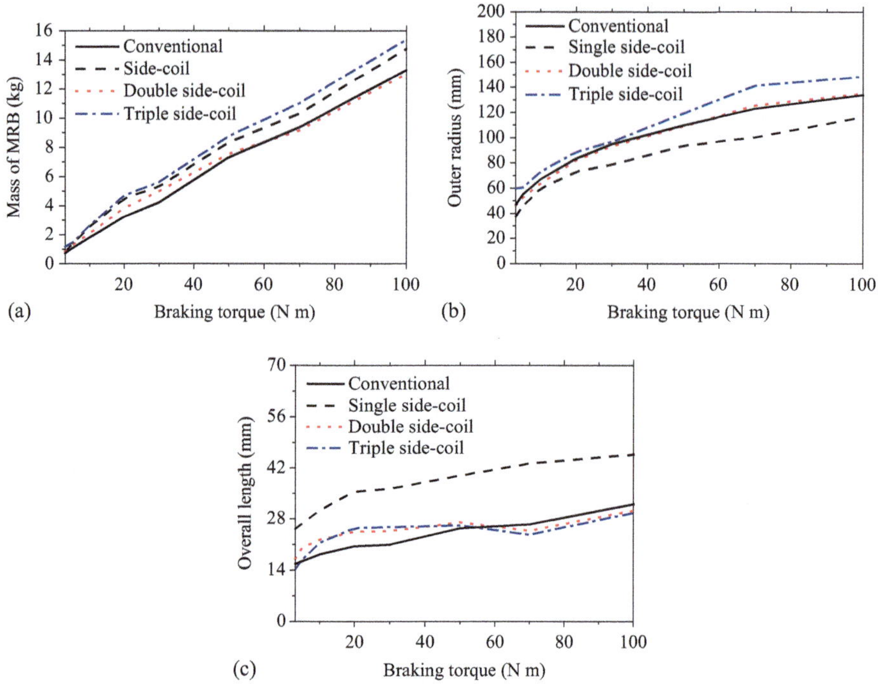

Figure 8.18 Optimal results of the MRBs when the power consumption is constrained to be smaller than that of conventional MRB, d = 1 mm: (a) mass vs. braking torque, (b) outer radius vs. braking torque and (c) overall length vs. braking torque

signal generated from the computer is provided to the current amplifier and the amplified step current from the amplifier is then applied to the coils of the MRBs.

Figure 8.22 shows step response of the single side-coil MRB prototype at the time of 0.5 s. Eight step currents are investigated which are 0.5, 0.75, 1.0, 1.25, 1.5, 1.75, 2.0 and 2.5 A. In Figure 8.22(a), the measured current applied to the coil is presented. It is observed from the figure that as steady state, the average measured currents are almost equal to the desired ones which are, respectively, 0.49, 0.746, 0.99, 1.24, 1.49, 1.756, 2.01 and 2.492 A. It is also observed that the steady values of the measured current are almost achieved at the time 0.7 s for all cases; thus, the current response time is around 0.2 s. From Figure 8.22(b), it is observed that the maximum braking torque of the prototype MRB at the applied current of 2.5 A is around 9.5 N m which is a bit smaller than that from the simulation (9.9 N m). The error is around 4% in this case. The main reason may come from inaccurate estimation of the friction torque, braking torque and loss of magnetic field to the ambient and at the contact between the magnetic parts of the MRBs. In addition, a slight difference between the design and the manufactured prototype is also a potential reason. At different values of the applied current, a well agreement between

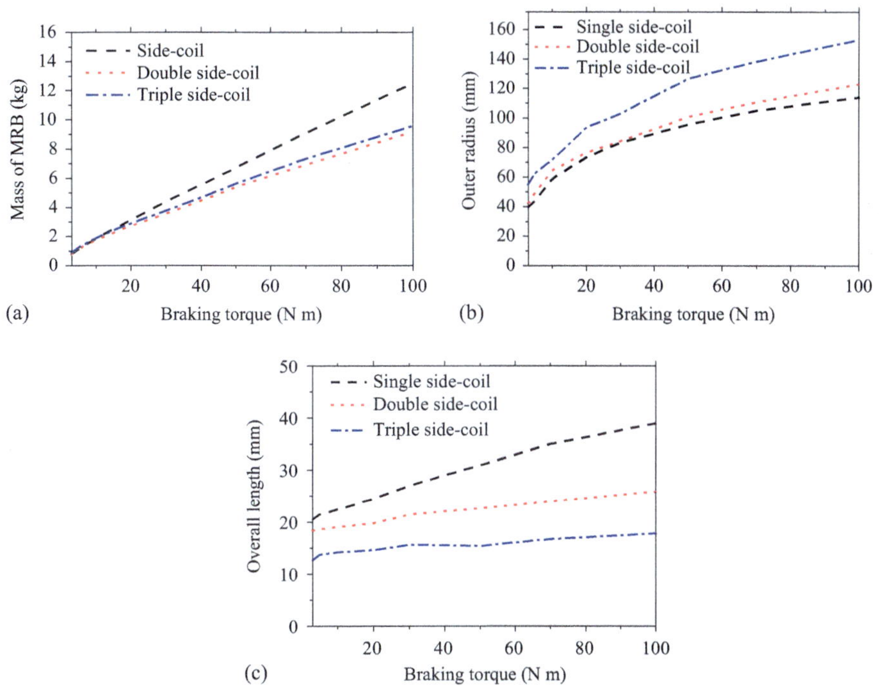

Figure 8.19 Optimal results of the multiple side-coil MRBs as a function of braking torque when the power consumption is constrained to be smaller than that of single side-coil MRB, d = 1 mm: (a) mass vs. braking torque, (b) outer radius vs. braking torque and (c) overall length vs. braking torque

Table 8.2 Parameters of the manufactured single and double side-coil MRB prototypes

Column heads	**Design parameters (mm)**	**Mass (kg)**	**Off-state torque (N m)**	**Fed power (W)/n_{turn}**
Single side-coil	$w_c = 2.8$; $h_c = 9.6$; $R_{ci} = 40.5$; $R = 56$; $L = 23$; $t_d = 8.05$; $R_d = 53.1$; $R_i = 21.5$	1.71	0.39	22/160
Double side-coil	$w_{c1} = 2.8$; $h_{c1} = 6.7$; $w_{c2} = 2.8$; $h_{c2} = 7.3$; $R_{c1i} = 28.8$; $R_{c2i} = 47.5$; $R = 59.8$; $L = 15.9$; $t_d = 4.3$; $R_d = 57.5$; $R_i = 17.3$	1.45	0.41	27.5/136

the manufactured and the simulated one is also achieved. The average braking torques at the applied currents of 0.5, 0.75, 1.0, 1.25, 1.5, 1.75, 2.0 A are, respectively, 1.56, 2.64, 3.75, 5.21, 6.65, 8.1, 8.7 N m while those from the simulation as shown in Figure 8.22(c) are 1.5, 2.6, 3.9, 5.3, 6.7, 7.9, 8.9 N m, respectively.

Table 8.3 Parameters of the triple side-coil MRB prototype

Column heads	Design parameters (mm)	Mass (kg)	Off-state torque (N m)	Fed power (W)/n_{turn}
Optimal design result	$w_{c1}=3.1$; $h_{c1}=8.2$; $w_{c2}=2.9$; $h_{c2}=9.1$; $w_{c3}=2.54$; $h_{c3}=9.7$; $R_{c1i}=56.4$; $R_{c2i}=79.6$; $R_{c3i}=101.1$; $R=116$; $L=15.7$; $t_d=4.4$; $R_d=113.4$; $R_i=32.1$	4.98	0.98	110/392
Actual parameters of the prototype	$w_{c1}=2.9$; $h_{c1}=8.2$; $w_{c2}=2.9$; $h_{c2}=9.1$; $w_{c3}=2.9$; $h_{c3}=9.7$; $R_{c1i}=56.4$; $R_{c2i}=79.6$; $R_{c3i}=101$; $R=116$; $L=15.8$; $t_d=4.4$; $R_d=113.2$; $R_i=32$	5.16	1.1	110/390

(a) (b) (c)

Figure 8.20 Prototypes of MRBs for experimental test: (a) single side-coil MRB, (b) double side-coil MRB and (c) triple side-coil MRB

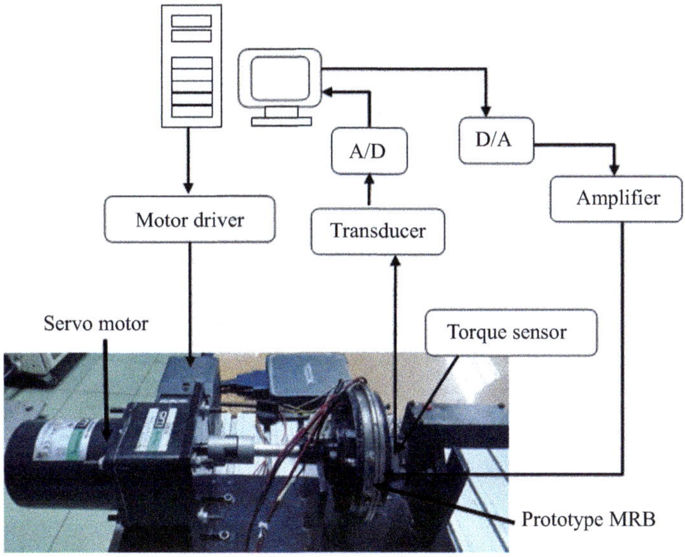

Figure 8.21 Experiment setup to test performance of the prototypes of the MRBs

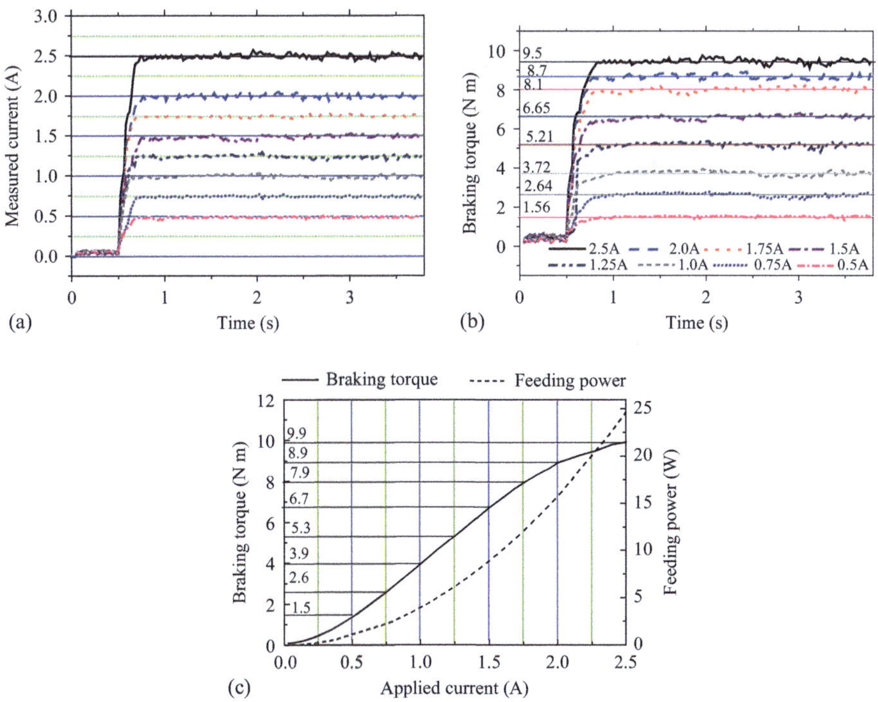

Figure 8.22 Step response of the single side-coil MRB prototype: (a) measured current, (b) measured torque and (c) simulated braking torque and feeding power vs. applied current

The corresponding errors are 4%, 1.54%, 3.85%, 1.7%, 0.75%, 2.53% and 2.25%. From Figure 8.22(b), it is also found that the steady values of the measured torque are almost achieved at the time 0.82 s for all cases; thus the torque response time is around 0.32 s. The larger response time of the measured torque compared to that of the measured current comes from the time response of the MRF, the hysteresis of mechanical system and the torque measurement.

Figure 8.23 shows step response of the double side-coil MRB prototype. From Figure 8.23(a), it is observed that as steady state, the average measured currents are almost equal to the desired ones which are, respectively, 0.482, 0.745, 0.984, 1.234, 1.48, 1.75, 2.02 and 2.484 A. The current response time is around 0.24 s which is a bit greater than that of the single side-coil MRB. This comes from larger time constant of the *R–L* circuit of the double side-coil MRB (the two coils at each side of the brake are connected in series in this case). From Figure 8.23(b), a well agreement between the manufactured and the simulated is also observed. The average braking torques at the applied currents of 0.5, 0.75, 1.0, 1.25, 1.5, 1.75, 2.0 and 2.5 A are, respectively, 1.26, 2.4, 3.6, 4.74, 6.14, 7.1, 8.13 and 10.04 N m while those from the simulation as shown in Figure 8.23(c) are 1.2, 2.3, 3.5, 4.7, 6.0, 7.2,

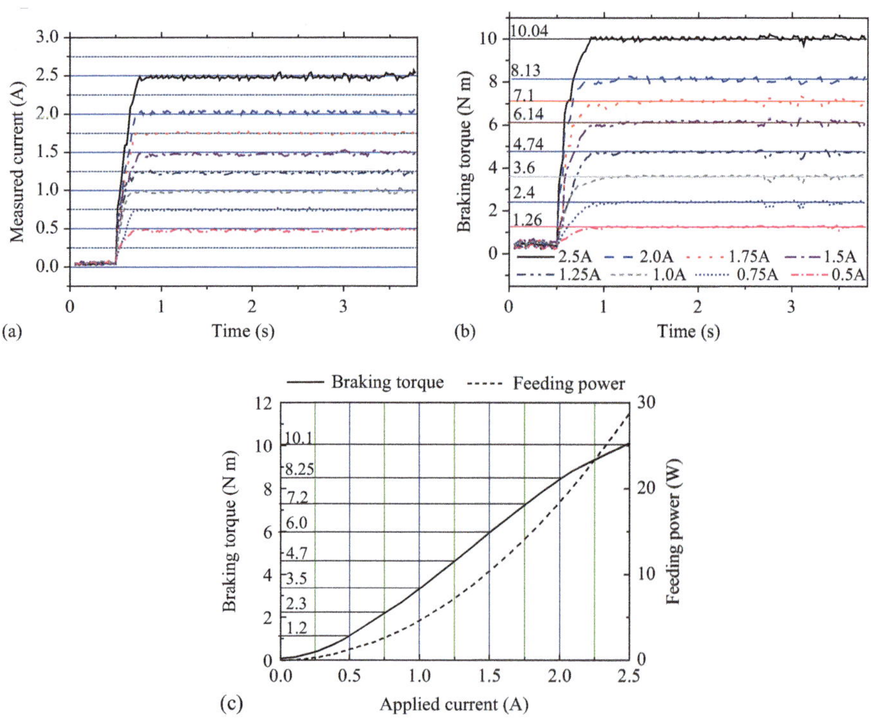

Figure 8.23 Step response of the double side-coil MRB prototype: (a) measured current, (b) measured torque and (c) simulated braking torque and feeding power vs. applied current

8.25 and 10.1 N m, respectively. The corresponding errors are 5%, 4.34%, 2.85%, 0.85%, 2.33%, 1.39%, 1.45% and 0.6%. From Figure 8.23(b), it is also found that the steady values of the measured torque are almost achieved at the time 0.84 s for all cases; thus the torque response time is around 0.34 s.

Figure 8.24 shows step response of the triple side-coil MRB prototype. The average measured currents observed from Figure 8.24(a) are, respectively, 0.488, 0.748, 0.994, 1.24, 1.49, 1.754, 2.01 and 2.486 A. The current response time is around 0.25 s which is a bit greater than that of the single and the double side-coil MRBs. The corresponding measured torques are 11.7, 18.9, 29.08, 40.0, 50.8, 57.73, 63.0 and 67.93 N m, while those from the simulation as shown in Figure 1.24(c) are 11.5, 18.9, 29.3, 39.8, 49.8, 57.9, 63.5 and 70.1 N m, respectively. The corresponding errors are 1.74%, 0.53%, 0.75%, 0.5%, 2.0%, 0.3%, 0.8% and 3.09%. In this case, the error is smaller than that in the previous cases. It is obvious that at high braking torque, the percentage error due to inaccurate estimation of friction torque and the measured data processing are reduced. The steady values of the measured torque are almost achieved at the time 0.88 s for all cases; thus, the torque response time is around 0.38 s.

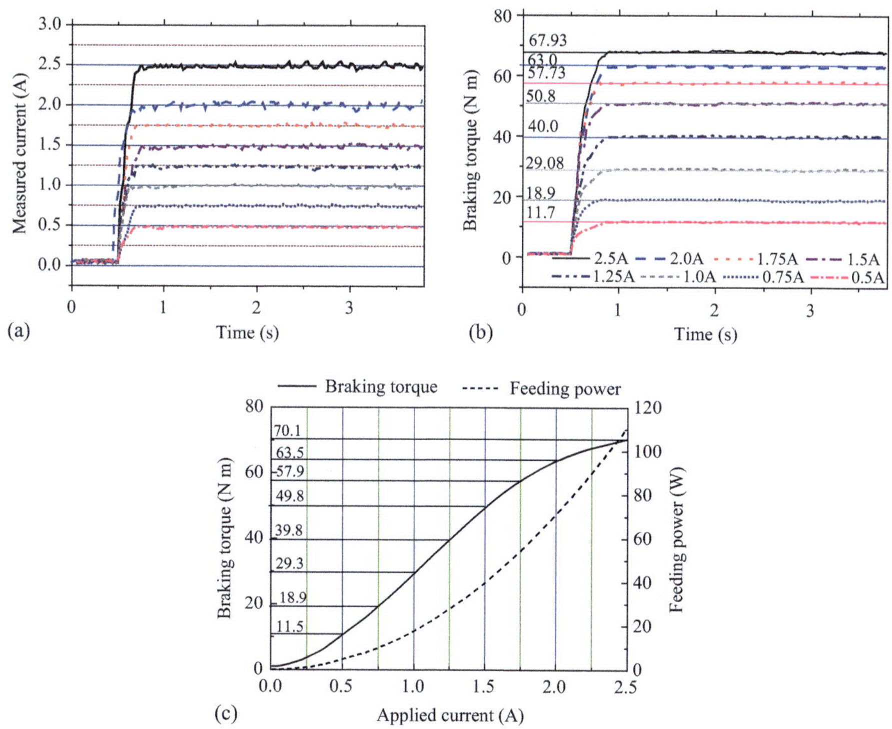

Figure 8.24 Step response of the triple side-coil MRB prototype: (a) measured current, (b) measured torque and (c) simulated braking torque and feeding power vs. applied current

8.5 Optimization of the side-coil MR brake with polygonal housing profile

In Section 8.4, optimization of the MRBs has been conducted; however, only rectangular shape of the housing profile was considered. From Figure 8.14, it is observed that magnetic flux density distributed in the brake housing is not uniform, especially at the corner of the housing profile. Therefore, the compact size of the MRBs can be significantly improved by optimization of the housing geometry to have more uniform distribution of magnetic flux density in the housing. In this section, the geometric optimization of the MRBs with polygonal housing profile determined by control points is conducted. The more control points of the brake envelop are, the more mass reduction of the brake can be obtained. However, the mass reduction of the brake tends to saturation when seven or more control points are considered [22]. Figures 8.25 and 8.26 show the geometry of the conventional and the single side-coil disc-type MRB with 7-control point polygonal housing profile. In these figures, the control point P_i is determined by its radius R_i and corresponding t_{hi}.

Figures 8.27 and 8.28 show the optimal solution of the conventional and the single side-coil MRB with 7-control point polygonal envelop when the braking torque is constrained to be greater than 10 N m. In this case, it is noted that in addition to the abovementioned design variables (the coil height h_c, the coil width w_c, the disc radius R_d, the disc thickness t_d, the outer radius of the MRB R and the inner radius of the coils in case of the single side-coil MRB R_c), the geometric parameters of the control point P_i (R_i, t_{hi}) are also considered as design variables. As shown in Figure 8.27(a) and (b), the optimization is converged after

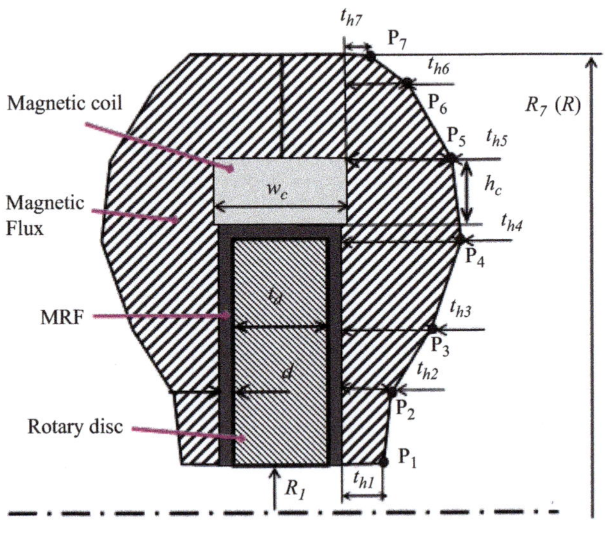

Figure 8.25 Geometry of the conventional MRBs with 7-control point envelop

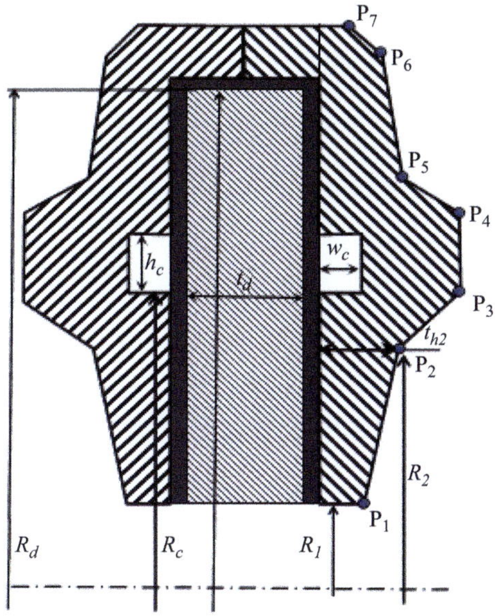

Figure 8.26 Geometry of the single side-coil MRBs with 7-control point envelop

30 iterations. At the optimum, the braking torque can reach up to 10 N m as constrained and the minimum mass is 1.44 kg and the values of design variables are (mm): $w_c = 6.9$, $h_c = 2.5$, $t_d = 4$, $R_d = 44$, $R_1 = 10.5$, $t_{h1} = 2$, $R_2 = 24.4$, $t_{h2} = 2.4$, $R_3 = 34$, $t_{h3} = 3.8$, $R_4 = 49.5$, $t_{h4} = 5.8$, $R_5 = 52.4$, $t_{h5} = 5.85$, $R_6 = 55.8$, $t_{h6} = 4.25$, $R(R_7) = 57.6$, $t_{h7} = 2$. In order to have a clear presentation of the MRB at the optimum, the magnetic distribution of the optimized disc-type MRB is presented in Figure 8.27(c) with numerical values of magnetic density at significant points. It is observed that the magnetic distribution in the housing is much for uniform than that in case of rectangular profile. For the single side-coil MRBs, from Figure 8.28(a) and (b), it is observed that the convergence occurs at the 25th iteration at which the braking torque can reach up to 10 N m as constrained and the minimum mass is 1.28 kg. At the optimum, the values of design variables are (mm) $w_c = 3.8$, $h_c = 4.3$, $t_d = 7.5$, $R_d = 44.5$, $R_c = 36$, $R_1 = 13$, $t_{h1} = 2$, $R_2 = 27.5$, $t_{h2} = 2.6$, $R_3 = 34$, $t_{h3} = 8$, $R_4 = 42$, $t_{h4} = 7.7$, $R_5 = 43.5$, $t_{h5} = 5.8$, $R_6 = 46.6$, $t_{h6} = 1.1$, $R(R_7) = 47.6$, $t_{h7} = 0.3$. The magnetic distribution of the optimized proposed MRB is presented in Figure 8.28(c) with numerical values of magnetic density at significant points. Again, a much more uniform distribution of magnetic flux density is observed in this case compared to the rectangular profile. From the results, it is found that by using polygonal housing profile, the magnetic flux density can be much more uniform and the mass of MRBs can be significantly reduced. It is also shown that the mass of single side-coil MRB is smaller than that in the case of the conventional one.

Figure 8.29 shows the mass and the off-state torque of the optimized MRBs at different values of the constrained braking torque. From Figure 8.29(a), it is found

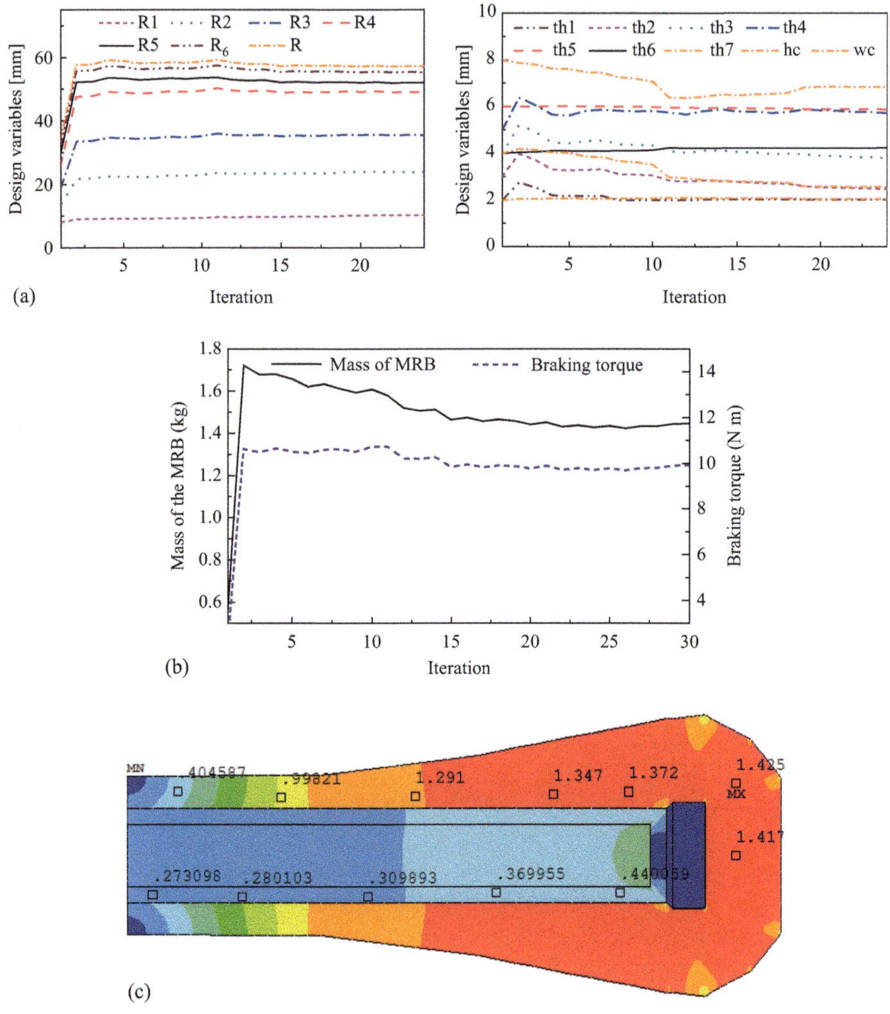

Figure 8.27 Optimization solution of the 7-segment polygon conventional brake, braking torque ≥ 10 N m: (a) design variables, (b) mass and braking torque and (c) magnetic flux density at the optimum

that the mass of the proposed MRB is always smaller than the conventional one at different values of the required braking torque. It is also observed from Figure 8.29(b) that the mass reduction percentage of the proposed MRBs compared to the conventional ones is around 10%–15%. The off-state torque of the optimized MRBs are also obtained and presented in Figure 8.29(c) at different value of the constrained braking torque. From the figure, it can be seen that the off-state torque of the proposed MRB is generally smaller than that of the conventional one. Especially, the off-state torque of

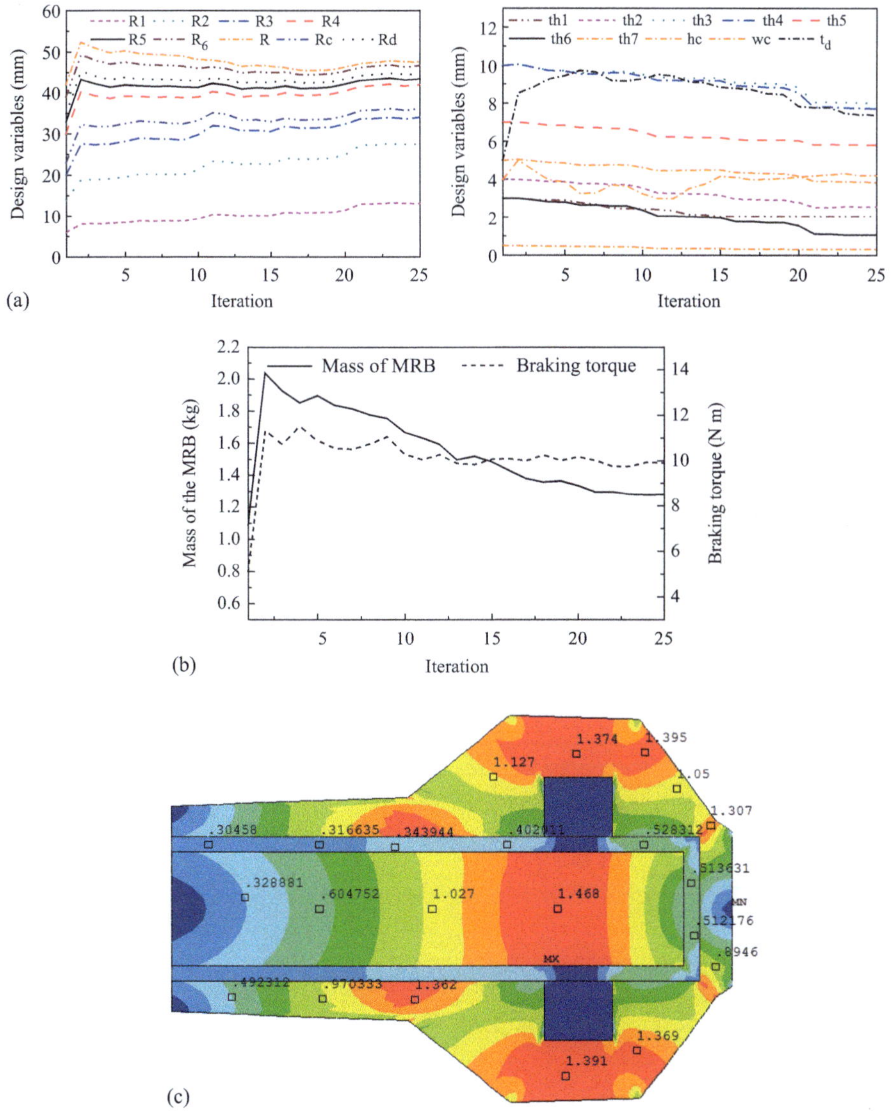

Figure 8.28 Optimization solution of the 7-segment polygon proposed brake, braking torque ≥ 10 N m: (a) design variables, (b) mass and braking torque and (c) magnetic flux density at the optimum

the proposed MRB with rectangular envelop is even though smaller than that of the conventional one with 7-control point polygonal envelop.

In order to evaluate the effects of the MRF duct size on performance characteristics of the proposed MRB, the optimal solution of the proposed and conventional MRBs, when the MRF duct size is 0.8 mm, are obtained and presented in Figure 8.30.

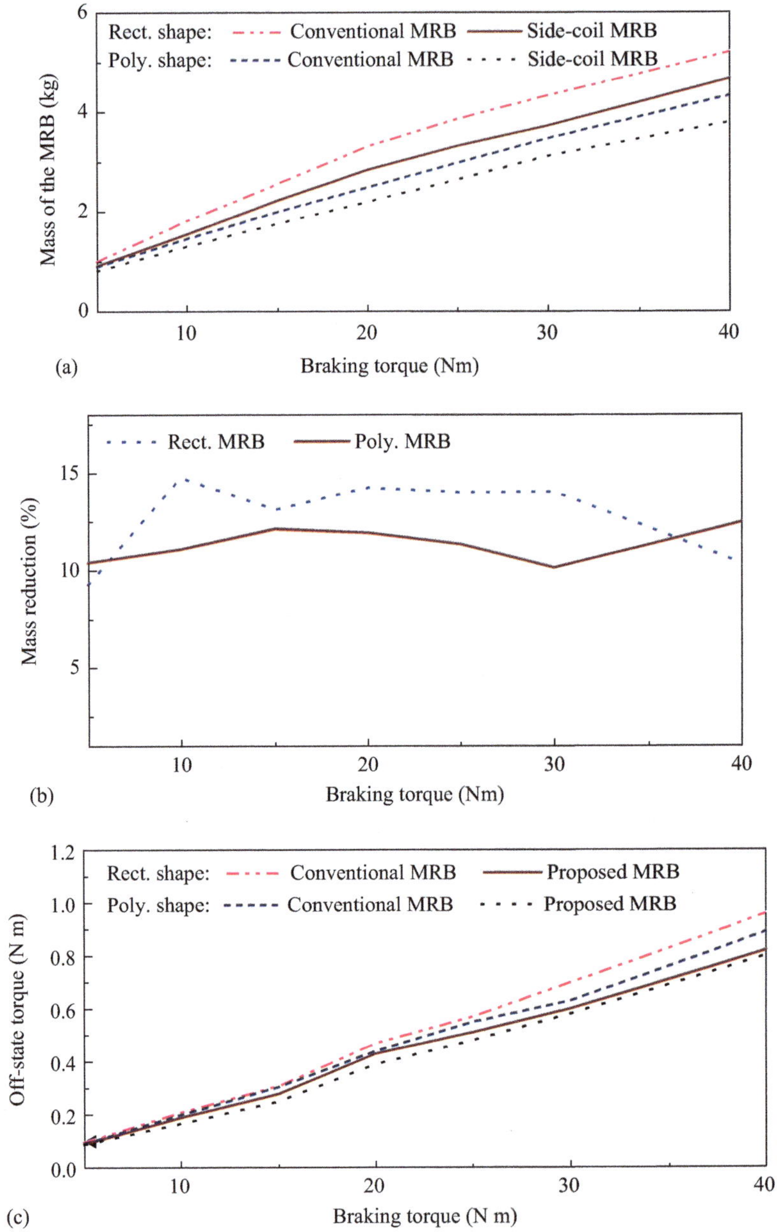

Figure 8.29 Optimal results of the MRBs as a function of maximum braking torque, d = 1 mm: (a) mass vs. braking torque, (b) mass reduction vs. braking torque and (c) off-state torque vs. braking torque

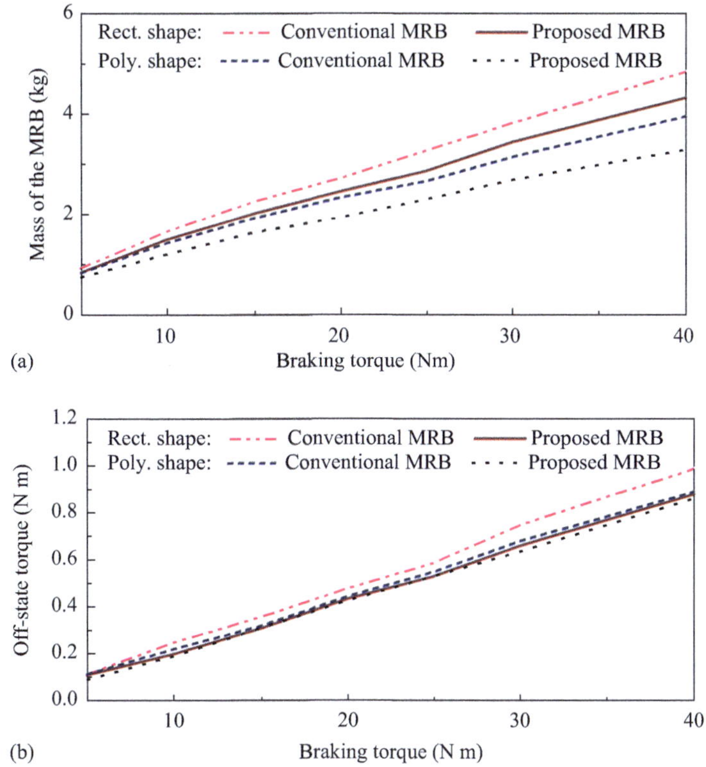

Figure 8.30 Optimal results of the MRBs as a function of maximum braking torque, d = 0.8 mm: (a) mass vs. braking torque and (b) off-state torque vs. braking torque

From the figure, it is found that by choosing small value of the MRF ducts, the mass of the optimized MRBs can be significantly reduced. However, the cost of manufacturing and maintenance by choosing small value of the MRF duct size is also an important issue that should be considered. It is also noted that a comparison between the conventional MRBs and the single side-coil MRBs in this case almost shows similar results to the previous case (MRF duct size is 1 mm).

8.6 Summary

In this chapter, an overview of MRB development was presented especially focused on development of configuration of MRBs. After the review, a new configuration of MRBs with magnetic coils placed on side housings (side-coil MRB) was introduced in Section 8.2. In Section 8.3, mathematic model of the MRBs such as braking torque, off-state torque was obtained based on Bingham plastic rheological behavior of the MRF. In Section 8.4, optimal design of the side-coil MRBs was conducted, in which the mass of the MRB was set as the objective function while the braking torque and power consumption are considered as constrained parameters (state variables). The results showed the advance of side-coil MRBs compared to conventional disc-type MRBs. The results also figured out the advantages and disadvantages of the single side-coil MRB compared to the multiple ones. To validate the optimal results, experimental works were conducted for three prototypes: single side-coil MRB (10 N m), double side-coil MRB (10 N m) and triple side-coil MRB (70 N m). The experimental results showed a very well agreement with simulated results. In Section 8.5, polygonal shape of the MRB housing profile was considered. The optimal results of the MRBs with 7-control point polygonal housing profile were conducted. The results showed that by using polygonal housing profile, the mass of the MRBs can be significantly reduced. Finally, it is remarked that the side-coil MRBs have many benefits in manufacturing, maintenance and operations compared to conventional one such as the place can be placed directly on the housing without a bobbin, the magnetic circuit can be checked, measured before the MRB is assembled and the MRB can be dissembled without removing the coils out of the housing.

References

[1] Rabinow J, 1951, Magnetic fluid torque and force transmitting device, *US patent* 2575360.

[2] Nguyen Q H and Choi S B, 2010, Optimal design of an automotive magnetorheological brake considering geometric dimensions and zero-field friction heat, *Smart Mater. Struct.* 19 (**11**), 1–10.

[3] Nguyen Q H and Choi S B, 2010, Optimal design of MR brake absorber for torsional vibration control, *Smart Mater. Struct.* 21 (**2**), 1–10.

[4] An J and Kwon D S, 2002, Haptic experimentation on a hybrid Active/Passive force feedback device, *In Proceedings of the IEEE International Conference on Robotics and Automation*, Washington, DC, 4217–4222.

[5] Ferre M, Galiana I, Wirz R and Tuttle N, 2011, Haptic device for capturing and simulating hand manipulation rehabilitation, *IEEE-ASME Trans. Mechatron.* 16 (**5**), 808–815.

[6] Li W H, Liu B, Kosasih P B and Zhang X Z, 2007, A 2-DOF MR actuator joystick for virtual reality applications, *Sens. Actuators, A* 137, 308–320.

[7] Huang J, Zhang J Q, Yang Y and Wei Y Q, 2002, Analysis and design of a cylindrical magnetorheological fluid brake, *J. Mater. Process. Technol.* 129, 559–562.

[8] Nguyen Q H, Nguyen P B and Choi S B, 2011, Optimal design of a hybrid MRB for haptic wrist application, *Proc. SPIE* 7977, 1–6.

[9] Nguyen Q H and Choi S B, 2012, Optimal design of a novel hybrid MR brake for motorcycles considering axial and radial magnetic flux, *Smart Mater. Struct.* 21 (**5**), 1–12.

[10] Nguyen Q H and Choi S B, 2012, Selection of magnetorheological brake types via optimal design considering maximum torque and constrained volume, *Smart Mater. Struct.* 21 (**1**), 1–11.

[11] Nguyen Q H, Lang V T and Choi S B, 2015, Optimal design and selection of magneto-rheological brake types based on braking torque and mass, *Smart Mater. Struct.* 24 (**6**), 1–10.
[12] Senkal D and Gurocak H, 2009, Compact MR-brake with serpentine flux path for haptics applications, *In World Haptics 2009*, Salt Lake City, UT, USA, March 18–20, 2009.
[13] Senkal D and Gurocak H, 2010, Serpentine flux path for high torque MRF brakes in haptics applications, *Mechatronics* 20, 377–383.
[14] Nguyen Q H, Nguyen N D and Choi S B, 2014, Optimal Design of a novel configuration of MR brake with coils placed on the side housings of the brake, *Proc. SPIE* 9059, 1–7.
[15] Nguyen Q H, Lang V T, Nguyen N D and Choi S B, 2015, Geometric optimal design of MR brake considering different shapes of the brake envelope, *Smart Matter. Struct.* 23 (**1**), 1–11.
[16] Nguyen N D, Le D T, Le D H and Nguyen Q H, 2018, Development of a new magnetorheological fluid based brake with multiple coils placed on the side housings, *Journal of Intelligent Material Systems and Structures* 29 (**20**), 734–748.
[17] Zubieta M, Eceolaza S, Elejabarrieta M J and Bou-Ali M M, 2009, Magnetorheological fluids: characterization and modeling of magnetization. *Smart Mater. Struct.* 18 (**9**), 1–6.
[18] EPS Division, 2006, Rotary seal design guide, Catalog EPS 5350, Parker Hannifin Corporation.
[19] Nguyen Q H, Han Y M, Choi S B and Wereley N M, 2007, Geometry optimization of MR valves constrained in a specific volume using the finite element method, *Smart Mater. Struct.* 16, 2242–2252.
[20] Nguyen Q H and Choi S B, 2008, Optimal design of vehicle MR damper considering damping force and dynamic range, *Smart Mater. Struct.* 18 (**1**), 1–10.
[21] Nguyen Q H and Choi S B, 2010, Optimal design of an automotive magnetorheological brake considering geometric dimensions and zero-field friction heat, *Smart Mater. Struct.* 19 (**11**), 1–10.
[22] Nguyen Q H, Lang V T, Nguyen N D and Choi S B, 2014, Geometric optimal design of MR brake considering different shapes of the brake envelope, *Smart Matter. Struct.* 23 (**1**), 1–11.

Chapter 9

Enhanced magnetic-sensing characteristics for application of magnetorheological elastomer

Miao Yu[1]

As a viscoelastic smart material with magnetic-control properties, magnetorheological (MR) elastomer (MRE) is prepared by solidifying the magnetic particles (hardness) into the viscoelastic polymer matrix (softness). MRE is a magnetic-sensing smart material whose mechanical property can be controlled by external magnetic field. MRE possesses advantages of both MR material and elastomer, such as rapid response, good reversibility, excellent polymer's properties and strong controllable capacity. It overcomes the disadvantages of sedimentation, poor stability and sealing problems of MR fluid (MRF). These advantages make MRE have broad application prospects in the fields of noise reduction, vibration attenuation and smart sensing. Consequently, it becomes a hotspot of MR materials in recent years. In this chapter, the MRE materials' preparation, characterization and application research are introduced. Finally, the future development of MREs is also prospected.

In the 1980s, people put forward the concept of smart material based on bionics. Meanwhile, the smart material was endowed with vital signs—be able to perceive the external environment, change its traits in real time to adapt the environment and meet its own development or even have the self-healing capabilities. Smart material is the frontier of interdisciplinary research and is also the natural product by the integration of information technology into materials science. Through several decades of development, smart material has become the research hotspot in the field of material science and engineering worldwide. Of necessity, it will become one of the most important material bases of information intelligent society in the future.

The viscoelastic smart material, which possesses magnetic-sensing properties, is developed from MRF, we called it magnetic-sensing viscoelastic materials (MSVMs). It contains MRF, MR plastomer, MR foam, MR gel and MRE. This chapter highlights the development of MRE material. As one of the most important member of MSVMs, MRE consists of a viscoelastic polymer matrix impregnated

[1]Key Lab for Optoelectronic Technology and Systems, Ministry of Education, College of Optoelectronic Engineering, Chongqing University, China

with ferromagnetic particles, typically spherical micron-sized magnetic particles [1–3]. It not only has the characteristics of energy storage and energy consumption as well as viscoelastic materials, also its physical properties (such as mechanical, electrical, optical, acoustic, thermal) can be adjusted continuously, rapidly and reversibly according to the applied magnetic field [4,5]. Compared with other smart materials, MRE has several advantages, such as significant magnetic effect, rapid response speed, high mechanical strength, simple processing and molding. Meanwhile, MRE manifests the superiority over MRF for its no settlement, good sealing performance and simple application structure [6–8]. Besides, its drive manner by noncontact space magnetic field extends MRE's application, such as biomedical, robot and aerospace.

As a consequence of above fascinating properties of MRE, great breakthroughs and fruitful achievements have been made in the fields of vibration and noise reduction, sensing, microfluidics and electromagnetic wave absorption. This chapter mainly focuses on the progress of material preparation, characterization, mechanisms model and application during the last few years. This summarizes the problems in recent research and application and thus analyzes its future prospects.

9.1 Overview of magnetorheological elastomer

The magnetic-sensing property of MRE is mainly manifested as its magnetic-control performance in the mechanical properties, such as the material's storage modulus and loss factor. Figure 9.1 shows the photo of silicone-rubber-based MRE and its magnetic-control modulus curve. In addition, based on the magnetic-induced interaction of magnetic particles by applied magnetic field, the electrical, thermal and optical properties of MRE can also produce rapid and reversible magnetic-control changes [9].

Due to the huge differences in the matrix, the MR mechanism of MRE is significantly different from its predecessor-MRF. The liquid MRF, in which magnetic particles can move freely in the matrix and form a chain or column structure parallel to the direction of applied magnetic field, showing an obvious phase

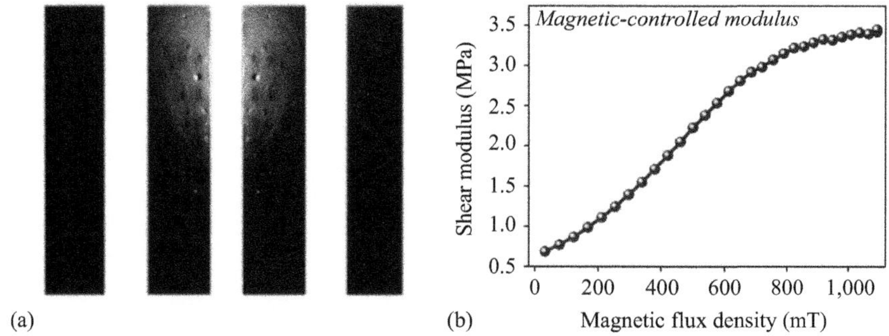

Figure 9.1 Photo (a) and magnetic-control modulus curve (b) of silicone rubber-based MRE

change characteristics. While, the solid MRE possesses a fixed structure in which the particles are solidified in the matrix; the particles' aggregation structure will not change significantly, but only tiny rotation and deviation of magnetic particles will be generated under an external magnetic field [10,11]. With increasing magnetic field intensity, the interaction between particles or between particles and matrix will be strengthened, showing higher rigidity rather than phase transition in the macroscopic. Besides, according to the working conditions, the MRF mainly works in the post yielding of the material and during its flow period. The chain structure of the magnetic particles in the MRF can be repeatedly destroyed and reconstructed. While the magnetic particles in MRE are bound to the solid matrix, its chain structure cannot be repaired autonomously after being damaged, so it usually works in the state before the yielding of material.

In 1995, by mixing silicone resin and iron powder, Shiga *et al.* prepared the magnetic viscoelastic gel, which is generally considered as the prototype of MRE [12]. After two decades of development, MRE has gradually formed four major research systems—material preparation, test characterization, mechanism research and device application. This chapter will introduce the research progress of MRE from these four aspects.

9.2 Material preparation of magnetorheological elastomer

During the preparation of MRE, its particles distribution can be regulated by magnetic field. As an example, Figure 9.2 shows the preparation of MRE based on polyurethane rubber [13]. The inserts 1, 2 and 3 of Figure 9.2 show the scanning electron microscopy (SEM) images of anisotropic MRE (prepared with magnetic field), isotropic MRE (prepared without magnetic field) and carbonyl iron powder (CIPs) (type: CN), respectively. It can be seen that by applying magnetic field in the prestructure process, the field-prepared MRE forms a distinct chain structure along the direction of magnetic field.

As we all know, MRE is mainly composed of matrix, magnetic particles and additives. In order to further improve the performance of MRE, researchers have made great efforts in material selection and modification from these three aspects.

9.2.1 Matrix

As a continuous phase in composite materials, the matrix adheres to the reinforced particles through the interface and transfers load to the particles in the form of shear force. MRE's mechanical strength, glass transition temperature, heat resistance and other properties are closely related to the matrix. In the matrix selection of MRE, researchers usually choose soft rubber elastomers, such as silicone rubber [14–16], natural rubber [17], polyurethane rubber [18–20], butadiene rubber [21,22], Polydimethylsiloxane (PDMS) [23], polyurethane/epoxy resin interpenetrating networks matrix [24] and hybrid rubber matrix [25,26]. Ubaidillah *et al.* analyzed the matrix types of MRE, as shown in Figure 9.3. It can be seen that silicone rubber matrix

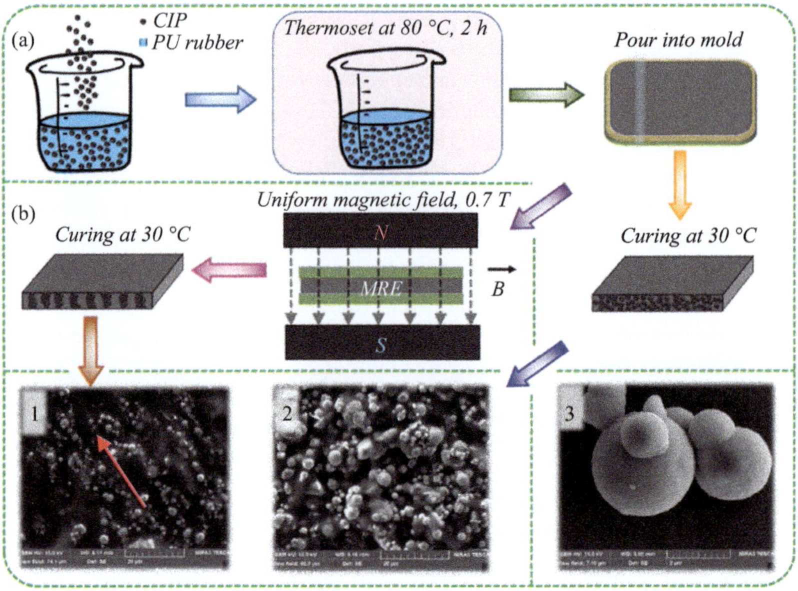

Figure 9.2 Preparation of MRE: (a) preparation process of isotropic MRE, (b) preparation process of anisotropic MRE. Insets 1, 2 and 3 show the SEM images of anisotropic MRE, isotropic MRE and CIPs (type CN), respectively [13]

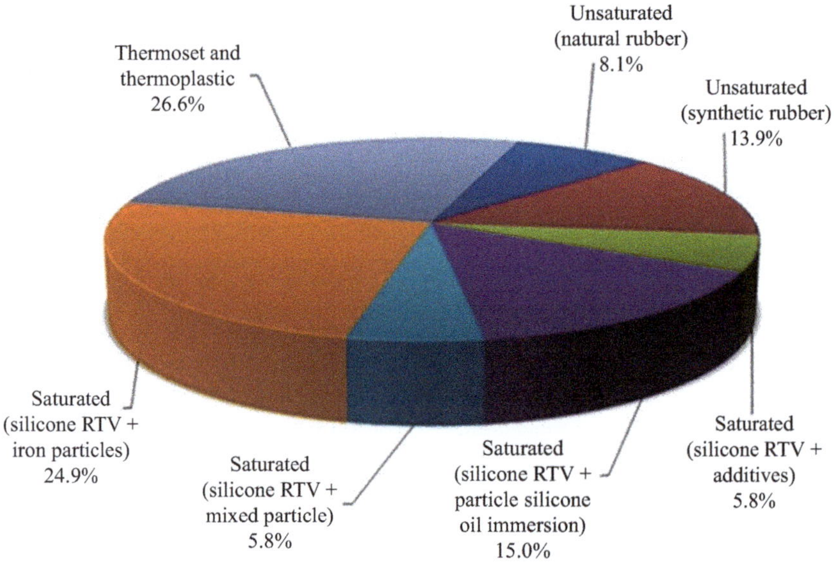

Figure 9.3 Mapping of MREs based on matrix material [9]

occupied half of the MRE [9]. The main reason is that the preparation process of silicone rubber is simple and the initial (zero-field) modulus is low, which ensures that MRE can obtain high MR effect.

9.2.2 Magnetic particles

Magnetic particles are the key point to obtain magnetic properties of MRE materials. In order to make MRE better adapt to engineering application requirements, the selection of magnetic particles is particularly important. With high saturation magnetization and low coercivity, carbonyl iron powder is an ideal choice [27]. In addition, nickel, cobalt powder and some hard magnetic particle powder such as barium ferrite, strontium ferrite, and rubidium ferroboron are also used as magnetic particles of MRE [28–31]. Yu *et al.* prepared MRE by filling hard magnetic particles. The results revealed that the novel MRE has remanence after being magnetized, and the bidirectional magnetic-modulus characteristics can be realized by its demagnetization and magnetization characteristics, as shown in Figure 9.4. When a magnetic field which opposite to the sample's magnetization direction is applied, the storage modulus of MRE decreases with the increase of magnetic field. While when the applied magnetic field is parallel to the sample's magnetization direction, the storage modulus of MRE increases sharply with increasing magnetic field [13].

Padalka *et al.* prepared MRE by using Fe, Co and Ni nanowires as the soft magnetic filler particles; it was found that the magnetic properties of MRE are closely related to the magnetic properties of the particles. Thereinto, the one which filled with Fe nanowires has the highest MR effect, while the one with Ni nanowires has the worst MR effect [31]. Moreover, the morphology and size of particles also have a great influence on MRE. As shown in Figure 9.5, Yu *et al.* prepared a dimorphic MRE incorporated with Fe nano-flakes modified carbonyl iron particles, which improved the material's MR effect and damping characteristics of the samples [32]. Li *et al.* prepared MRE containing two kinds of carbonyl iron powders with different particle sizes, and they found that the mixing of magnetic particles with different particle sizes could make the material obtain greater MR effect [33].

Ubaidillah *et al.* counted the shape and size distribution of magnetic particles used in the preparation of MRE. From Figure 9.6, it can be seen that most of current researchers prefer to use the spherical magnetic particles (diameter: 0–10 μm) as the doping particles [9].

As the polarity of matrix and magnetic particle differ greatly, the weak interfacial strength restricts the enhancement of MRE's mechanical strength. Especially, in order to obtain high MR effect, researchers often need to increase the volume fraction of magnetic particles. However, it leads to more serious agglomeration problems and interface defects. In the study of MRE interface modification, Qiao *et al.* used titanate coupling agent to modify the surface characteristics of CIP; it improved the combination between particles and poly(styrene-b-ethylene-ethylenepropylene-b-styrene) (SEEPS) matrix and also significantly improved the MR effect of MRE [34]. Jiang *et al.* modified the surface of CIP by using Span80, and which improved the relative MR effect of MRE [35]. Wang *et al.* employed

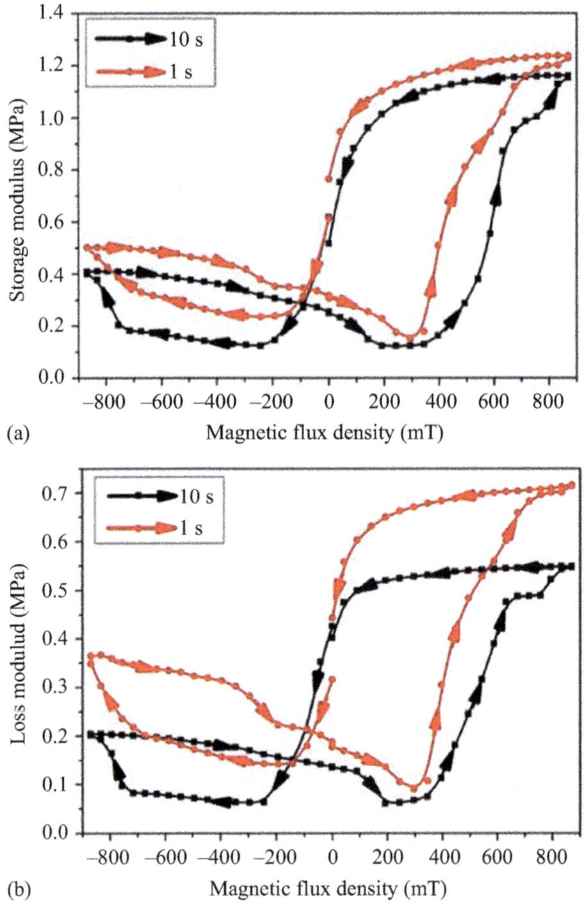

Figure 9.4 Modulus-magnetic field curves of magnetic MRE: (a) storage modulus and (b) loss modulus [13]

silane coupling agent to modify the CIP, and the results showed that the surface modification could improve the dispersion characteristic of particles in the matrix [36]. In addition, Behrooz *et al.* improved the corrosion resistance of MRE by surface coating [37]. In Yu Miao's research group, tetramethoxysilane and polyaniline were used to coat carbonyl iron powder, respectively, which improved the MR effect and reduced the Payne effect of MRE [38,39]. As shown in Figure 9.7, the interface modification not only reduces the interface defects but also improves the particle-agglomeration phenomenon. Actually, the chain-like structure formed in the prestructure process of MRE is also a kind of ordered particle-agglomeration phenomenon, and which will inevitably affect the material strength of MRE. Therefore, interface modification and reasonable prestructure method are both crucial to the mechanical properties of MRE.

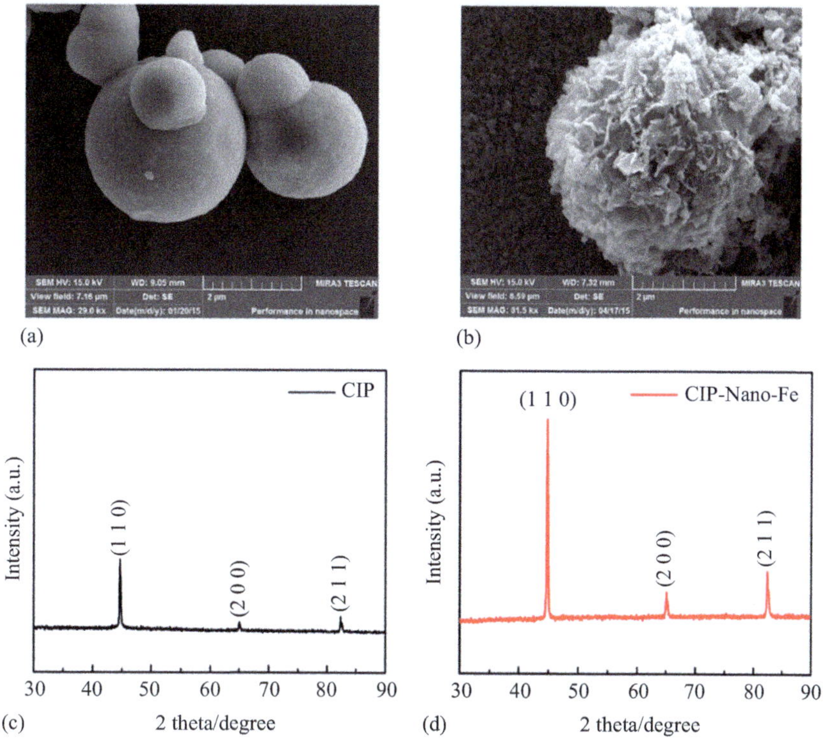

Figure 9.5 The SEM and XRD of CI particles before and after coated with Fe nanostructures: (a) and (b) the SEM micrograph of CIP and CIP-Nano-Fe; (c) and (d) the XRD patterns of CIP and CIP-Nano-Fe [32]

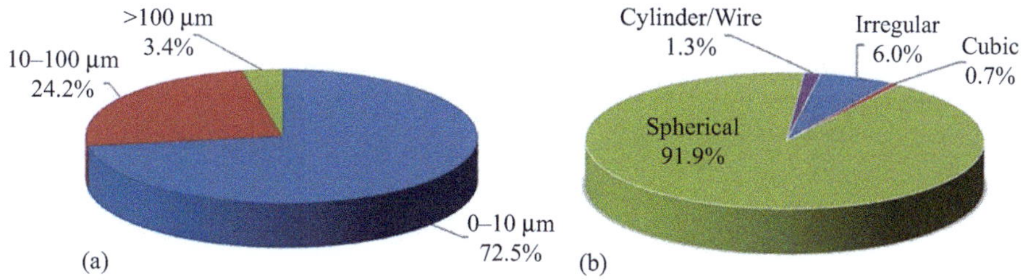

Figure 9.6 Mapping of MREs based on magnetizable particle: (a) based on particle size and (b) based on particle shape [9]

9.2.3 Additives

Improved additives or additional enhanced particles are also an effective means to improve the magnetomechanical properties of MRE. Based on this, the reinforcing agent mainly employed at present are silicon carbide particles [40,41], nanometer

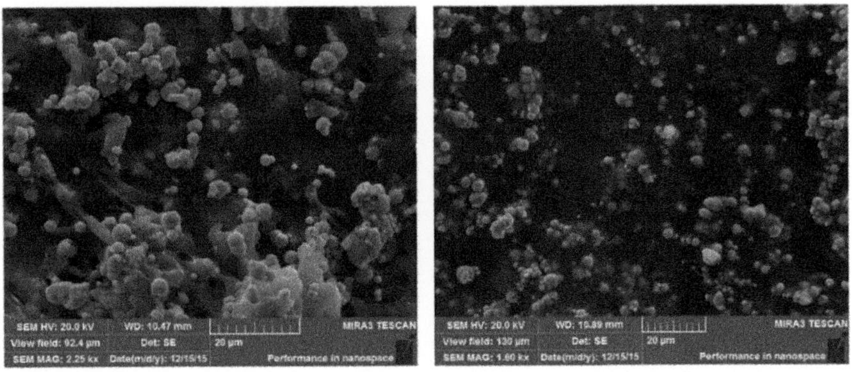

(a) (b)

Figure 9.7 SEM images of (a) unmodified MREs and (b) modified MREs [38,39]

ferric oxide [42], carbon black [43], carbon nanotubes [44] and so on. From the whole perspective of MRE material development, currently prepared MRE materials often fail to take into account both mechanical strength and MR effect. Low MR effect or mechanical strength limits the MRE's engineering application value. Taken together, the MR effect should not be the only evaluation index of MRE. According to specific application background, producing MRE materials which possess comprehensive performance and could better satisfy the requirements should be the main research priority in the future.

9.3 Test characterization of magnetorheological elastomers

The magnetic-control mechanical properties are the most widely studied magnetic properties of MRE. At present, the most commonly used test instrument is the rheometer of Anton Paar GmbH. The instrument is equipped with a magnetorheology device, which can realize the oscillatory shear test of MRE under different magnetic flux densities [45–47]. The physical diagram of rheometer and the schematic illustration of testing part are shown in Figure 9.8. Besides, Gong *et al.* equipped an electromagnet module in the traditional dynamic mechanical analyzer, so as to test the dynamic mechanical properties of MRE [40].

Kallio *et al.* and Koo *et al.* studied the effects of strain amplitude and frequency on the dynamic compressive mechanical properties of MRE [48,49]. As a typical viscoelastic material, MRE's static viscoelasticity also shows an obvious dependence on the magnetic field. Qi *et al.* conducted a systematic study on the creep recovery and stress relaxation characteristics of MRE and found that the magnetic field can reduce the creep of MRE and limit its recovery and stress relaxation at the same time [50,51]. Zhu *et al.* proposed a Maxwell behavior model with time constant λ to describe the response characteristic of MRE, and they

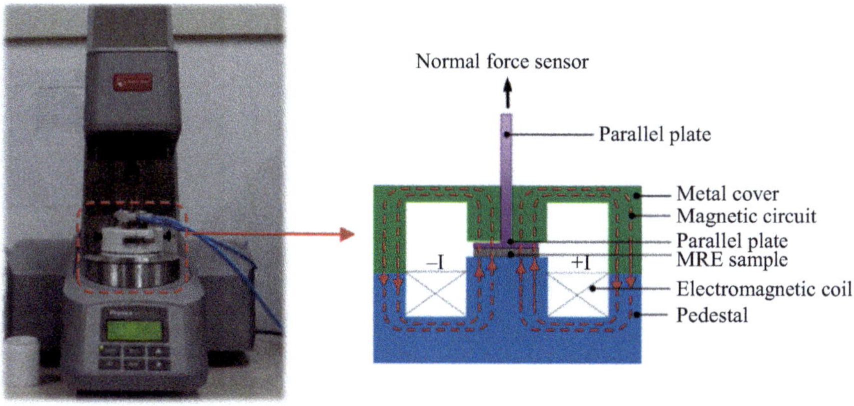

Figure 9.8 The photo and the schematic diagram of the Anton Paar MCR 301 parallel plates rheometer

tested and analyzed the magnetic response time of MRE under compression mode. They found that the anisotropic MRE possessed a better responsive performance (response time was about 12 ms) over isotropic one (response time was about 26 ms), as shown in Figure 9.9 [52]. Besides the intrinsic importance for contributing to the response evaluation and effective design of MRE device, the study may conduce to the optimal design of controller for MRE control system.

In order to better promote the application of MRE in practical engineering, the stability has been paid more and more attention by researchers. Thermogravimetric analysis and differential scanning calorimetry were used to study the heat resistance of MRE [24,53]. In addition, the fatigue property of MRE has also attracted more and more attention recently [17,40,54].

In terms of microstructure characterization of MRE, the most commonly used means at present are SEM, transmission electron microscopy, atomic force microscopy and digital holographic interferometry. The computer tomography (CT) can reconstruct the three-dimensional internal structure of material. More and more researchers have studied the distribution and movement behavior of particles inside MRE using this method [10,55–57], as shown in Figure 9.10.

The conductivity of ferromagnetic particles in MRE is several orders of magnitude higher than that of the matrix, and the unique chain structure inside the anisotropic MRE can form a conductive pathway, so it brings unique electrical characteristics to MRE. Tian *et al.* doped graphene into MRE and found that the doped MRE's resistance would change with the applied magnetic field [58]. Bica prepared the conductive MRE by adding graphene particles and studied the effects of magnetic field and pressure on its conductivity [59]. Recently, Fan *et al.* studied the piezoresistive effect of MRE. The results showed that the electrical properties of MRE mainly relate to the internal particle structure and compressive load [60]. Besides, Gong *et al.* conducted an in-depth study on the conductive MRE; they prepared the strain-sensitive conductive MRE with sponge [61] and studied the

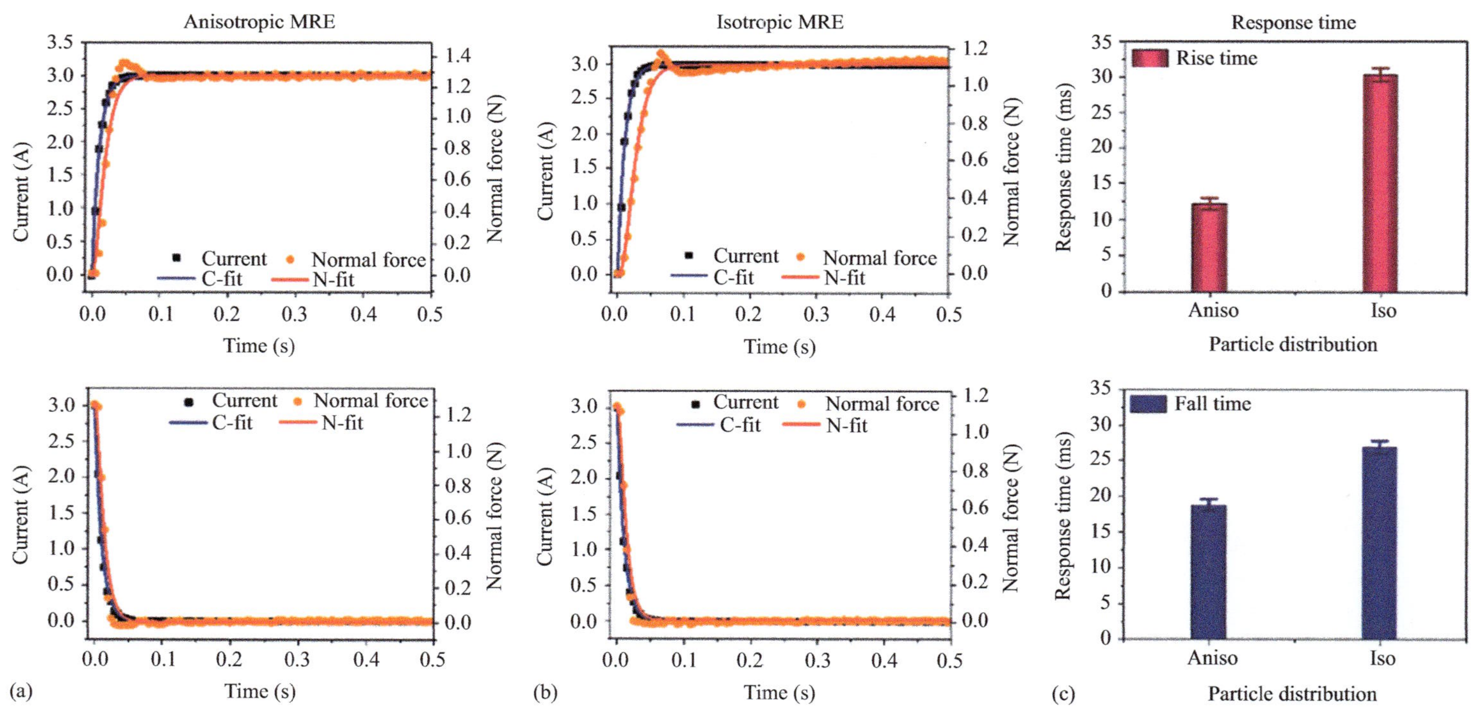

Figure 9.9 *Response time in rise and fall edge under different particles distribution: (a) anisotropic MRE, (b) isotropic MRE and (c) dependence of the response time on the particle distribution [52]*

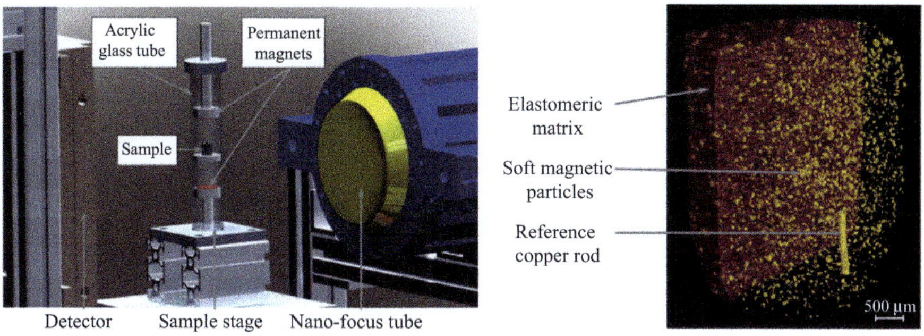

Figure 9.10 Experimental setup of Xμ-CT and the tomogram of the MRE sample [10]

coupling relationship between the impedance characteristics and dynamic mechanical characteristics of MRE under fatigue load [62]. The above research highlights the application prospect of conductive MRE in sensing field.

The magnetic particles in MRE will produce interaction force under magnetic field. Through the coupling effect between magnetic particles and elastic matrix, MRE will deform obviously under applied magnetic field. Martin *et al.* studied the magnetostriction of MRE in 2004, and the maximum shrinkage deformation was measured by optical method, the result reached to 10,000 ppm [63,64]. Guan *et al.* used strain gauges to measure the magnetostriction of columnar MRE; the results showed that the magnetostriction effect was 184 ppm [65]. Gong *et al.* used digital holographic interferometry to analyze the two-dimensional deformation of MRE under magnetic field. It was observed that both contraction and tensile deformation existed in MRE, and the deformation state was mainly related to the particle distribution state [66]. Qi *et al.* studied the magneto-induced deformation and response time of MRE thin films by using laser displacement sensor; the response time was found to be within 50 ms [67]. Recently, Feng *et al.* studied the magneto-induced deformation of MRE thin films by using laser Doppler velocimeter and digital-image-processing method, the system schematic diagram and test results of digital image processing method are shown in Figure 9.11 [68].

In addition to above properties, the researchers found that MRE also has unique magnetic properties in optical, thermal, acoustic and electromagnetic shielding. Zhang *et al.* used nano-Fe_3O_4 particles doped into PDMS matrix to prepare thin film MRE. It was found that the transmittance of MRE under 800 Gs magnetic field decreased by about 8.61% compared with the one under zero field, and the anisotropic MRE had a better transmittance than the isotropic one [69]. This work provides the possibility for MRE to develop magnetically controlled optical devices. Wu *et al.* tested the thermal conductivity of polyurethane MRE and found that the thermal conductivity of anisotropic MRE is higher than that of isotropic MRE, the reason is that the internal particle chain structure can be regarded as a heat conduction channel to improve the thermal conductivity [70]. Chen *et al.*

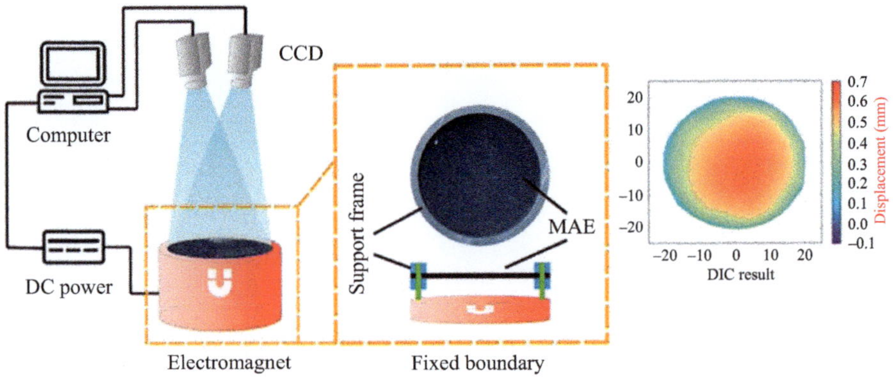

Figure 9.11 Schematic illustration of the digital image correlation (DIC) method and the DIC result [68]

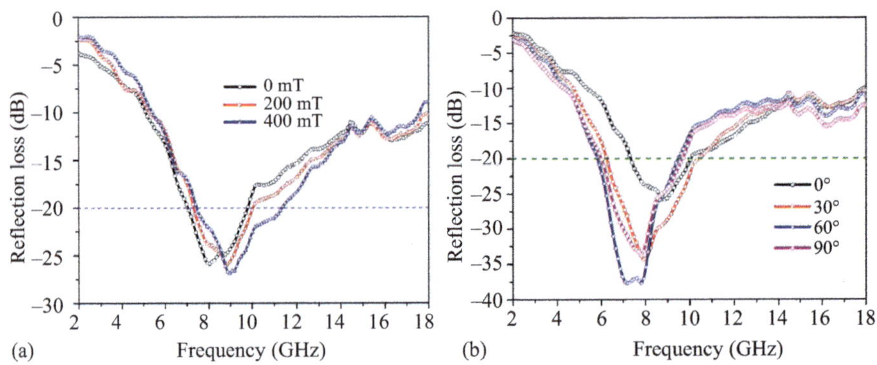

Figure 9.12 Reflection loss curves of the MRE with the different (a) applied magnetic field intensity and (b) angle [76]

investigated the acoustic characteristics of MRE under different magnetic fields, they found that the resonance frequency increased with the increase of magnetic field [71]. Faramarz and Xu *et al.* also studied the tunable band gap characteristics of the phononic crystals composed of MRE [72–75]. Besides, Yu *et al.* studied the electromagnetic wave absorption characteristics of MRE, as shown in Figure 9.12; they found that MRE's absorbing property is closely related to the magnetic field intensity and the spacing and angle of MRE's internal chain structure [76]. The test characterization of MRE is the basis of application, and by various advanced instruments to fully dig the MRE's magnetic-control characteristics is the only way to promote the development of MRE research.

9.3.1 Mechanism of magnetorheological elastomers

At present, the research on MRE magnetic-sensing mechanism mainly includes two aspects: micro and macro magnetic-sensing mechanism. The MRE microscopic theoretical model generally simplifies the particles as magnetic dipoles and obtains the MRE's magnetic modulus under magnetic field by calculating the change of particles interaction energy. Among them, the most representative ones are the dipole model established by Jolly of LORD Company [77] and the coupling field model proposed by Shen [78]. In order to improve the accuracy of the dipole model, many researchers have introduced more influencing factors into the calculation model [33,77,79]. Thereinto, Chen established the finite column length model and predicted the magnetic characteristics of MRE in shear mode [80]. However, the magnetic dipole model ignores the influence of particle and matrix forces on MRE's viscoelasticity. Based on the theory of magnetic elasticity, many scholars proposed the constitutive model of MRE viscoelastic by introducing the magnetoelastic strain energy density function to describe the coupling relationship between MRE internal mechanical energy and magnetic field energy. Such models can reflect the effect of the interaction between the matrix and ferromagnetic particles on the viscoelasticity of MRE. According to the Cauchy stress invariant theory, Dorfmann *et al.* considered MRE as a hyperelastic material and ignored the internal electric polarization strength. By using the continuous medium theory, they deduced the constitutive equation of isotropic MRE, and the relation between MRE stress and external magnetic field under different magnetic field directions and deformation conditions was obtained [81–83]. Borcea *et al.* established a comprehensive model of isotropic MRE by using the principle of minimum potential energy and obtained its relationship between stress and strain [5]. Based on the finite deformation, Rudykh *et al.* deduced the occurrence conditions of MRE's macroscopic instability. By regarding the internal chain structure of MRE as multilayer microstructure, they established a precise mechanical model to describe the relation between MRE's stiffness and external magnetic field [84]. Danas proposed the transversely isotropic energy density function based on the theoretical basis of MRE finite strain, which can describe the mechanical properties of MRE under different preloads, particle chain directions and external magnetic fields [85].

The micro magnetic-sensing mechanism is usually based on certain ideal conditions, but actually, the particle size, distribution state, interface structure, matrix characteristics and so on all will affect the magnetic properties. Therefore, there is a lack of unified micro magnetic-sensing to describe MRE's magnetic properties perfectly. For practical applications, the establishment of macroscopical mechanics model has more practical application value. At present, the macro-mechanical model of MRE is mainly based on the traditional viscoelastic parameterized and non-parameterized macro-model. For example, Chen *et al.* described the viscoelastic properties of MRE such as stiffness and damping by using spring element and viscous pot [80], as shown in Figure 9.13.

Li Weihua's research group has established a four-parameter model that characterizes MRE under dynamic loading conditions and obtained the values of

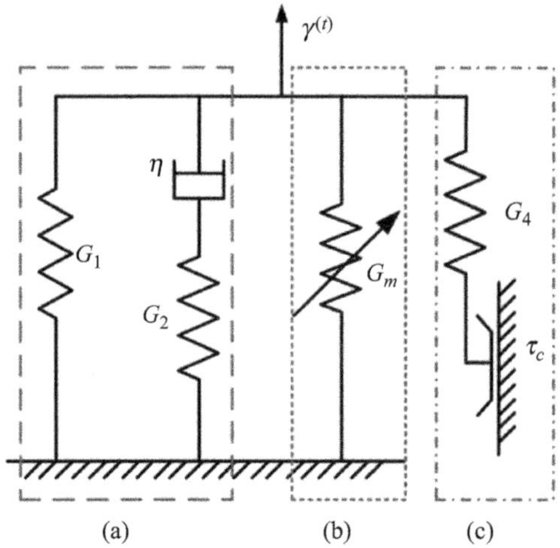

Figure 9.13 MRE magnetic viscoelastic model based on the spring and dashpot (a) the viscoelasticity of the polymer composite, (b) the magnetic-field-induced mechanical properties, and (c) interfacial slippage between the matrix and the particles. [80]

each parameter through experimental data fitting [86]. In addition, they studied the creep characteristics of MRE, and used the four-parameter model to discuss the MRE's creep characteristics [87]. Norouzi *et al.* proposed an improved Kelvin–Voigt viscoelastic model, which can describe the stress–strain relationship of MRE under certain excitation conditions, and the parameters of the model would not change with the change of external load and magnetic field [88]. According to the microstructure and macroscopic mechanical test results of MRE, Xin *et al.* proposed a model combining traditional viscoelastic with magnetic-sensing viscoelastic model, which can simultaneously describe the relationship between MRE's viscoelastic and strain, frequency and magnetic field [89]. Recently, Cantera *et al.* reviewed the mechanism model of MRE's magnetic sensing characteristics in detail and classified the theoretical models from the aspects of particles interaction mechanism, magnetoelastic theory, magneto-viscoelastic theory, phenomenological model and so on [90]. They believed that the theoretical model studies on the characteristics (creep, stress relaxation, fatigue, durability, etc.) of MRE over a long period of time are needed to be conducted in future studies, which can help to push MRE into practical engineering applications.

9.3.2 Application research of magnetorheological elastomer

Based on the excellent magnetic-control mechanical properties of MRE, researchers have designed and developed numerous intelligent vibration absorbers, which have achieved a lot of applications in the fields of civil engineering, automobile, precision machining and so on [91–97]. Now many research groups have made important contributions in promoting the application of MRE in the field of vibration control [98–104]. Figure 9.14 shows a typical laminated MRE lateral vibration isolator and its magnetically controlled shift–frequency curve.

Figure 9.15 shows a hybrid MRE–MRF isolation mount, which was designed based on the characteristics of MRE's controllable stiffness and MRF's controllable damping. It realized the independent controllability of stiffness and damping of vibration isolator [105].

Considering variable amplitude and frequency vibration of some isolated systems, Fu *et al.* design an active/semi-active hybrid isolator using MRE isolator and piezoelectric stack actuator to suppress the wide-frequency vibration [106], the structure is shown in Figure 9.16. It can be switched to active isolator, semi-active one and passive one according to the spectrum signature of the excitation. Besides, an active and semi-active fuzzy switch controller is designed for the hybrid isolator, which is independent of the system model, self-tuning and robust. The results showed that the hybrid isolator, combined with the fuzzy switch controller, can attenuate full frequency vibration effectively.

Aiming at the micro-vibration problem in precision machining or measuring platform system, Fu *et al.* also proposed a semi-active fuzzy control method to solve the complex nonlinear problem of MRE micro-vibration isolation system. In addition, a nonparametric auto regression neural network was used to model the MRE isolator [107,108], as shown in Figure 9.17.

In addition to intelligent vibration damping applications, in recent years, applications based on magnetostrictive, magnetoelectrics, magneto-optics and other

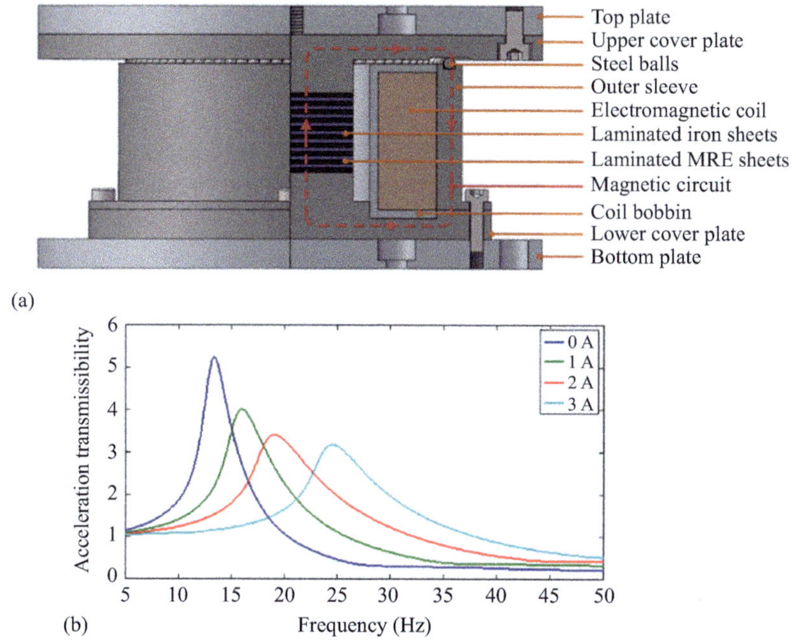

Figure 9.14 The (a) structure and (b) transfer characteristic of laminated MRE isolation bearing

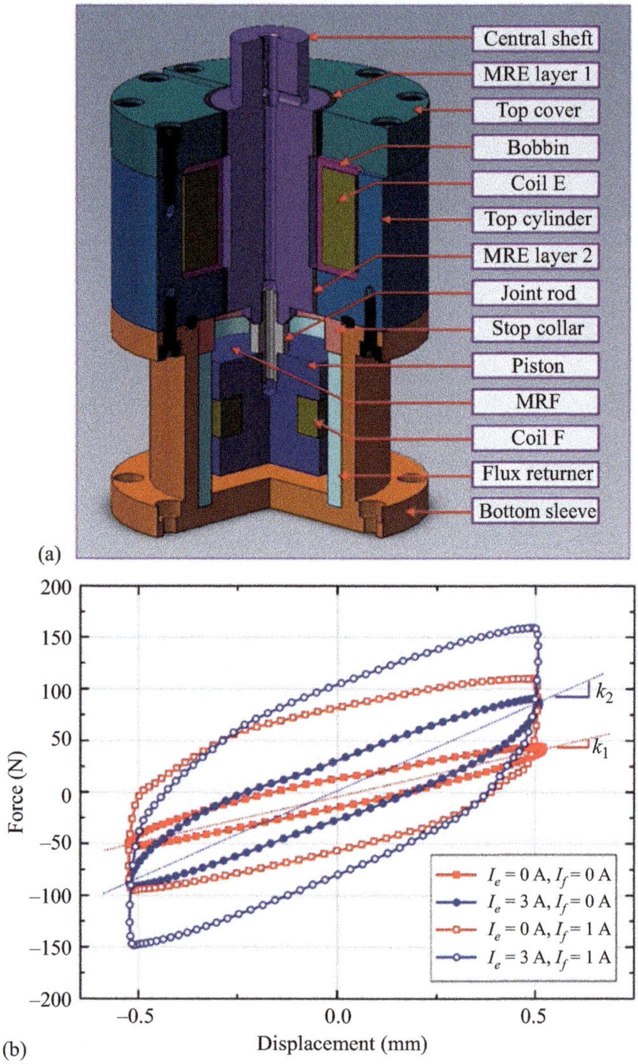

Figure 9.15 The structure (a) and force–displacement (b) of laminated-MRE isolation bearing [105]

characteristics of devices emerge in endlessly. For example, based on the excellent magnetic deformation characteristics of MRE, researchers have developed a series of magnetic controlled flexible actuators and have carried out a series of applied research in the fields of soft robots, sensors, microfluidics and so on [109–112]. Recently, Li *et al.* reported a microfluidic control system based on MRE magnetostrictive, which successfully realized the controlled microfluidic system by magnetic field [113]. Aiming at the application of MRE in soft robot, Schmauch *et al.* prepared anisotropic MRE membrane and developed various flexible actuators [114].

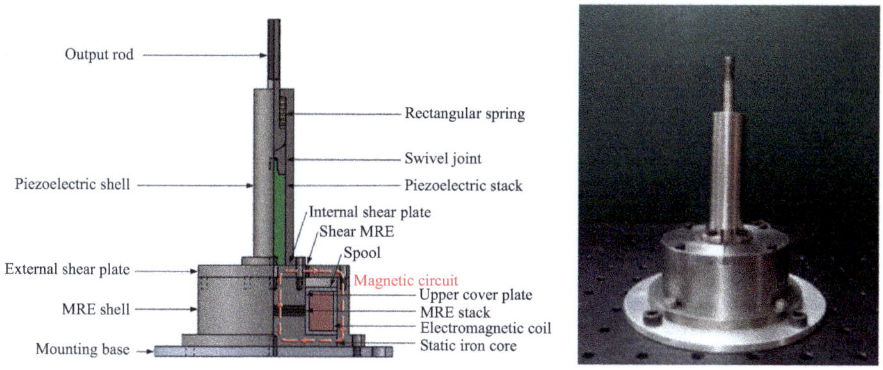

Figure 9.16 The structure of the active/semi-active isolator using MRE isolator and piezoelectric stack actuator [106]

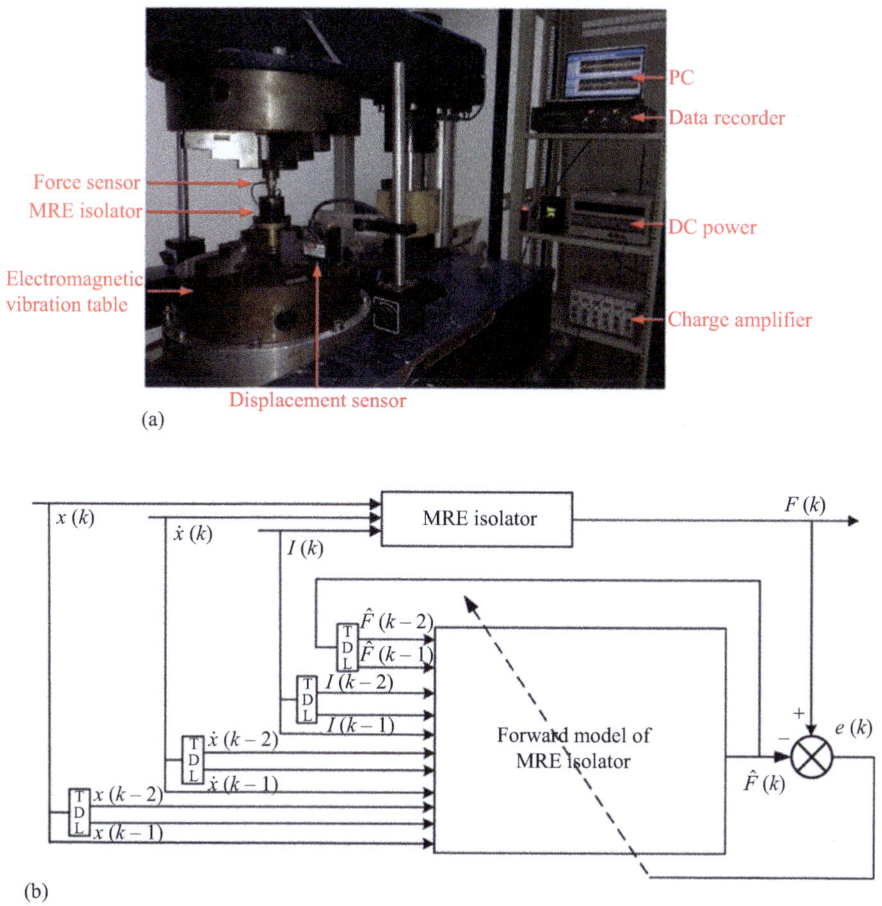

Figure 9.17 (a) Test system of MRE device and (b) self-regressive neural network modeling of MRE vibration isolation system [107,108]

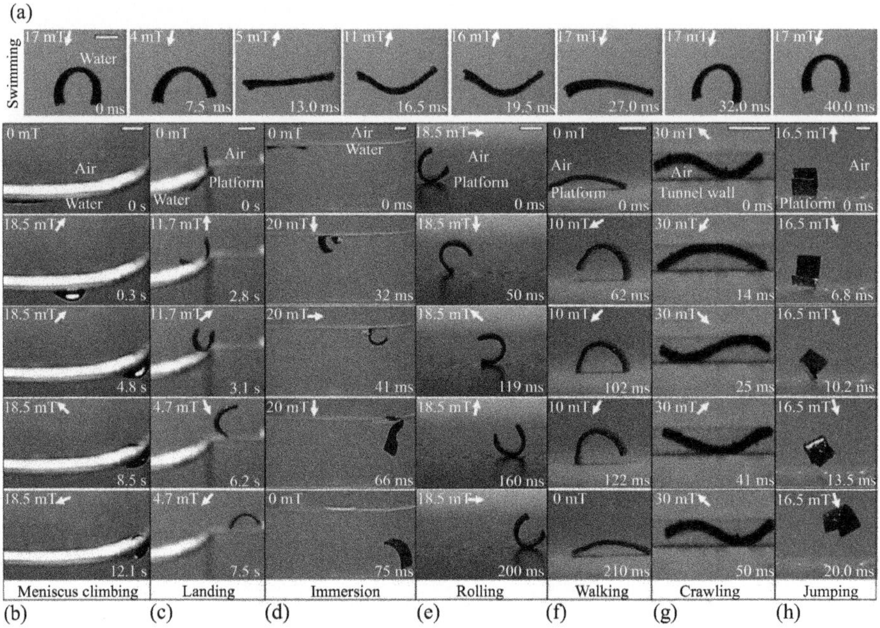

Figure 9.18 Different kinds of actuating behavior of soft robot [115]

Gong *et al.* prepared magnetic composite film by using MRE and polyvinylidene fluoride, and by using the composite film, they made some flexible claws. The claw could capture and release small objects by controlling the applied magnetic field. Besides, they also studied the change of the induced charge of the actuator during the operation and made a preliminary attempt to monitor the deformation by real-time feedback [110]. Lately, Hu *et al.* reported a small soft-bodied robot made of silicone rubber and rubidium–iron–boron particles in the journal "Nature." As shown in Figure 9.18, a series of actions such as crawling, grasping and jumping of the soft robot were realized by designing the corresponding time-varying magnetic field. The results proved that it has considerable application prospects in medical filed [115].

Qi *et al.* designed and fabricated a magnetic field sensor with self-energizing characteristics; it was based on the sensitive magnetic deformation characteristics of MRE film combined with friction nano-generator. As shown in Figure 9.19, the sensor can realize the monitoring of uniform magnetic field under passive condition [67].

In addition, based on the magnetoelectric properties of MRE, Bica fabricated a sensor by using MRE with conductive graphite sheet [116]. Electrical characteristics are the hotspot of recent MRE research, and the development of sensors based on their electrical characteristics will become an important branch of the application of MRE.

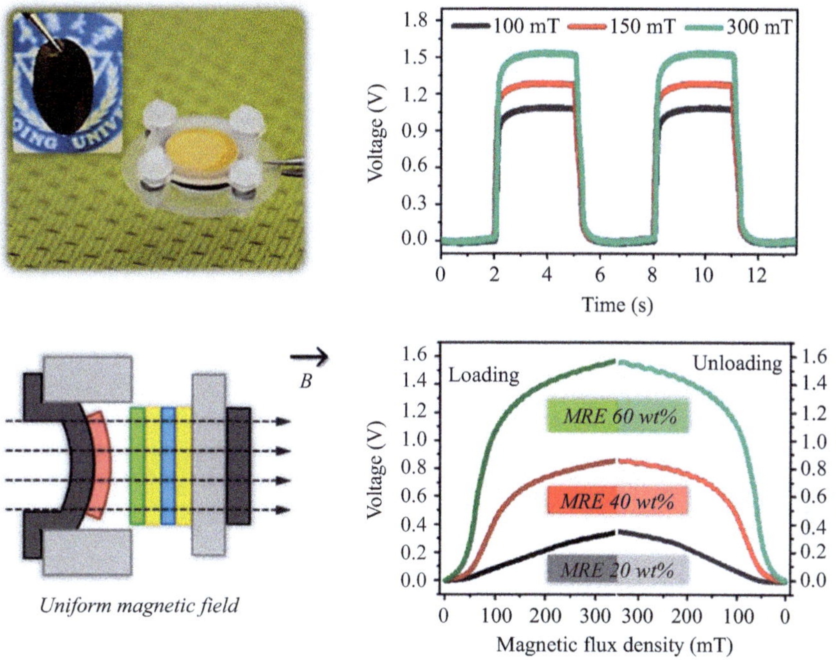

Figure 9.19 Self-power magnetic-field sensor based on MRE film and its electrical response characteristics [67]

9.4 Conclusions and prospect

The emergence of MSVM has strongly promoted the development of smart materials, opened up a broader space for the design of magnetic-controlled smart devices and provided new ideas for solving engineering problems. Although some breakthroughs have been made in the study of the materials, many theoretical and technical problems need to be solved urgently in order to realize its popularization of engineering applications. First, the magnetic-induced properties of MRE under low magnetic field still need to be improved. It is necessary to improve its stability (such as durability and weatherability) over long timescales, by simultaneously taking account of its static characteristics and magnetic-induced effects. Second, the material research and device research are mostly in a split state. It is the key to promote MRE's engineering application by coordinating and optimizing the material characteristics and device parameters for specific application objects and environments. Finally, MRE has great application prospects in the fields of optics, thermology, acoustics and electromagnetism shield. However, the related research is just getting started, the related material preparation and device design need to be further studied. In summary, MRE is a cross-disciplinary research direction of material, physics, information and other disciplines. In order to realize the large-scale engineering application of magnetic-sensitive viscoelastic materials as soon as possible, we need joint efforts from researchers in all fields.

References

[1] Carlson, J. D., and Jolly, M. R. 2000. MR fluid, foam and elastomer devices. Mechatronics, 10, 555–569.
[2] Davis, L. 1999. Model of magnetorheological elastomers. Journal of Applied Physics, 85, 3348–3351.
[3] Gong, X. L., Zhang, X. Z., and Zhang, P. Q. 2005. Fabrication and characterization of isotropic magnetorheological elastomers. Polymer Testing, 24, 669–676.
[4] Bellan, C., and Bossis, G. 2002. Field dependence of viscoelastic properties of MR elastomers. International Journal of Modern Physics B, 16, 2447–2453.
[5] Borcea, L., and Bruno, O. 2001. On the magneto-elastic properties of elastomer–ferromagnet composites. Journal of the Mechanics and Physics of Solids, 49, 2877–2919.
[6] Ginder, J. M., Nichols, M. E., Elie, L. D., and Clark, S. M. 2000. Controllable-stiffness components based on magnetorheological elastomers. Smart Structures and Material 2000: Smart Structures and Integrated Systems, 3985, 418–425.
[7] Hoang, N., Zhang, N., and Du, H. 2011. An adaptive tunable vibration absorber using a new magnetorheological elastomer for vehicular powertrain transient vibration reduction. Smart Materials and Structures, 20, 015019.
[8] Jeong, U. C., Yoon, J. H., Yang, I. H., *et al.* 2013. Magnetorheological elastomer with stiffness-variable characteristics based on induced current applied to differential mount of vehicles. Smart Materials and Structures, 22, 115007.
[9] Sutrisno, J., Purwanto, A., and Mazlan, S. A. 2015. Recent progress on magnetorheological solids: materials, fabrication, testing, and applications. Advanced Engineering Materials, 17, 563–597.
[10] Gundermann, T., and Odenbach, S. 2014. Investigation of the motion of particles in magnetorheological elastomers by X-μCT. Smart Materials and Structures, 23, 105013.
[11] Schumann, M., Borin, D. Y., Huang, S., Auernhammer, G. K., Muller, R., and Odenbach, S. 2017. A characterisation of the magnetically induced movement of NdFeB-particles in magnetorheological elastomers. Smart Materials and Structures, 26, 095018.
[12] Shiga, T., Okada, A., and Kurauchi, T. 1995. Magnetroviscoelastic behavior of composite gels. Journal of Applied Polymer Science, 58, 787–792.
[13] Yu, M., Qi, S., Fu, J., and Zhu, M. 2015. A high-damping magnetorheological elastomer with bi-directional magnetic-control modulus for potential application in seismology. Applied Physics Letters, 107, 111901.

[14] Agirre-Olabide, I., Elejabarrieta, M. J., and Bou-Ali, M. M. 2015. Matrix dependence of the linear viscoelastic region in magnetorheological elastomers. Journal of Intelligent Material Systems and Structures, 26, 1880–1886.

[15] Kimura, T., Umehara, Y., and Kimura, F. 2012. Magnetic field responsive silicone elastomer loaded with short steel wires having orientation distribution. Soft Matter, 8, 6206–6209.

[16] Yu, M., Fu, J., Ju, B. X., Zheng, X., and Choi, S. B. 2013. Influence of x-ray radiation on the properties of magnetorheological elastomers. Smart Materials and Structures, 22, 125010.

[17] Lokander, M., Reitberger, T., and Stenberg, B. 2004. Oxidation of natural rubber-based magnetorheological elastomers. Polymer Degradation and Stability, 86, 467–471.

[18] Ju, B. X., Yu, M., Fu, J., Zheng, X., and Yang, Q. 2013. Study on the properties of porous magnetorheological elastomers under shock effect. 13th International Conference on Electrorheological Fluids and Magnetorheological Suspensions (ERMR2012), 412.

[19] Slawinski, G., Miedzinska, D., Niezgoda, T., and Boczkowska, A. 2012. Experimental investigations of MREs behavior under the cyclic load. Environmental Degradation of Engineering and Materials Engineering and Technologies, 183, 163–168.

[20] Wei, B., Gong, X. L., and Jiang, W. Q. 2010. Influence of polyurethane properties on mechanical performances of magnetorheological elastomers. Journal of Applied Polymer Science, 116, 771–778.

[21] Sun, T. L., Gong, X. L., Jiang, W. Q., Li, J. F., Xu, Z. B., and Li, W. H. 2008. Study on the damping properties of magnetorheological elastomers based on cis-polybutadiene rubber. Polymer Testing, 27, 520–526.

[22] Zhang, W., Gong, X. L., Sun, T. L., Fan, Y. C., and Jiang, W. Q. 2010. Effect of cyclic deformation on magnetorheological elastomers. Chinese Journal of Chemical Physics, 23, 226–230.

[23] Li, W. H., and Nakano, M. 2013. Fabrication and characterization of PDMS based magnetorheological elastomers. Smart Materials and Structures, 22, 055035.

[24] Yu, M., Qi, S., Fu, J., Yang, P. A., and Zhu, M. 2015. Preparation and characterization of a novel magnetorheological elastomer based on polyurethane/epoxy resin IPNs matrix. Smart Materials and Structures, 24, 045009.

[25] Ge, L., Gong, X. L., Fan, Y. C., and Xuan, S. H. 2013. Preparation and mechanical properties of the magnetorheological elastomer based on natural rubber/rosin glycerin hybrid matrix. Smart Materials and Structures, 22, 115029.

[26] Zhao, X. Y., Yang, J. N., Zhao, D. T., *et al.* 2015. Natural rubber/nitrile butadiene rubber/hindered phenol composites with high-damping properties. International Journal of Smart and Nano Materials, 6, 239–250.

[27] Farshad, M., and Benine, A. 2004. Magnetoactive elastomer composites. Polymer Testing, 23, 347–353.

[28] Koo, J.-H., Dawson, A., and Jung, H.-J. 2012. Characterization of actuation properties of magnetorheological elastomers with embedded hard magnetic particles. Journal of Intelligent Material Systems and Structures, 23, 1049–1054.
[29] Kramarenko, E. Y., Chertovich, A. V., Stepanov, G. V., *et al.* 2015. Magnetic and viscoelastic response of elastomers with hard magnetic filler. Smart Materials and Structures, 24, 035002.
[30] Stepanov, G. V., Chertovich, A. V., and Kramarenko, E. Y. 2012. Magnetorheological and deformation properties of magnetically controlled elastomers with hard magnetic filler. Journal of Magnetism and Magnetic Materials, 324, 3448–3451.
[31] Yu, M., Yang, P., Fu, J., Liu, S., and Choi, S.-B. 2016. A theoretical model for the field-dependent conductivity of magneto-rheological gels and experimental verification. Sensors and Actuators A: Physical, 245, 127–134.
[32] Yu, M., Zhu, M., Fu, J., Yang, P. A., and Qi, S. 2015. A dimorphic magnetorheological elastomer incorporated with Fe nano-flakes modified carbonyl iron particles: preparation and characterization. Smart Materials and Structures, 24, 115021.
[33] Li, W. H., and Zhang, X. Z. 2010. A study of the magnetorheological effect of bimodal particle based magnetorheological elastomers. Smart Materials and Structures, 19, 035002.
[34] Qiao, X. Y., Lu, X. S., Li, W. H., *et al.* 2012. Microstructure and magnetorheological properties of the thermoplastic magnetorheological elastomer composites containing modified carbonyl iron particles and poly (styrene-*b*-ethylene-ethylene propylene-*b*-styrene) matrix. Smart Materials and Structures, 21, 115028.
[35] Jiang, W. C., Yao, J. J., Gong, X. L., and Chen, L. 2008. Enhancement in magnetorheological effect of magnetorheological elastomers by surface modification of iron particles. Chinese Journal of Chemical Physics, 21, 87–92.
[36] Wang, Y. L., Hu, Y. A., Chen, L., *et al.* 2006. Effects of rubber/magnetic particle interactions on the performance of magnetorheological elastomers. Polymer Testing, 25, 262–267.
[37] Behrooz, M., Sutrisno, J., Zhang, L. Y., Fuchs, A., and Gordaninejad, F. 2015. Behavior of magnetorheological elastomers with coated particles. Smart Materials and Structures, 24, 035026.
[38] Chen, D., Yu, M., Zhu, M., Qi, S., and Fu, J. 2016. Carbonyl iron powder surface modification of magnetorheological elastomers for vibration absorbing application. Smart Materials and Structures, 25, 115005.
[39] Yu, M., Qi, S., Fu, J., Zhu, M., and Chen, D. 2017. Understanding the reinforcing behaviors of polyaniline-modified carbonyl iron particles in magnetorheological elastomer based on polyurethane/epoxy resin IPNs matrix. Composites Science and Technology, 139, 36–46.
[40] Wang, Y., Gong, X. L., Yang, J., and Xuan, S. H. 2014. Improving the dynamic properties of MRE under cyclic loading by incorporating silicon carbide nanoparticles. Industrial and Engineering Chemistry Research, 53, 3065–3072.

[41] Yang, J., Gong, X. L., Zong, L. H., Peng, C., and Xuan, S. H. 2013. Silicon carbide-strengthened magnetorheological elastomer: preparation and mechanical property. Polymer Engineering and Science, 53, 2615–2623.
[42] Yang, P., Yu, M., and Fu, J. 2016. Ni-coated multi-walled carbon nanotubes enhanced the magnetorheological performance of magnetorheological gel. Journal of Nanoparticle Research, 18, 61.
[43] Nayak, B., Dwivedy, S. K., and Murthy, K. S. R. K. 2015. Fabrication and characterization of magnetorheological elastomer with carbon black. Journal of Intelligent Material Systems and Structures, 26, 830–839.
[44] Li, R., and Sun, L. Z. 2011. Dynamic mechanical behavior of magnetorheological nanocomposites filled with carbon nanotubes. Applied Physics Letters, 99, 131912.
[45] Ge, L., Xuan, S. H., Liao, G. J., Yin, T. T., and Gong, X. L. 2015. Stretchable polyurethane sponge reinforced magnetorheological material with enhanced mechanical properties. Smart Materials and Structures, 24, 037001.
[46] Jung, H. S., Kwon, S. H., Choi, H. J., Jung, J. H., and Kim, Y. G. 2016. Magnetic carbonyl iron/natural rubber composite elastomer and its magnetorheology. Composite Structures, 136, 106–112.
[47] Tian, T. F., Zhang, X. Z., Li, W. H., Alici, G., and Ding, J. 2013. Study of PDMS based magnetorheological elastomers. 13th International Conference on Electrorheological Fluids and Magnetorheological Suspensions (ERMR2012), 412, 012038.
[48] Kallio, M., Lindroos, T., Aalto, S., Järvinen, E., Kärnä, T., and Meinander, T. 2007. Dynamic compression testing of a tunable spring element consisting of a magnetorheological elastomer. Smart Materials and Structures, 16, 506–514.
[49] Koo, J.-H., Khan, F., Jang, D.-D., and Jung, H.-J. 2010. Dynamic characterization and modeling of magneto-rheological elastomers under compressive loadings. Smart Materials and Structures, 19, 117002.
[50] Qi, S., Yu, M., Fu, J., Li, P. D., and Zhu, M. 2016. Creep and recovery behaviors of magnetorheological elastomer based on polyurethane/epoxy resin IPNs matrix. Smart Materials and Structures, 25, 015020.
[51] Qi, S., Yu, M., Fu, J., and Zhu, M. 2017. Stress relaxation behavior of magnetorheological elastomer: Experimental and modeling study. Journal of Intelligent Material Systems and Structures, 29, 205–213.
[52] Zhu, M., Yu, M., Qi, S., and Fu, J. 2018. Investigations on response time of magnetorheological elastomer under compression mode. Smart Materials and Structures, 27, 055017.
[53] Fuchs, A., Sutrisno, J., Gordaninejad, F., Caglar, M. B., and Yanming, L. 2010. Surface polymerization of iron particles for magnetorheological elastomers. Journal of Applied Polymer Science, 117, 934–942.
[54] Zhang, W., Gong, X. L., Jiang, W. Q., and Fan, Y. C. 2010. Investigation of the durability of anisotropic magnetorheological elastomers based on mixed rubber. Smart Materials and Structures, 19, 085008.

[55] Borin, D., Günther, D., Hintze, C., Heinrich, G., and Odenbach, S. 2012. The level of cross-linking and the structure of anisotropic magnetorheological elastomers. Journal of Magnetism and Magnetic Materials, 324, 3452–3454.
[56] Günther, D., Borin, D. Y., Günther, S., and Odenbach, S. 2011. X-ray micro-tomographic characterization of field-structured magnetorheological elastomers. Smart Materials and Structures, 21, 015005.
[57] Wang, Y., Xuan, S., Dong, B., Xu, F., and Gong, X. 2015. Stimuli dependent impedance of conductive magnetorheological elastomers. Smart Materials and Structures, 25, 025003.
[58] Tian, T., Li, W., and Deng, Y. 2011. Sensing capabilities of graphite based MR elastomers. Smart Materials and Structures, 20, 025022.
[59] Bica, I., Anitas, E. M., Bunoiu, M., Vatzulik, B., and Juganaru, I. 2014. Hybrid magnetorheological elastomer: influence of magnetic field and compression pressure on its electrical conductivity. Journal of Industrial and Engineering Chemistry, 20, 3994–3999.
[60] Fan, Y., Liao, C., Xie, L., and Chen, X. 2018. Piezo-capacitive behavior of magnetically structured particles-based conductive polymer with high sensitivity and wide working range. Journal of Materials Chemistry C, 6, 5401–5411.
[61] Ge, L., Gong, X. L., Wang, Y., and Xuan, S. H. 2016. The conductive three dimensional topological structure enhanced magnetorheological elastomer towards a strain sensor. Composites Science and Technology, 135, 92–99.
[62] Wang, Y., Xuan, S., Ge, L., Wen, Q., and Gong, X. 2016. Conductive magnetorheological elastomer: fatigue dependent impedance-mechanic coupling properties. Smart Materials and Structures, 26, 015004.
[63] Martin, J. E., Anderson, R. A., and Gulley, G. 2004. Magnetostriction of field-structured composites. Materials and Devices for Smart Systems, 785, 433–438.
[64] Martin, J. E., Anderson, R. A., Read, D., and Gulley, G. 2006. Magnetostriction of field-structured magnetoelastomers. Physical Review E, 74, 051507.
[65] Guan, X. C., Dong, X. F., and Ou, J. P. 2008. Magnetostrictive effect of magnetorheological elastomer. Journal of Magnetism and Magnetic Materials, 320, 158–163.
[66] Gong, X. L., Liao, G. J., and Xuan, S. H. 2012. Full-field deformation of magnetorheological elastomer under uniform magnetic field. Applied Physics Letters, 100, 211909.
[67] Qi, S., Guo, H., Chen, J., *et al.* 2018. Magnetorheological elastomers enabled high sensitive self-powered tribo-sensor for magnetic field detecting. Nanoscale, 10, 4745–4752.
[68] Feng, J., Xuan, S., Lv, Z., Pei, L., Zhang, Q., and Gong, X. 2018. The magnetic-field-induced deformation analysis of magnetoactive elastomer film by means of DIC, LDV, and FEM. Industrial and Engineering Chemistry Research, 57, 3246–3254.

[69] Zhang, Q. S., Peng, B., Li, D., Yang, Y., and Liu, Y. L. 2014. The effect of film microtexture and magnetic field on transparency of Fe_3O_4-PDMS nanocomposite films. IEEE Photonics Technology Letters, 26, 2181–2184.

[70] Wu, J. K., Gong, X. L., Fan, Y. C., and Xia, H. S. 2010. Anisotropic polyurethane magnetorheological elastomer prepared through in situ polycondensation under a magnetic field. Smart Materials and Structures, 19, 105007.

[71] Chen, X., Xu, X. C., Ai, S. G., Chen, H. S., Pei, Y. M., and Zhou, X. M. 2014. Active acoustic metamaterials with tunable effective mass density by gradient magnetic fields. Applied Physics Letters, 105, 071913.

[72] Bayat, A., and Gordaninejad, F. 2015. Band-gap of a soft magnetorheological phononic crystal. Journal of Vibration and Acoustics, 137, 011011.

[73] Bayat, A., and Gordaninejad, F. 2015. Dynamic response of a tunable phononic crystal under applied mechanical and magnetic loadings. Smart Materials and Structures, 24, 065027.

[74] Xu, Z., and Wu, F. 2015. Elastic band gaps of magnetorheological elastomer vibration isolators. Journal of Intelligent Material Systems and Structures, 26, 858–864.

[75] Xu, Z., Wu, F., and Guo, Z. 2013. Shear-wave band gaps tuned in two-dimensional phononic crystals with magnetorheological material. Solid State Communications, 154, 43–45.

[76] Yu., M., Yang., P., Fu., J., Liu., S., and Qi., S. 2016. Study on the characteristics of magneto-sensitive electromagnetic wave-absorbing properties of magnetorheological elastomers. Smart Materials and Structures, 25, 085046.

[77] Gong, X., Zhang, X., and Zhang, P. 2005. Fabrication and characterization of isotropic magnetorheological elastomers. Polymer Testing, 24, 669–676.

[78] Shen, Y., Golnaraghi, M. F., and Heppler, G. R. 2004. Experimental research and modeling of magnetorheological elastomers. Journal of Intelligent Material Systems and Structures, 15, 27–35.

[79] Han, Y., Hong, W., and Faidley, L. E. 2013. Field-stiffening effect of magneto-rheological elastomers. International Journal of Solids and Structures, 50, 2281–2288.

[80] Chen, L., and Jerrams, S. 2011. A rheological model of the dynamic behavior of magnetorheological elastomers. Journal of Applied Physics, 110, 013513.

[81] Bustamante, R., Dorfmann, A., and Ogden, R. 2007. A nonlinear magnetoelastic tube under extension and inflation in an axial magnetic field: numerical solution. Journal of Engineering Mathematics, 59, 139–153.

[82] Bustamante, R., Dorfmann, A., and Ogden, R. 2011. Numerical solution of finite geometry boundary-value problems in nonlinear magnetoelasticity. International Journal of Solids and Structures, 48, 874–883.

[83] Dorfmann, A., and Ogden, R. 2003. Magnetoelastic modelling of elastomers. European Journal of Mechanics—A/Solids, 22, 497–507.

[84] Rudykh, S., and Bertoldi, K. 2013. Stability of anisotropic magnetorheological elastomers in finite deformations: a micromechanical approach. Journal of the Mechanics and Physics of Solids, 61, 949–967.
[85] Danas, K., Kankanala, S., and Triantafyllidis, N. 2012. Experiments and modeling of iron-particle-filled magnetorheological elastomers. Journal of the Mechanics and Physics of Solids, 60, 120–138.
[86] Li, W. H., Zhou, Y., and Tian, T. F. 2010. Viscoelastic properties of MR elastomers under harmonic loading. Rheologica Acta, 49, 733–740.
[87] Li, W., Zhou, Y., Tian, T., and Alici, G. 2010. Creep and recovery behaviors of magnetorheological elastomers. Frontiers of Mechanical Engineering in China, 5, 341–346.
[88] Norouzi, M., Sajjadi Alehashem, S. M., Vatandoost, H., Ni, Y. Q., and Shahmardan, M. M. 2016. A new approach for modeling of magnetorheological elastomers. Journal of Intelligent Material Systems and Structures, 27, 1121–1135.
[89] Xin, F.-L., Bai, X.-X., and Qian, L.-J. 2016. Modeling and experimental verification of frequency-, amplitude-, and magneto-dependent viscoelasticity of magnetorheological elastomers. Smart Materials and Structures, 25, 105002.
[90] Cantera, M. A., Behrooz, M., Gibson, R. F., and Gordaninejad, F. 2017. Modeling of magneto-mechanical response of magnetorheological elastomers (MRE) and MRE-based systems: a review. Smart Materials and Structures, 26, 023001.
[91] Deng, H. X., and Gong, X. L. 2008. Application of magnetorheological elastomer to vibration absorber. Communications in Nonlinear Science and Numerical Simulation, 13, 1938–1947.
[92] Eem, S. H., Jung, H. J., and Koo, J. H. 2011. Application of MR elastomers for improving seismic protection of base-isolated structures. IEEE Transactions on Magnetics, 47, 2901–2904.
[93] Kavlicoglu, B., Wallis, B., Sahin, H., and Liu, Y. M. 2011. Magnetorheological elastomer mount for shock and vibration isolation. Active and Passive Smart Structures and Integrated Systems, 7977, 79770Y.
[94] Li, W. H., Zhang, X. Z., and Du, H. P. 2012. Development and simulation evaluation of a magnetorheological elastomer isolator for seat vibration control. Journal of Intelligent Material Systems and Structures, 23, 1041–1048.
[95] Li, Y. C., Li, J. C., Tian, T. F., and Li, W. H. 2013. A highly adjustable magnetorheological elastomer base isolator for applications of real-time adaptive control. Smart Materials and Structures, 22, 095020.
[96] Liao, G. J., Gong, X. L., Xuan, S. H., Kang, C. J., and Zong, L. H. 2012. Development of a real-time tunable stiffness and damping vibration isolator based on magnetorheological elastomer. Journal of Intelligent Material Systems and Structures, 23, 25–33.
[97] Tu, J. W., Tu, B., Mei, S. T., Huang, L., and Xu, J. Y. 2014. Research on new type MRE isolator and its mechanical model. Materials Research Innovations, 18, 552–558.

[98] Behrooz, M., Wang, X. J., and Gordaninejad, F. 2014. Modeling of a new semi-active/passive magnetorheological elastomer isolator. Smart Materials and Structures, 23, 045013.
[99] Behrooz, M., Wang, X. J., and Gordaninejad, F. 2014. Performance of a new magnetorheological elastomer isolation system. Smart Materials and Structures, 23, 045014.
[100] Eem, S.-H., Koo, J.-H., and Jung, H.-J. 2018. Feasibility study of an adaptive mount system based on magnetorheological elastomer using real-time hybrid simulation. Journal of Intelligent Material Systems and Structures. Available at: https://doi.org/10.1177/1045389X18754347.
[101] Fu, J., Lai, J. J., Liao, G. Y., Yu, M., and Bai, J. F. 2018. Genetic algorithm based nonlinear self-tuning fuzzy control for time-varying sinusoidal vibration of a magnetorheological elastomer vibration isolation system. Smart Materials and Structures, 27, 085010.
[102] Ni, Y. Q., Ying, Z. G., and Chen, Z. H. 2010. Magneto-rheological elastomer (MRE) based composite structures for micro-vibration control. Earthquake Engineering and Engineering Vibration, 9, 345–356.
[103] Opie, S., and Yim, W. 2011. Design and control of a real-time variable modulus vibration isolator. Journal of Intelligent Material Systems and Structures, 22, 113–125.
[104] Yang, J., Sun, S. S., Tian, T. F., *et al.* 2016. Development of a novel multi-layer MRE isolator for suppression of building vibrations under seismic events. Mechanical Systems and Signal Processing, 70–71, 811–820.
[105] Xing, Z. W., Yu, M., Sun, S. S., Fu, J., and Li, W. H. 2016. A hybrid magnetorheological elastomer-fluid (MRE-F) isolation mount: development and experimental validation. Smart Materials and Structures, 25, 015026.
[106] Fu, J., Li, P. D., Liao, G. Y., Lai, J. J., and Yu, M. 2018. Active/semi-active hybrid isolation system with fuzzy switching controller. Journal of Intelligent Material Systems and Structures, 29, 101–115.
[107] Fu, J., Li, P. D., Wang, Y., Liao, G. Y., and Yu, M. 2016. Model-free fuzzy control of a magnetorheological elastomer vibration isolation system: analysis and experimental evaluation. Smart Materials and Structures, 25, 035030.
[108] Fu, J., Liao, G. Y., Yu, M., Li, P. D., and Lai, J. J. 2016. NARX neural network modeling and robustness analysis of magnetorheological elastomer isolator. Smart Materials and Structures, 25, 125019.
[109] Behrooz, M., and Gordaninejad, F. 2016. A flexible micro fluid transport system featuring magnetorheological elastomer. Smart Materials and Structures, 25, 025011.
[110] Feng, J., Xuan, S., Ding, L., and Gong, X. 2017. Magnetoactive elastomer/PVDF composite film based magnetically controllable actuator with real-time deformation feedback property. Composites Part A: Applied Science and Manufacturing, 103, 25–34.
[111] Fuhrer, R., Schumacher, C. M., Zeltner, M., and Stark, W. J. 2013. Soft iron/silicon composite tubes for magnetic peristaltic pumping: frequency-dependent pressure and volume flow. Advanced Functional Materials, 23, 3845–3849.

[112] Gao, W., Wang, L. L., Wang, X. Z., and Liu, H. Z. 2016. Magnetic driving flowerlike soft platform: biomimetic fabrication and external regulation. ACS Applied Materials and Interfaces, 8, 14182–14189.
[113] Tang, S. Y., Zhang, X., Sun, S., *et al.* 2018. Versatile microfluidic platforms enabled by novel magnetorheological elastomer microactuators. Advanced Functional Materials, 28, 1705484.
[114] Schmauch, M. M., Mishra, S. R., Evans, B. A., Velev, O. D., and Tracy, J. B. 2017. Chained iron microparticles for directionally controlled actuation of soft robots. ACS Applied Materials and Interfaces, 9, 11895–11901.
[115] Hu, W., Lum, G. Z., Mastrangeli, M., and Sitti, M. 2018. Small-scale soft-bodied robot with multimodal locomotion. Nature, 554, 81.
[116] Bica, I. 2011. Magnetoresistor sensor with magnetorheological elastomers. Journal of Industrial and Engineering Chemistry, 17, 83–89.

Chapter 10

Multi-scale modeling on tensile modulus of magnetorheological elastomers

Shiwei Chen[1,2] *and Xiaojie Wang*[1]

Magnetorheological (MR) elastomers (MREs) are made by embedding magnetic particles (micron size) in the polymer matrix. The magnetizing particles in the matrix would interact with each other under an external magnetic field and consequently induce deformation and modulus changes of MREs. With these unique controllable properties, MREs could be possible candidates for a variety of applications, such as in soft actuators [1], vibration absorber [2,3] and magnetic sensor [4,5]. Using MREs in protective systems for civil structures has the advantage of no-leakage and non-sedimentation of ferrous particles since they are solid materials [6].

MREs differ in their compositions and synthesis. In preparing MREs, if the particles are uniformly blended with the liquid matrix, and cured without magnetic field, then an isotropic MRE material is produced. However, if the magnetic field is applied during the curing process, the particles will be organized in chain structure or more complex structure. Thus, an anisotropic MRE is obtained.

As shown in Figure 10.1, the MRE in which particles are randomly distributed in the matrix is named uniformed MRE having isotropic properties. The MRE cured under magnetic field is generally called structured MRE with chain-like structures. Experimental results showed that the structured MREs have larger field-responsive modulus than those uniformed MREs [7–9].

The general theoretical work for modeling of the magnetoelastic response of particle-filled solids is based on two approaches: (1) the micromechanical approach based on the multi-scale analysis of spatial distribution of particles, and (2) the "direct" method using conservation laws of continuum mechanics.

Similar to MR fluids, the field-dependent properties of MREs result from the magnetic dipole interactions between particles within the host material. Most micromechanical models were developed to understand whether particle or chains behavior of MR fluids could be used for modeling of MR elastomers. Jolly *et al.* [10] employed a simple dipole model of particle interactions within a chain to

[1]Institute of Advanced Manufacturing Technology, Hefei Institutes of Physical Science, Chinese Academy of Sciences, People's Republic of China
[2]Institute of Architectural Engineering, Chongqing University of Science and Technology, People's Republic of China

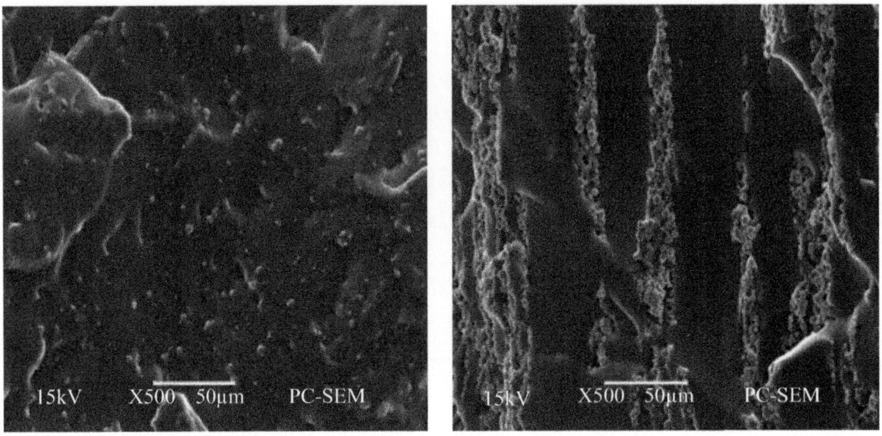

(a) (b)

Figure 10.1 Microscopic structure of MRE: (a) uniformed and (b) structured

predict the shear modulus of an MRE. The relationship between particle polarization and average composite flux density was established by considering the particle saturation theory. In order to "fit" experimental data, they proposed two adjustable parameters, which accounts for unmodeled multipolar magnetic interactions and gaps between adjacent particles. Davis [11] utilized a decoupled model from his prior theoretical work on MR fluids to analyze the magnetomechanical properties of MREs in zero and saturated magnetic fields. He predicted that the optimum particle volume fraction for the largest percentage change of shear modulus at saturation is 27% (by volume). However, the magnetoelastic behavior of MRE composites involves a much more complicated mechanical and magnetic interaction than that of nonmagnetic composites; therefore, a fully coupled magnetoelastic model was proposed by Yin and Sun [12] considering pair-wise particle interactions for magnetic and elastic fields embedded in infinite domain. Yin and Sun [13] further developed a new model with multiple in-plane chains to evaluate the tensile and shear modulus of MRE. In their model, they chose the Green function method to analyze the magneto-induced stress between particles considering the influence of particles' size on MR effect. Zhu *et al.* [14] also took into account the interaction between the particle chains in the plane and introduced another multiple chains model to study the shear modulus of MREs. Han *et al.* [15] assumed that the particle chains are zigzag in the structured MREs. Then they used COMSOL Multiphysics software to analyze the tensile modulus, shear modulus and viscoelastic effect of MREs with zigzag chain structure.

The continuum mechanics approach for modeling of MREs is to couple the equations of magnetic field and equations of deformation field. The principle of minimum potential energy is used to find the magnetomechanical free energy equations, which describes the constitutive relation of MREs. Borcea and Bruno

[16] studied the macroscopic magnetomechanical behavior of composite materials comprised random, statistically homogeneous distributions of ferromagnetic inclusions embedded in an elastic matrix. Unlike Jolly *et al.* [10] and Davis's work [11], Borcea and Bruno's model took into account fully coupled interparticle magnetoelastic interactions and predicted overall deformations through a stress–strain relation under an applied stress and magnetic field. The advantage of this model is that it does not use empirical magnetization expressions; instead, the distributions of magnetization in the MRE composite are obtained from the basic minimum energy principles of magnetoelasticity. A phenomenological model was presented for an MRE by considering the full system of constitutive equations in an electromagnetic field [17]. For an isotropic incompressible MRE, a four parameter model for shear modulus was obtained. Both theoretical studies [16,17] were based on the assumption that the MREs are isotropic and elastic materials, and the particle volume fraction is small.

The structured MREs are directional composite material systems. Since they are cured under an applied magnetic field, they have larger field-responsive modulus than those cured at zero magnetic field. The structures of the ferrous particles within field-cured elastomers aligned with the field direction results in an asymmetric stress state under deformation. The microstructure of particle aggregation plays an important role in determining the macroscopic properties of MREs. Kankanala and Triantafyllidis [18], and Dorfmann and Ogden [19] applied the restricted incompressible isotropic hyper-elastic free energy equations for the constitutive relation of MRE to obtain theoretical solutions for the unidirectional stretch and pure shear of classical mechanics problems. The former considered the magnetization intensity **M** as basis variables, while the latter employed magnetic flux density **B**. Based on the work in [18,19], Bustamante [20] treated structured MRE as a kind of transverse isotropic material. He introduced structural tensors into the free energy equations and obtained theoretical solutions for simple shear as well as the tension and torsion problems of structured MRE. Later, Danas *et al.* [21] applied the Kankanala and Triantafyllidis' theory [18] to investigate the magnetostrictive phenomenon of chain-like MRE both theoretically and experimentally. Saxena *et al.* [22] proposed a novel transverse isotropic viscoelastic energy density equation for the structured MRE. They considered the damping characteristics as a function of deformation gradient and multiplicative decomposition of magnetic flux density. This transverse isotropic viscoelastic model can describe the MRE behavior with a time-dependent magnetic field and a time-dependent strain. All the continuum mechanics approaches have achieved reasonable solutions in understanding the MR effect of MRE in the shear modulus under different working conditions, but the limitation still exists in revealing the relationship among the tensile modulus of MRE, the spatial configuration of MRE structures and the particle volume fractions.

A multi-scale analysis based on continuum-mechanics and microscopic theoretical approaches has been developed by Ivaneyko *et al.* [23–25] in order to address the influence of spatial configuration of particles on the mechanical properties of structured MREs. Ivaneyko *et al.*'s work assumed that the particles in

MRE are organized in simple cubic lattice [23], body-centered cubic and hexagonal close-packed lattice [24]. Then the uniform, chain-like and layer-like structures of MREs were investigated according to these particle spatial configurations. Their results have shown that the field-dependent behavior of MREs is related with the distribution of the magnetic particles. MREs can demonstrate either uniaxial expansion or contraction along the magnetic field, and the tensile modulus can be positive or negative, depending on the structures of magnetic particles inside MREs. Although Ivaneyko *et al.*'s model [25] can provide excellent explanations on the shear behavior of structured MRE relating with particles' spatial distribution, some results about the negative tensile modulus were contradictory to the experimental data [21,26]. In addition, they did not consider the influence of anisotropy of zero-field tensile modulus on magneto-induced modulus of structured MREs when calculating the tensile modulus in the absence of magnetic field. Moreover, their theoretical work ignored the influence of magnetic field on tensile modulus in other two directions, which should be investigated when considering the particle spatial distributions on the magnetic field strength and its effect on the tensile modulus changes.

In this chapter, the micromechanical model will be incorporated into the macroscopic constitutive laws derived from the equivalent effective medium theory by considering the anisotropy of these particular materials. Micromechanics approach has been one of the effective tools in describing mechanics-related behavior of composites. MREs are two-phase composites containing chain- or column-structured magnetic particles. The magnetomechanical behaviors of MRE should be well described within the framework of micromechanics. This chapter will focus on investigating the relationship among tensile modulus, magnetization and particle volume fraction of the structured MRE under a uniform magnetic field. In Section 10.1, we established a three parameter representative volume element (RVE) model for the layer and chain-structured MREs and derived the MRE's tensile modulus in absence of magnetic field by using Mori–Tanaka method and evaluated the solutions in comparing with finite element method (FEM) model. In Section 10.2, in terms of the magneto-induced stress in the RVE, we proposed a model for the tensile modulus of anisotropic MRE under a uniform magnetic field; then the proposed model was evaluated by comparing with both the energy method and FEM. In Section 10.3, we analyzed the anisotropic of the MRE's tensile modulus for layer and chain-like structured MREs. Conclusions are given in Section 10.4.

10.1 Tensile modulus in the absence of magnetic fields

10.1.1 Three parameters of RVE model

MRE is a type of composite material with heterostructure [27]. In this chapter, we build a representative volume element (RVE) to describe the microstructure of MRE as shown in Figure 10.2. It is assumed that magnetic particles have the same

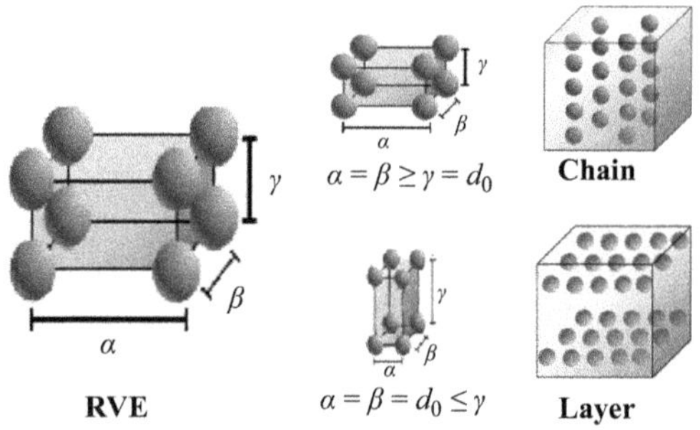

Figure 10.2 Three parameters RVE model of MREs

radius and are tightly bonded to the matrix. Meanwhile, both magnetic particles and polymer matrix are considered as linear elastic materials.

Considering the modeling method on particles' spatial microstructure referred to [23–25], we use α, β, γ to define the difference of particles' microstructure in the present model. In this proposed model, α, β, γ represent the space along x, y and z direction between particles, respectively. It is assumed that interval d_0 between neighboring particles is fixed in the structured MRE. And we can design the MRE spatial structures by altering the values of α, β, γ, such as $\alpha = \beta \geq \gamma = d_0$ representing the chain structure of MRE, while $\alpha = \beta = d_0 \leq \gamma$ representing the layer structure of MRE. The particle volume fraction c_p can be obtained as follows:

$$c_p = \frac{4\pi r^3}{3\alpha\beta\gamma} \tag{10.1}$$

where r in this chapter is the spheres radius.

10.1.2 Theoretical solutions

In this subsection, the tensile modulus of MREs in the absence of magnetic field are theoretically analyzed with respect to the particle volume fractions and spatial structures. Provided that MRE is in infinitesimal deformation and the deformation is in linear elastic stage, the tensile modulus is equivalent to Young's modulus. As shown in Figure 10.3, we consider MRE as a type of fiber-reinforced composite, and the layer and chain structures can be portrayed by altering the fiber-reinforced direction and spatial distribution.

The reinforced fiber, which is composed of the matrix and magnetic particles that are homogeneously dispersed, is a kind of isotropic particle-reinforced composite. Therefore, we use Mori–Tanaka method to evaluate the equivalent tensile modulus E_f of reinforced material. Then, the layer structure and chain structure zero-field tensile modulus E_ε can be obtained by combining the tensile modulus of fiber-reinforced and matrix materials.

According to the stress–strain relationship, the arithmetic mean symbol is defined as $\langle\cdot\rangle = V^{-1} \oint dV$. There are two phases inside fiber: matrix phase and

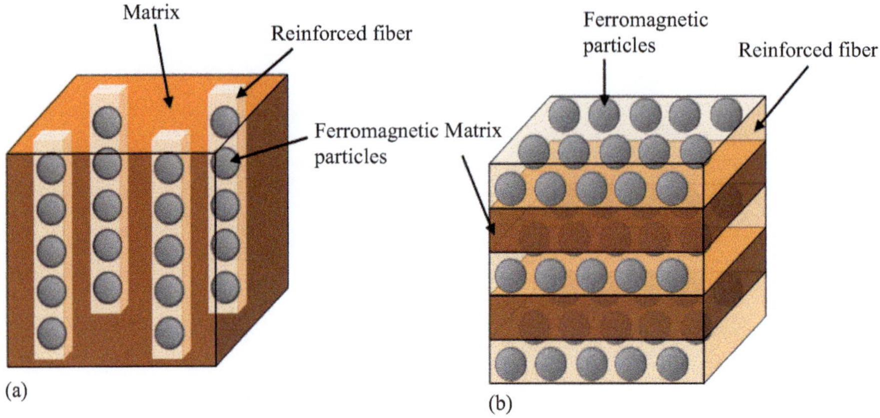

Figure 10.3 Fiber-reinforced structure of MRE: (a) chain and (b) layer

inclusion phase, in which the average stress tensor $\langle \boldsymbol{\sigma} \rangle$ and the average strain tensor $\langle \boldsymbol{\varepsilon} \rangle$ of each element satisfy the following equations:

$$\langle \boldsymbol{\sigma} \rangle = \bar{\mathbf{C}} \langle \boldsymbol{\varepsilon} \rangle \tag{10.2}$$

$$\langle \boldsymbol{\sigma} \rangle_r = \mathbf{C}_r \langle \boldsymbol{\varepsilon} \rangle_r \tag{10.3}$$

where $\bar{\mathbf{C}}$ is the equivalent elastic tensor of an element, in (10.3), $\mathbf{C}_r$ is elastic tensor of r phase. In this chapter, when $r = 0$, $\mathbf{C}_0$ stands for elastic tensor of matrix phase, when $r = 1$, $\mathbf{C}_1$ is elastic tensor of inclusion phase. Similarly, $\langle \boldsymbol{\sigma} \rangle_r$ is the average stress tensor of the r phase and $\langle \boldsymbol{\varepsilon} \rangle_r$ is the average strain tensor of the r phase.

According to the definition of arithmetic mean symbol and the structure of reinforced fiber, the average stress tensor of an element can be also described as (10.4), which is the sum of average stress of each phase

$$\langle \boldsymbol{\sigma} \rangle = \frac{1}{V} \oint \boldsymbol{\sigma} dV = \sum_{r=0}^{1} \frac{V_r}{V} \frac{1}{V_r} \oint \boldsymbol{\sigma} dV = \sum_{r=0}^{1} c_r \langle \boldsymbol{\sigma} \rangle_r \tag{10.4}$$

where $V = d_0^3$ is the total volume of reinforced fiber, V_r is the relative volume of phase r, $c_r = V_r/V$ is the volume fraction that phase r accounts for. It should be noted that c_0 represents the matrix phase volume fraction of MRE and c_1 represents the inclusion phase volume fraction of MRE. The relation between the average strain of each phase and the far-field boundary condition is

$$\langle \varepsilon \rangle = \mathbf{N}_r \bar{\varepsilon} \tag{10.5}$$

where $\bar{\varepsilon}$ is the uniform strain on the unit cell boundary, $\mathbf{N}_r$ is the local tensor of the r phase. Its physical significance roots in influence coefficient of the far-field stress on the r phase. $\mathbf{N}_0$ represents the localized relation tensor of matrix and $\mathbf{N}_1$ is the localized relation tensor of inclusion phase. It satisfied

$$c_0\mathbf{N}_0 + c_1\mathbf{N}_1 = \mathbf{I} \tag{10.6}$$

where $\mathbf{I}$ characterize an identity matrix. Equations (10.2)–(10.5) reduce to the equivalent elastic modulus of reinforced layer:

$$\bar{\mathbf{C}} = \mathbf{C}_0 + \sum_{r=0}^{1}(c_r(\mathbf{C}_1 - \mathbf{C}_0)\mathbf{N}_r) \tag{10.7}$$

In terms of the Mori–Tanaka method [28], we define that the strain on a particular inclusion is the average strain of matrix, and the inclusion phase local tensor $\mathbf{N}_1$ can expressed as

$$\mathbf{N}_1 = (\mathbf{I} + \mathbf{P}_1(\mathbf{C}_1 - \mathbf{C}_0))^{-1}\left[c_0\mathbf{I} + c_1(\mathbf{I} + \mathbf{P}_1(\mathbf{C}_1 - \mathbf{C}_0))^{-1}\right] \tag{10.8}$$

where tensor $\mathbf{P}_1$ is related to the inclusions shape and its component could be expressed as (10.9), when spherical particles are assumed.

$$P^1_{ijkl} = \frac{10\delta_{ij}\delta_{kl} + 9(2\mathbf{G}_0 + \mathbf{K}_0)\left(\delta_{ij}\delta_{kl} + \delta_{ij}\delta_{kl} - (2/3)\delta_{ij}\delta_{kl}\right)}{(30\mathbf{G}_0(4\mathbf{G}_0 + 3\mathbf{K}_0)} \tag{10.9}$$

where $\mathbf{G}_0$ and $\mathbf{K}_0$ are the shear modulus and volume modulus of matrix phase, respectively, δ_{ij} represents the Kronecker tensor. The subscripts i, j, k and l satisfied Einstein notation, both represent i, j, k and l followed by rotation in turns of number 1, 2, 3, and also represent the x-, y-, z-axis of Cartesian coordinates in the linear space. By substituting (10.8) and (10.6) into (10.7), we can derive equivalent volume modulus and equivalent shear modulus. These two independent moduli can be expressed as follows:

$$\bar{G} = \mathbf{G}_0 + \frac{10\mathbf{G}_0(4\mathbf{G}_0 + 3\mathbf{K}_0)(\mathbf{G}_1 - \mathbf{G}_0)}{10\mathbf{G}_0(4\mathbf{G}_0 + 3\mathbf{K}_0) + 12c_0(3\mathbf{G}_0 + \mathbf{K}_0)(\mathbf{G}_1 - \mathbf{G}_0)}c_1 \tag{10.10}$$

$$\bar{K} = \mathbf{K}_0 + \frac{3\mathbf{G}_0(4\mathbf{G}_0 + 3\mathbf{K}_0)(\mathbf{K}_1 - \mathbf{K}_0)}{3\mathbf{G}_0(4\mathbf{G}_0 + 3\mathbf{K}_0) + 9c_0(\mathbf{K}_1 - \mathbf{K}_0)}c_1 \tag{10.11}$$

where $\mathbf{G}_1$, $\mathbf{K}_1$ characterize the shear modulus and volume modulus of magnetic particles, respectively. c_1 is the volume fraction that magnetic particles occupying in the reinforced fiber, which is expressed as follows:

$$c_1 = \frac{4\pi r^3}{3d_0^3} \tag{10.12}$$

Then, the MRE tensile modulus $\bar{E}_{\perp}$ which is perpendicular to the reinforced fiber and $\bar{E}_{\parallel}$ which is parallel to the reinforced fiber are represented as the following equations:

$$\bar{E}_{\perp} = \bar{E}_f c_f + E_m(1 - c_f) \tag{10.13}$$

$$\bar{E}_{\parallel} = \left((1 - c_f) + c_f\left(\frac{E_m}{\bar{E}_f}\right)\right)^{-1} \tag{10.14}$$

where $\bar{E}_f = 9\bar{G}\bar{K}/(3\bar{K} + \bar{G})$ is the equivalent elastic modulus of reinforced fiber, c_f is the volume fraction of the fibers in the matrix, which is expressed as follows:

$$c_f = \frac{d_0^3}{\alpha\beta\gamma} \tag{10.15}$$

10.1.3 FEM solutions

We apply COMSOL Multiphysics V4.1. to evaluating equivalent elastic modulus of MRE at multiple scales. The FEM model is shown in Figure 10.4.

We use $\mathbf{U} = (u,v,w)$ to describe the displacement field of RVE. The boundary conditions are set up where $u = 0$ for the boundary S_{fead}, $v = 0$ for the boundary S_{abcd}, $w = 0$ for the boundary S_{abhe}. There is a continuous periodic boundary conditions (PBC) module in COMSOL which can be utilized to setup the PBC, where S_{fead} is continuous to S_{hbcg}, S_{abcd} is continuous to S_{ehgf}, S_{abhe} is continuous to S_{dcgf}.

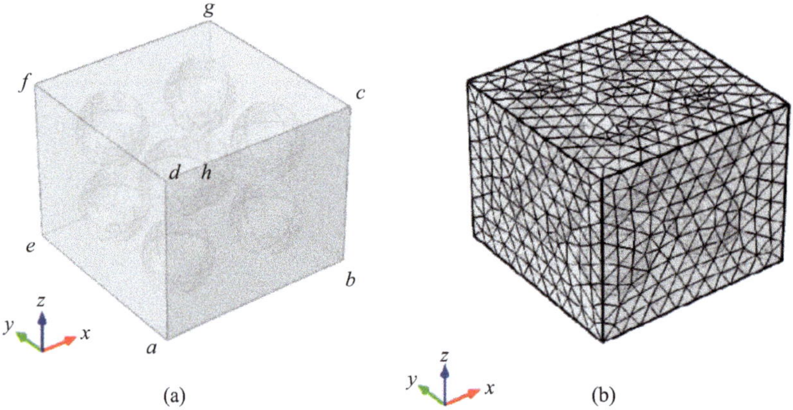

Figure 10.4 Finite-element model of RVE: (a) model and (b) mesh

To stretch RVE along the x-axis, a supplementary displacement boundary condition $u = \varepsilon_x l_{ab}$ is required to set up on S_{hbcg}. Thus, an infinitesimal strain ε_{x-x} occurs in the x direction. Thanks to the approximate incompressibility of MRE, the stretch also produces strain ε_{y-x}, ε_{z-x} in the y, z direction.

Hence, the strain energy density of volume unit v_f could be obtained as the following equation [29]:

$$v_f = \frac{1}{2}E_z\varepsilon_z^2 + \frac{1}{2}E_y\varepsilon_y^2 + \frac{1}{2}E_x\varepsilon_x^2 \tag{10.16}$$

Then, the average stress $\langle\sigma\rangle_{x-x}$ which is produced by the stretch on RVE in the x direction can be described as (10.17):

$$\langle\sigma_{x-x}\rangle = \frac{\partial v_{f-x}}{\partial\varepsilon_{x-x}} = E_x\varepsilon_{x-x} + \varepsilon_{y-x}E_y\frac{\partial\varepsilon_{y-x}}{\partial\varepsilon_{x-x}} + \varepsilon_{z-x}E_z\frac{\partial\varepsilon_{z-x}}{\partial\varepsilon_{x-x}} \tag{10.17}$$

When the displacement excitation ε_{x-x} is small enough in (10.18), we derive (10.18) according to the definition of Poisson's ratio.

$$\frac{\partial\varepsilon_{y-x}}{\partial\varepsilon_{x-x}} = \frac{\partial v_{xy}\varepsilon_{x-x}}{\partial\varepsilon_{x-x}} = \varepsilon_{x-x}\frac{v_{xy}(\varepsilon_{x-x}) - v_{xy}(0)}{\varepsilon_{x-x}} + v_{xy}(0) = v_{xy}(\varepsilon_{x-x}) \tag{10.18}$$

where v_{xy}, which is a function of strain ε_{x-x}, represents the Poisson ratio of material in x–y direction, which also means the cross-sectional shrinkage rate in the y direction when MRE suffers from the stretch in the x direction. Generally, tensor of the Poisson ratio v_{ij} could be expressed as follows:

$$v_{ij}(\varepsilon_{i-i}) = \frac{d\varepsilon_{j-i}}{d\varepsilon_{i-i}}(i,j = 1,2,3) \tag{10.19}$$

Particularly, when the ε_{i-i} is infinitesimal strain, the following equation could be derived.

$$v_{ij}(\varepsilon_{i-i}) = \frac{\varepsilon_{j-x}(2\varepsilon_{i-i}) - \varepsilon_{j-x}(\varepsilon_{i-i})}{\varepsilon_{i-i}}(i,j = 1,2,3) \tag{10.20}$$

By substituting (10.18) into (10.19), (10.21) could be derived.

$$\langle\sigma_{x-x}\rangle = E_{\varepsilon x}\varepsilon_{x-x} + v_{xy}(\varepsilon_{x-x})v_{xy}(0)E_{\varepsilon y}\varepsilon_{z-z} + v_{xz}(\varepsilon_{x-x})v_{xz}(0)E_{\varepsilon z}\varepsilon_{x-x} \tag{10.21}$$

Similarly, when RVE is stretched along the y, z direction, the infinitesimal strains ε_{y-y} and ε_{z-z} are created. Then (10.22) and (10.23) could be derived as the following:

$$\langle\sigma_{y-y}\rangle = v_{yx}(\varepsilon_{y-y})v_{yx}(0)E_{\varepsilon x}\varepsilon_{y-y} + E_{\varepsilon y}\varepsilon_{y-y} + v_{yz}(\varepsilon_{y-y})v_{yz}(0)E_{\varepsilon z}\varepsilon_{y-y} \tag{10.22}$$

$$\langle\sigma_{z-z}\rangle = v_{zx}(\varepsilon_{x-x})v_{zx}(0)E_{\varepsilon x}\varepsilon_{z-z} + v_{zy}(\varepsilon_{x-x})v_{zy}(0)E_{\varepsilon y}\varepsilon_{z-z} + E_{\varepsilon z}\varepsilon_{z-z} \tag{10.23}$$

By considering (10.21)–(10.23), the zero-field equivalent Young's modulus $E_{\varepsilon i}$ are derived.

$$\begin{bmatrix} E_{\varepsilon x} \\ E_{\varepsilon y} \\ E_{\varepsilon z} \end{bmatrix} = \begin{bmatrix} \varepsilon_{x-x} & v_{xy}(\varepsilon_{x-x})v_{xy}(0)\varepsilon_{x-x} & v_{xz}(\varepsilon_{x-x})v_{xz}(0)\varepsilon_{x-x} \\ v_{yx}(\varepsilon_{y-y})v_{yx}(0)\varepsilon_{y-y} & \varepsilon_{y-y} & v_{yz}(\varepsilon_{y-y})v_{yz}(0)\varepsilon_{y-y} \\ v_{zx}(\varepsilon_{z-z})v_{zx}(0)\varepsilon_{z-z} & v_{zy}(\varepsilon_{z-z})v_{zy}(0)\varepsilon_{z-z} & \varepsilon_{z-z} \end{bmatrix}^{-1} \begin{bmatrix} \langle\sigma_{x-x}\rangle \\ \langle\sigma_{y-y}\rangle \\ \langle\sigma_{z-z}\rangle \end{bmatrix} \tag{10.24}$$

Overall, average stress and strains in three directions of RVE under different working conditions are obtained, after the simulation on the unidirectional stretch in the x, y, z direction. The Poisson ratio of MRE is also obtained by introducing the strain data under different working conditions into (10.20). And to bring the Poisson ratio we acquired and average stress into (10.24), we can get the zero-field Young's modulus $E_{\varepsilon i}$.

10.1.4 Results

Using the parameters from Table 10.1 for simulation, Figure 10.5 shows the tensile modulus of MRE in the absence of magnetic field for the layer and chain-structured MREs obtained by using both theoretical and FEM methods.

Table 10.1 Simulation parameter

Parameter	Value	Parameter	Value
Particle tensile modulus	200 GPa	Matrix Poisson's ratio	0.49 μm
Particle Poisson's ratio	0.29	Spheres radius r	1.3 μm
Matrix tensile modulus	0.5 MPa	Particle spacing d_0	3 μm Body text

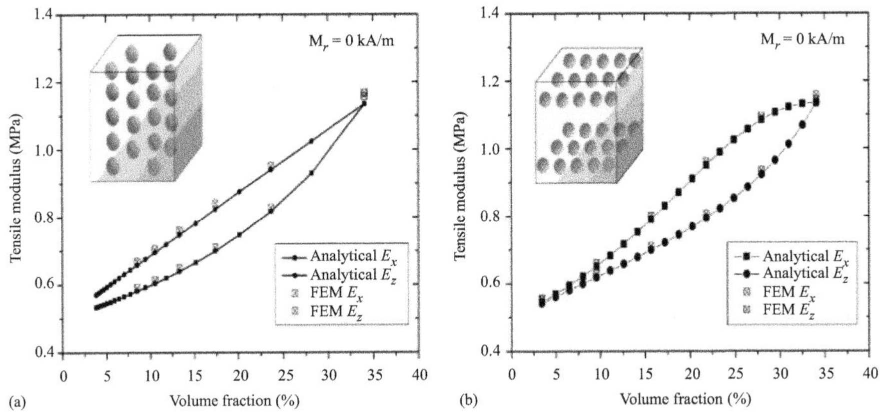

Figure 10.5 Finite-element model of RVE: (a) model and (b) mesh

We can see that the theoretical and FEM results match with each other very well. Both E_x and E_z increase with the particles' fraction, and the tensile modulus E_x parallel to the reinforced fiber is always larger than E_z which is the tensile modulus perpendicular to the fiber. As the particles' fraction increases, E_z initially increases more than E_x. When the particles' fraction reaches 20.17% or even higher, E_x start to increase more than E_z and when the volume fraction reaches at 34.08%, i.e., $\gamma = \beta = \alpha = d_0$, E_x is completely equal to E_z. It means that particles are homogeneously distributed and MRE can be regarded as an isotropic material at this higher particle volume fraction.

Figure 10.5(b) shows that tensile modulus of layer structure MRE alters with particles' volume fraction. Different with the chain-structured MRE, the tensile modulus E_x for the layer structure MRE is less than E_z.

10.2 Tensile modulus under applied magnetic fields

When the magnetic field is applied on MRE, MRE will become stiffer or softer than usual. It is mainly due to the particle interactions in micromechanics. In this section, we will discuss how the particle volume fraction and their microstructures affect the MRE macro-properties in terms of the field-induced tensile modulus.

When the RVE is under the far-field strain, which is caused by stretched along MRE in the $i(i = 1,2,3 = x,y,z)$ direction, the deforming stress is created inside it. At the same time, the magnetic particles attract and repulse each other when the magnetic field H is applied, and the magneto-induced stress is generated along the i direction. Thus, the deforming stress and magneto-induced stress constitute the total stress σ_i. In this study, we define the tensile modulus in the y direction E_i of MRE as variation of the total stress σ_i which is changed stress ε_i in the i direction. Assuming small deformation, E_i can be described as follows:

$$E_i = \frac{\partial \sigma_i}{\partial \varepsilon_i} = \frac{\partial(\sigma_{\varepsilon i} + \sigma_{mi})}{\partial \varepsilon_i} = E_{\varepsilon i} + \frac{\partial \sigma_{mi}}{\partial \varepsilon_i} \tag{10.25}$$

where $E_{\varepsilon i}$, which is also called pure tensile modulus in this chapter, represents MRE's tensile modulus in the absence of magnetic defined in Section 10.1. From (10.25), it is known that if the magneto-induced stress of MRE in the presence of magnetic field is obtained, combined with the results from Section 10.1.4, the tensile modulus will be subsequently obtained. In Section 10.2.1, we will first analyze the dependence of particles on magnetization and the magneto-induced stress which is the result of interactions among particles. In Sections 10.2.2 and 10.2.3, we will then discuss tensile modulus in two independent directions of transverse isotropic MRE.

10.2.1 Magnetic-induced stress

When the MRE is in the static uniform magnetic field in the z direction, the magnetic particles are magnetized, which satisfies the Frohlich–Kennelly equation [30]:

$$\mathbf{M}_r = \frac{M_s \chi \mathbf{H}_0}{M_s + \chi \mathbf{H}_0} \tag{10.26}$$

where $\mathbf{M}$ represents the magnetization intensity, M_s is the saturation magnetization, χ is the initial magnetic susceptibility. Each particle is assumed to be a magnetic dipole, so the magnetic moment of the i particle is $\mathbf{m}_i = V_i\mathbf{M}_r$, where V_i is the particle volume. According to the magnetic dipole theory [31], the magnetic intensity at particle i that particle j creates can be described as

$$\tilde{\mathbf{B}}_{ij} = \frac{\mu_0}{4\pi\|\mathbf{r}_{ij}\|^3}\left(3\left(\mathbf{m}_j \cdot \mathbf{r}_{ij}\right)\mathbf{r}_{ij} - \mathbf{m}_j\right) \tag{10.27}$$

where μ_0 is the permeability of vacuum. $\mathbf{r}_{ij}$ is the displacement vector from i to j. $\mathbf{F}_i$ is the force on the magnetic dipole i.

$$\mathbf{F}_i = -\mathbf{m}_i \cdot \nabla(\mathbf{B}_0 + \sum_{i \neq j} \tilde{\mathbf{B}}_{ij}) = -\mathbf{m}_i \cdot \sum_{i \neq j} \tilde{\mathbf{B}}_{ij} \tag{10.28}$$

where $\mathbf{B}_0$ is the uniform magnetic field, therefore $\nabla\mathbf{B}_0 = 0$. $\sum_{i \neq j} \tilde{\mathbf{B}}_{ij}$ is the total magnetic intensity at particle i. By substituting (10.27) into (10.28), assuming that all the particles have the same magnetization intensity, which is $\mathbf{m}_i = \mathbf{m}_j = \mathbf{m}$, so that the three components of $\mathbf{F}_i$ can be described as the following equations:

$$F_{ix} = -\frac{3\mu_0\|\mathbf{m}\|^2}{4\pi\|\mathbf{r}_{ij}\|^7}\left(r_x\|\mathbf{r}_{ij}\|^2 - 5r_z^2 r_x\right) \tag{10.29}$$

$$F_{iy} = -\frac{3\mu_0\|\mathbf{m}\|^2}{4\pi\|\mathbf{r}_{ij}\|^7}\left(r_y\|\mathbf{r}_{ij}\|^2 - 5r_z^2 r_y\right) \tag{10.30}$$

$$F_{iz} = -\frac{3\mu_0\|\mathbf{m}\|^2}{4\pi\|\mathbf{r}_{ij}\|^7}\left(3r_z\|\mathbf{r}_{ij}\|^2 - 5r_z^3\right) \tag{10.31}$$

Figure 10.6 describes the RVE load conditions under an applied magnetic field.

According to Figure 10.6, the RVE under magnetic field is equivalent to Figure 10.6(c), and its magneto-induced stress can be described as follows:

$$\begin{aligned} \sigma_{mx} &= -F_{ix}/\beta\gamma \\ \sigma_{my} &= -F_{iy}/\alpha\gamma \\ \sigma_{mz} &= -F_{iz}/\alpha\beta \end{aligned} \tag{10.32}$$

FEM method can also be used to evaluate the magento-induced stress according to work in Section 10.1.3. MFNC module of COMSOL Multiphysics is adopted to analyze the magnetic distribution after the magnetization of MRE.

$$\mathbf{F}_i^{\mathrm{Maxwell}} = \int_{\partial\Omega} \boldsymbol{\sigma}^{\mathbf{M}} \cdot \mathbf{n}ds = \int_{\partial\Omega} \mu_0\left[\mathbf{H}\mathbf{H}^{\mathrm{T}} - \frac{1}{2}|\mathbf{H}|^2\mathbf{I}\right] \cdot \mathbf{n}ds \tag{10.33}$$

where $\boldsymbol{\sigma}^{\mathbf{M}}$ is the Maxwell stress tensor, $\mathbf{I}$ is the second rank unit tensor, $\mathbf{n}$ characterizes the normal vector on particles' boundary. After the evaluation, the surface force is used as a force boundary condition in the finite-element mechanical analysis so that the magneto-induced stress on the RVE is obtained.

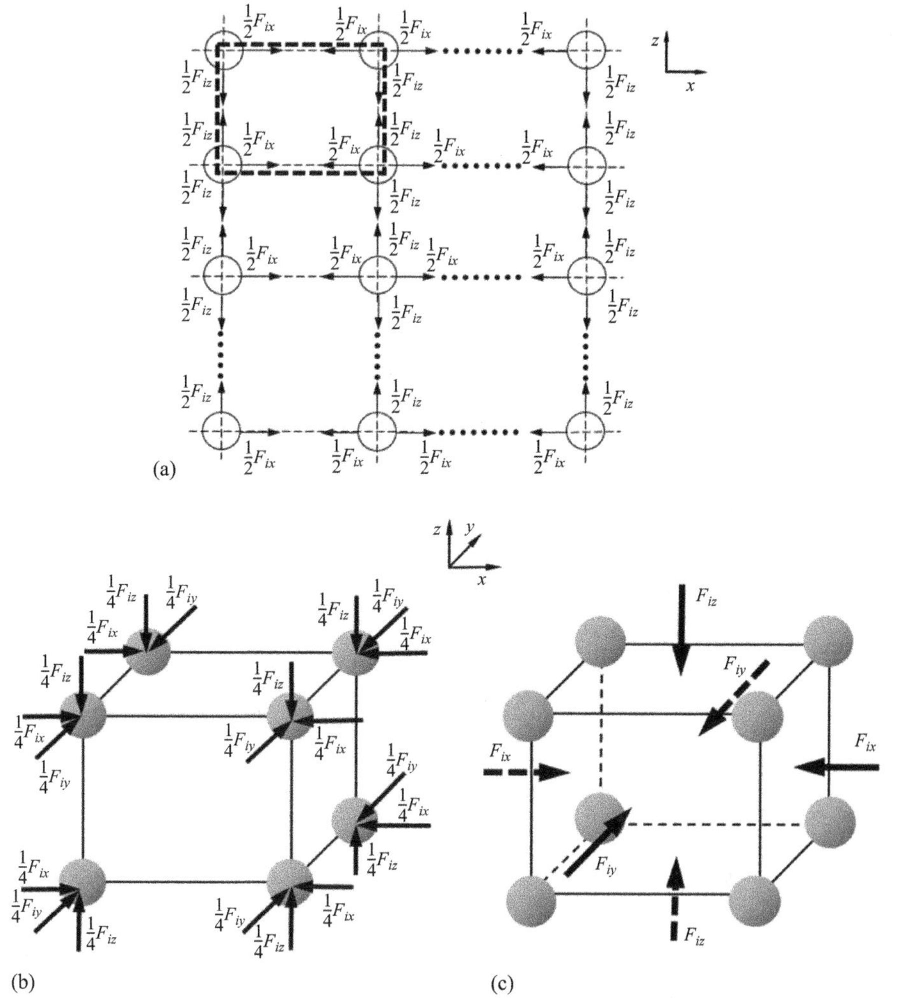

Figure 10.6 RVE force analysis: (a) whole MRE, (b) RVE and (c) RVE equivalent force

10.2.2 The x, y direction tensile modulus under magnetic field

Structured MRE can be considered as transverse anisotropic in this chapter. When the uniform magnetic field is applied in the z direction, the tensile modulus in the x and y direction are equivalent. In this subsection, tensile modulus in the x and y direction will be discussed. According to the definition of tensile modulus, the stretch on MRE in the x direction makes RVE under a far-field strain $\varepsilon_{\varepsilon x}$. So distance vector between particles j and i is shown as

$$\mathbf{r}_{ij} = r_{ijx0}(1 + \varepsilon_{\varepsilon x})\boldsymbol{i} + r_{ijy0}(1 + \varepsilon_{y-\varepsilon x})\boldsymbol{j} + r_{ijz0}(1 + \varepsilon_{z-\varepsilon x})\boldsymbol{k} \tag{10.34}$$

where r_{ijx0}, r_{ijy0}, r_{ijz0} are the initial distances in the x, y, z directions between particles j and i. Particularly, when take the i particles as the origin of coordinates, (10.25) can derived as

$$\mathbf{r}_j = r_{jx0}(1 + \varepsilon_{\varepsilon x})\mathbf{i} + r_{jy0}(1 + \varepsilon_{y-\varepsilon x})\mathbf{j} + r_{jz0}(1 + \varepsilon_{z-\varepsilon x})\mathbf{k} \tag{10.35}$$

where r_{jx0}, r_{jy0}, r_{jz0} are the initial coordinates in the x, y, z directions of particle j, if the magnetic field is applied, and the far-field strain $\varepsilon_{\varepsilon x}$ stays unchanged, the total magneto-induced stress in the x direction can be described as (10.36) due to the incompressibility of MRE and obeying the principle of linear superposition.

$$\tilde{\sigma}_{mx} = \sigma_{mx} + \sigma_{x-my} + \sigma_{x-mz} \tag{10.36}$$

where equivalent magneto-induced stress $\tilde{\sigma}_{mx}$ is composed of three parts. The first one is the stress σ_{mx} which is produced by the interaction between particles in the x direction. The second is the stress σ_{x-my} in the x direction which is caused by the interaction between particles in the y direction. The explanation of this phenomenon is that magnetostrictive strain ε_{my} that occurs on RVE in the y direction will produce magnetostrictive strain ε_{x-my} in the x direction, owing to the incompressibility of MRE. Because of the effect of far-field strain $\varepsilon_{\varepsilon x}$, this magneto-induced strain ε_{x-my} is restricted, and then the stress σ_{x-my} occurs. The third part is the stress σ_{x-mz} in the x direction caused by particles' interaction in the z direction. They can be expressed as

$$\begin{aligned} \sigma_{x-my} &= E_{\varepsilon x}\varepsilon_{x-my} \\ \sigma_{x-mz} &= E_{\varepsilon x}\varepsilon_{x-mz} \end{aligned} \tag{10.37}$$

Thanks to the incompressibility of MRE, ε_{x-my} and σ_{my} have the following relationship:

$$\left(1 + \varepsilon_{x-my}\right)\left(1 + \varepsilon_{z-my}\right)\left(1 + \frac{\sigma_{my}}{E_{\varepsilon y}}\right) = 1 \tag{10.38}$$

where ε_{x-my} is the magnetostrictive strain in the z direction caused by the magneto-induced stress. According to the definition of Poisson's ratio, ε_{x-my} satisfies the following equation:

$$\varepsilon_{x-my} = \frac{\nu_{23}}{\nu_{21}}\varepsilon_{z-my} = \frac{E_{\varepsilon z}}{E_{\varepsilon x}}\varepsilon_{z-my} \tag{10.39}$$

Considering (10.28)–(10.30), generally, according to the symmetry, given that x, y, $z = i$, j, k, the magneto-induced stress σ_{i-mj} in the i direction caused by the magneto-induced strain σ_{mj} in the j direction can be described as

$$\sigma_{i-mj} = -\frac{1}{2}(E_{\varepsilon i} + E_{\varepsilon k}) + \frac{1}{2}\sqrt{(E_{\varepsilon i} + E_{\varepsilon k})^2 - \frac{4E_{\varepsilon i}E_{\varepsilon k}\sigma_{mj}}{E_{\varepsilon i} + \sigma_{mj}}} \tag{10.40}$$

Thanks to the incompressibility of MRE, $\varepsilon_{x\text{-}my}$ and σ_{my} have the following relationship:

$$\varepsilon_{i-mj} = -\frac{1}{2}(1 + E_{\varepsilon k}/E_{\varepsilon i}) + \frac{1}{2}\sqrt{(1 + E_{\varepsilon k}/E_{\varepsilon i})^2 - \frac{4E_{\varepsilon k}\varepsilon_{\varepsilon j}}{E_{\varepsilon i}(1 + \varepsilon_{\varepsilon j})}} \tag{10.41}$$

According to the definition of tensile modulus in this study in (10.16), the tensile modulus of MRE in the x direction is written as the following:

$$E_x = E_{\varepsilon x} + E_{mx} = E_{\varepsilon x} + \frac{\partial(\sigma_{mx} + \sigma_{mx-y} + \sigma_{mx-z})}{\partial \varepsilon_{\varepsilon x}} \tag{10.42}$$

E_{mx} in (10.33) is defined as magneto-induced modulus that is generated in applied magnetic field. Considering (10.29)–(10.31), (10.34), (10.40) and (10.41), the equation is derived and simplified as below:

$$E_{mx} = \mathbf{C} \cdot \sum_{j=2}^{N} \mathbf{A}_j \mathbf{T}_j \tag{10.43}$$

where $\mathbf{C}$ is the function of magneto-induced stress—it indicates the influence of magneto-induced stress σ_{m} on magneto-induced modulus E_{mx}, which is called magneto-stress weighting vector and could be obtained as the following:

$$\begin{aligned}
\mathbf{C}(1,1) &= -\frac{3\mu_0 \|\mathbf{m}\|^2}{4\pi\alpha\gamma} \\
\mathbf{C}(2,1) &= \frac{-3\mu_0\|\mathbf{m}\|^2 E_{\varepsilon x} E_{\varepsilon y} E_{\varepsilon z}}{4\pi\gamma\beta\sqrt{(E_{\varepsilon x} + E_{\varepsilon z})^2 (E_{\varepsilon y} + \sigma_{my})^4 + 4E_{\varepsilon x}E_{\varepsilon y}\sigma_{my}(E_{\varepsilon y} + \sigma_{my})^3}} \\
\mathbf{C}(3,1) &= \frac{-3\mu_0\|\mathbf{m}\|^2 E_{\varepsilon x} E_{\varepsilon y} E_{\varepsilon z}}{4\pi\alpha\beta\sqrt{(E_{\varepsilon x} + E_{\varepsilon y})^2 (E_{\varepsilon z} + \sigma_{mz})^4 + 4E_{\varepsilon x}E_{\varepsilon y}\sigma_{my}(E_{\varepsilon z} + \sigma_{mz})^3}}
\end{aligned} \tag{10.44}$$

where $\mathbf{A}_j$ in (10.34) is the function of particles' spatial coordinates; it is a matrix of particles' structure, indicating the influence of spatial coordinates of particle j on magneto-induced modulus. After the superposition, it also can be regarded as the influence of microstructure of particles on magneto-induced modulus and be expressed as the following:

$$\begin{aligned}
\mathbf{A}_j(1,1) &= \|\mathbf{r}_j\|^4 - 5r_{jy}r_{jx}\|\mathbf{r}_j\|^2 - 5\|\mathbf{r}_j\|^2 r_{jz}^2 + 35r_{jy}^2 r_{jx} r_{jy} \\
\mathbf{A}_j(1,2) &= -5r_{jx}^2\|\mathbf{r}_j\|^2 + 35r_{jx}^2 r_{jz}^2 \\
\mathbf{A}_j(1,3) &= -15r_{jx}r_{jz}\|\mathbf{r}_j\|^2 + 35r_{jz}^3 r_{jx} \\
\mathbf{A}_j(2,1) &= -5r_{jy}r_{jx}\|j\|^2 + 35r_{jx}r_{jz}^2 r_{jy} \\
\mathbf{A}_j(2,2) &= \|\mathbf{r}_j\|^4 - 5r_{jy}^2\|\mathbf{r}_j\|^2 - 5\|\mathbf{r}_j\|^2 r_{jz}^2 + 35r_{jy}^2 r_{jz}^2 \\
\mathbf{A}_j(2,3) &= -15r_{jy}r_{jz}\|\mathbf{r}_j\|^2 + 35r_{jz}^3 r_{jy} \\
\mathbf{A}_j(3,1) &= -15r_{jx}r_{jz}\|\mathbf{r}\|_j^2 + 35r_{jx}r_{jz}^3 \\
\mathbf{A}_j(3,2) &= -15r_{jz}^2\|\mathbf{r}\|_j^2 + 35r_{jx}^2 r_{jz}^2 \\
\mathbf{A}_j(3,3) &= 3\|\mathbf{r}_j\|^4 - 15r_{jx}r_{jz}\|\mathbf{r}_j\|^2 - 15\|\mathbf{r}_j\|^2 r_{jz}^2 + 35r_{jz}^3 r_{jy}
\end{aligned} \tag{10.45}$$

Besides, $\mathbf{T}_j$ in (10.34) is the function of far-field strain $\varepsilon_{\varepsilon x}$, which demonstrates the influence of far-filed strain on magneto-induced modulus, also called far-filed strain weighting vector and could be obtained as following:

$$\begin{aligned} \mathbf{T}_j(1,1) &= r_{jx0} \\ \mathbf{T}_j(2,1) &= r_{jy0}\frac{E_{\varepsilon z}}{\sqrt{(E_{\varepsilon y}+E_{\varepsilon z})^2(1+\varepsilon_x)^4+4E_{\varepsilon z}E_{\varepsilon y}\varepsilon_x(1+\varepsilon_z)^3}} \\ \mathbf{T}_j(3,1) &= r_{jz0}\frac{E_{\varepsilon x}}{\sqrt{(E_{\varepsilon y}+E_{\varepsilon z})^2(1+\varepsilon_x)^4+4E_{\varepsilon z}E_{\varepsilon y}\varepsilon_x(1+\varepsilon_z)^3}} \end{aligned} \tag{10.46}$$

From (10.35)–(10.37), we can see that magneto-induced modulus can be described as a result of coupling of $\mathbf{A}_j$, $\mathbf{B}_j$, $\mathbf{C}$, or it can be said that magneto-induced modulus is determined by magneto-induced stress, particle spatial structure and far-field strain. Since $\mathbf{C}<0$ and $\mathbf{B}_j>0$ ($i=1,2,3$), and the particles' structure matrix $\mathbf{A}_j$ can either be positive or negative; the positive and negative properties of magneto-induced modulus are directly determined by the particle spatial coordinates that induce the spatial structure of particles.

Particularly, if we ignore the effect of anisotropy of pure tensile modulus on far-field strain and magneto-induced stress, that means $E_{\varepsilon x}=E_{\varepsilon y}=E_{\varepsilon z}$, and considering the $\varepsilon_{\varepsilon x}=0$, the proposed model will reduce to the "energy" method. In the "energy" method, the total magneto-induced stress as the first derivative of magnetic energy was considered. The elastic modulus was calculated by using the following equations:

$$E_x = E_{\varepsilon x} + \frac{\partial^2 v_m}{\partial \varepsilon_x^2} \tag{10.47}$$

where v_m is the magnetic energy intensity, whose expression is as

$$v_m = \frac{1}{V}\sum_{i=1}^{N}\sum_{j=1ui}^{N}\tilde{\mathbf{B}}_{ij}\cdot\mathbf{m}_i \tag{10.48}$$

To compare the proposed method with previous studies, we choose the particle volume fraction of 17.04%, which means that the MRE turns out to be layer structure if $\alpha=\beta=d_0=0.5\gamma$. The influence of magnetization on the elastic modulus E_x in the x direction was analyzed. The similar analysis by using FEM and the "energy" method were also carried out. In the simulation, the saturation magnetization M_s is 1,520 kA/m, and the permeability of polymer equals to the permeability of vacuum μ_0 at 1.256e−6 H/m. Figure 10.7 presents the results of relative tensile modulus $E_x/E_{\varepsilon x}$ changes with strain and magnetic fields by using the three methods for layer-structured MREs. $E_{\varepsilon x}$ represents the MRE's tensile modulus at zero magnetic field.

The results of three methods are much closer with each other as shown in Figure 10.7. Meanwhile, the magnetic-field-induced nonlinear behavior of MRE is

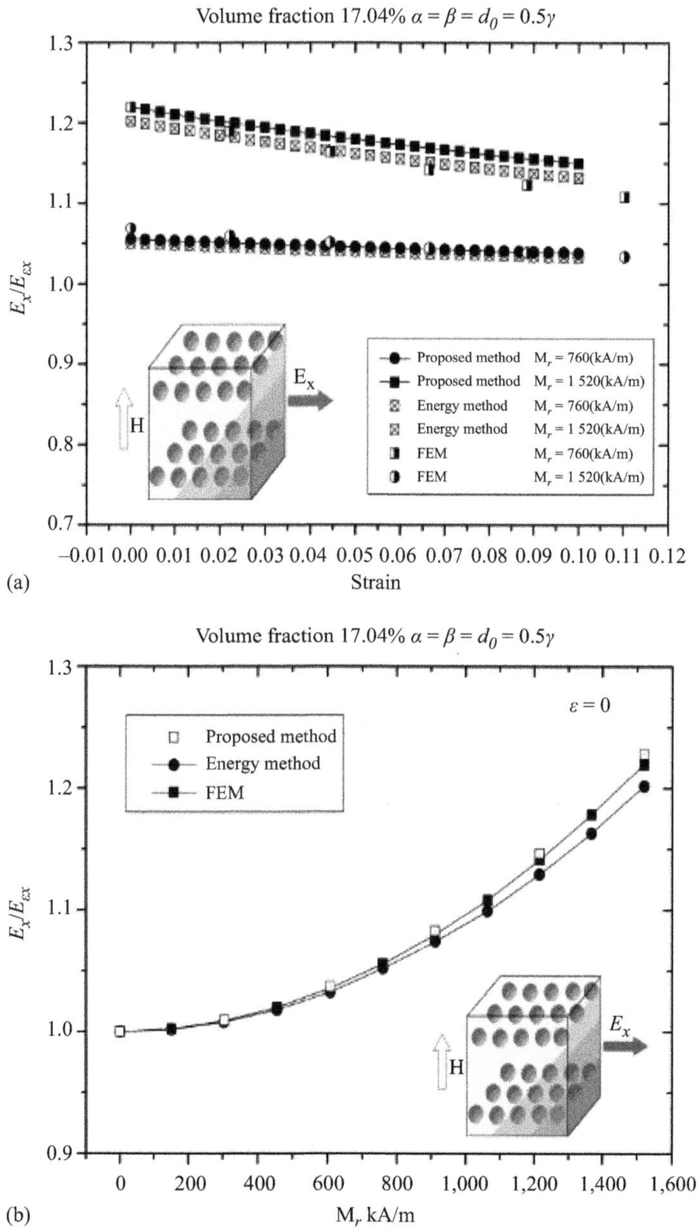

Figure 10.7 Magneto mechanic coupling simulation of layer structure at (a) different strains and (b) different magnetizations

clearly demonstrated in the tensile modulus, which is a function of strain. This phenomenon was also observed in experiment [32]. The nonlinear tensile modulus under a magnetic field is caused by the changes of particle interaction force which is due to the strain changes.

Figure 10.7(b) shows the tensile modulus changes with magnetization intensity at zero strain. The results indicate that the tensile modulus E_x in the x direction increases quadratically with magnetization intensity $\mathbf{M}_r$.

Besides, Figure 10.8 shows the FEM simulation results. An RVE of MRE with layer structure is established in Figure 10.8(a), containing 17.04% particles in it. Figure 10.8(b) shows the Mises stress contour plot with $\mathbf{M}_r = 1{,}520$ kA/m, when the RVE is saturated and not strained.

The red arrow is the magnetic flux density of magnetic particles in RVE. Figure 10.8(c) and (d) shows that the Mises stress contour plot with $\varepsilon_y = 0.1$. RVE in Figure 10.8(c) is in the absence of magnetic field, while RVE in Figure 10.8(b) is not.

As shown in Figure 10.8, the particles are magnetized under the magnetic field, and magnetostrictive stress occurs. The distance of particles changes after the MRE is strained, which subsequently influence the magnetostrictive stress. This phenomenon results in toughly stretching MRE in the y direction because the Young's modulus E_y is enhanced by the magnetic-field intensity.

Figure 10.9 compares three methods for the relative tensile modulus of MRE with layer structure varying with the particles' volume fraction, under the magnetic field of $\mathbf{M}_r = 1{,}520$ and 760 kA/m in the z direction, respectively.

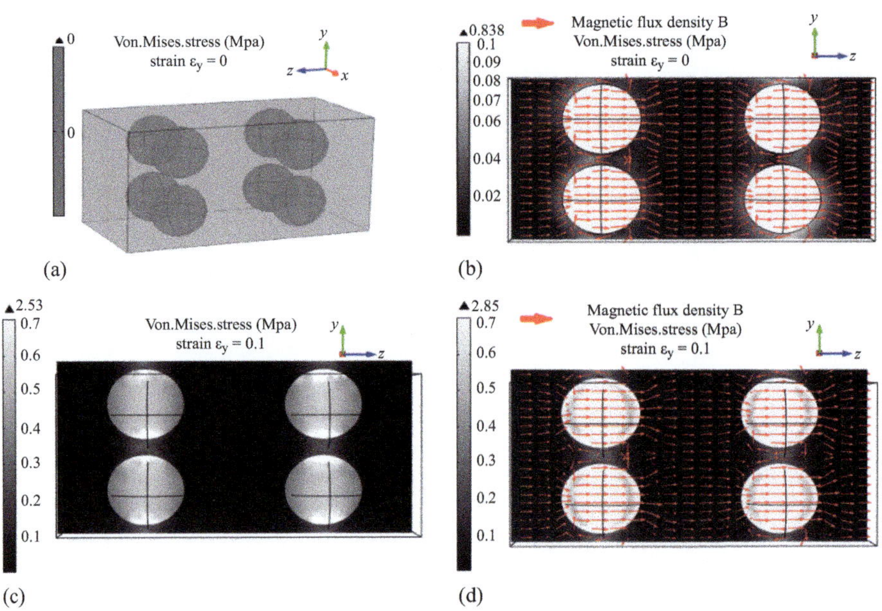

Figure 10.8 FEM simulation results: (a) FEM model, (b) $\varepsilon_y = 0$ with magnetized, (c) $\varepsilon_y = 0.1$ and (d) $\varepsilon_y = 0.1$ with magnetized

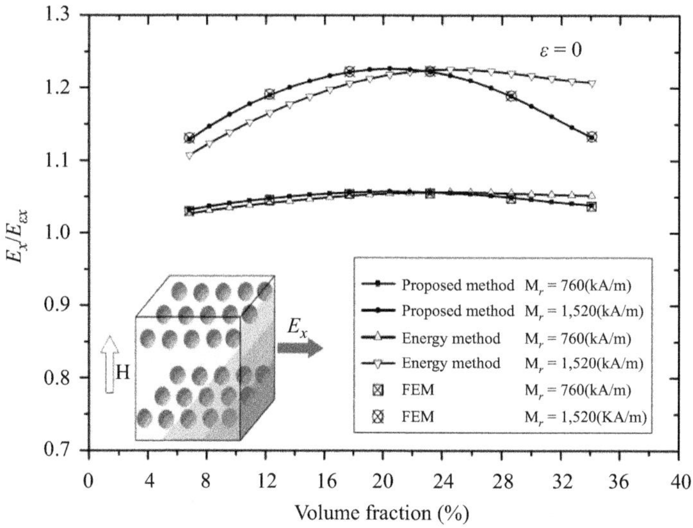

Figure 10.9 *$E_x/E_{\varepsilon x}$ of MRE with layer structure varies with particles' volume fraction*

As depicted in Figure 10.9, the method proposed in this chapter is closer to the FEM solutions than "energy" method, for a whole range of particle volume fractions. Since the "energy" method does not take into account the effect of magnetic energy releasing in different directions as well as the anisotropy of zero-field tensile modulus, as a result, all the energy is released along the side with E_x, which leads to predict a larger tensile modulus at higher particle volume fractions, while a smaller tensile modulus at lower particle volume fractions. All the results show that there is an optimal particle volume fraction in which the relative tensile modulus $E_x/E_{\varepsilon x}$ has the most changes under an applied magnetic field. And it will reach the highest value of 122.7% when particles' volume fraction is 20.2%.

Using the proposed method, the results of relative tensile modulus $E_x/E_{\varepsilon x}$ changes with strain, and magnetic fields for chain structure MRE in the x direction are shown in Figure 10.9.

Figure 10.10(a) shows that the tensile modulus changes with strain in the x direction under two magnetic fields of $\mathbf{M}_r = 1{,}520$ and 760 kA/m, respectively. Different from the results for the layer structure MRE as shown in above, the tensile modulus for the chain structure of MRE in the x direction decreases in the presence of magnetic field. And the more MRE is strained, the more tensile modulus decreases.

Figure 10.10(b) shows the relative tensile modulus of chain-structured MRE under two magnetic fields for different particles' volume fractions. When the particle volume fraction is less than 23.0%, the magneto-induced stress is mainly provided by the attraction force between particles in the z direction. And the attraction force increases as the deformation increases, which causes stretching

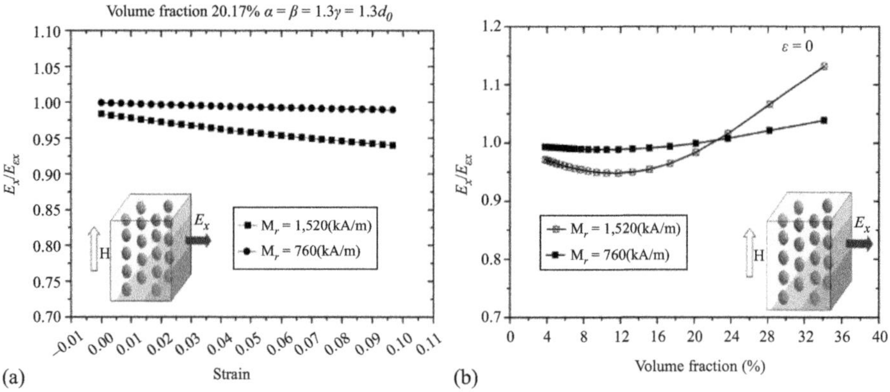

Figure 10.10 Relative tensile modulus $E_x/E_{\varepsilon x}$ for the chain structure MRE in the x direction (a) at different strains and (b) at different volume fractions

MRE in the x direction more easily when particles are being magnetized, resulting in $E_x/E_{\varepsilon x}<1$. However, when the particle volume fraction is over 23.0%, the distance between particles becomes closer. The magneto-induced stress is mainly provided by the repulsion force of particles in the x, y directions. With the deformation increasing, the repulsion force in the x directions declines, which makes the expansion in the x direction more difficult, resulting in magneto-induced positive tensile modulus referring to $E_x/E_{\varepsilon x}>1$.

The distance in the y direction between particles decreases and the repulsion force increases with the increasing stress, resulting in an increase in the absolute value of σ_{my}, and a decline in its numerical value. Meanwhile, the distance in the x direction between particles starts to increase, which increases the numerical value of σ_{mx}. Given that the particles in the z direction are too far to attract each other, the σ_{mz} almost remains the same. Thus, the total magneto-induced stress of RVE grows with the strain, and the elastic modulus in the x direction is enhanced.

10.2.3 The z direction tensile modulus under magnetic fields

In this subsection, we apply the same method to study the tensile modulus of MRE in the z direction under an applied magnetic field. According to (10.25), the MRE is stretched in the z direction, and thus the far-field strain ε_z occurs. The position vector of particle j in RVE can derived as

$$\mathbf{r}_j = r_{jx}(1+\varepsilon_{x-\varepsilon z})\mathbf{i} + r_{jy}(1+\varepsilon_{y-\varepsilon z})\mathbf{j} + r_{jz}(1+\varepsilon_{\varepsilon z})\mathbf{k} \tag{10.49}$$

And the tensile modulus in the z direction of RVE is expressed as following:

$$E_z = E_{\varepsilon x} + \frac{\partial(\sigma_{mz} + \sigma_{mz-x} + \sigma_{mz-y})}{\partial \varepsilon_{\varepsilon x}} \tag{10.50}$$

where E_{mz} in (10.50) is defined as magneto-induced modulus in the z direction which is generated in the presence of magnetic field. Considering (10.29)–(10.31), (10.25), (10.31), (10.32), the equation is derived and simplified as follows:

$$E_{mz} = \tilde{\mathbf{C}} \cdot \sum_{j=2}^{N} \mathbf{A}_j \tilde{\mathbf{T}}_j \tag{10.51}$$

In the comparison between (10.43) and (10.51), tensile moduli in these threes directions have the same particles' structure matrix $\mathbf{A}_j$. If the far-field strain direction is changed, the magneto-stress weighting vector $\tilde{\mathbf{C}}$ and far-field strain vector $\tilde{\mathbf{T}}_j$ are also changed, and $\tilde{\mathbf{C}}$ could be expressed as follows:

$$\begin{aligned}
\mathbf{C}(1,1) &= \frac{-3\mu_0 \|\mathbf{m}\|^2 E_{\varepsilon x} E_{\varepsilon y} E_{\varepsilon z}}{4\pi\gamma\beta\sqrt{(E_{\varepsilon x}+E_{\varepsilon z})^2(E_{\varepsilon x}+\sigma_{mx})^4 + 4E_{\varepsilon x}E_{\varepsilon y}\sigma_{my}(E_{\varepsilon x}+\sigma_{mx})^3}} \\
\mathbf{C}(2,1) &= \frac{-3\mu_0 \|\mathbf{m}\|^2 E_{\varepsilon x} E_{\varepsilon y} E_{\varepsilon z}}{4\pi\alpha\beta\sqrt{(E_{\varepsilon x}+E_{\varepsilon y})^2(E_{\varepsilon y}+\sigma_{my})^4 + 4E_{\varepsilon x}E_{\varepsilon y}\sigma_{my}(E_{\varepsilon y}+\sigma_{my})^3}} \\
\mathbf{C}(3,1) &= -\frac{3\mu_0 \|\mathbf{m}\|^2}{4\pi\alpha\gamma}
\end{aligned} \tag{10.52}$$

Besides, the far-field strain vector $\tilde{\mathbf{T}}_j$ in (10.49) could be obtained as the following:

$$\begin{aligned}
\tilde{\mathbf{T}}_j(1,1) &= r_{jx0} r_{jy0} \frac{E_{\varepsilon z}}{\sqrt{(E_{\varepsilon y}+E_{\varepsilon z})^2(1+\varepsilon_z)^4 + 4E_{\varepsilon z}E_{\varepsilon y}\varepsilon_z(1+\varepsilon_z)^3}} \\
\tilde{\mathbf{T}}_j(2,1) &= r_{jy0} \frac{E_{\varepsilon z}}{\sqrt{(E_{\varepsilon y}+E_{\varepsilon z})^2(1+\varepsilon_z)^4 + 4E_{\varepsilon z}E_{\varepsilon y}\varepsilon_z(1+\varepsilon_z)^3}} \\
\tilde{\mathbf{T}}_j(3,1) &= r_{jz0}
\end{aligned} \tag{10.53}$$

By considering (10.49), the tensile modulus of MRE E_z is derived. The results of tensile modulus in the z direction of MRE with layer structure are given in Figure 10.9.

The relative tensile modulus $E_z/E_{\varepsilon z}$ of MRE in the z direction under different strains is shown in Figure 10.11(a). The results indicate that the tensile modulus in the z direction of MRE with layer structure increases as the intensity of magnetic field increases. However, the tensile modulus is almost constant as the strain increases.

Figure 10.11 shows the relative tensile modulus $E_z/E_{\varepsilon z}$ of MRE with layer structure change with particles' volume fraction for two different magnetic field strengths. When the particle volume fraction is below 25.9%, the magneto-induced stresses in the z direction are almost independent with particle volume fractions. When the particles' volume fraction is higher than 25.9%, the tensile modulus E_z decreases in the presence of magnetic field. Further increasing the particle volume

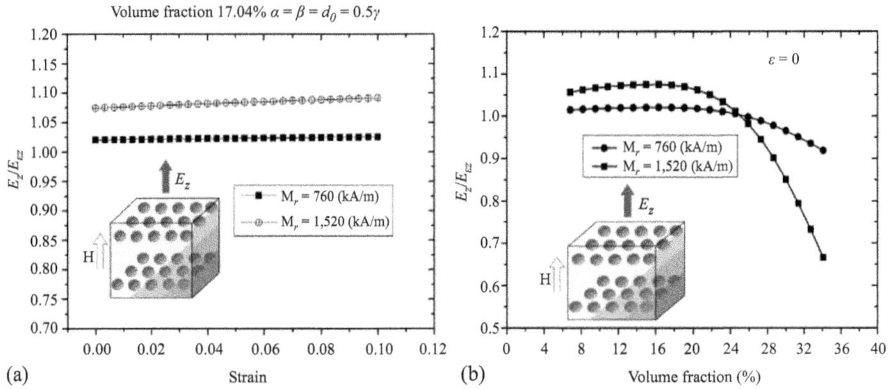

Figure 10.11 Tensile modulus in the z direction of layer structure MRE (a) at different strains and (b) at different volume fractions

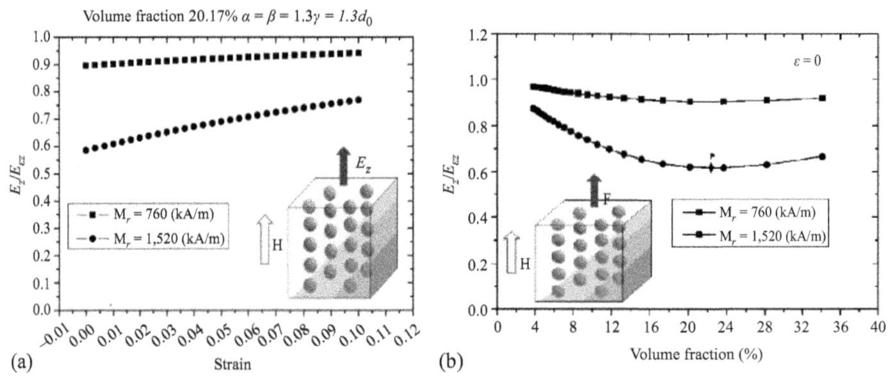

Figure 10.12 Tensile modulus in the z direction of chain structure MRE (a) at different strains and (b) at different volume fractions

fraction would cause the magneto-induced stresses E_z decrease more under the same magnetic field.

Similar results of the relative tensile modulus $E_z/E_{\varepsilon z}$ of MRE with chain structure shown in Figure 10.12.

The results indicate that the tensile modulus of MRE with chain structure decreases as the intensity of magnetic field increases and remains constant as the strain increases under magnetic field strength of 760 kA/m.

The relative tensile modulus $E_z/E_{\varepsilon z}$ of MRE with chain structure as a function of particle volume fraction is illustrated in Figure 10.12(b). The relative changes of the tensile modulus E_z decrease with particle fractions. The changes reach its maximum value at the particle volume fraction of 23.7% (the point P as shown in Figure 10.12(b)).

For both layer and chain structures of MREs, our theoretical results show that the relative tensile modulus $E_z/E_{\varepsilon z}$ is less than 1 in most cases. It means that the modulus E_z decreases in the presence of magnetic field which is called negative MR effect. This phenomena has also been reported by previous researches in the theoretical study of chain-like structured MREs [20,22].

The tensile moduli E_Z of structured MREs have been investigated experimentally [32]. The results show that when the tensile strain is between 0.1 and 0.2, the structured MREs demonstrate the negative MR effect. This effect decays as the particle volume fraction reduces, which conforms to our study.

10.3 Anisotropy of the structured MRE

The particles are orderly distributed, which offers anisotropy to structured MRE not only in the presence of magnetic field but also in the absence of field. For the same particle volume fraction, our studies show that the MRE specimen with layer or chain structure has different field-induced tensile moduli under the same magnetic field. In this section, we will discuss how the structured particles generate the anisotropy of MREs, and how the particles' distribution affects their field-induced properties in tensile modulus.

10.3.1 Tensile modulus in different direction

In this subsection, we will discuss about the anisotropy of MREs by showing the ratio of E_z/E_X for both chain and layer structures as a function of magnetic field strength and particle fractions. The magnetic fields are applied in the z direction. $E_z/E_X = 1$ represents the isotropy of MREs; and the larger the value of E_z/E_X deviating from 1, the higher the anisotropy of MREs.

Figure 10.13 shows the E_z/E_X of layer structure and chain structure MREs changes with magnetization intensity for three different particle volume fractions. As shown in Figure 10.13, the ratio of E_z/E_X for both chain and layer-structured MREs decreases with magnetization intensity.

Figure 10.13 shows how the anisotropic behavior of MRE changes with particle volume fractions and magnetic fields. Under zero magnetic field, the anisotropy of MREs is created by the particle structures inside the elastomers during curing process. It can be seen that the ratio of E_z/E_X for chain-structured MREs reaches maximum at 17.04% of particles' volume fraction, while layer-structured MRE reaches its minimum at 23.3%. The reinforced directions of MRE with chain-like and layer-like structure are the opposite. When the particles are fully magnetized under an applied magnetic field, the anisotropy of magnetized-structured MRE has nothing to do with particles' spatial configuration and is dependent on the magnetization intensity uniformly along the z direction. Besides, as particles' volume fraction increases, the anisotropic properties of MREs for both chain and layer structures are enhanced with applied magnetic field.

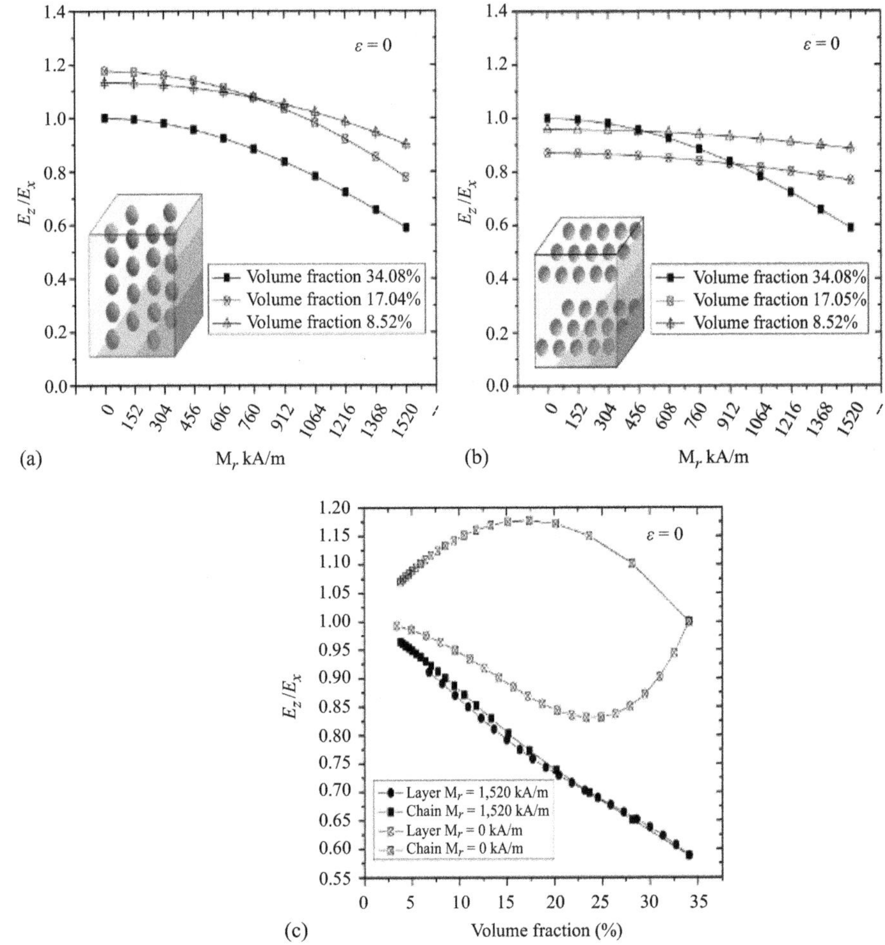

Figure 10.13 Anisotropy of MRE: (a) chain structure, (b) layer structure and (c) at different volume fractions

10.3.2 Tensile modulus in the different anisotropy microstructure

In this subsection, we will discuss the influence of particles' structure on magnetorheology effect of MREs in tensile modulus. Figure 10.14(a) and (b) shows the comparisons of tensile modulus E_x and E_z for the layer-structured and chain-structured MREs with the same particle volume fractions that change with magnetization intensity. Three different particle volume fractions are chosen in these studies. We can see that the difference between layer-structured and chain-structured MRE disappears with high particle volume fraction of 34.08%. The chain structure has higher tensile modulus E_z in the z direction than layer structure at the same particle volume fraction in the absence of the magnetic field. In the

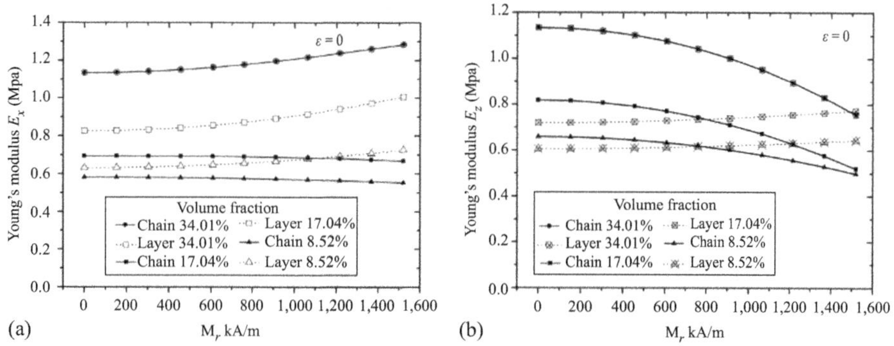

Figure 10.14 Tensile modulus of chain structure MRE at different structures: (a) E_x and (b) E_z

chain structure, the influence of magnetization on E_z is negative, while it is positive in the layer structures.

With a low particle volume fraction, the tensile modulus E_x of layer-structured MREs increases with magnetic field while tensile modulus E_x of chain-structured MREs decreases.

Figure 10.14 demonstrates that there is no difference between layer and chain structure at high volume fraction. However, when the particle fraction is low, MRE with layer structure has a bigger MR effect in the x direction, while MRE with chain structure has a stronger effect in z direction. That is to say, when the magnetic field is applied to the direction that particles are relatively closer, the MRE will have a stronger MR effect.

10.4 Conclusions

We have investigated the tensile modulus of structured MRE to understand its anisotropic properties in terms of spatial configuration of particles, particle volume fraction and magnetization density. A micromechanical model incorporated into the macroscopic constitutive laws was derived from the equivalent effective medium theory by considering the anisotropy of structured MREs. A three-parameter microscopic RVE was constructed to describe the microstructure of MREs in which the magnetic particle were arranged in the matrix forming layer or chain structure. By considering the transverse isotropic of pure tensile modulus, we established a relationship between tensile modulus of magnetic-induced stress and mechanical strain, which can derive tensile modulus of structured MRE in two directions under uniform magnetic field. The theoretical analysis showed that the field-response effect of tensile modulus of MRE was dependent on the magneto-induced stress, magneto-induced spatial structure and far-field strain.

Considering the particle interaction force in the presence of magnetic field under tensile stress, we found that when the magneto-induced stress decreases with mechanic strain, the positive MR effect is generated, while the magneto-induced stress increases with mechanical strain, the negative MR effect is generated. Simulation results show that there is an optimal particles' volume fraction for structured MRE, in which the changing rate of tensile modulus is the largest in one direction.

We evaluated the anisotropy of MREs by showing the ratio of E_z/E_x for both chain and layer structures as a function of magnetic field strength and particle fractions. The anisotropy of MRE with chain structure is most evident with the particles loading of 17.04%, in the absence of magnetic field and the anisotropy of MRE with layer structure reaches peak with the particles loading of 23.3%. When the magnetic field is applied on, the anisotropy of MRE with chain structure is weakened and one with layer structure is enhanced with the particles loading lower than 10.1%. When the particles' volume fraction is higher than 10.1%, the anisotropy of both layer structure and chain structure of MRE is reinforced.

References

[1] Böse H, Rabindranath R, Ehrlich J. Soft magnetorheological elastomers as new actuators for valves. Journal of Intelligent Material Systems and Structures, 2012, 23(9): 989–994.

[2] Ginder J M, Nichols M E, Elie L D, *et al.* Controllable-stiffness components based on magnetorheological elastomers. SPIE's 7th Annual International Symposium on Smart Structures and Materials. International Society for Optics and Photonics, 2000: 418–425.

[3] Behrooz M, Wang X, Gordaninejad F. Performance of a new magnetorheological elastomer isolation system. Smart Materials and Structures, 2014, 23(4): 045014.

[4] Wang X, Gordaninejad F, Calgar M, *et al.* Sensing behavior of magnetorheological elastomers. Journal of Mechanical Design, 2009, 131(9), 091004.

[5] Sutrisno J, Purwanto A, Mazlan S A. Recent progress on magnetorheological solids: materials, fabrication, testing, and applications. Advanced Engineering Materials, 2015, 17(5): 563–597.

[6] Chen S, Wang X, Zhang Z, *et al.* Optimal design of laminated-MRE bearings with multi-scale model. Smart Materials and Structures, 2016, 25(10): 105037.

[7] Bellan C, Bossis G. Field dependence of viscoelastic properties of MR elastomers. International Journal of Modern Physics B, 2002, 16(17n18): 2447–2453.

[8] Wu J, Gong X, Fan Y, *et al.* Anisotropic polyurethane magnetorheological elastomer prepared through in situ polycondensation under a magnetic field. Smart Materials & Structures, 2010, 19(10): 105007.

[9] Varga Z, Filipcsei G, Zrínyi M. Magnetic field sensitive functional elastomers with tunable elastic modulus. Polymer, 2006, 47: 227–233.

[10] Jolly M R, Carlson J D, Muñoz B C, *et al.* The magnetoviscoelastic response of elastomer composites consisting of ferrous particles embedded in a polymer matrix. Journal of Intelligent Material Systems and Structures, 1996, 7(6): 613–622.
[11] Davis L C. Model of magnetorheological elastomers. Journal of Applied Physics, 1999, 85(6): 3348–3351.
[12] Yin H M, Sun L Z. Magnetoelastic modelling of composites containing randomly dispersed ferromagnetic particles. Philosophical Magazine, 2006, 86(28): 4367–4395.
[13] Yin H, Sun L. Magnetoelasticity of chain-structured ferromagnetic composites. Applied Physics Letters, 2005, 86(26): 261901.
[14] Zhu Y, Gong X, Dang H, *et al.* Numerical analysis on magnetic-induced shear modulus of magnetorheological elastomers based on multi-chain model. Chinese Journal of Chemical Physics, 2006, 19(2): 126–130.
[15] Han Y, Zhang Z, Faidley L A E, *et al.* Microstructure-based modeling of magneto-rheological elastomers. SPIE Smart Structures and Materials+ Nondestructive Evaluation and Health Monitoring. International Society for Optics and Photonics, 2012: 8342: 83421B.
[16] Borcea L, Bruno O. On the magneto-elastic properties of elastomer–ferromagnet composites. Journal of the Mechanics and Physics of Solids, 2001, 49(12): 2877–2919.
[17] Brigadnov I A, Dorfmann A. Mathematical modeling of magneto-sensitive elastomers. International Journal of Solids and Structures, 2003, 40(18): 4659–4674.
[18] Kankanala S V, Triantafyllidis N. On finitely strained magnetorheological elastomers. Journal of the Mechanics and Physics of Solids, 2004, 52(12): 2869–2908.
[19] Dorfmann A, Ogden R W. Nonlinear magnetoelastic deformations. The Quarterly Journal of Mechanics and Applied Mathematics, 2004, 57(4): 599–622.
[20] Bustamante R. Transversely isotropic nonlinear magneto-active elastomers. Acta Mechanica, 2010, 210(3–4): 183–214.
[21] Danas K, Kankanala S V, Triantafyllidis N. Experiments and modeling of iron-particle-filled magnetorheological elastomers. Journal of the Mechanics and Physics of Solids, 2012, 60(1): 120–138.
[22] Saxena P, Hossain M, Steinmann P. Nonlinear magneto-viscoelasticity of transversally isotropic magneto-active polymers. Proceedings of the Royal Society A: Mathematical, Physical and Engineering Science, 2014, 470 (2166): 20140082.
[23] Ivaneyko D, Toshchevikov V P, Saphiannikova M, *et al.* Magneto-sensitive elastomers in a homogeneous magnetic field: a regular rectangular lattice model. Macromolecular Theory and Simulations, 2011, 20(6): 411–424.
[24] Ivaneyko D, Toshchevikov V, Borin D, *et al.* Mechanical properties of magneto-sensitive elastomers in a homogeneous magnetic field: theory and experiment. Macromolecular Symposia, 2014, 338: 96–107.

[25] Ivaneyko D, Toshchevikov V, Saphiannikova M, *et al.* Mechanical properties of magneto-sensitive elastomers: unification of the continuum-mechanics and microscopic theoretical approaches. Soft Matter, 2014, 10(13): 2213–2225.

[26] Elkins J. Oxidative stability in surface modified magnetorheological elastomers. Diss. Masters Thesis, 2005. University of Nevada, Reno.

[27] Eshelby J D. The determination of the elastic field of an ellipsoidal inclusion, and related problems. Proceedings of the Royal Society of London. Series A. Mathematical and Physical Sciences, 1957, 241(1226): 376–396.

[28] Mori T, Tanaka K. Average stress in matrix and average elastic energy of materials with misfitting inclusions. Acta Metallurgica, 1973, 21(5): 571–574.

[29] Mura T. Micromechanics of defects in solids. The Hague/Boston/London: Martinus Nijhoff Publishers, 1982.

[30] Park B J, Song K H, Choi H J. Magnetic carbonyl iron nanoparticle based magnetorheological suspension and its characteristics. Materials Letters, 2009, 63(15): 1350–1352.

[31] Rosensweig R E. Ferrohydrodynamics. New York: Cambridge University Press, 1985.

[32] Schubert G, Harrison P. Large-strain behaviour of magneto-rheological elastomers tested under uniaxial compression and tension, and pure shear deformations. Polymer Testing, 2015, 42: 122–134.

Chapter 11

Experimental study and mathematical model on different magnetorheological elastomers

Zhao-Dong Xu[1], Jun-Tao Zhu[2], and Ying-Qing Guo[3]

Magnetorheological (MR) elastomer (MRE) is a kind of smart material, which exhibits rapid variations in their rheological property such as viscosity and modulus when subjected to different magnetic fields [1–4]. The physical phenomena seen in MREs are very similar to those in MR fluids, where the particles try to arrange themselves in the direction of the magnetic field [5,6,28], that is, to form an anisotropic ordered pre-configuration such as chains or more complex structures, which are solidified into a polymer medium. The advantage of MREs over MR fluids is that magnetic particles do not undergo sedimentation; therefore, these MREs have both the MR effect and good mechanical performance. MREs generally consist of three main components: matrix, magnetic particles and additives. Some literatures show that the MR effect of MREs is relative to the matrix. The relative MR effect was higher in the softer elastomers, which indicate that the matrix has influence on the MR effect [25]. Therefore, in order to improve the MR effects of MRE, these soft matrices, such as silicon rubber and gels, were selected [7,8]. In these reports, the maximum increment in the shear modulus due to the MR effect is about 50%–60% of the zero-field modulus, depending on the matrix elastomer [8,9]. However, those MREs based on the soft matrices are usually ill-suited for most load-bearing applications due to their low strength and reduced fatigue life. Natural rubber (NR), excellent in mechanical property, is usually used as the matrix for fabricating MREs [3,10]. Several kinds of viscoelastic materials based on nitrile butadiene rubber (NBR) and silicone rubber (SR) matrices were studied. The results indicated that NBR matrix VE materials have a high energy dissipation capacity, whereas SR matrix VE materials have stable performance under different working conditions [11,12]. Bromobutyl rubber was adopted as the matrix material because of its high damping capacity, and the magnetoviscoelasticity properties of the MRE specimens are investigated with respect to different magnetic fields, displacement amplitudes and frequencies under sinusoidal

[1]Key Laboratory of C&PC Structures of the Ministry of Education, Southeast University, China
[2]Civil Engineering School, Zhengzhou University, China
[3]Mechanical and Electronic Engineering College, Nanjing Forestry University, China

loadings. The experimental results show that the MREs have a high loss factor which can reach to 0.682 [11–13]. On the other hand, the damping capability is greatly influenced by the effect of vibration control, which can reduce the peak responses of the structures. So the damping property of MREs was studied by researchers. These researches indicated that the influences of strain amplitude, frequency and magnetic field on the damping capacities of MREs are small [14,15]. Carbonyl iron power is generally used as the filler material to fabricate MREs because of its high permeability, saturation magnetization and low remnant magnetization. The effect of particle shape, size and volume fraction on the performance of MREs has been investigated [16,17]. It was identified that the content and size of the magnetic filler play an important contribution in the shear modulus. Additives are usually used to improve the property of MREs, i.e., to improve the particles or matrix adhesion, alter the property of the matrix (e.g., its hardness) and prevent the elastomer from oxidizing [16].

So far, most of the research has been focused on experimental approach to study the performance of MREs under external magnetic field. Few are about theoretical aspect; therefore, it is an important problem how to describe the performance of MREs in the form of mathematic models. In these studies, a few models were proposed in analyzing the field-dependent modulus of MREs based on simple chain mechanisms and particle dipole interactions. Originally, simply considering the dipole interactions in the same direction as external field and assuming that the magnetic interaction only exists between the two nearby dipoles, the modulus change of the material is calculated [18]. A point-dipole model is proposed, where the magnetic interaction occurs only between the adjacent particles within a chain [19]. And another model calculated the shear increment by using finite element analysis, which was for isolated single chains of periodically spaced dipoles [20]. Taking into account the fully coupled magnetoelastic interactions, the model calculated from the basic minimum energy principle of magnetoelasticity was developed [21]. A direct theory of magnetoelastic membranes based on the three-dimensional theory is applicable to conventional elastomers magnetized through infusion with uniformly dispersed ferrous particles [22]. The other models described as a relationship between force–displacement and stress–strain, which is a behavior of macro mechanics. In this aspect, very few mathematical models are available to predict their property. A mathematical model to represent the stress–strain relationship of the MRE was derived by taking into account all dipole interactions in a chain [10]. A constitutive model was derived based on pseudo-elasticity theory, which is the inclusion of two variables in the energy function in order to separately capture the stress softening and residual strain effects [23]. And a four-parameter linear viscoelastic model was proposed, which could predict the dynamic mechanical property of MRE performance under various working conditions [24]. A microphysical model based on the assumption of the chi-square distribution of the distance between adjacent ferromagnetic particles is proposed, which can eliminate the error generated by the assumption of the uniform distribution and describe the MR effect more exactly. Based on the proposed microphysical model, the magnetoviscoelasticity parameter model is modified to describe the dynamic properties of MREs [13].

This study concerns the method for fabricating MREs based on different carbonyl iron powder proportions in two different matrices. A total of 12 MRE samples, divided into two types, were manufactured. The physical property and the dynamic mechanical property tests were carried out to compare the performance of the specimens under different loading cases, particularly, the viscoelastic property of the two different kinds of MREs will be discussed. In addition, a magnetic-induced viscoelastic parameter model is verified by comparing these experimental and numerical results, which can describe the viscoelastic property of MREs.

11.1 MR elastomers fabrication

11.1.1 Selection of materials

Matrices: The studies of some available literatures on fabricating and testing of MREs show that mainly network polymer matrices have been used, i.e., silicon rubber, NR and polyurethane gels [25,26]. These studies experimentally indicate that the high MR effects can be obtained by using the soft matrices; therefore, the soft matrices, such as silicon rubbers and gels, are the best selection for fabricating MREs. However, as a smart material used in engineering vibration control, it must have a capability of deformation resistance of the structures under excited loadings; therefore, these soft matrices with worse mechanical property are not suitable to be applied in structural vibration control. So, for the purpose to apply the MREs in area of structural vibration control, the NR with good mechanical property was selected to fabricate the MREs in our previous work [27]; however, the damping property of which is not good for its application. As we all know, damping can effectively reduce the peak of structures vibration responses, so, we should make effort to improve the damping property of MREs. According to the author's previous works [28,29], Brominated Isoprene Isobutylene Rubber (BIIR) is selected as the matrix for fabricating MREs, which has a high damping property compared with other rubbers in our researches.

Magnetic fillers: The carbonyl iron particles are commonly used to fabricate MREs with the advantages of low residual magnetism, high magnetic permeability and economy. The shape of carbonyl iron particle is usually a spherical form, whose characteristic makes it easily move in the matrix when curing under external magnetic field. The size of the particles also influences the MR effect of MREs, which reported in the literature [16]. The mechanical property of MREs decreases although the MR effect can increase with the increment of particle size. Considering these factors, the carbonyl iron powder, type, R2F, with a mean particle size of approximately 3.30 μm is selected as magnetic fillers to fabricate the MREs.

Additives: The additives used in our tests include reinforcing agent, vulcanizing agent, plasticizer and accelerator, which can improve the mechanical property of MREs. For instance, the reinforcing agent, i.e., high abrasion furnace black (type: N330, average size is 27–34 nm, supplied by Shanghai Kai Nuo Chemical Co., LTD) and semi-reinforcing purpose furnace black, can improve the tearing strength of the synthetic rubber, and the plasticizer (Stearic acid, type: 800, supplied by Shanghai Zhi Tai Chemical Co., LTD) can modify the matrix and increase the wettability of the carbonyl iron powder.

Based on the quantity of tests, the recipes used in this study are decided. The recipe of MREs with NR matrix mainly includes 100 phr NR, 19 phr reinforcing agent (semi-reinforcing purpose furnace black), 11.5 phr vulcanizing agent (ZnO and S), 13 phr plasticizer (stearic acid and 46# machine oil), 2 phr accelerator (promoter Z), in which phr is the weight parts based on per hundred parts of rubber. The recipe of MREs with BIIR matrix mainly includes 100 phr BIIR, 45 phr reinforcing agent (high abrasion furnace black), 13.5 phr vulcanizing agent (ZnO, MgO and WS curing resin), 15 phr plasticizer (TOTM and stearic acid), 21 phr accelerator (2203LC hydrocarbon resin).

11.1.2 Procedure of MRE specimens' fabrication

The MRE test samples are manufactured in the style of sandwich, which can be seen in Figure 11.1. In order to ensure the uniformity of magnetic field, the plates of the specimens are made of aluminum and the permanent magnet is used, the face area of which is larger than that of the MREs sample. The manufacturing of MRE specimens consists of mixing, preforming configuration and curing. Take the MRE sample with BIIR, for example, first, 100 phr BIIR, 94.9 phr additives and the carbonyl iron particles are mixed homogeneously using a two-roll mill, type, XK-400. Then, the mixture and the aluminum plates are put into a mold for preforming under an external magnetic field of 100 mT, generated by the high-temperature-resistant permanent magnet (seen in Figure 11.2), at 135 °C for 15 min. Finally, the specimen is vulcanized on a plate vulcanization machine, type, TH-6009. The vulcanization process is operated under a pressure of approximately 15 MPa and at 165 °C for 30 min. In this test, in order to investigate the property of MREs with BIIR and the differences on performance between the two kinds of MREs

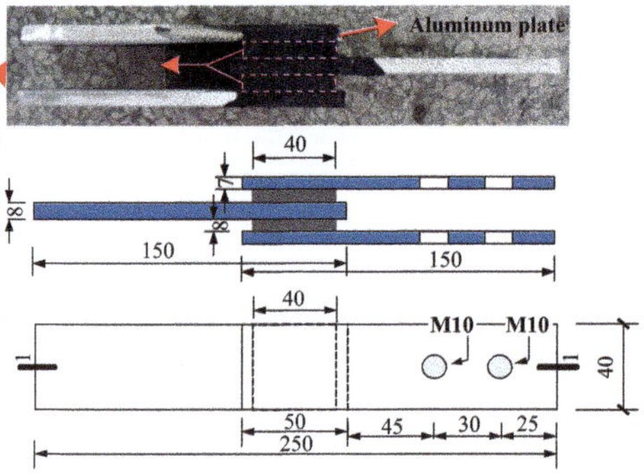

(a) (b)

Figure 11.1 MRE test specimens: (a) MRE device Photo and (b) structure diagram of the MRE device

Figure 11.2 Manufacturing setup

(MREs with BIIR and NR), a total of four MRE test samples are manufactured, which are, respectively, named BIIR60 and BIIR70, compared to the test samples NR50, NR60, NR70 and NR80 in previous work [27].

11.2 Performance tests of MREs

11.2.1 Test setup and procedure

The physical property of the MRE samples were tested, the indexes of which mainly include fracture strength, elongation at break and hardness. The testing instruments used in this test are universal-material-testing machine (type, DXLL-5000, made in China) and shore hardness tester (type, LX-A, made in China), seen in Figure 11.3.

A measuring stand making it possible to examine the effect of the magnetic field on the performance of the MREs was designed and built. The test specimens are subjected to cyclic shearing at set constant magnetic field intensities. The quasi-static and dynamical property of MRE samples were tested using an MTS hydraulic pulsator, type, PA-100, with a maximum force of $\pm$100 kN. The magnetic field on both sides of the tested instrument, ranging from 0 up to 300 mT, is generated by a magnetic field apparatus. The magnetic field apparatus consists of permanent magnets and aluminum boxes, the magnetic field strength of which is changed by changing the distance of the two boxes and the quantity of the magnets, and the magnetic field strength is measured by a teslameter. The experimental setup is shown in Figure 11.4. With the material test system, the displacement control mode is employed to measure the dynamic mechanical property of MRE specimen.

The physical property test is conducted under national rubber industry standard. For quasi-static property testing, the stretching rate usually is set below 1 mm/min (the height of test samples is 8 mm; therefore, the shear rate approximately is 0.017/s), and

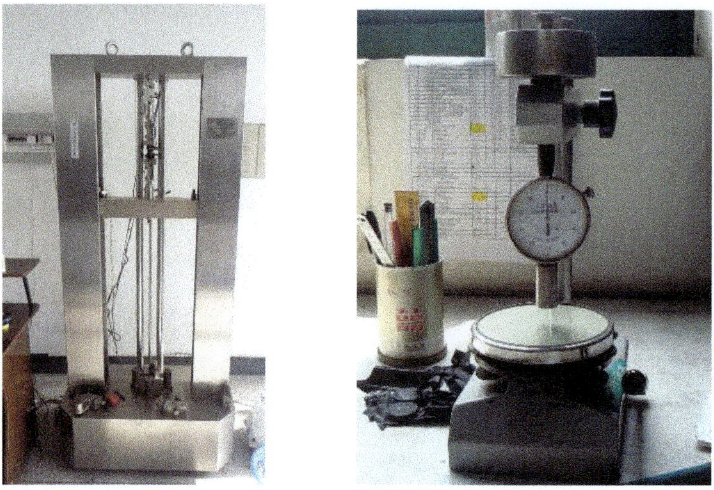

(a) (b)

Figure 11.3 Testing instruments for physical and mechanical property test: (a) universal material testing machine and (b) shore hardness tester

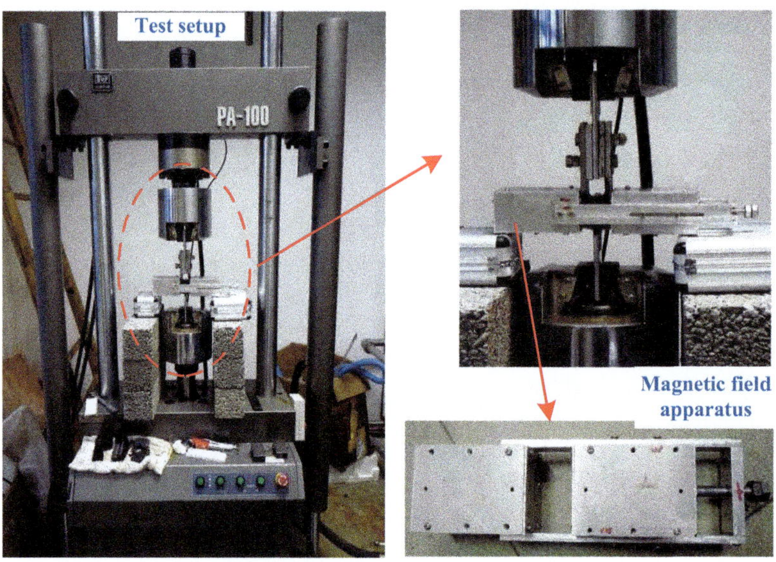

Figure 11.4 Experimental setup

the value of shear deformation will be set as 0.4 for the purpose invoicing shear damage of MREs or bond failure between MRE and aluminum plates. The shear stress and shear strain test results under a constant magnetic field of 300 mT are recorded by the data-acquisition system.

For dynamic mechanical property testing, sinusoidal excitations with constant displacement amplitudes and driving frequency are applied to the MRE specimens under different magnetic fields, and the resultant time-dependent forces are then recorded. In tests, three typical strain amplitudes of 12.5%, 25% and 50% (the dimensions of MRE parts of the specimen is 40 mm × 40 mm × 8 mm, and the shear area is 40 mm × 40 mm), corresponding to the three displacement amplitudes of 1, 2 and 4 mm, and four typical frequencies (1, 2, 5 and 10 Hz) as well as a number of magnetic fields are selected. The test program is designed and shown in Table 11.1. It consists of three stages with different displacement amplitudes, and 10 cycles are recorded for each set of frequency during each stage. Two identical samples are made for each of the types of MREs and the loading program is applied to them at four magnetic field strength values: 0, 100, 200 and 300 mT. For each loading case, the real-time force and displacement signals are recorded by the data-acquisition system and then used to reconstruct the force–displacement hysteresis loop.

11.2.2 Results and analysis

11.2.2.1 Physical property test

The main physical property test results of MRE samples are listed in Table 11.2. It can be seen from Table 11.2 that the fracture strength and elongation at break of the MRE samples with NR are all higher than that of the samples with BIIR, which indicates that the physical property of samples with NR is better than that of the samples with BIIR. In addition, the physical property of the samples with the same matrix decreases steadily with an increase in the fraction of carbonyl iron powder; on the contrary, the hardness increases with the increment of the carbonyl iron

Table 11.1 MRE devices test program

Order	Displacement (mm)	f (Hz)	H (mT)
1	1	1, 2, 5, 10	0, 100, 200, 300
2	2	1, 2, 5, 10	0, 100, 200, 300
3	4	1, 2, 5	0, 100, 200, 300

Table 11.2 Physical property results of MRE

Matrix	Carbonyl iron powder (phr/%)	Curing condition	Fracture strength (MPa)	Elongation at break (%)	Shore-A hardness
BIIR	60	165 °C × 30 min	4.8	490	51
	70	–	3.7	422	57
NR	50	150 °C × 20 min	13.4	600	42
	60	–	11.4	520	49
	70	–	9.1	450	55
	80	–	7	400	60

powder. It means that the addition of carbonyl iron powder results in a decrease of the physical property of MREs and an increase of the hardness. For the reason that the magnetic particles are taking the fact of filling agent, when the filling agent is too high, the interparticle spaces could be left unfilled, which would result in much worse strength parameters of the composite. Actually, the lowest value of the fracture strength in the test is 3.7 MPa, the MRE sample with BIIR matrix and with the proportion of 70% of carbonyl iron powder, which still can meet the request of strength for applications.

11.2.2.2 Quasi-static property test

For the purpose of investigating whether the MREs exists MR effects, the quasi-static property test is processed. Each type of MRE sample is tested according to the preset test procedure. Take samples NR70 and BIIR70, for example, Figure 11.5 shows the stress–strain curves of the two samples at two extreme magnetic field strength values: 0 and 300 mT. The slope of the stress–strain curve is the shear modulus of the MRE materials. It can be seen from these figures that the shear modulus obtained at the magnetic field of 300 mT is higher than that of 0 mT, which demonstrates that the MREs exhibit MR effects. The shear stress shows a linear relationship with the shear strain when the magnetic field is 0 mT. This demonstrates that the MREs have the linear elastic characteristic without external magnetic field. On the other hand, the relationship of the shear stress and strain expressed obvious nonlinear property when the shear strain is below 0.1 and with the magnetic field of 300 mT. This means that the nonlinear property of MREs is caused by the external magnetic field and shear strain, that is to say, the MR effect is a function with magnetic field and shear strain.

11.2.2.3 Dynamic mechanical property test

In order to study the dynamic mechanical property of MRE, these MRE samples are tested under harmonic loadings with various displacement amplitudes (1, 2 and 4 mm) and frequencies (1, 2, 5 and 10 Hz). Originally, the experimental force–displacement traces are recorded, and only one cycle finally is taken for the

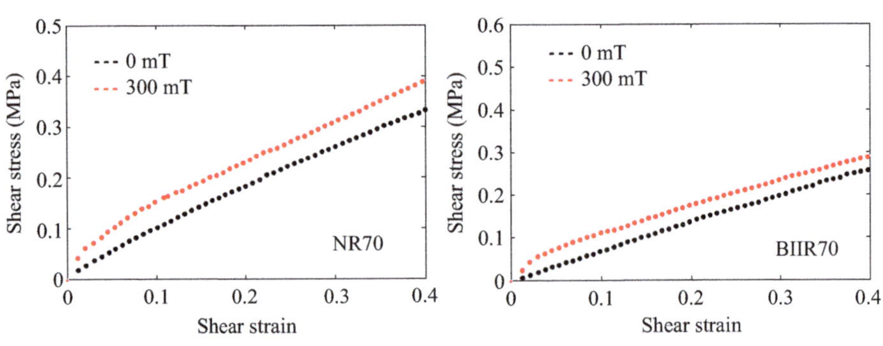

Figure 11.5 Comparison of the stress–strain curves at external magnetic field

calculations in order to make sure that the values represented only the steady state. Figure 11.6 shows the force–displacement relationships of the two kinds of MRE samples (take NR70 and BIIR70, for example) at the constant displacement amplitude of 2 mm, frequency of 10 Hz but at various magnetic fields from 0 to 300 mT. From this figure we can see that the hysteresis loops of the two kinds of MRE can shape ellipses; and the slope of the major axis of the elliptical loops varies with the magnetic field. The experimental force F increases with the increment of magnetic field strength values, while the displacement amplitude is constant. This means that because of the changes in the material, a greater force is required to maintain the same level of displacement, that is to say the stiffness of MREs varies with the magnetic fields. The areas of the elliptical loops increase slightly with the increment of the magnetic fields, which demonstrates that the damping capacity of the MRE samples is a function of the applied magnetic fields. In addition, the difference between the two kinds of MRE is that the areas of the elliptical loops of BIIR70 are larger than that of NR70, which indicates that the damping of BIIR70 is higher than NR70, that is to say, the damping property of BIIR70 is better than NR70. Figure 11.7 shows the shear storage modulus G_1 and loss factors η of these MRE samples with different matrices at various magnetic fields. It can be seen from this figure that the values of G_1 and η increase with the increment of the magnetic fields, for example, the value G_1 of BIIR70 increases from 0.79 to 0.985 MPa with the growth of 24.7% when the magnetic flux is increased from 0 to 300 mT, comparing with NR70, the value of which increases from 0.906 to 1.162 MPa with a growth of 28.3%. This result demonstrates that MRE materials have controllable mechanical property. Obviously, the value η of BIIR70 reaches to 0.682 at magnetic field of 300 mT, which is extremely higher than that of the MRE samples with NR. This advantage means that the MREs with BIIR, compared to MREs with NR, have a better damping capability in mitigating vibration applications.

Figure 11.8 shows the force–displacement curves of the MRE samples, NR60 and BIIR60, at the constant displacement amplitude of 2 mm, magnetic fields of 200 mT but at various frequencies from 1 to 10 Hz. It can be seen from these

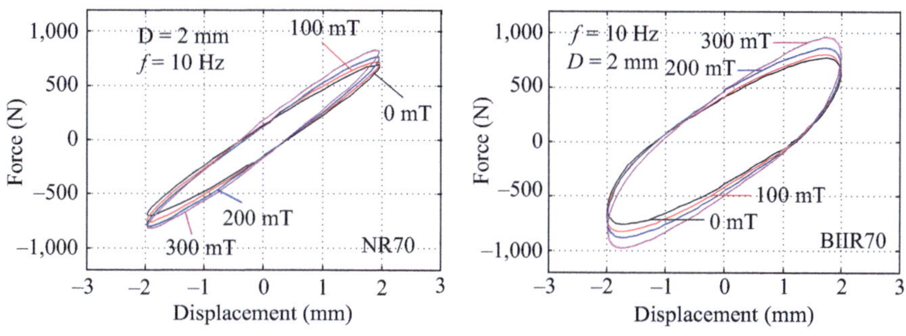

Figure 11.6 Force–displacement curves of MRE devices for different magnetic fields

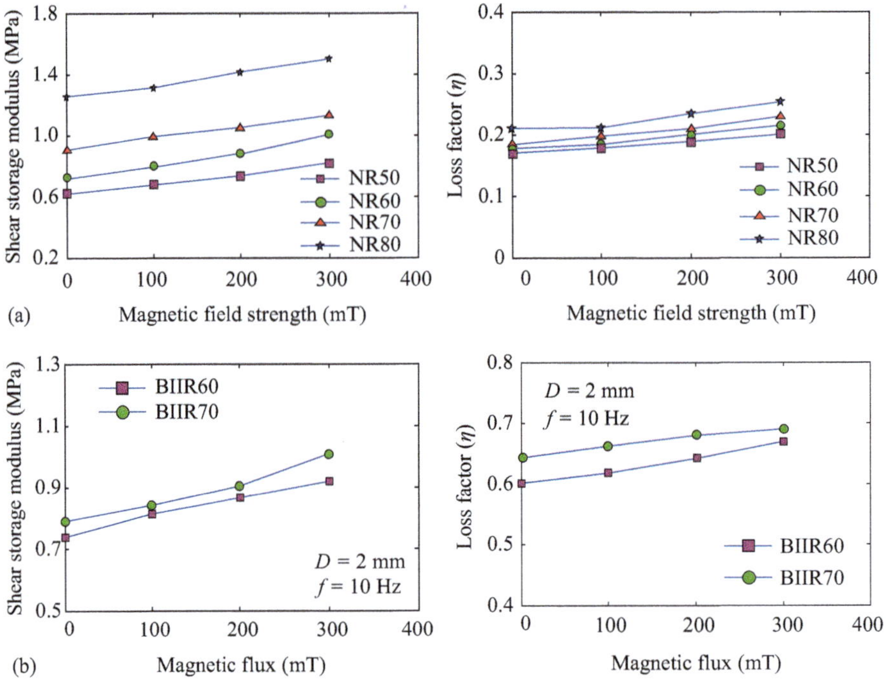

Figure 11.7 Shear storage modulus G_1 and loss factor η for different magnetic fields: (a) G_1 and η of MRE devices with NR matrix and (b) G_1 and η of MRE devices with BIIR matrix

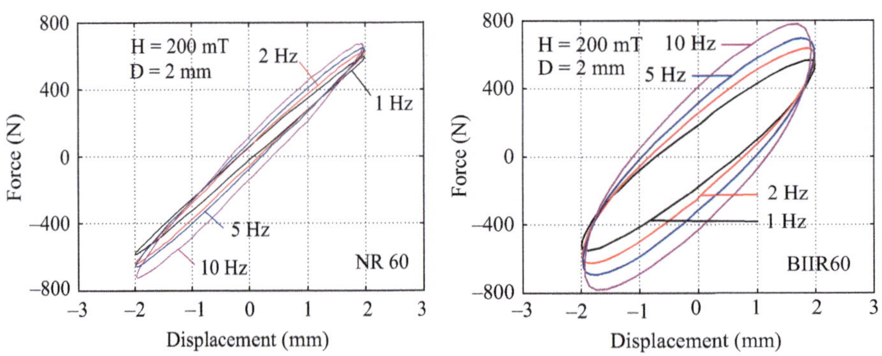

Figure 11.8 Force–displacement curves of MRE devices for different frequencies

figures that the changes in slope of the main axis of hysteresis loops are not significant, whereas the loops area increase markedly, which demonstrate that the damping capacity of the MRE specimens is greatly influenced by the excitation frequencies. As presented in Figure 11.9, the values G_1 and η of these MRE

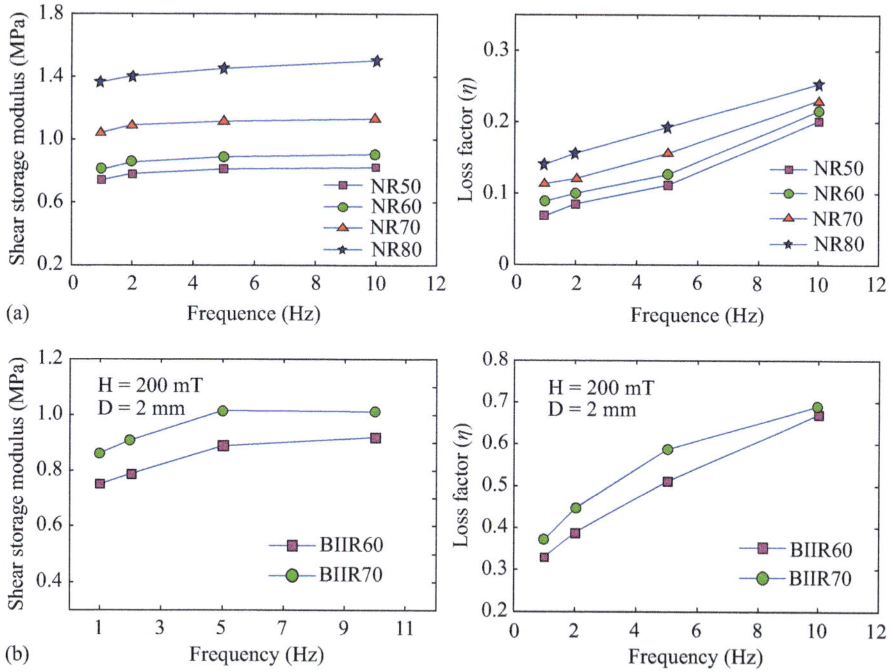

Figure 11.9 Shear storage modulus G_1 and loss factor η for different frequencies: (a) G_1 and η of MRE devices with NR matrix and (b) G_1 and η of MRE devices with BIIR matrix

samples all increase with the increment of excitation frequencies, especially the values of η, for example, the value of η of BIIR60 increases from 0.325 to 0.673 when the frequency increases from 1 to 10 Hz. This is because the fraction times between particles in the matrix increase when the working frequency becomes larger. Furthermore, the slope of the main axis increases steadily when the frequency inputs change from 1 to 5 Hz, while the slope of the main axis increases slightly when the frequency inputs change from 5 to 10 Hz. This implies that the stiffness of the specimen shows a slowly increasing trend with the frequency. In addition, at each testing frequency level, the values G_1 and η of the MRE samples all generally increase with the weight fraction of the particles.

Figure 11.10 shows the force–displacement curves of NR70 and BIIR70 at the constant frequency of 2 Hz, magnetic field of 200 mT but at various displacement amplitudes. It can be seen from Figure 11.10 that the slope of the main axis of these elliptical loops obviously change with the increment of the displacement amplitudes, which means that the stiffness of MRE specimens decreases when the displacement amplitudes increase from 1 to 4 mm. Figure 11.11 shows the curves of G_1 and η of these MRE samples changed with displacement amplitudes. It can be seen from these figures that the values of G_1 and η, especially G_1, greatly decrease with the increment of the displacement amplitudes, which demonstrates

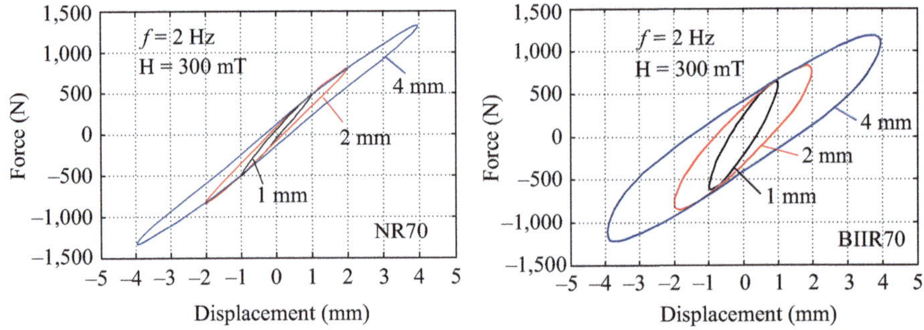

Figure 11.10 Force–displacement curves of MRE devices for various displacement amplitudes

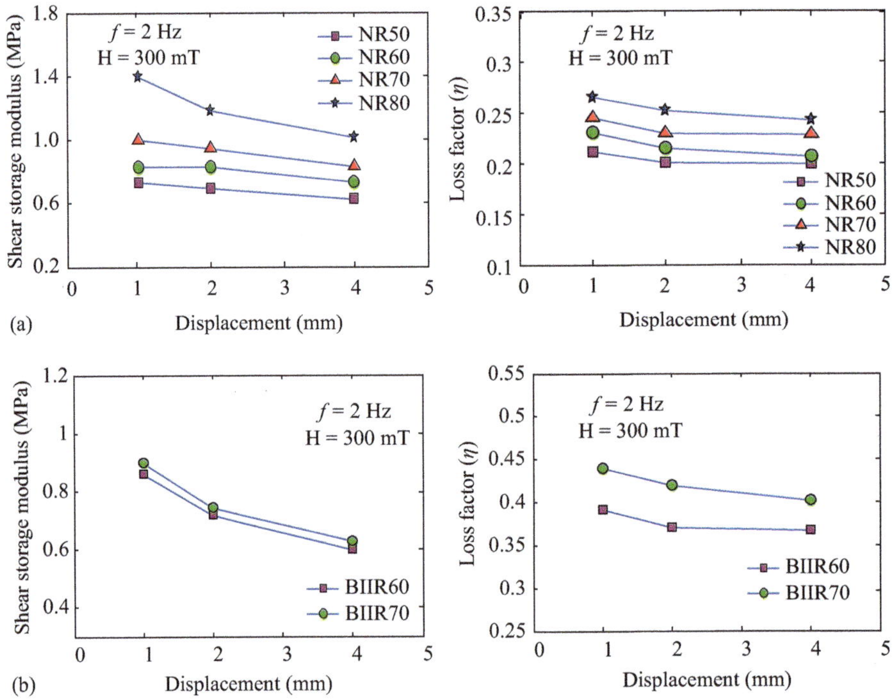

Figure 11.11 Comparison of shear storage modulus and loss factor for various displacement amplitudes: (a) G_1 and η of MRE devices with NR matrix and (b) G_1 and η of MRE devices with BIIR matrix

that the mechanical property of MREs is very sensitive to the strain amplitude. This is because the magnetic effect mainly depends on the distance between magnetic particles; therefore, the interactions between particles in MREs sharply decrease with the increment of displacement amplitude.

11.3 Numerical simulations

11.3.1 The microphysical model based on chi-squared distribution

Generally, the difficulty in the modeling procedure of MREs is how to describe the distribution of magnetic particles in mathematics precisely and concisely. From the scanning electron microscope (SEM) photographs of MRE [3], as shown in Figure 11.12, the particles align in a chain structure under the action of magnetic field, and moreover, the more intense the magnetic field, the more the particles tend to form compact chains. These facts suggest that the distance between particles appears to be inclined to partial distribution.

When simplifying ferromagnetic particles as dipoles, as shown in Figure 11.13, the interactive force between two poles, $\mathbf{m}_1$ and $\mathbf{m}_2$, can be formulated as

$$F = -\frac{3\mu_0}{4\pi r^4}[3(\mathbf{m}_1 \cdot \mathbf{r}_0)(\mathbf{m}_2 \cdot \mathbf{r}_0) - \mathbf{m}_1 \cdot \mathbf{m}_2] \tag{11.1}$$

where $\mathbf{m}_0$ is the vacuum permeability, $\mathbf{r}_0$ is the direction vector, r is the distance of the two dipoles, and $\mathbf{m}_1$ and $\mathbf{m}_2$ are the magnetic moments of dipoles.

According to (11.1), interaction F between two dipoles is approximately inversely proportional to high order of distance r, that is, F decreases sharply with an increase in r. So it can be concluded that the interaction among particles in the chain structure is determined by the adjacent distance, namely, the smallest distance.

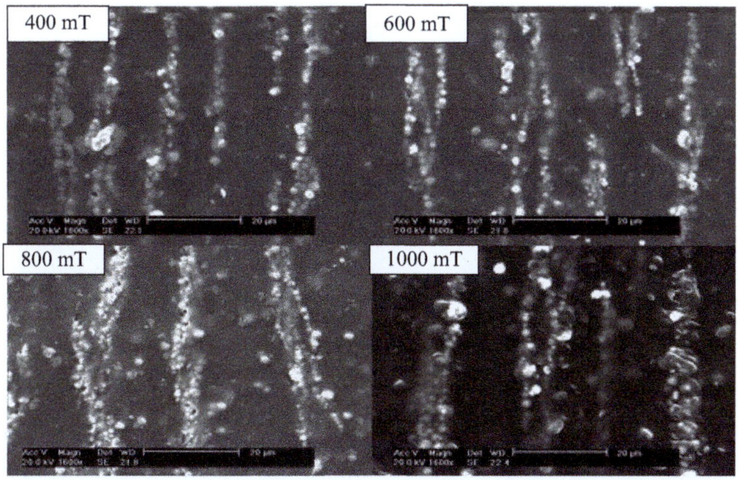

Figure 11.12 Particle distributions with different magnetic field intensities

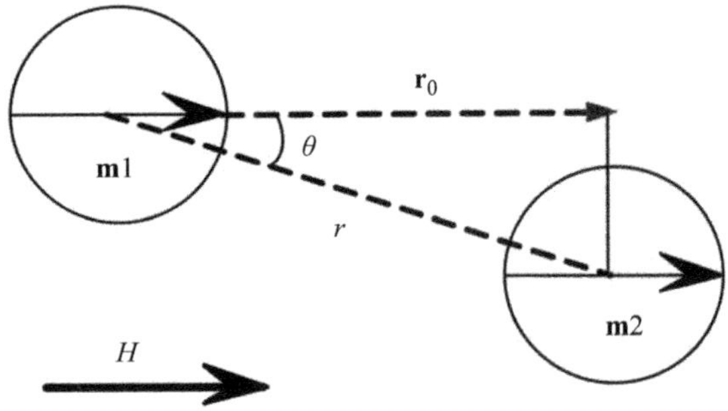

Figure 11.13 Schematic of dipolar model

Based on the above analysis, the main hypotheses are summarized as follows:

1. Ferromagnetic particles have the same size and the same dipole moment.
2. The interaction exists just between adjacent dipoles, ignoring other dipoles' action.
3. Due to the action of magnetic field during curing, the distance between adjacent dipoles meets chi-squared distribution, rather than the uniform distribution.

To consider two particles under the action of magnetic field **H**, the two particles are magnetized as two dipoles to produce magnetic moments $\mathbf{m}_1$ and $\mathbf{m}_2$, respectively. The interaction energy of the two dipoles is

$$E_{12} = \frac{1}{4\pi\mu_0\mu_1}\left[\frac{\mathbf{m}_1 \cdot \mathbf{m}_2 - 3(\mathbf{m}_1 \cdot \mathbf{e}_\mathbf{r})(\mathbf{m}_2 \cdot \mathbf{e}_\mathbf{r})}{r^3}\right] \tag{11.2}$$

where $\mathbf{m}_1$ is the relative permeability of the medium and $\mathbf{e}_\mathbf{r}$ is the unit vector of **r**. When $\mathbf{m}_1$ and $\mathbf{m}_2$ have the same direction and quantity, (11.2) can be simplified as

$$E_{12} = \frac{|m|^2(1 - 3\cos^2\theta)}{4\pi\mu_0\mu_1|r|^3} = \frac{|m|^2\left(1 - 3\left(r_0^2/r_0^2 + x^2\right)\right)}{4\pi\mu_0\mu_1\left(r_0^2 + x^2\right)^{3/2}} \tag{11.3}$$

Since the shear strain can be defined as $\gamma = x/r_0$ and the magnetic moment can be formulated with dipole moment magnitude per unit particle volume J_p as $|m| = J_p V_i = J_p (1/6)\pi d^3$, the interaction energy of the two dipoles is also formulated as

$$E_{12} = \frac{\pi d^6(\gamma^2 - 2) \cdot J_p^2}{144\mu_0\mu_1 r_0^3 \cdot (1 + \gamma^2)^{5/2}} \tag{11.4}$$

where d is the diameter of particles. To characterize the relative distance between adjacent particles, the random variable h ($h \geq 0$) is introduced, and therefore, r_0 can be expressed as $r_0 = (1 + h) \cdot d$. Obviously, the real distance is enlarged with the increase in h, and $h = 0$ means adjacent particles contact each other. Substituting r_0 with $(1 + h) \cdot d$, (11.4) can be modified as

$$E_{12} = \frac{(\gamma^2 - 2) \cdot J_p^2 \pi d^3}{144\mu_0\mu_1 (1 + h)^3 \cdot (1 + \gamma^2)^{5/2}} \tag{11.5}$$

Assuming $h \sim \chi^2(n)$

$$h \sim \chi^2(n) : f(h) = \begin{cases} 0 & h < 0 \\ \dfrac{h^{n/2-1}}{2^{n/2} \cdot e^{h/2} \cdot \Gamma(n/2)} & h \geq 0 \end{cases} \tag{11.6}$$

where n is the expectation of h. Distribution parameter n can reflect the average level of adjacent particles' distance, that is, the smaller the n, the more compactly the particles distribute. E_{12} can be regarded as a function of h, and the expectation of E_{12} is

$$\begin{aligned} E(E_{12}) &= \int_{-\infty}^{+\infty} f(h) \cdot E_{12} \cdot dh = \frac{(\gamma^2 - 2) \cdot J_p^2 \pi d^3}{144\mu_0\mu_1 (1 + h)^3 \cdot (1 + \gamma^2)^{5/2}} \cdot \int_{-\infty}^{+\infty} f(h) \cdot \frac{1}{(1 + h)^3} \cdot dh \\ &= \frac{(\gamma^2 - 2) \cdot J_p^2 \pi d^3}{144\mu_0\mu_1 (1 + h)^3 \cdot (1 + \gamma^2)^{5/2}} \cdot g(n) \end{aligned} \tag{11.7}$$

where

$$g(n) = \frac{e^{1/2}}{32 \cdot 2^{n/2}} \cdot \begin{bmatrix} n^2 \cdot (4 \cdot \Gamma(1 - n/2, 1/2) - 4 \cdot \Gamma(-n/2, 1/2) + \Gamma(-1 - n/2, 1/2)) + \cdots \\ n \cdot (-24 \cdot \Gamma(1 - n/2, 1/2) + 8 \cdot \Gamma(-n/2, 1/2) + 2 \cdot \Gamma(-1 - n/2, 1/2)) + \cdots \\ 32 \cdot \Gamma(1 - n/2, 1/2) \end{bmatrix}$$

This expression describes the influence on MR effect, which the intensity of particles' distribution makes. The functional relationship between n and $g(n)$ is shown in Figure 11.14, from which it can be seen that $g(n)$ always decreases with n, but the rate of decent tends to be 0; it is noteworthy that when $n \geq 6$, the value of $g(n)$ is close to 0, that is, particles in MRE distribute so sparsely that the MR effect can be ignored.

The expectation of E_{12} times total number of particles n_p, and then, the average interaction energy density of the elastomer can be formulated as

$$U = \frac{n_p \cdot E(E_{12})}{V_e} = \frac{\phi \cdot (\gamma^2 - 2) \cdot J_p^2}{24\mu_0\mu_1 \cdot (1 + \gamma^2)^{5/2}} \cdot g(n) \tag{11.8}$$

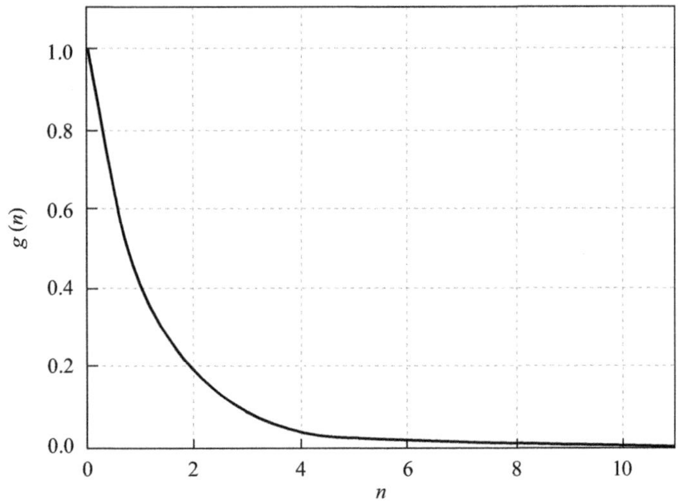

Figure 11.14 g(n)–n curves

in which ϕ is the volume fraction of the particle, $\phi = (4/3)\pi \cdot ((r^3)/(2r+2\omega)^3)$. Consequently, the magnetic-induced shear stress can be derived from (11.8)

$$\sigma_m = \frac{\partial}{\partial \gamma} U = \frac{\phi \cdot (4-\gamma^2) \cdot \gamma \cdot J_p^2}{8\mu_0\mu_1 \cdot (1+\gamma^2)^{7/2}} \cdot g(n) \tag{11.9}$$

And then, the magnetic-induced shear modulus G_m is

$$G_m = \frac{\phi \cdot (4-\gamma^2) \cdot J_p^2}{8\mu_0\mu_1 \cdot (1+\gamma^2)^{7/2}} \cdot g(n) \tag{11.10}$$

To describe the macro mechanics behavior of MREs, that is, the relationship of force–displacement or stress–strain, it is necessary to propose a macro model based on (11.10), in which G_m can be used as the magnetic-induced part.

11.3.2 Magnetoviscoelasticity parametric model

From these experimental results, we can see that the MRE material exhibits a feature that its modulus and damping capability are both magnetic field dependent, and the behavior of MRE without external magnetic field processes the viscoelastic property; generally speaking, the mechanical behaviors of MRE can be described as magnetic-induced viscoelastic property. So, a magnetoviscoelasticity parametric model is developed by us to predict these characteristics, as shown in Figure 11.15 [27], which is extended based on the viscoelastic fractional derivative model [30]. In this model, a spring element, a fractional derivative dashpot element, a nonlinear spring element and an additional analogs dashpot element are paralleled together. The parameters of these mechanical elements are G_0, τ, $G_m(H)$, μ and α. G_0 is the

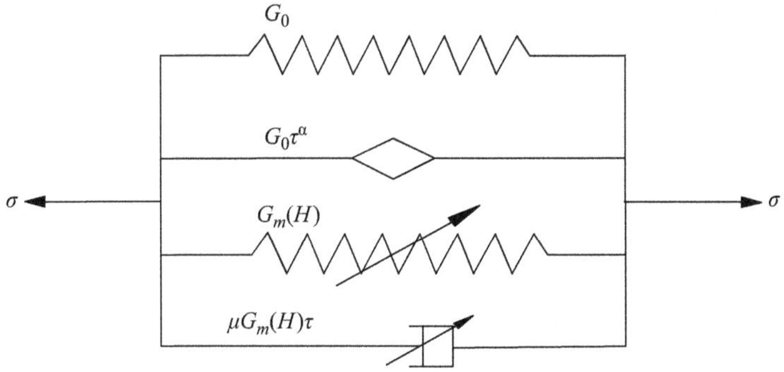

Figure 11.15 Magnetic-induced viscoelastic parameter model for MRE

shear modulus of MRE without magnetic field, τ is a parameter related to time, $G_m(H)$ is the modulus generated from the MR effects and μ is a parameter determined by experimental results. α is a fraction number and the range of α is 0–1, and in this model, α is the arbitrary order derivative, which is the right difference between Newton dashpot and Abel dashpot. G_0 and $G_0\tau\alpha$ form a fractional derivative model, which mainly deals with the viscoelastic characteristic of matrix, while $G_m(H)$ represents the field dependence of modulus, and $\mu G_m(H)\tau$ deals with the field-dependence damping capability.

Suppose the input strain is ε, while the output stress is σ; the constitutive equation can be expressed as

$$\sigma(t) = G_0\varepsilon(t) + G_0\tau^\alpha D^\alpha[\varepsilon(t)] + G_m(H)\varepsilon(t) + \mu G_m(H)\tau\dot{\varepsilon}(t) \tag{11.11}$$

where α is the fractional order. In order to get the complex modulus G^*, the Fourier transformation is undertaken for (11.11), then

$$G^* = \frac{\sigma(\omega)}{\varepsilon(\omega)} = G_0 + G_0\tau^\alpha(i\omega)^\alpha + G_m(H) + i\mu G_m(H)\tau\omega = G_1(\omega) + iG_2(\omega) \tag{11.12}$$

where G_1 and G_2 are real and imaginary parts of the complex modulus, respectively. The $i^\alpha = \cos(\alpha\pi/2) + i\sin(\alpha\pi/2)$ is undertaken for (11.12), then the storage modulus G_1 and loss modulus G_2 can be obtained

$$G_1 = G_m(H) + G_0\left(1 + \tau^\alpha\omega^\alpha \cos\frac{\alpha\pi}{2}\right) \tag{11.13}$$

$$G_2 = \mu G_m(H)\omega\tau + G_0\tau^\alpha\omega^\alpha \sin\frac{\alpha\pi}{2} \tag{11.14}$$

and the loss factor can be calculated

$$\eta = \frac{G_2}{G_1} = \frac{\mu G_m(H)\omega\tau + G_0\tau^\alpha\omega^\alpha \sin(\alpha\pi/2)}{G_m(H) + G_0(1 + \tau^\alpha\omega^\alpha \cos(\alpha\pi/2))} \tag{11.15}$$

where ω is the driving frequency.

According to the experiment in this chapter, the excite input is displacement u, which is a harmonic loading.

$$u(t) = u_0 \sin(\omega t) \tag{11.16}$$

So, the force–displacement relationship can be expressed as

$$F(t) = u_0 \frac{n_v A_s}{t_s} \sqrt{G_1^2 + G_2^2} \sin(\omega t + \varphi) \tag{11.17}$$

where φ is the phase angle difference between the input and the output, which can be calculated as $\varphi = \tan^{-1}(G_2/G_1)$, that is, $\varphi = \tan^{-1}\eta$. n_v is the number of the shear plane of MRE specimens. A_s and t_s are area and height of the shear plane.

11.3.3 Parameter identification

In the model, there are five parameters, i.e., G_0, τ, α, μ and $G_m(H)$, four of which should be identified by using the experimental data, and one of these parameters, $G_m(H)$, is determined by experimental results or microphysical model [18,19,21]. In this experiment, the input is displacement $u(t)$, and the force $F(t)$ can be received by the equipment. The relationship of $u(t)$ and $F(t)$ can be described as (11.17), which will be used to identify the parameters for the model. So, the four parameters are estimated on the base of the least squares method to minimize the error between the model-calculated force of FM and the experimental result of FE. The error in the model is represented by the objective function J given by

$$J = \sum_{j=1}^{N} \sum_{i=1}^{n} \left(FM_j(i) - FE_j(i) \right)^2 \tag{11.18}$$

where N is the number of the experimental case and n is the experimental numbers of the hysteresis loops. FM_j is the force value calculated by (11.17). This optimization is finished by compositing program in the software MATLAB®.

The four parameters of different MRE samples are identified from some experimental case. Taking the MRE sample BIIR60, for example, the four parameters are estimated by using the least squares method on the base of the experimental data, which are obtained from the experimental case of different frequency, displacement and a constant magnetic fields, then we compare the results between the model and other experimental data by the fitting results. The identified results are shown in Figure 11.16, and the values of identified parameters are $G_0 = 0.265$ MPa, $\tau = 0.58$ s, $\alpha = 0.18$ and $\mu = 0.15$. $G_m(H)$ can be calculated by microphysical model or received from the experimental data, so it does not need to identify. So, one kind of MRE can get one set of parameters. It can be seen from Figure 11.16 that the calculating values are in good agreement with the experimental data, especially, the fitting result of G_1 is better than that of loss factor η for the reason that the shear storage modulus G_1 changing with the magnetic-induced modulus $G_m(H)$ is determined by the experimental results, but the loss factor η is a complicated parameter determined by several conditions. For the other sample BIIR70,

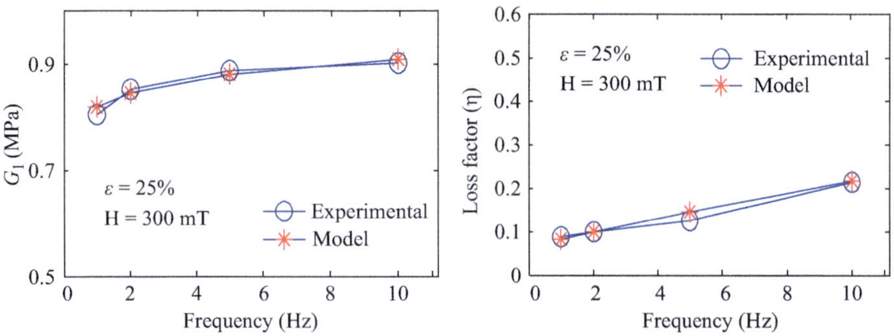

Figure 11.16 The identified results of parameters of NR60

following the same method, these four parameters, G_0, τ, α and μ are also estimated at a constant strain amplitude of 25%, and the identified results are $G_0 = 0.285$ MPa, $\tau = 0.62$ s, $\alpha = 0.20$ and $\mu = 0.12$, respectively. The identified results of the previous work and this chapter are listed in Table 11.3. It can be seen from the identified results that the parameters G_0, τ and α increase steadily with the increment of the proportion of particles, and the parameters τ and α related to the loss modulus are also increased with the changing of particle proportions. These results demonstrate the original modulus G_0 and the damping capability influenced by the proportions of magnetic particles.

11.3.4 Numerical simulation results

The shear modulus G_1 and loss factor η of sample BIIR60 are calculated by using the parameters estimated from the system identification process, seen in Figure 11.17. From the figure we can see that the parameters of mechanical property G_1 and η calculated by experimental data and numerical simulation are in good agreement. For example, the experimental values of G_1 can increase from 0.788 up to 0.92, 0.132 MPa increment at the strain amplitude of 25% and the driving frequency of 10 Hz, and the error between experimental values and model simulation is no more than 3%. Obviously, the shear storage modulus G_1 of experimental and numerical increases steadily when the magnetic flux increases from 0 to 300 mT. It does not like the shear storage modulus G_1, the loss factor η increases with the magnetic fields slightly. These results demonstrated again that the MRE can work as variable stiffness specimens.

The force–displacement curves of the MRE samples with BIIR are reconstructed by using the parameters estimated from the system-identification process, which are compared with the experimental data curves, seen in Figure 11.18. It can be seen from these figures that the curves plotted by experimental data and numerical simulation can form elliptical loops and are in good agreement. The areas and the slopes of the main axis of the elliptical loops, reconstructed by numerical simulation, increase with the increment of the magnetic fields, which

Table 11.3 The results of identified parameters

Samples	Parameters			
	G_0 (MPa)	τ (s)	α	μ
NR50	0.37	0.40	0.10	0.024
NR60	0.41	0.43	0.11	0.015
NR70	0.46	0.49	0.13	0.003
NR80	0.54	0.51	0.15	0.001
BIIR60	0.265	0.58	0.18	0.15
BIIR70	0.285	0.62	0.20	0.12

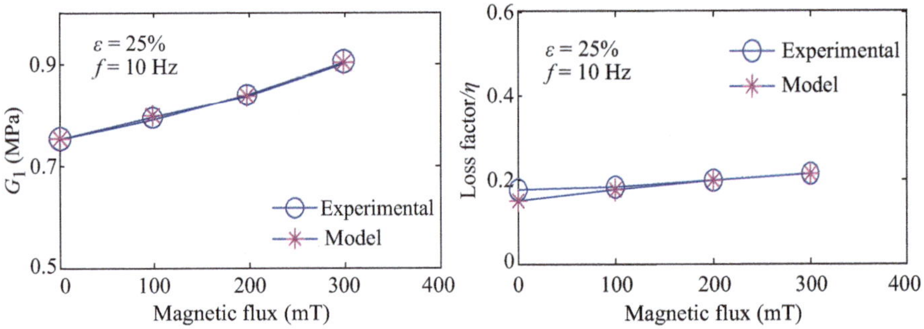

Figure 11.17 The comparison of NR60 between experimental and model-calculated results

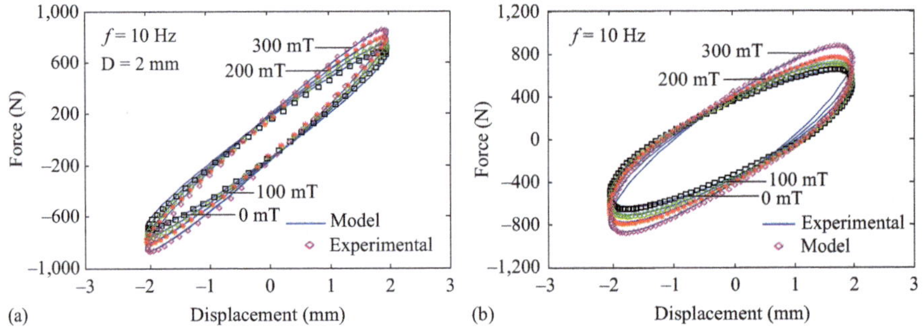

Figure 11.18 Comparisons between experimental data with model simulated results: (a) NR70 and (b) BIIR 70

demonstrates that the stiffness and damping of MRE specimens can change with the magnetic fields. And the comparison results demonstrate that the model can simulate the experimental data well, that is to say, the proposed model can simulate the stiffness and damping characteristics of MREs.

11.4 Conclusion

In this chapter, a kind of new MRE is developed with the BIIR matrix, a high damping property rubber. In order to investigate the performance of the new kind of MRE and find out the difference with the MRE based on NR matrix, the physical and dynamic mechanical properties of these samples are tested. In addition, the proposed magnetoviscoelasticity parametric model is introduced, and the parameters are identified for this kind of MRE, then the force–displacement curves obtained from model simulation and experimental data are compared with the numerical simulation. The following conclusions can be obtained from the experimental and numerical analyses:

1. The result of physical property test shows that the MRE samples with NR matrix have better physical property than that of samples with BIIR matrix, and the minimum value of the fracture strength of MRE sample with BIIR is no less than 3.7 MPa.
2. The experimental curves of the MRE specimens show that the areas and the slope of the main axis of those loops vary with the magnetic field, which reflects variable damping and stiffness property of MREs. The G_1 and η steadily increase with the increment of magnetic fields and driving frequencies but decrease with the strain amplitudes. In addition, the value η of the MREs with BIIR matrix obviously higher than that of MREs with NR matrix.
3. The comparison between the experimental data and numerical results demonstrates that the proposed microphysical model based on the chi-square distribution may calculate the MR effect of MREs exactly, and the magnetoviscoelasticity parametric model based on the microphysical model can describe the MRE's dynamic performance effectively.

Acknowledgments

The study was financially supported by the National Natural Science Foundation of China (51878355), National Science Fund for Distinguished Young Scholars with Grant (51625803), the Program of Chang Jiang Scholars of Ministry of Education, Ten Thousand Talent Program of Leading Scientists and the Program of Jiangsu Province Distinguished Professor. These supports are gratefully acknowledged.

References

[1] Carlson, J.D., Jolly, M.R. (2000). "MR fluid, foam and elastomer devices." *Mechatronics*, 10(4–5): 555–569.

[2] Fuchs, A., Zhang, Q., Elkins, J., Gordaninejad, F., and Evrensel, C. (2007). "Development and characterization of magnetorheological elastomers." *Journal of Applied Polymer Science*, 105: 2497–2508.

[3] Chen, L., Gong, X.L., and Li, W.H. (2007). "Microstructures and viscoelastic property of anisotropic magnetorheological elastomers." *Smart Materials and Structures*, 16(6): 2645–2650.

[4] Popp, K.M., Kroger, M., Li, W.H., Zhang, X.Z., and Kosasih, P.B. (2010). "MRE property under shear and squeeze modes and applications." *Journal of Intelligent Material Systems and Structures*, 21(15): 1471–1477.

[5] Xu, Z.D., Shen, Y.P., and Guo, Y.Q. (2003)."Semi-active control of structures incorporated with magnetorheological dampers using neural networks." *Smart Materials and Structures*, 12(1): 80–87.

[6] Hou, C.Y. (2008). "Fluid dynamics and behavior of nonlinear viscous fluid dampers." *Journal of Structural Engineering*, 134(1): 56–63.

[7] Wilson, M.J., Fuchs, A., and Gordaninejad, F. (2002). "Development and characterization of magnetorheological polymer gels." *Journal of Applied Polymer Science*, 84(14): 2733–2742.

[8] Rao, V.P., Maniprakash, S., Srinivasan, S.M., and Srinivasa, A.R. (2010). "Functional behavior of isotropic magnetorheological gels." *Smart Materials and Structures*, 19: 085019.

[9] Wang, Y.L., Hu, Y., Chen, L., *et al.* (2006). "Effects of rubber/magnetic particle interactions on the performance of magnetorheological elastomers." *Polymer Testing*, 25(2): 262–267.

[10] Shen, Y., Golnaraghi, M.F., and Heppler, G.R. (2004). "Experimental research and modeling of magnetorheological elastomers." *Journal of Intelligent Material Systems and Structures*, 15(1): 27–35.

[11] Xu, Z.D., Liao, Y.X., Ge, T., and Xu, C. (2016)."Experimental and theoretical study on viscoelastic dampers with different matrix rubbers." *Journal of Engineering Mechanics*, 142(8): 04016051.

[12] Xu, Z.D., Suo, S., and Lu, Y. (2016). "Vibration control of platform structures with magnetorheological elastomer isolators based on an improved SAVS law." *Smart Materials and Structures*, 25(6): 065002.

[13] Xu, Z.D., Suo, S., Zhu, J.T., and Guo, Y.Q. (2018). "Performance tests and modeling on high damping magneorheological elastomers based on bromobutyl rubber." *Journal of Intelligent Material Systems and Structures*, 29(6): 1025–1037.

[14] Sun, T.L., Gong, X.L., Jiang, W.Q., Li, J.F., Xu, Z.B., and Li, W.H. (2008). "Study on the damping property of magnetorheological elastomers based on cis-polybutadiene rubber." *Polymer Testing*, 27(4): 520–526.

[15] Zajac, P., Kaleta, J., Lewandowski, D., and Gasperowicz, A. (2010). "Isotropic magnetorheological elastomers with thermoplastic matrices: structure, damping property and testing." *Smart Materials and Structures*, 19(4): 045014.

[16] Lokander, M., Stenberg, B. (2003). "Improving the magnetorheological effect in isotropic magnetorheological rubber materials." *Polymer Testing*, 22(6): 677–680.

[17] Stepanov, G.V., Abramchuk, S.S., Grishin, D.A., Nikitin, L.V., Kramarenko, E.Y., and Khokhlov, A.R. (2007). "Effect of a homogeneous magnetic field on the viscoelastic behavior of magnetic elastomers." *Polymer*, 48(2): 488–495.

[18] Shiga, T., Okada, A., and Kurauchi, T. (1995). "Magnetoviscoelastic behavior of composite gels." *Journal of Applied Polymer Science*, 58(4): 787–92.

[19] Jolly, M.R., Carlson, J.D., and Muňoz, B.C. (1996). "A model of the behavior of magnetorheological materials." *Smart Materials and Structures*, 5(5): 607–614.

[20] Davis, L.C. (1999). "Model of magnetorheological elastomers." *Journal of Applied Physics*, 85(6): 3348–3351.

[21] Borcea, L., Bruno, O. (2001). "On the magneto-elastic property of elastomer–ferromagnet composites." *Journal of the Mechanics and Physics of Solids*, 49 (12): 2877–2991.

[22] Steigmann, D.J. (2004). "Equilibrium theory for magnetic elastomers and magnetoelastic membranes." *International Journal of Non-Linear Mechanics*, 39: 1193–1216.

[23] Dorfmann, A., Ogden, R.W. (2004). "A constitutive model for the Mullins effect with permanent set in particle-reinforced rubber." *International Journal of Solids and Structures*, 41: 1855–1878.

[24] Li, W.H., Zhou, Y., and Tian, T.F. (2010). "Viscoelastic property of MR elastomers under harmonic loading." *Rheologica Acta*, 49(7): 733–740.

[25] Lokander, M., Reitberger, T., and Stenberg, B. (2004). "Oxidation of natural rubber-based magnetorheological elastomers." *Polymer Degradation and Stability*, 86(3): 467–471.

[26] Boczkowska, A., Awietjan, F.A. (2009). "Smart composites of urethane elastomers with carbonyl iron." *Journal of Material Science*, 44(15): 4104–4111.

[27] Zhu, J.T., Xu, Z.D., and Guo, Y.Q. (2012). "Magnetoviscoelasticity parametric model of an MR elastomer vibration mitigation device." *Smart Materials and Structures*, 21: 075034.

[28] Xu, Z.D., Guo, Y.Q. (2006). "Fuzzy control method for earthquake mitigation structures with magnetorheological dampers." *Journal of Intelligent Material and Systems*, 17(10): 871–881.

[29] Xu, Z.D. (2007). "Earthquake mitigation study on viscoelastic dampers for reinforced concrete structures." *Journal of Vibration and Control*, 13(1): 29–45.

[30] Meral, F.C., Royston, T.J., and Magin, R. (2010). "Fractional calculus in viscoelasticity: An experimental study." *Communications in Nonlinear Science and Numerical Simulation*, 15(4): 939–945.

Chapter 12

Promising application of magnetorheological elastomer

Ubaid Ubaidillah[1] *and Endra Dwi Purnomo*[1]

Initiated in 1983 by Rigbi and Jilken [1], soft magnetic particles in micron size were embedded in viscoelastic type polymers. At the current stage, they were not considered as a smart material since the performance investigation did not reveal the change of intrinsic properties influenced by external magnetic fields. Later on, in 1995–96, the exploration on their smart properties began to be investigated.

Viscoelastic-matrix-based smart material called magnetorheological (MR) materials have been studied intensively in the last 20 years. Researchers have developed specific terminologies for various features. The term magnetoactive elastomers have been used by several research groups [1–5]. Other researchers have coined the terms ferrogels [6–8], magneto-active polymers [9], elastomer–ferromagnetic composites [10], MR visco-elastomers [11], soft magnetic elastomers [12–14], ferromagnetic elastomer composites [10], and magnet filler-polymer composites or magnetic polymer (Magpols) [15]. Most researchers use the term MR elastomer (MRE) referred to their functional behavior that is rheological properties. Therefore, the term "MRE" will be used as a reference name in this chapter. Till October 2018, we have detected at least 953 scientific articles found in Scopus Indexing Engine as the publication growth can be seen in the following chart (see Figure 12.1).

MRE is a type of magnetically actuated smart material in which their rheological properties such as storage and loss moduli, loss tangent (loss factor/damping factor), as well as physical properties, such as thermal conductivity, electrical capacitance, sound absorption properties, magnetostriction, and resistivity, can be altered in a fraction of milliseconds by applying magnetic fields [16]. Both rheological and physical responses of the MRE are referred to in the foundation of devices' design and development. For instance, the tunable rheological properties are relevant for vibration absorber since the field-dependent modulus affects both stiffness and damping characteristics. Meanwhile, the physical properties are applicable for sensory development according to the change of resistance, capacitance, and thermal reluctance.

[1]Mechanical Engineering Department, FT-UNS, Indonesia

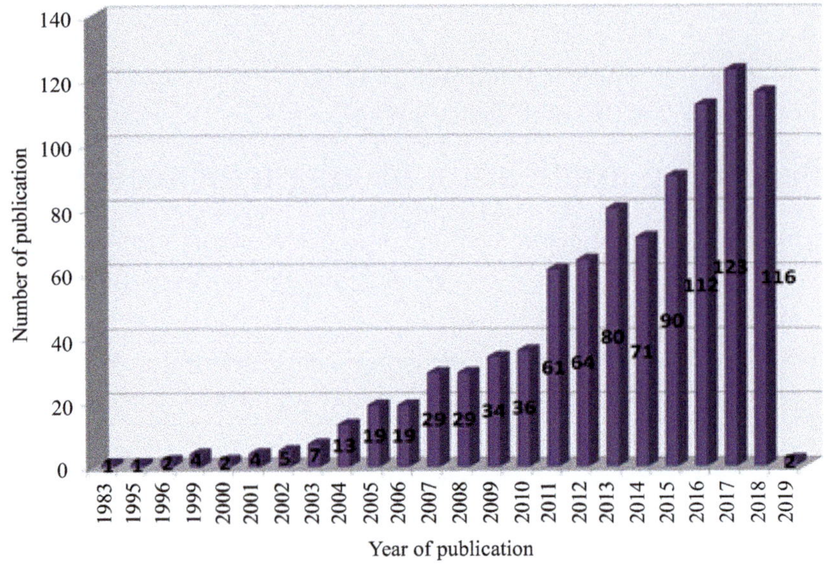

Figure 12.1 Publication growth of MREs since 1983, data logged from Scopus™

Implementations of MRE in vibration isolation, sensors, as well as actuators have been patented and disseminated in several papers. Typical applications in vibration absorber include variable stiffness bushing [17,18], propeller shaft [19], variable spring rate [20], and prosthetic leg [21]. MRE applications for sensory use are tire pressure control [22] and MEMS [23]. Meanwhile, solicitation in the active actuator was releasable attachment [24] and active morphing composites [15,25]. Various MRE-based equipment until the latest finding in 2018 will be provided later.

This chapter explored the existing application of MRE-based devices which were proposed in several field applications. It covers the design parameter and general performances of the devices. Some applications which have been published through patents as well as scientific articles were highlighted in the last two decades. Working principle and prototype performances are also delivered as well. To ease reader in structuring the various application of MRE, we classified practical application into three areas of semi-active, active, and sensory systems. However, the discussion here is limited to the basic working principle of the device without further elaboration with their control strategies. This is because control system in MRE-based devices has a broad coverage. Besides, the fundamental technique in controlling MRE-based devices is similar to that of controlling MR-fluid-based equipment.

12.1 Semi-active system

Basically, conventional system of vibration absorber consists of inertial element (mass), spring element, as well as damping element [26]. It is proven undoubtedly simple, economic, and reliable choices for vibration absorber in the last few decades. Elastomeric material here plays two important elements those are spring and damping in a single body. It is because elastomer has two distinguish properties, namely, storing and damping at the same time. This device protects structural damage in civil structure, machine elements, and so on, caused by external vibration by matching both external and natural frequencies [26]. However, during application, conventional elastomeric vibration absorber has some limitations such as narrow band of effective frequency, not effective for frequency changing, and not available for detuning. Therefore, researchers move to either active or semi-active vibration absorber to overcome the limitation [27].

MRE-based semi-active vibration absorber offers some advantages compared to the active one such as low-power consumption and better fail-safe in nature. Active vibration absorber provides better absorption and adaptability for wideband frequency. However, this type needs higher energy consumption and consequently costing. Semi-active vibration absorber is a good alternative because when the device is facing a failure problem, it goes back becoming passive system. So far, the application of MRE in semi-active system has been penetrated deeply in vibration area.

MRE as semi-active vibration absorber has been used for both variable stiffness and damper since their storage and loss modulus called viscoelasticity can be customized as per desired. The application of vibration absorber is generally based on MRE working modes such as compression, tension, linear shearing, torsional shearing, as well as bending. In vibration area, MRE has been used as tunable vibration absorber in seismic structures, automotive, medical, robotic, and machining devices. This subchapter provides a brief explanation of their working principles.

12.1.1 Generic tunable vibration absorber

In some occasion, MRE-based semi-active vibration absorber is also mentioned as active vibration absorber. However, the term "active" here does not mean it can actively actuate force to overcome external disturbance as in active system discussed previously. Active means the properties of MRE-based vibration absorber can be manipulated actively depending on the desired condition [28,29]. These vibration absorbers are sometimes called dynamic vibration absorbers, active–passive vibration absorbers [30] or adaptive tuned vibration absorbers (ATVA) [31–33], vibration insulator [34,35], and impact element [36].

The basic configuration of MRE-based vibration absorber can be classified into two types, which are single and multilayers. Single layer type is referred to vibration absorber employing a layer of MRE as controllable stiffness and damping material. This type is commonly applied for limited space of vibration mounting with low electromagnetic capacity. Note that the flux density needed to magnetize MRE is higher than commonly used in MRF. The carbonyl iron trapped within elastomeric matrix has very limited space for movement. Therefore, to achieve strong attraction, they need higher magnetization. Single layer also has low thickness of MRE because the lower the thickness, the better the magnetic-field penetration. The limitation of deformation space, single layer MRE is suitable for low-amplitude vibration.

Schematic and prototype of various single layer MRE-based vibration absorber are depicted in Figure 12.2 [37–42]. Working mode of MRE that is usually adopted

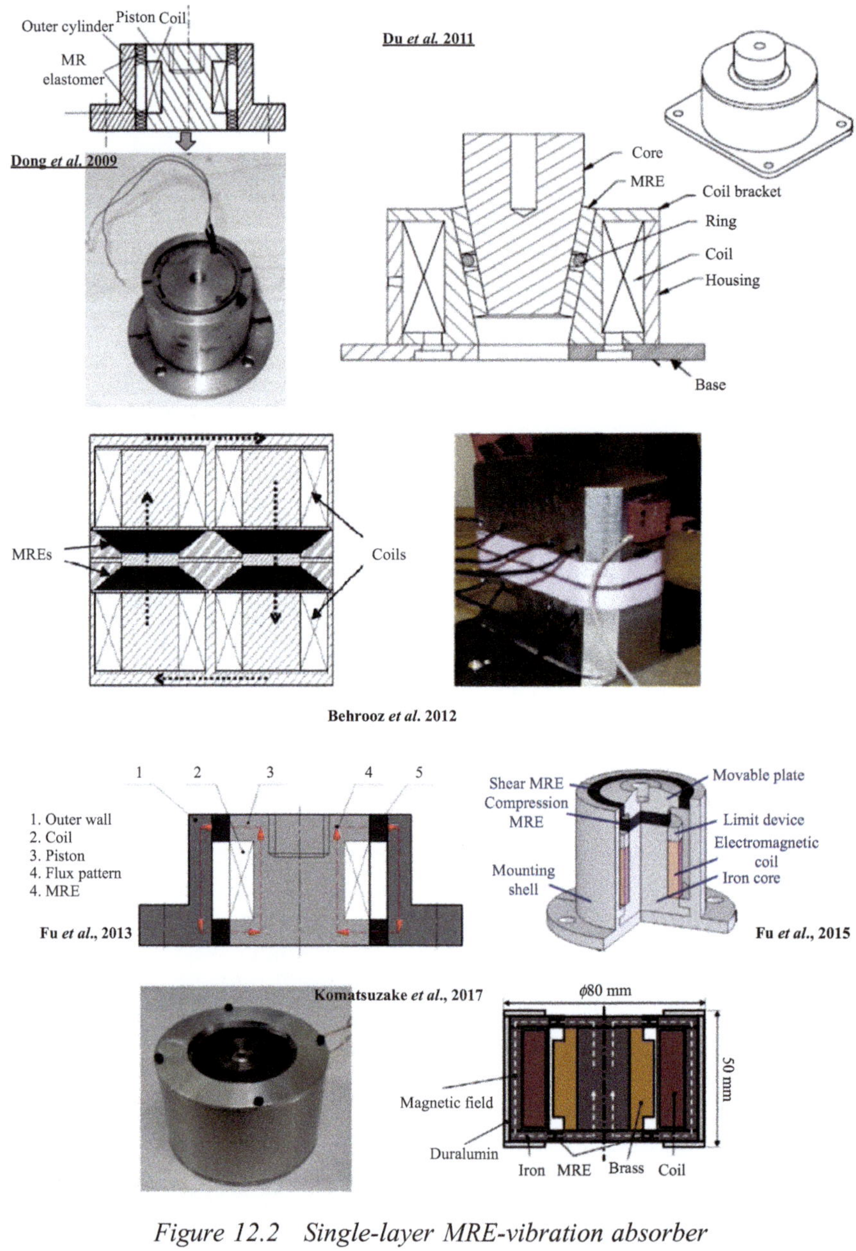

Figure 12.2 Single-layer MRE-vibration absorber

for vibration absorber in single layer is shear mode. The MR effect is more pronounced in this mode compared to the compression. Shear storage modulus of the MRE is strongly affected by the magnetic flux flown perpendicularly across the carbonyl iron. Frequency shifting under applying magnetic fields was found about 100–200 Hz [38] and 40–50 Hz [43]. Meanwhile, Du *et al.* [42] achieved stiffness increment from about 3,000–9,500 N/m at current variation of 0–3 A. The change of stiffness becomes a key role of natural frequency shifting. To avoid fatal destruction, the natural frequency of the system should be unequal with the external disturbances.

One of the most famous techniques for mitigating earthquake effects is isolating structure and its content from potentially harmful ground movement especially at the frequency where the structure is most influenced [44,45]. A massive utilization of passive base isolation has been becoming great attention in Japan, China, and the United States over the past two decades [44]. Practical implementation of seismic base isolation needs optimum design which is referred to the amplitude and frequency ranges of designate earthquake. Therefore, a seismic isolator is usually appropriate for only one earthquake. Thus, this condition forced researchers to develop seismic isolator that coverable to wideband of earthquake.

In passive system, the existing design has a larger size than the common machinery isolator. It is because a giant load should be supported by the isolator. In civil engineering application, a seismic isolator or absorber should have a low radial stiffness and high vertical stiffness simultaneously [27,44]. In lateral direction, seismic isolator should have capability to protect the superstructure from external destructive vibration. At the same time, the vertical stiffness is responsible for supporting the gravity loads as well as other vertical superstructure loads.

A conventional seismic base isolater is constructed of a bulky rubber with inserted metal layers inside. Laminated or sandwich structure offers high vertical stiffness and high flexibility in lateral direction. This arrangement is also proven to protect rubber bulging failures. The steel layers are vulcanized together with the rubber resulting in a strong capacity for supporting large axial load. The rubber used in seismic isolator does not change its properties once installed. Therefore, it cannot adapt to various earthquake condition. MRE is the most appropriate candidate for solving this issue.

Apart from single layer MRE-based vibration absorber, multilayer has been chosen due to its feasibility for large-scale structure such as high-rise building, bridge, flyover road, and so on. The other consideration is that the multilayer type has been proven in supporting super structure. Introducing MRE as the replacement of conventional rubber for seismic isolator can solve the existing problem. However, the major problem in MRE-based vibration absorber is the magnetic circuit which often fails magnetizing thick MRE. Therefore, the addition steel layers inside the bulky MRE would be increasing the magnetic permeability of the sandwich.

In 2009, Li *et al.* [44] introduced MRE-based seismic isolator having large capacity suitable for super structure. It has a diameter of 232 mm and thickness of 276 mm. The MRE and steels were arranged sandwich like with diameter of

140 mm and thickness of 137 mm. The isolator can support 280 kg vertical loading. The configuration of multilayer seismic absorber can be seen in Figure 12.3. The MRE was made of silicon rubber and carbonyl iron particle. The 2 mm thickness of 47 MRE layers were bonded particularly between 46 pieces steel plates with 1 mm thickness. Magnetic circuit employed in the prototype was using 1.2 mm copper wire, 3,100 wire turns with 32 Ω resistant.

Large diameter of the sandwich becomes a challenge in providing uniform magnetic flux across the laminated structure. Thanks to the steel yoke as outer wall, the magnetic field is able to looping around steel structures. During characterization, the seismic isolator was located on the shaker table and the other edge was permanently connected to the load cell. This configuration is carried out to explore its basic characteristic force versus both displacement and velocity. The shaker can be adjusted to its shaking amplitude and frequency. At a set of experiment, a particular current was supplied onto the coils. From the experiment, it was found that the change of stiffness at maximum current of 5 A, 0.5–3 Hz excitation frequency, was about 33%–45%. Moreover, the damping be tuned from 7.8% to 11.2%. The change of lateral stiffness was about 38%. This prototype has exhibited the feasibility and opportunity for smart seismic isolation system.

Another group also proposed multilayer seismic isolator, but it has different approach compared to Li *et al.* [44]. Since the developed prototype can only be tuned in positive increment, in 2015, Sun *et al.* [46–48] developed a different approach to achieve both positive and negative modulus changes. A piece of permanent magnet was located in between layers. The permanent magnet provides initial magnetization to the MRE sandwich. The DC current here can change its

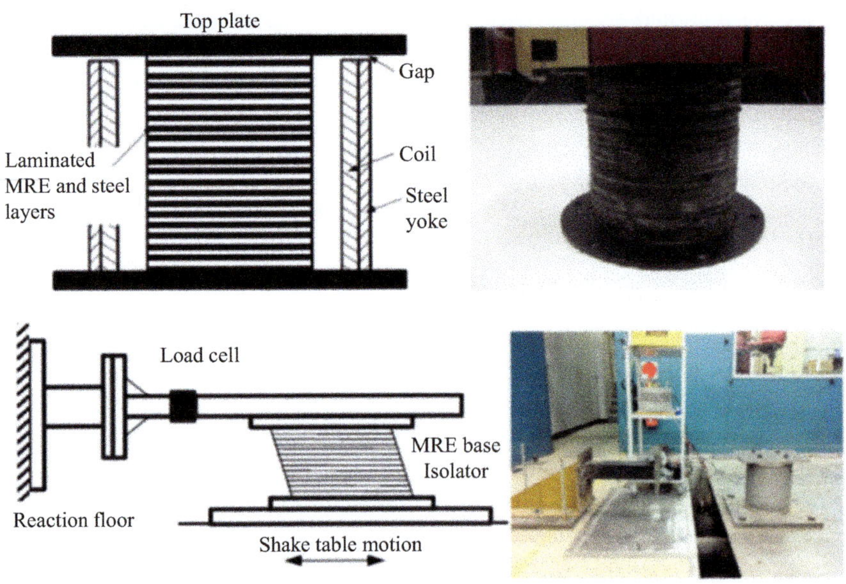

Figure 12.3 Multilayer MRE-based seismic isolator by Li et al. [44,45]

magnitude and direction. When a certain current is applied to the coil, it generates magnetic flux in which the direction of pole will be following right hand rule. If the magnetic flux has the same poles with the permanent magnet, the magnetic flux strengthen each other resulting in a higher complex modulus. However, if the magnetic flux across the sandwich element has the different direction, the magnetic flux becomes weak, and that is why the negative change of sandwich layer can be achieved. Besides, changing the complex modulus in positive and negative amounts also represents the change of natural frequency. It means that the hybrid magnetic system can widen the range of frequency shifting.

Fabrication method of the prototype by Sun *et al.* [48] has the same way to that of Li *et al.* [44]. Even the MRE matrix was made of silicon room temperature vulcanization (RTV) and carbonyl iron particles. Here, the MRE was made in layers and then bonded onto steel layers. The permanent magnet was located in the middle of the sandwich [48], while the other prototype was in each end of the sandwich [46]. The configurations have the same role. Based on the experiment, the magnitude and phase of transmissibility have the wider frequency shifting from about 3–19 Hz at applied current from −4 to 4 A, interval of 1 A. The prototype has proven working for vibration case in low frequencies and high amplitudes. Compared to the single layer as well as passive isolator which was also provided in the study, multilayer seismic isolator demonstrated superiority in terms of vibration attenuation. Prototype by Sun *et al.* can be seen in Figure 12.4.

Regarding fabrication technique, silicon-based MRE is indeed the simplest way for curing and made in layers [26]. However, it has some drawbacks for the real application. Silicon has the main problem of bubble trap when it is cured in a bulky form. It needs vacuum process for degassing. Besides, the manual bonding method is not suitable for real application. Common rubber bearing or seismic isolator containing steel layers have a different fabrication method. Wahab *et al.* [27] fabricated the following standard way for preparing seismic isolator. They used natural rubber as the MRE matrix and carbonyl iron particles for the filler. The full-scaled MRE seismic isolator is portrayed in Figure 12.5.

At the first stage, the natural rubber compound was mixed with the carbonyl iron using double roll mill becoming flat mixtures having thickness of 1 mm. The steel plates were sprayed with epoxy glue as binding agent. Both mixture and plate were arranged like a sandwich placed in a hot mold. The raw sandwich was then pressed at a certain pressure, temperature, and curing time. The vulcanization parameters were first determined using moving dies rheometer. Unfortunately, Wahab *et al.* only investigated the compression response of the seismic isolator. Based on compression test, the axial stiffness changed as the increase of magnetic flux generated by the coil. At compression stroke of 1.5 mm, the compression force varied from 8.6 to 9.9 kN under applied current of 0–3 A. It means that the magnetic flux can increase the stiffness about 14.5%.

Generation of magnetic flux in multilayer type seismic isolator has been becoming a major issue till recent time. Trying to solve this issue, two groups have proposed a different electromagnetic circuit. Gu *et al.* [49] and Przybylski *et al.* [50] proposed multicoil and Halbach array, respectively. Not only the magnetic

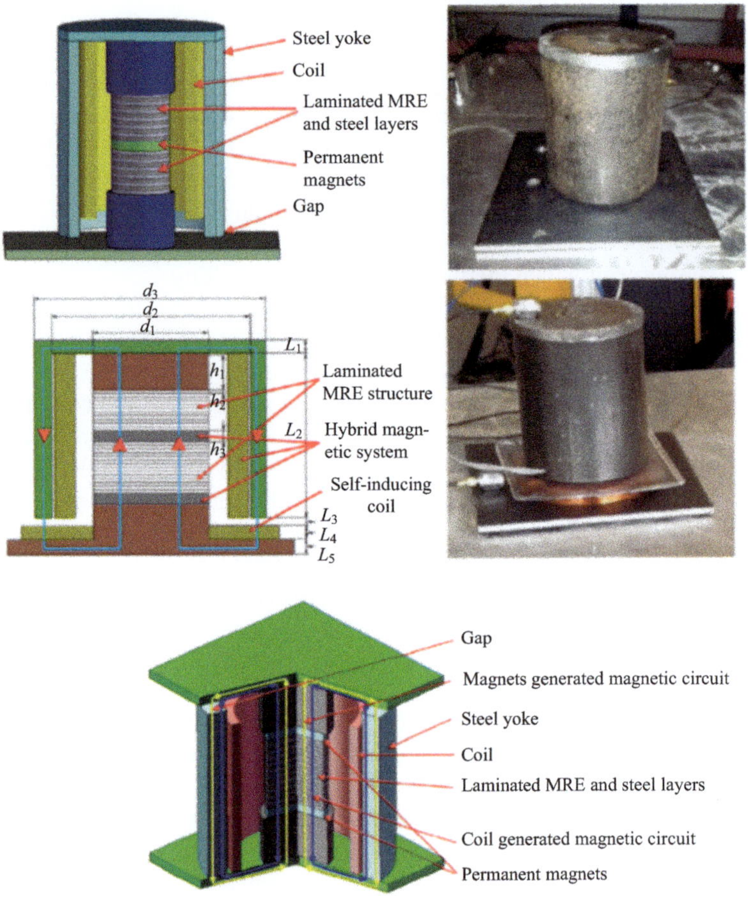

Figure 12.4 Multilayer MRE seismic isolator by Sun et al. [46–48]

flux generation issue, the time response was also the main focus of the study by Gu *et al.* [49]. By manipulating the driving current to the coil, the time response was examined under step driving current or the so-called transient condition. Three ways of driving current were tested, namely, series, parallel, and field quenching. The series configuration represents a single large coil. From Figure 12.6, it can be clearly seen that the parallel and field-quenching strategy have better transient response compared to the series configuration in terms of rising and falling currents. The fast transient response provide advantages for real-time control implementation for seismic protection.

Since the conventional coil configuration was difficult to induced high magnetic fields, Halbach array was introduced to overcome this drawback. This electromagnetic configuration has been proven in various electric motor, refrigerator, and so on. Its unique construction provides light weight and more efficient systems.

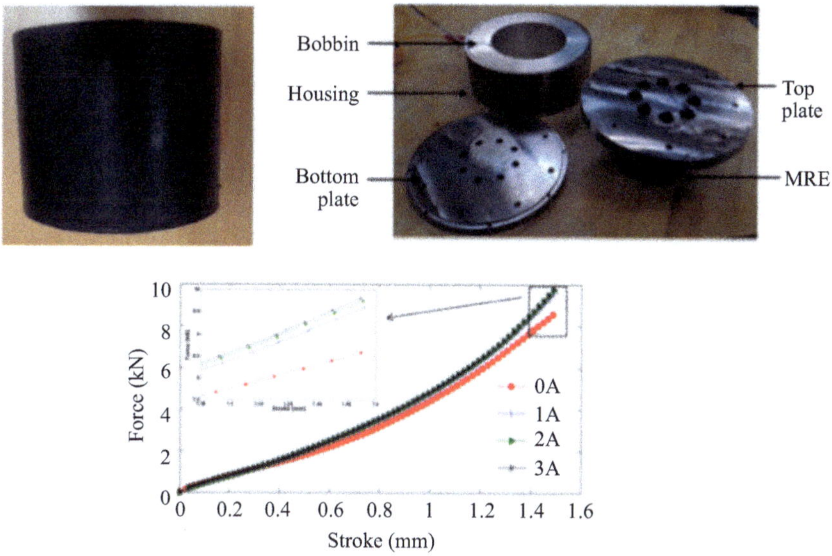

Figure 12.5 Full-scaled MRE seismic isolator by Wahab et al. [27]

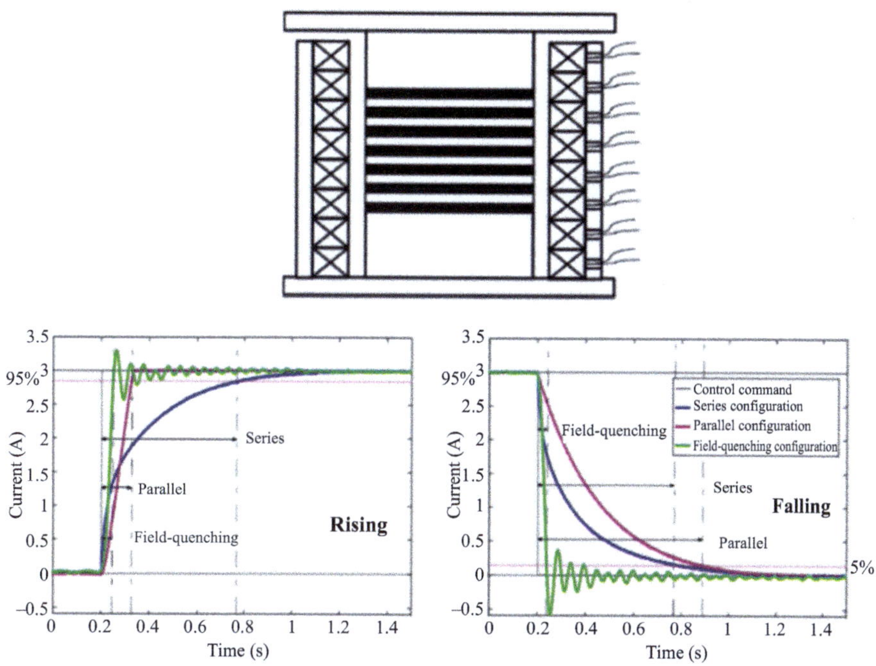

Figure 12.6 Multicoil multilayer MRE seismic isolator by Gu et al. [49]

The Halbach array system in this isolator was managed by rotating an array of magnets around the device to change the angle of the magnetic field vector induced inside the laminated structure or by rotating two or more overlapping Halbach arrays to change the value of the magnetic field induced inside the array [50].

This array consumes much less power than a common magnetic coil because it only uses electrical power to control and rotate the arrays. Moreover, it can also tune the device and freeze it into a position where it will be tuned properly in the case of an upcoming seismic event combined with a loss of power [50]. The configuration of Halbach-array-based MRE multilayer seismic isolator is depicted in Figure 12.7. From the figure, it can be seen that the configuration of laminated MRE steel follows a generic laminated structure. The sandwich structure has diameter of 30 mm employing 15 layers MRE having 1.67 mm, 16 layer steels 1 mm thickness, and 4 mm thick neodymium magnet. A magnetic field is generated by a neodymium magnet placed inside the layered structure, and the circular dipolar Halbach array surrounding the isolator. This system can provide a magnetic flux from 300 to 90 mT depending on how the Halbach array is rotated around the isolator. The rotation of Halbach array was taken over by an electric motor. Therefore, such isolator needs no external power. However, this configuration was

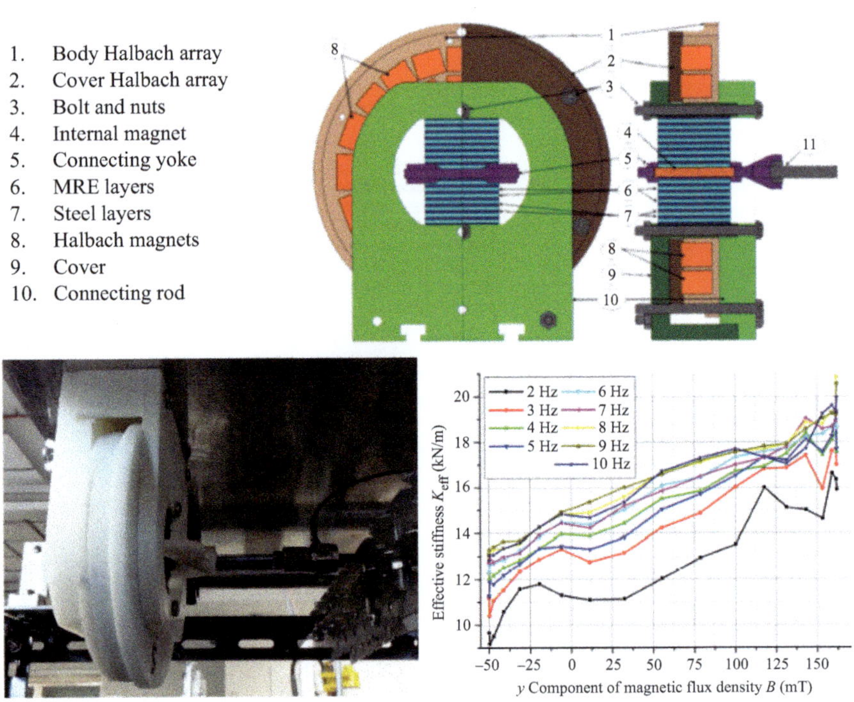

Figure 12.7 Multilayer MRE seismic isolator using Halbach array [50]

applied only in small size MRE sandwich. During experiment, the amplitude and frequency were varied under increasing magnetic flux. Overall results showed that the effective stiffness increased linearly with the increment of magnetic fields.

12.1.2 Automotive application

Automotive systems can adopt adaptive vibration absorber to replace the passive absorber for a better engine vibration, ride, and handling performance. The penetration of MREs in automotive area has been acknowledged in some patents and publication such as seat absorber [51], tunable slip yoke [19], bushing [17,18], spring pad/stopper [22], and rubber spring [52,53].

Tuning the stiffness constant of a rubber bush on a longitudinal axle is meaningful for improving the mentioned performance as perceived by the passengers. A specific problem occurs during the vehicle transmission shift, which needs a hard rubber bush or a relatively high stiffness constant. On the other hand, optimum ride quality requires a low stiffness constant rubber bush. The variation of rubber bush spring rate can be realized by developing smart rubber that covers such requirements. The MREs bush, patented by Watson in 1997, is a novel technique for tuning the stiffness constant, which allows for increasing passenger satisfaction as a result of a decrement in the dissatisfaction associated with braking noise, vibration and harshness [17]. The placement of the MREs bush in the longitudinal suspension link is mainly proposed for overcoming disturbance in a longitudinal direction. The state of MREs bush stiffness depends on the electrical current supplied by the power regulator as a function of the output signal from the control unit. Longitudinal measurement sensors provide signals that are processed in the control unit to determine the appropriate electrical current due to a particular control reference point [54]. Ginder *et al.* [18] also reported their work regarding MRE-based bushing. Both prototypes have the same MRE-based bushing structure. The configuration of MRE bush can be seen in Figure 12.8.

Vibration attenuation in the vehicle driveline, namely, tunable slip yoke was also introduced in US20036623364B2 [19] by Badolato and Pawlowski. The driveline is one of the main sources of noise and vibration, which plays an essential role in the comfort evaluation of the vehicle. The vibration is caused by the non-aligned driveline axes between the transmission and the rear drive axle, which is interconnected through a universal joint. The slip yoke damper, as shown in Figure 12.9, is adapted to the driveline assembly, and the stiffness can be tuned to different frequencies to dampen the vibration. This device consists of an outer case, an inner case, and a section between the outer and inner cases. The MREs is housed in the section between the two cases. The magnetic means are capable of being rotated either toward the inner house or away from the inner house such that the elastic modulus of the MREs is either increased or decreased.

Experimental investigation of an adaptive tune vibration absorber for the driveline shaft was also reported by Hoang *et al.* [32,55,56]. They proposed an ATVA for solving drive train resonant problem during transient or acceleration. By using

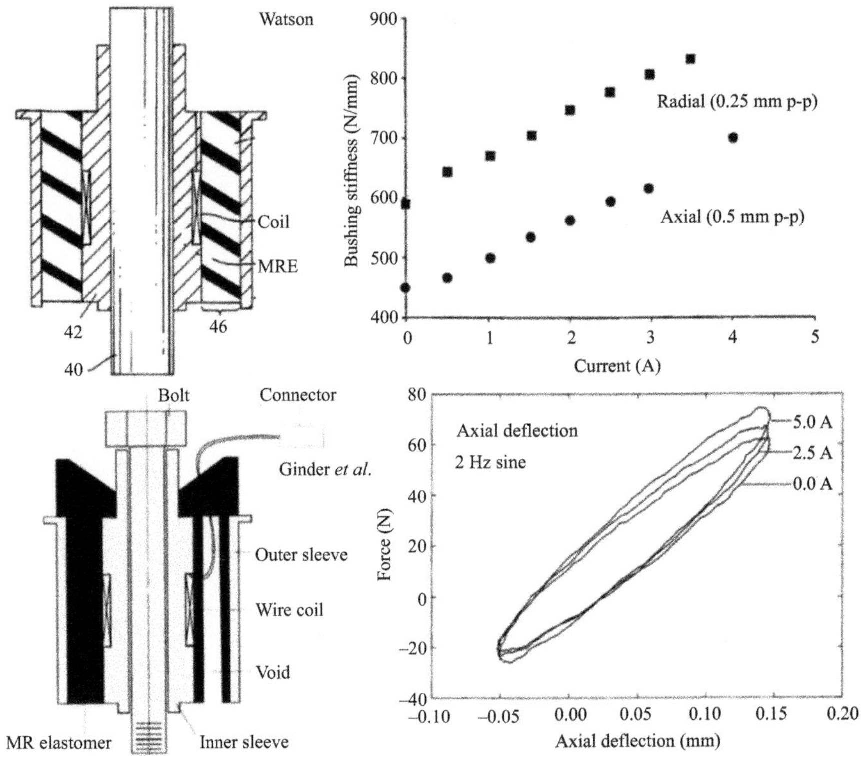

Figure 12.8 MRE bushing patented by Watson [17] and presented by Ginder et al. [18]

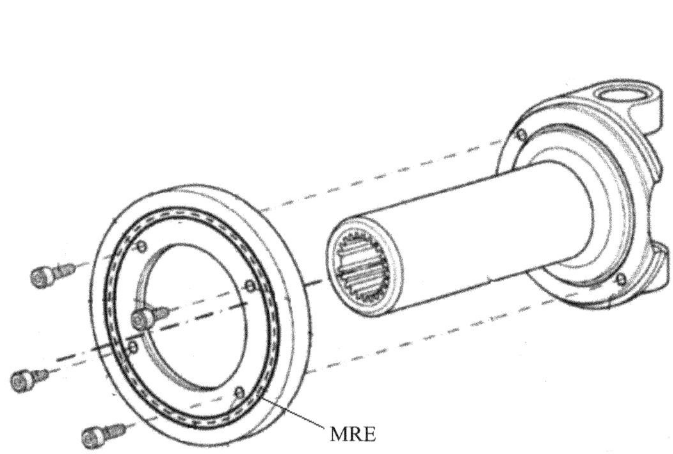

Figure 12.9 MRE-based tunable slip yoke [19]

this device, the power train frequency can be shifted far away from the excitation frequency so that the transient response is reduced significantly [56]. Configuration of vibration absorber for power train can be seen in Figure 12.10.

The conventional metal spring often possesses a constant spring stiffness requiring the vehicle operator to compromise vehicle handling or operator comfort. Improving the ride performance through the spring element is still limited to an adjustable air spring (manual or controlled). In US8210547B2 [20] and US2013/0087985A1 [53], the inventors offered a tunable spring stiffness using MRE. These two types of apparatus are relatively novel approaches for tuning the spring element through its base material. The stiffness constant of the MREs spring is adjusted in response to the intensity of the magnetic field. The application of MREs in the air spring was also proposed and patented by Li and Ding [52] The spring stiffness in their work can be adjusted by either air pressure or the tunable properties of the spring material.

Figure 12.11 illustrates the MREs spring assembly, which consists of a combination between the conventional strut suspension and the MREs element. The MREs element forms a cylinder with the first end connected to the spring and the

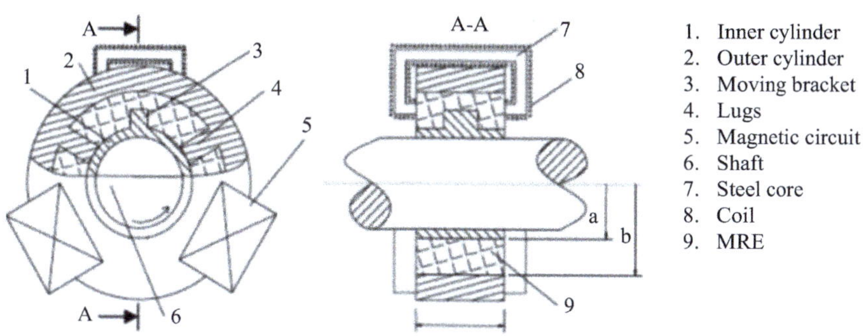

Figure 12.10 Powertrain or driveline MRE-based absorber [56]

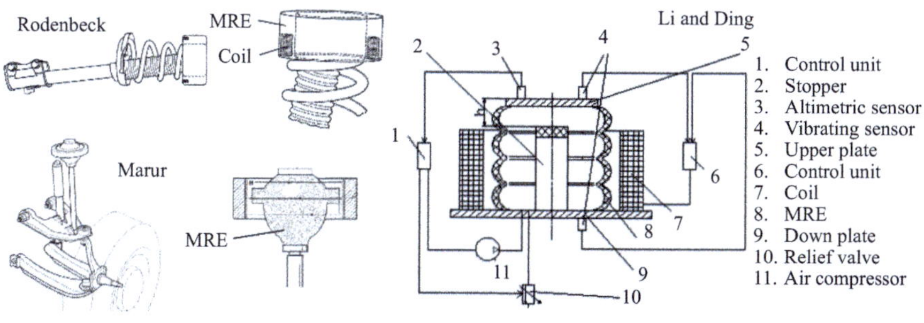

Figure 12.11 Tunable stiffness spring based on MRE

second end to the vehicle body or chassis. More than one assembly of coil and core are immersed within the MREs as an electromagnet generator. The stiffness of the strut assembly increases when the MREs are magnetized. The lack of magnetization makes this configuration act as a passive system with a single stiffness constant.

Apart from Rodenbeck [20], the suspension system replaces the helical spring with a MREs spring. The system includes a MREs element and an electromagnet with the ability to provide high flux density. The MREs element is made of natural and synthetic rubber as a composite matrix and micron size particle, such as iron, nickel, cobalt, and manganese. Curing of the MREs element is under the presentation of a magnetic flux to achieve anisotropic conditions since the spring effect is only unidirectional. The construction seems to be simple; however, due to the relatively thick shape of MREs, the magnetic flux generated should sufficiently stiffen the MREs element even though it will consume more energy. Optimization of the micron-sized particles makes the MREs spring work in an optimum manner. This approach to the shape design and matrix materials will also improve the performance of the device, such as a wide range of stiffness tuning ability.

12.1.3 Medical application

MR-based devices have also penetrated into medical rehabilitation since 1999 in the form of artificial prosthetic knees [57]. The patented prosthetic knee uses an MR fluid damper for tunable damping. However, the use of MR fluid damper only provides advantage in terms of controllable damping. Nevertheless, stiffness parameter in such device is also beneficial for the user. The need for adjustable stiffness and damping embedded in the prosthetic knee arises from the fact that conventional prosthetic and orthotic devices cannot provide an appropriate level of stiffness for users based on the activity level [58]. Passive devices cannot adapt the feeling range of motion of a limb's natural movement, such as the natural stride of a human foot. High stiffness is needed in a hard strike, e.g., running and jumping while low stiffness is necessary for comfort at rest or a soft strike or moving casually, e.g., walking [59]. The configuration of MRE-based prosthetic leg is portrayed in Figure 12.12.

The active orthotic and prosthetic devices, which are controllable and adaptable to the user's activity level and are capable of being controlled either manually or automatically by responding to signals that represent the current activity level of the user, are important because the operating characteristics of the device can be changed in real time [21]. Currently, MREs are applied on the prosthetic devices to simplify the design and were patented in 2013 by Thorsteinsson *et al.* [58]. This MRE component is both manually and automatically tunable to vary the stiffness of the shock absorption member via the application of a magnetic flux [58]. According to Jonsdottir *et al.*, the fabricated MRE-based prosthetic knee did not achieve optimum results since it needed high density of magnetic flux. So far, the change of MRE stiffness was about 37% under 0.3 T in which such amount of magnetic flux needed high current supplied to the coil.

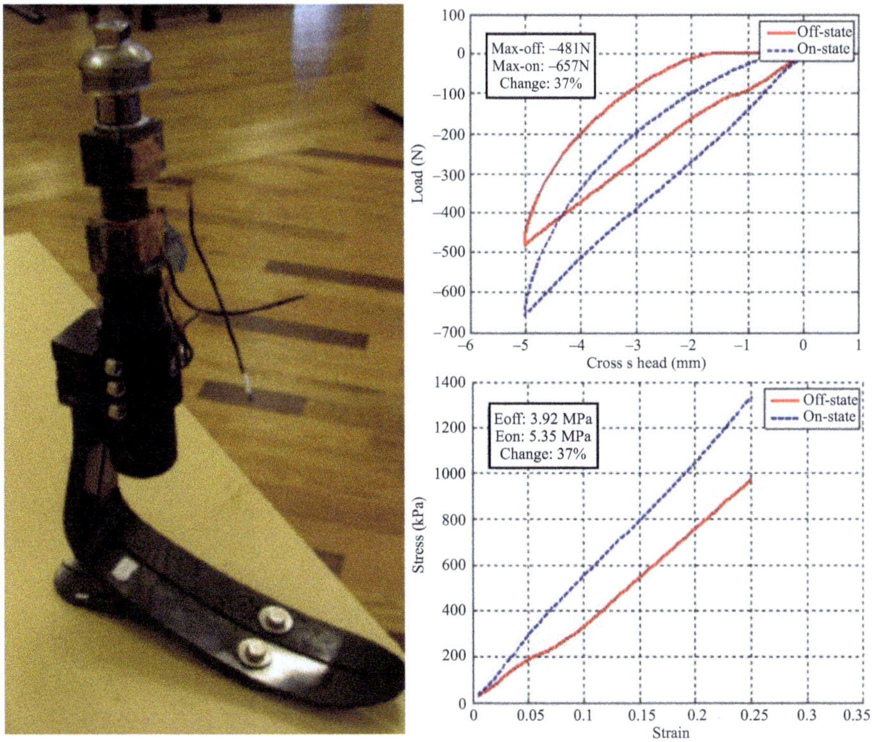

Figure 12.12 Prosthetic leg utilizing MRE [58–60]

12.1.4 Robotic application

In 2017, Christie *et al.* [61] introduced MRE for robotic application as could be seen in Figure 12.13. The main function of MRE in this device was also variable stiffness for C-shaped leg hexapod robot. This breakthrough was proposed to replace rotational spring in C-shaped leg proposed by Huang *et al.* [62]. The torsional spring of the prototype that typically represents an approximation of the compliant C-shaped leg stiffness was replaced by a torsional joint which could be adjusted to its stiffness by means of a permanent magnet and electromagnet within the structure [61]. Owing to the distinguishing characteristic of MRE, it has potential to lead advantageous performance of robotic locomotion with greater energy efficiency.

In the proposed design, the permanent magnet enhances the MRE resulting in a higher complex modulus compared to that of free magnetic fields. The main parts of the MRE-based joint are MRE, electromagnetic coil, permanent magnet, and low carbon steel magnetic circuit formed by the top and base covers and the outer yoke. The structure of low carbon steel allows the magnetic flux flow in an appropriate manner. Once the electromagnet was energized, the complex modulus can be either

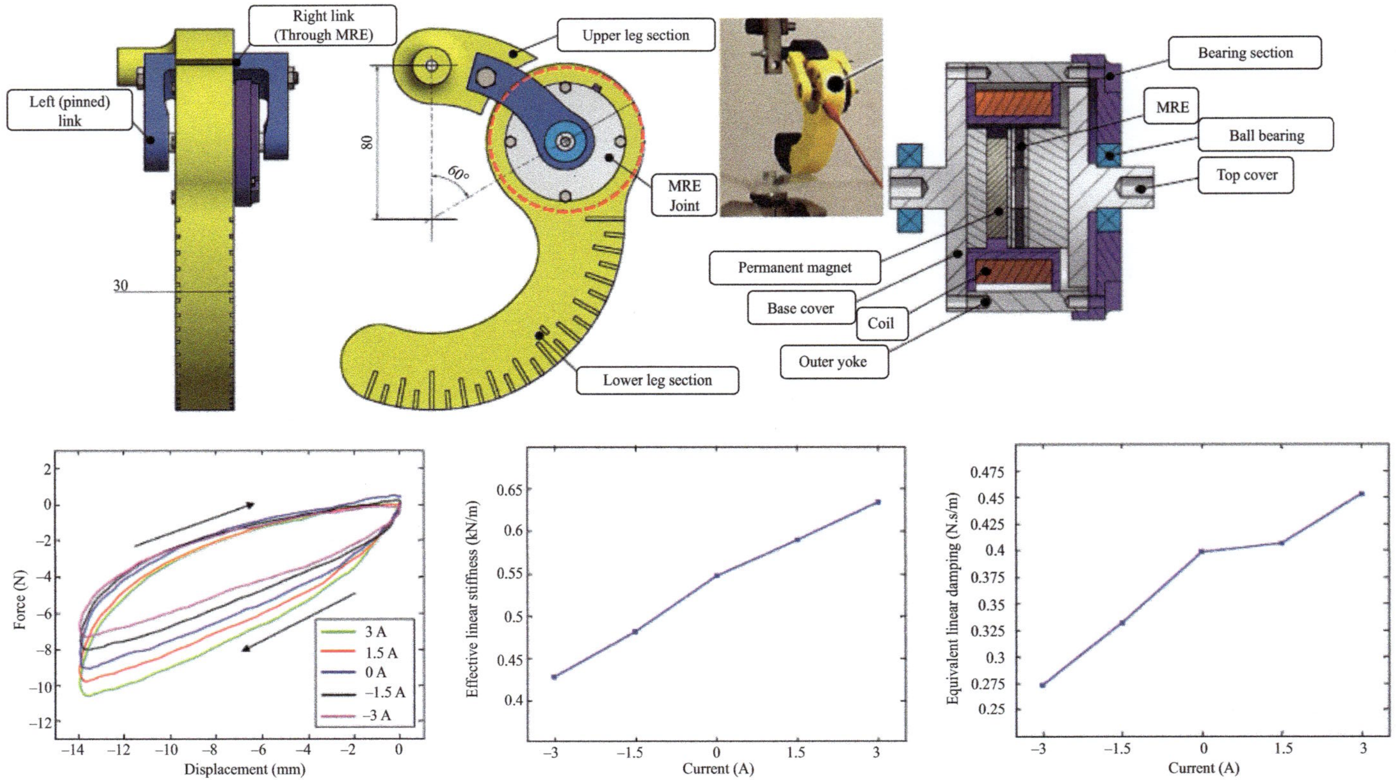

Figure 12.13 MRE for robotic leg by Christie et al. [61]

weakened or strengthened depending on the current flow. Based on the experiment, it can be seen that both effective linear stiffness and equivalent linear damping increases following the increment of applied current. The total alteration of stiffness and damping shifting was recorded about 48% referring to the lowest value of achievable stiffness and damping constants [61].

12.1.5 Machining application

A robotic machining usually experiences chattering phenomenon on the machining tools such as boring bar and milling cutter. It is because the robotic machining has a lower structural stiffness compared to the conventional computer numerical control (CNC) machines. The chattering phenomenon in conventional CNC machines is often compensated by their weight. Moreover, the entire robotic machining structures also vibrate during light machining process. Therefore, the chattering phenomenon in robotic machining has been becoming interest of research in recent years [63]. So far, this issue has been treated by incorporating passive and active chatter compensation technologies such as changing robot configuration, process behavior, system structures including stiffness and damping, as well as active force control.

To overcome drawbacks in passive and active chatter compensation technologies, Yuan *et al.* [63] proposed an MRE-based vibration absorber for robotic machining system. With the employment of MRE, the absorber has capability of controlling the vibration through stiffness and damping tuning. Figure 12.14 shows the configuration of vibration absorber.

Shaking test was first carried out to identify the capability of frequency and phase shifting. The proposed absorber has shifting frequency ranging from 9.7 to 15.5 Hz under current variation of 0–1.8 A. The prototype was then integrated to the robotic arm after the shaking test. However, the experiment did not focus on

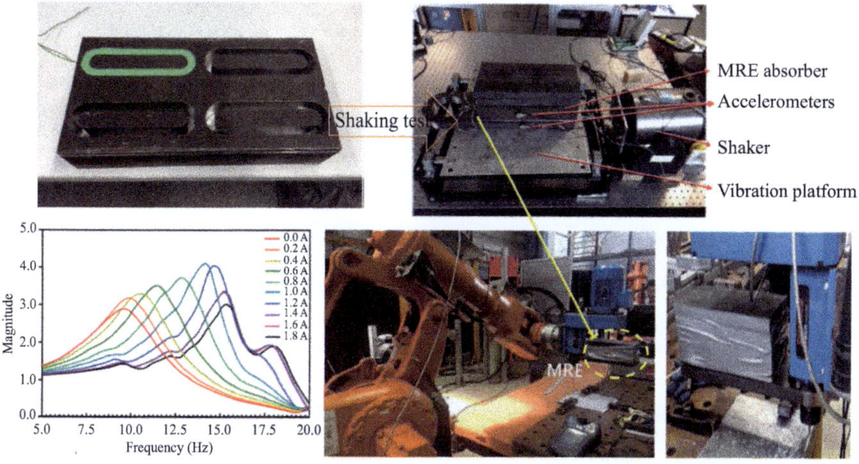

Figure 12.14 MRE-based vibration absorber for robotic machining [63]

controlling absorber at the moment. It only validated the effectiveness of using MRE absorber in a milling process under various cutting condition. Therefore, the experimental result only measured cutting force, chatter frequency, and power spectral density of the system. Experimental setup showed that if the natural frequency met the target frequency, the chatter could be absorbed effectively. Based on this finding, a set of control strategy would be effectively matching the natural and target frequencies.

12.2 Active system

When an MRE has a soft and highly elastic matrix, it will respond to the external magnetic field by changing the shape following the magnetic field's strength and pattern. The reaction can either lengthen or shorten from the initial shape which means reversible process. This advantage is useful for some application such as artificial muscle, pumping, valve, active attachment, even morphing airfoil. The last mentioned term (morphing airfoil) seems interesting, but it is still not realized yet.

12.2.1 Releasable attachment

The right side application depicted in Figure 12.15 is a remote controllable attachment that has been patented in US6877193B2A [24]. The releasable fastener system shown in the illustration encompasses a loop and a hook section. The loop section includes a support, which is made of magnetic metal as well as a loop material fixed on one side. The hook section consists of a magnetic metal plate as a

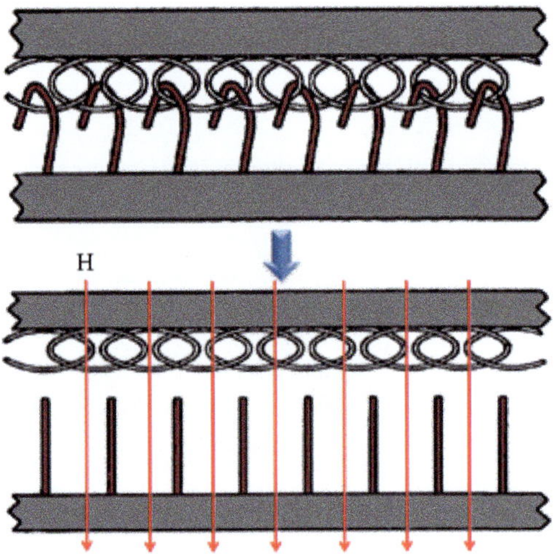

Figure 12.15 Releasable attachment by Ottaviani et al. [24]

support and some closely spaced upright hook elements extending from the counterpart direction.

If the hook and loop sections are pressed together, they interlock to form a releasable engagement. The resulting joint created by the engagement is relatively resilient to shear and tension forces as well as weak in peel strength forces. MREs, which provide a change in the shape orientation, are used as the basic material of the hook components. In this manner, the shape orientation of the hook components can be remotely changed to provide a decrease or magnification in the shear or tension-off forces also to provide adjustable damping capabilities to the fastener system.

12.2.2 Artificial muscle

Figure 12.16 referred to the implementation of actuation for the artificial muscle that has been achieved by Nguyen *et al.* [15]. In the report, the term Magpol was used as a replacement of MREs. The Magpol consists of highly elastic materials (silicone type) and soft magnetic particles. This smart composite has the high responsiveness and is capable of remote and contactless actuation. Some actuation modes can be achieved such as elongation, contraction, deflection, shear, torsion, rotation, and coiling. Both homogeneous and nonhomogeneous magnetic fields can be applied to induce the Magpol to get various actuation responses. The position or shape alteration can be controlled depending on the magnetic field intensity. However, the application of Magpol as artificial muscle by taking advantages from deformation modes is still limited on the electromagnetic field generation. The magnitude and direction of magnetic fields play an important role in achieving the desired mode.

12.2.3 Microfluidic and peristaltic pumping

Wonderful ideas came from Fuhrer *et al.* [64] and Tang *et al.* [65] which utilized MRE for pumping system for microfluidic system. Microfluidic systems allow for

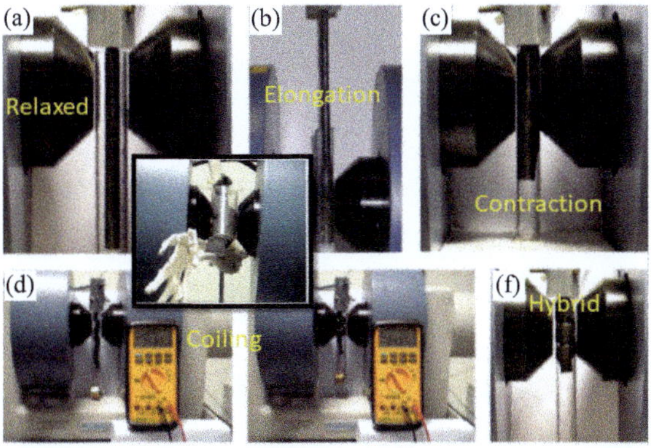

Figure 12.16 Magnetic polymer by Nguyen et al. [15]

precise control over flow variables such as velocity, shear stress, temperature, and concentration of chemicals in microscale, making them very useful for handling small volumes of samples and reagents for various complex analysis and reactions [65]. So far, microfluidic pumping system needs very complicated tiny structure and is consequently hard to fabricate resulting in a noneconomic pricing. Pumping system employing MRE has been proven free from rotating part and minimum noise generated.

In 2013, Fuhrer *et al.* introduced application of MRE for peristaltic pumping. The proposed peristaltic pump utilized a set of MRE tubes. Silicon RTV was used as the main matrix of MRE tubes which carry carbonyl iron particles. The fabrication process of MRE tube was simple molding. Each part has its own electromagnet for actuation. The shape of inner tube is designed as shown in Figure 12.17, thus the MRE tube can be easily deformed. The magnetic flux deformed the MRE tube so that the tube flattens. In a series of MRE tubes, the schedule of magnetic fields energizing could result in a useful driving force to transport the fluid inside an MRE tube.

Apart from peristaltic pump, Tang *et al.* proposed a membrane-type pumping system as can be seen in Figure 12.18. This prototype does not use electromagnet for actuation but permanent magnets. Five magnets were radially arrayed at the bottom side of MRE membrane. To energize the membrane, a motor was needed to rotate the permanent magnet. Therefore, the actuation rate of MRE membrane exactly depends on the angular speed of motor. The higher the angular speed, the faster the pumping rate. The pumping system needs thin MRE in which it would become another challenge for material preparation.

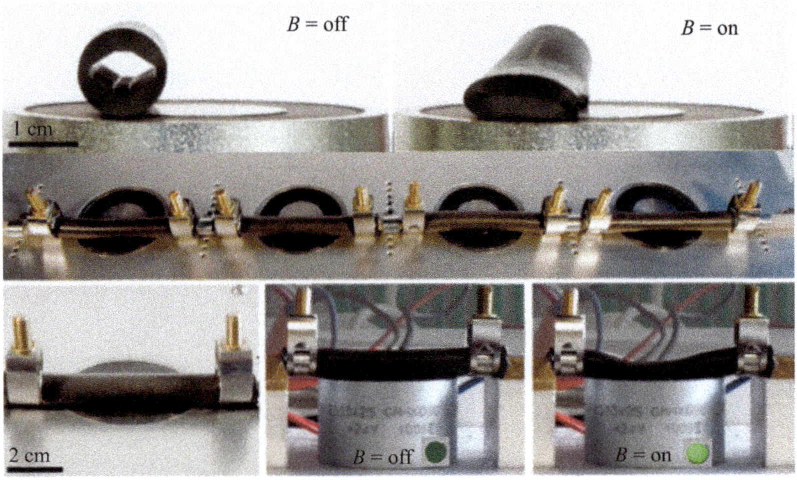

Figure 12.17 Peristaltic MRE-based pump by Fuhrer et al. [64]

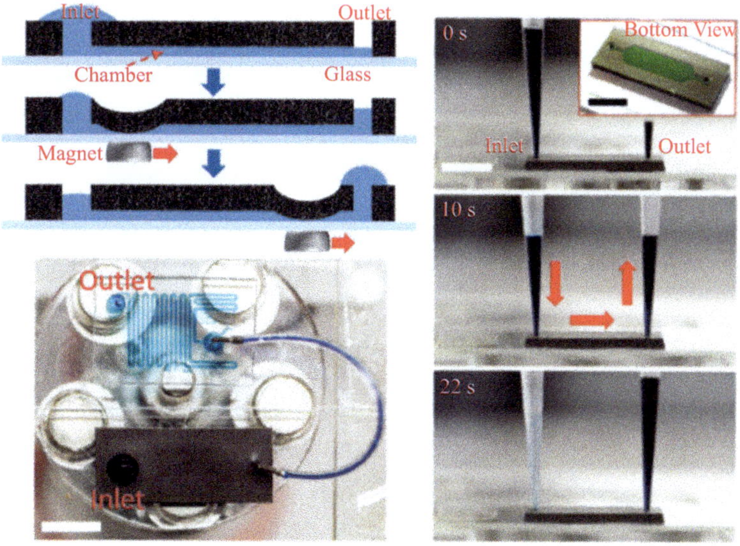

Figure 12.18 Microfluidic MRE-based pump by Tang et al. [65]

12.2.4 Valve system

The stretching capability of MRE under inhomogeneous magnetic fields was utilized by Böse *et al.* [66] for developing valve based on MRE. When the magnetic fields were removed, it went back relaxing to its original shape. Such actuation needs very soft and elastic polymer. Here, the strength of magnetic attraction acting on the soft-magnetic particles and elastic polymer force strongly influences the degree of actuation. The magnetic fields attract soft magnetic particles resulting in displacement from their original position. On the other hand, the elastic polymer network reacts to the magnetic attraction and try to restore the particles into their initial position.

The configuration of MRE-based valve is depicted in Figure 12.19. The valve was equipped with ring-shaped MRE. In normal condition (without magnetic fields), the ring-shaped MRE has a certain air gap. The magnetic fields generated by coil could attract the MRE and radially deformed resulting in a closed gap. Expansion of the MRE would block the air flow through the gap. During valve performance test, the prototype was installed in a closed system air flowing at 0.36 m^3/h. Two kinds of sensors were used those were pressure sensor and air flow meter. Based on the experiment, when a certain current was applied into the coil, the flow rate decreased due to the effect of reducing air gap. The higher the current applied to the coil, the lower the air gap resulting in a lower flow rate. Meanwhile, the maximum pressure increased as the increase of applied current. Based on the working principle, the state of matrix plays important roles in stretching and relaxing stages. The softer matrix would ease stretching, and more elastic matrix leads to faster relaxing stages.

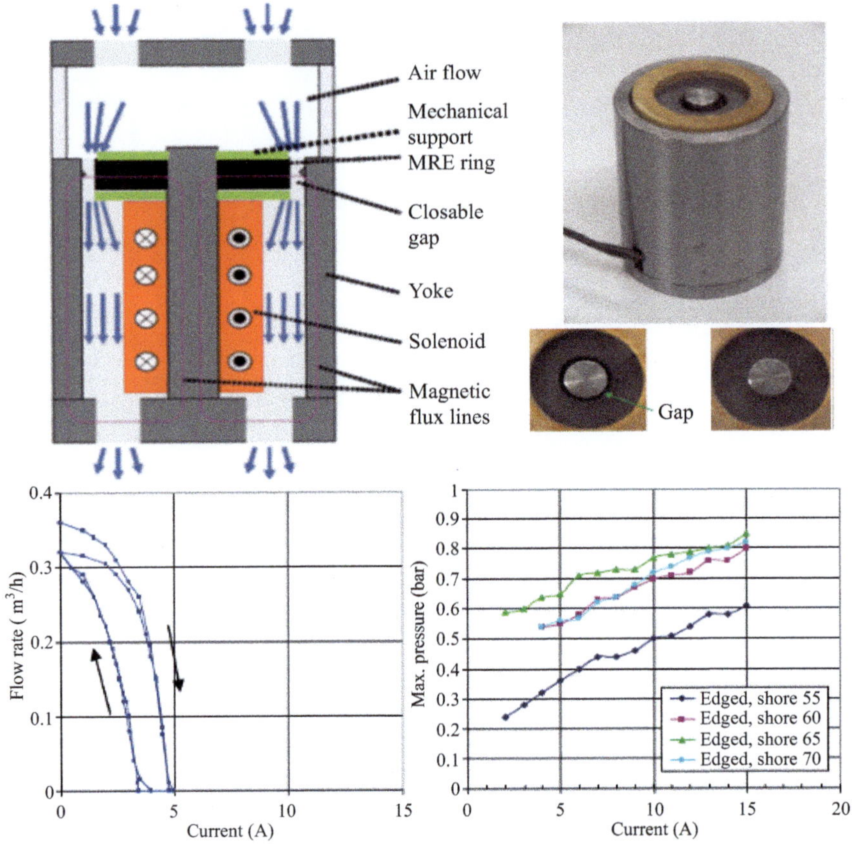

Figure 12.19 MRE-based valve by Böse et al. [66]

12.3 Sensory system

Capability of sensing in MRE has been providing a promising implementation. For instance, the use of MRE for tunable vibration absorber is not only purely for energy storing and dissipation, but also the device can monitor the condition [67]. This mechanism is usually known as self-sensing capability. This is because the distance between particles inside this smart material is not only influencing the mechanical but also its electrical properties [68]. Here, the change of dimension due to the effect of external forces and magnetic stimuli affects its electrical properties, namely, magnetostriction and magnetoresistance. This fact has been investigated and proved by some researchers. This advantage took an important role in the further MRE-based sensors development. Early in 1999, Bednarek found a giant magnetostriction in ferromagnetic powder embedded in silicon matrix. The length ratio of all samples increased along with the increment of flux density as can be seen in Figure 12.20. Here, the silicon matrix did not interact practically with the

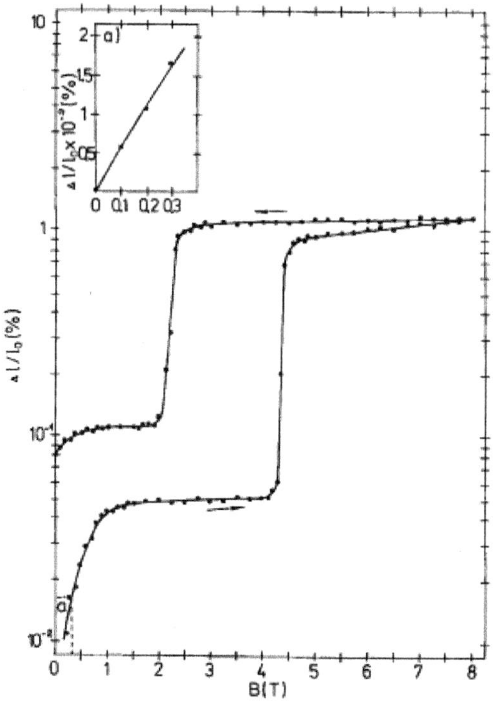

Figure 12.20 Giant magnetostriction found by Bednarek in 1999 [67,68]

magnetic flux because of very low susceptibility. Therefore, the magnetostriction is only influenced by interparticle attraction and mutual interaction of the magnetic particles. The magnetic flux induced the magnetic dipole in the particle causing those particle interactions.

From Figure 12.20, we can see that the increase of flux density obviously affected the length ratio of the material. The curves implied a positive magnetostriction phenomenon. The applied flux density induced the magnetic dipole moments within the particles. Even, the mode of both increment and decrement magnetic flux variation influences the length ratio values. Two curves exhibit the hysteresis phenomenon.

12.3.1 Impedance and magnetoresistance

Ten years later after Bednarek [x], Wang *et al.* characterized sensing behaviors of MRE which consisted of impedance and magnetoresistance. They found significant finding in impedance and resistant changes of such materials which were strongly depended on the applied force and magnetic fields. The schematic diagram, samples, magnetoresistance, as well as impedance spectroscopy results are depicted in Figure 12.21. The measurement diagram was important for establishing relationship between electrical properties and external stimuli. The

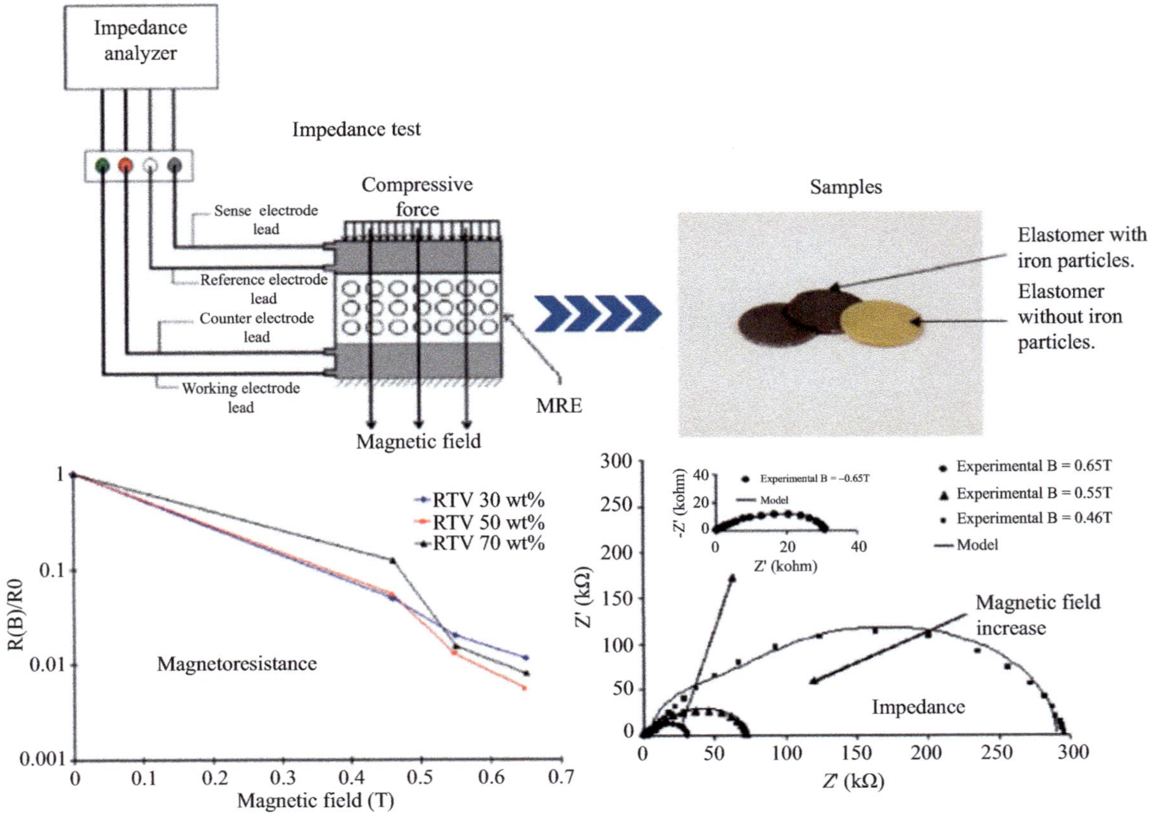

Figure 12.21 Magnetoresistance MRE sensor by Wang et al. [69]

material was placed in between two electrodes made of steel. The external force was applied to deform the sample at various magnetic fields across the sample. Some sensors were used to measure the deflection and force, namely, LVDT and load sensor. These sensors logged all data which was further used in the determination of stress and strain mechanical properties. Meanwhile, the electrical properties those were impedance and resistance were measured using impedance spectroscopy. The samples had round and thin shapes. During measurement, the magnetic field increased sequentially. Detail of experimental setup can be referred in Wang *et al.* [69].

The significant alteration in impedance under magnetic field variation was found during the whole testing frequency ranges. Magnitude and phase of the impedances were reduced as a result of applied external magnetic flux. The effect of magnetic flux, however, had less effect on the impedances at higher frequencies. This fact was further known as magnetoimpedance effect (MI effect). Moreover, the electrical properties of these silicon-based MREs are also shown in Figure 12.21. It can be seen that the increase of applied magnetic flux was followed by the significant reduction of resistance at all samples. The decrease of resistance was predicted due to microstructure interaction resulting in a closer interparticle distance as the effect of magnetic flux. The phenomenon also affected the impedance of MRE in which the increase of magnetic flux. The complex form of impedance can also be seen in Figure 12.21.

Another magnetoresistor was also proposed by Bica [70] in 2011 in better appearance (see Figure 12.22). In his study, silicon-based MRE was prepared

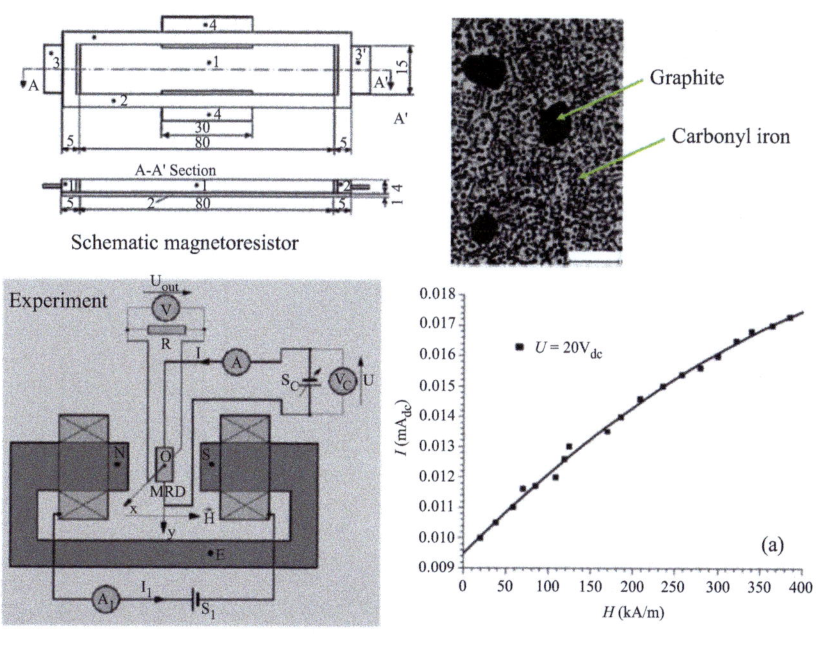

Figure 12.22 Magnetoresistor by Bica [70]

with additional graphite. It is known that silicon is nonelectric conductive materials. Therefore, graphite plays an important role in changing to conductive materials. During experimental test, the MRE acts as variable resistor. The control voltage is set fix, while the output voltage and current intensity is measured in the presence and absence of magnetic fields. Current intensity enhanced as the increase of magnetic field is under constant control current. It showed that the resistivity of the sample changed with the change of magnetic field. The magnetic dipole moment strongly depended on the fields. Here, the dipole moment changed its orientation so that there are dipole interactions between neighboring iron particles.

12.3.2 Magneto-induced capacitor

In the same year, Bica *et al.* [71] also treated MRE as dielectric material to investigate the potential development of capacitor. This application took advantages from the magneto-controlled electrical properties. As can be seen in Figure 12.23, a piece of silicon-based MRE was fixedly glued between two parallel copper plates

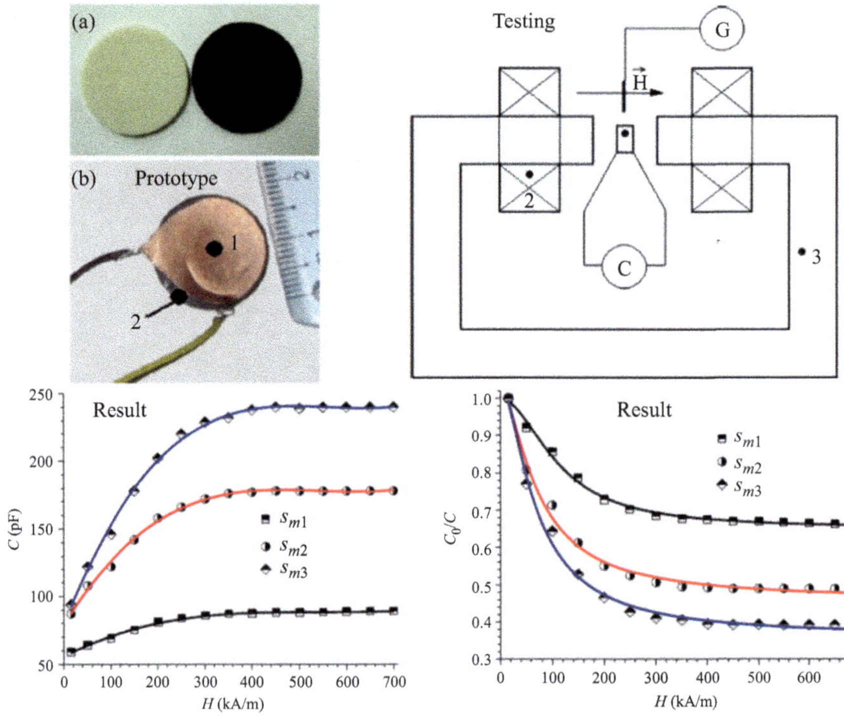

Figure 12.23 MRE-based capacitor by Bica et al. [71]

becoming a piece of capacitor. The capacitor was then located between two electromagnetic cores. Sets of capacitance data were taken under various magnetic field intensities. The procedures were also undertaken for the control capacitor that was pure silicon.

It has introduced three MRE-based capacitor with different carbonyl iron particle loads (three types MRE in the experiment). Referring to Figure 12.23, Bica *et al.* found two trends in each single curve. The capacity increased monotonically in exponential trend until particular field intensity and then saturated at the increase of field intensity. Besides, the higher loading of carbonyl iron particles led to the higher degree of increment on the first region (before saturation). Here, the change of capacitance also depends on the magnetic dipole interaction between particles and their orientation when experiencing magnetization.

12.3.3 Displacement sensor

Another creative idea for application of MRE in sensory system is displacement sensor. The concept is also similar to the previously discussed sensor. It uses the magnetoresistance phenomenon and its relation to the strain alteration. Ausanio *et al.* [72] develop MRE-based displacement sensor in 2014. The team introduces nickel particle inside silicon matrix instead of iron particle. The nickel enabled high saturation value at lower magnetization field. Therefore, greater total strain in longitudinal direction was obtained at the same Young modulus when it was magnetized with a fixed magnetization gradient. The larger longitudinal strain produced obvious resistant change in the sample.

The prototype of MRE-based displacement sensor is depicted in Figure 12.24. During experiment, the MRE sample was equipped with contact capillary wire

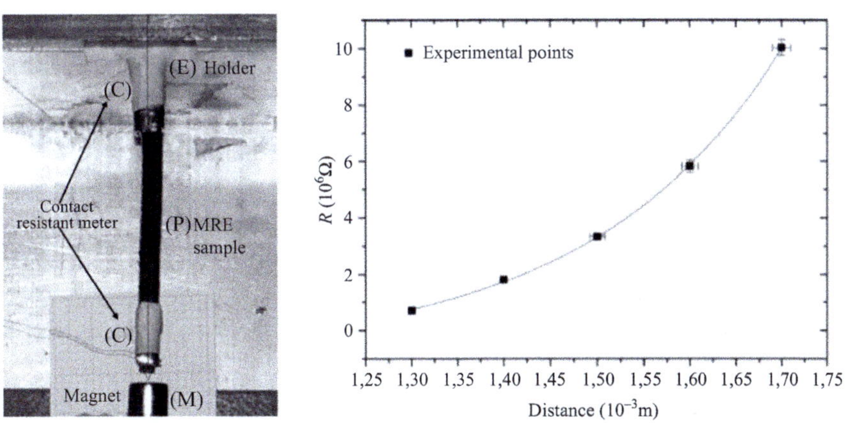

Figure 12.24 MRE-based displacement sensor [72]

for resistant meter connection. It was clamped in vertical longitudinal direction. Permanent magnet was located in the other edge having a certain distance. It can be seen that the resistance strongly depends on the distance between magnet and sensing core. This device is a potential for monitoring the displacement of any non-magnet on which the permanent magnet is set. Thanks to the giant resistant changes which were induced by a moderate gradient magnetic induction.

Three years later, Ding *et al.* [73] proposed a multifunctional strain sensor utilizing highly elastic yarn. They introduced silver nanowires (AgNWs) for the excellent electric properties. Here, the MRE based on silicon was used for protecting shell and wrapped the AgNWs coated lastex yarn. The magnetic composite fiber sensor is illustrated in Figure 12.25 in which this sensor has out an standing function [73]. It can measure not only the strain but also the magnetic and electric properties. Therefore, it is potentially used for magnetic–mechanic–electric wearable sensor. In mechanical fields, the sensor can be used to measure tension and bending, while in electrical fields, the resistivity change also apparently represented the electric parameters. This sensor is also magnetically responsive in which it is potential for portable magnetic fields measurement.

The sensing mechanism can be explained as follows: the microstructure change of AgNWs network plays important roles in sensing properties of the sensor. Total resistant of sensor is contributed by contact resistant of AgNWs joints and their inherent resistant. The strain applied on the sensor makes the embedded AgNWs in MRE stretch and slide. Therefore, the number of joints in AgNWs network reduces and the nanowire distance increases. As a result, there is growth resistance as the effect of applied strain. Besides, the magnetic field induction influences the magnetic dipole interaction between iron particles inside the silicon matrix. As intensively studied before, the magnetic dipole interaction also

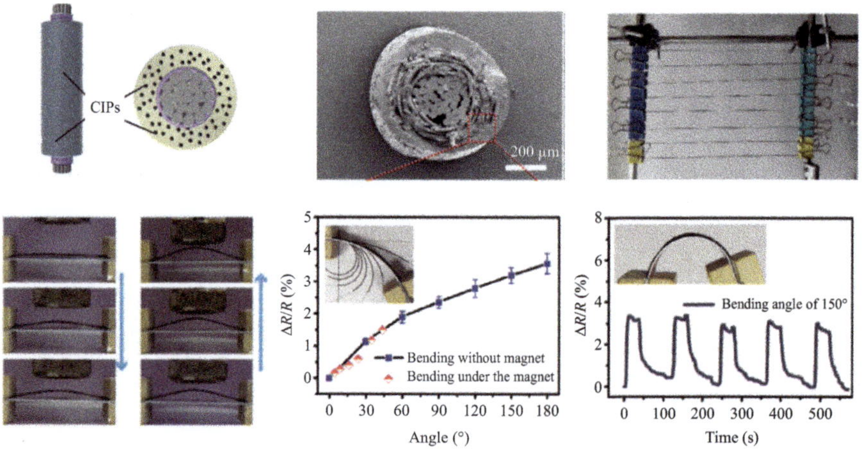

Figure 12.25 Magnetic composite fabric sensor [73]

influences the resistivity. Magnetic field across the sample is also detected from the strain growth in the sample.

12.3.4 Tire pressure sensor

In US8176958B2, Rodenbeck *et al.* proposed a dynamic tire pressure control, as shown in Figure 12.26, which incorporated MREs in a compact wheel assembly [20]. It consists of MREs assembly embedded within a wheel chamber. The MREs assembly has the primary function of adjusting the tire pressure within the chamber between the rim and the tire assembly when a magnetic field is applied. The MREs becomes stiffer and changes its shape resulting in an increment of cavity air pressure. The feedback control system communicates with a tire pressure sensor. When the feedback control system receives a feedback signal from the condition sensor that a tire pressure is outside a preset range, the feedback control system may produce a control signal to control an output of the electromagnet to activate the MREs element.

12.3.5 Magnetic field sensor

The implementation of MREs in magnetic field sensor or magnetometer applies magnetostriction phenomenon. In 2012, Du and Chen [23] introduced MRE for magnetometer based in this phenomenon as can be seen in Figure 12.27. The MRE was incorporated with silicon cup that was usually utilized in piezoelectric sensors. The MRE layers were attached to the silicone cup or silicone sensitivity diaphragm embedded with a piezoresistive Wheatstone bridge. The interaction between the MREs layers and magnetic fields generated a deflection on the sensitivity diaphragm. The interactions altered the piezoresistance and unbalance the Wheatstone bridge. The sensor has an excellent linearity in sensing the magnetic

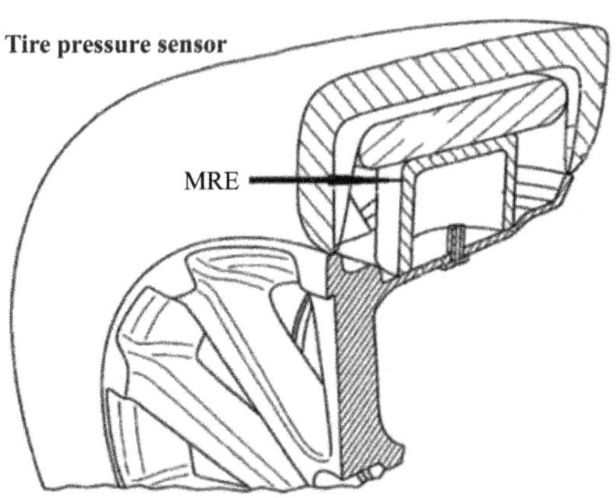

Figure 12.26 MRE-based tire pressure sensor [22]

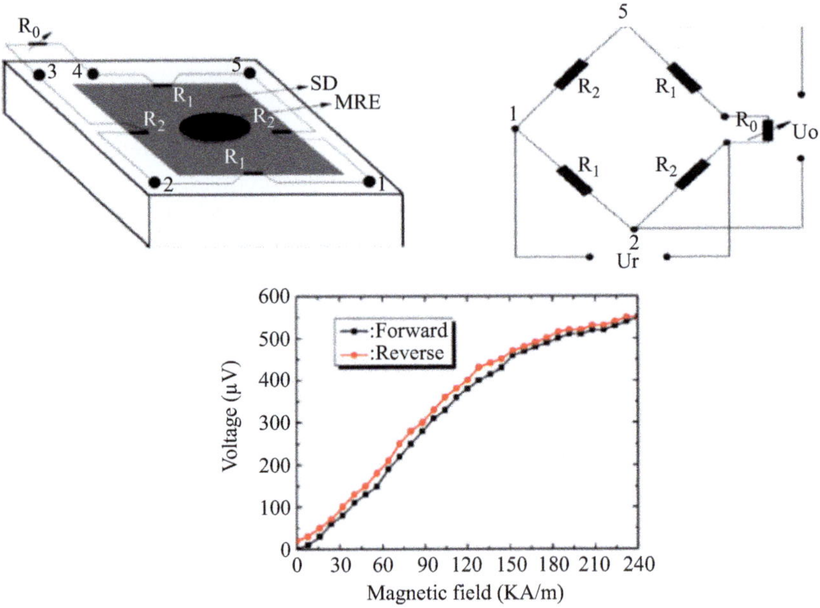

Figure 12.27 Magnetometer-based MRE [23]

field. This sensor is easily integrated into electronic circuit due to its simplicity, low cost, and low power.

12.3.6 Tactile sensor

In robotic application, tactile sensor is considered as an important part for system interaction with other unknown objects. For complex surface objects, robot needs a soft and flexible sensor to detect them all. So far, there were a lot of flexible tactile sensor, but they have a problem for measuring three axes forces due to their structures. In recent year, Kawasetsu *et al.* [74] introduced an MRE-based tactile sensor which could measure three axis forces without deteriorating durability, sensitivity, as well as simpler structure than the existing one.

The configuration of MRE-based tactile sensor can be found in this tactile sensor that applies basic working mechanism of magnetic induction on the coil which strongly depends on the core shape. Figure 12.28 shows the configuration and MRE-based tactile sensor. Here, MRE acts as the main core in which the deformation applied on the MRE will affect the inductance of four coils. The change of inductance is supported by aggregation of iron particles inside the MRE. The distance between deformed MRE and coil is proportional to the inductance changes.

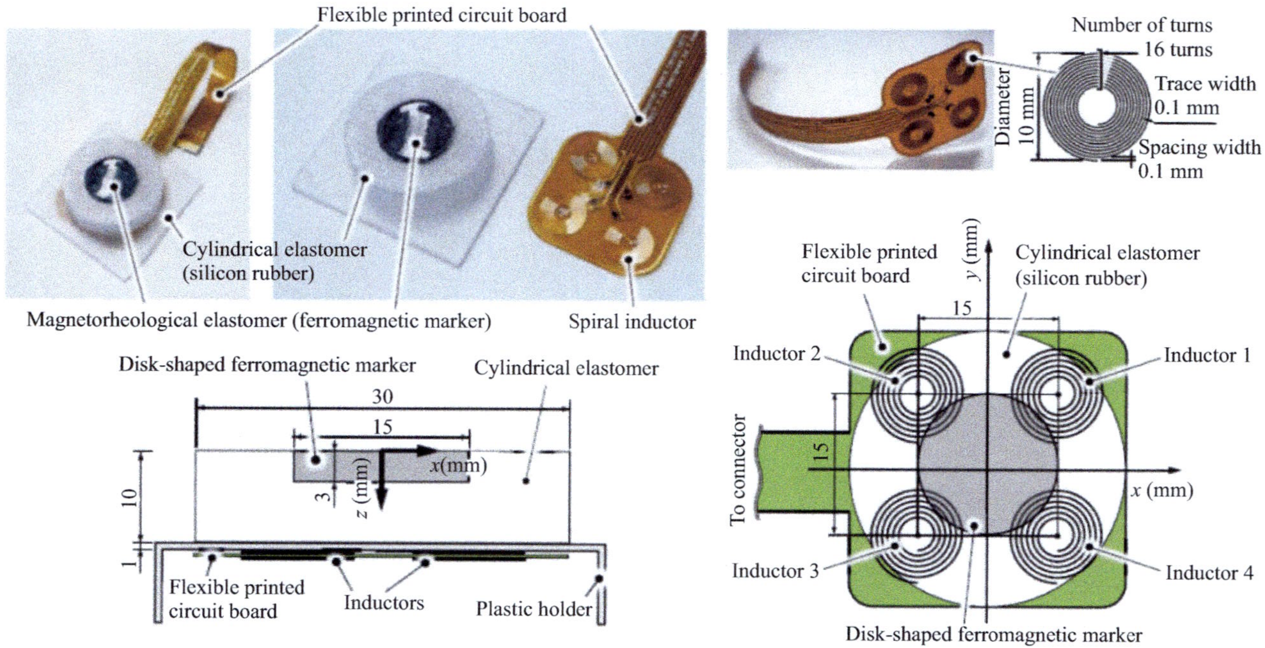

Figure 12.28 Three-dimensional MRE-based tactile sensor

References

[1] M. Farshad and A. Benine, "Magnetoactive elastomer composites," *Polym. Test.*, vol. 23, no. 3, pp. 347–353, 2004.

[2] G. Bossis, C. Abbo, S. Cuttilas, S. Lacis, and C. Metayer, "Electroactive and electrostructured elastomers," *Int. J. Mod. Phys. B*, vol. 15, no. 6 & 7, pp. 564–573, 2001.

[3] M. Mayer, R. Rabindranath, J. Börner, E. Hörner, A. Bentz, J. Salgado, *et al.*, "Ultra-soft PDMS-based magnetoactive elastomers as dynamic cell culture substrata," *PLoS One*, vol. 8, no. 10, p. e76196, 2013.

[4] G. V. Stepanov, A. V. Chertovich, and E. Y. Kramarenko, "Magnetorheological and deformation properties of magnetically controlled elastomers with hard magnetic filler," *J. Magn. Magn. Mater.*, vol. 324, no. 21, pp. 3448–3451, 2012.

[5] A. Stoll, M. Mayer, G. J. Monkman, and M. Shamonin, "Evaluation of highly compliant magneto-active elastomers with colossal magnetorheological response," *J. Appl. Polym. Sci.*, vol. 131, no. 2, 39793, 2014.

[6] T. Y. Liu, S. H. Hu, D. M. Liu, and S. Y. Chen, "Magnetic-sensitive behavior of intelligent ferrogels for controlled release of drug," *Langmuir*, vol. 22, no. 14, pp. 5974–5978, 2006.

[7] J. A. Galicia, F. Cousin, E. Dubois, O. Sandre, V. Cabuil, and R. Perzynski, "Static and dynamic structural probing of swollen polyacrylamide ferrogels," *Soft Matter*, vol. 5, pp. 2614–2624, 2009.

[8] M. Zrinyi, L. Barsi, and A. Bu, "Deformation of ferrogels induced by nonuniform magnetic fields," *J. Chem. Phys.*, vol. 104, no. 21, pp. 8750–8756, 1996.

[9] Y. Han, W. Hong, and L. E. Faidley, "Rate dependent finite deformation of magneto-active polymers," in *Behavior and Mechanics of Multifunctional Materials and Composites*, 2011, vol. 7978, 797819.

[10] L. Borcea and O. Bruno, "On the magneto-elastic properties of elastomer – ferromagnet composites," *J. Mech. Phys. Solids*, vol. 49, pp. 2877–2919, 2001.

[11] Z. Ying, H. Chen, and Y. Ni, "Magnetorheological visco-elastomer and its application to suppressing microvibration of sandwich plates," in *The 3rd International Conference on Smart Material and Nanotechnology in Engineering*, 2012, vol. 8409, p. 840918.

[12] O. V. Stolbov, Y. L. Raikher, and M. Balasoiu, "Modelling of magnetodipolar striction in soft magnetic elastomers," *Soft Matter*, vol. 7, no. 18, p. 8484, 2011.

[13] Y. L. Raikher and O. V. Stolbov, "Numerical modeling of large field-induced strains in ferroelastic bodies: a continuum approach," *J. Phys.: Condens. Matter*, vol. 20, no. 20, p. 204126, 2008.

[14] P. Melenev, Y. Raikher, G. V. Stepanov, V. Rusakov, and L. Polygalova, "Modeling of the field-induced plasticity of soft magnetic elastomers," *J. Intell. Mater. Syst. Struct.*, vol. 22, no. 6, pp. 531–538, 2011.

[15] V. Q. Nguyen, A. S. Ahmed, and R. V. Ramanujan, "Morphing soft magnetic composites," *Adv. Mater.*, vol. 24, no. 30, pp. 4041–4054, 2012.

[16] Y. C. Li, J. C. Li, W. H. Li, and H. Du, "A state-of-the-art review on magnetorheological elastomer devices," *Smart Mater. Struct.*, vol. 23, no. 12, p. 123001, 2014.
[17] J. R. Watson, "Method and Apparatus for Varying the Stiffness of a Suspension Bushing," US005609353A, 1997.
[18] J. M. Ginder, M. E. Nichols, L. D. Elie, and S. M. Clark, "Controllable-stiffness components based on magnetorheological elastomers," in *Smart Structures and Integrated System*, 2000, vol. 3985, pp. 1–8.
[19] A. R. Badolato and R. P. Pawlowski, "Tunable Slip Yoke Damper Assembly," US006623364B2, 2003.
[20] P. D. Rodenbeck, "Active Magneto-Rheological Spring Assemblies and Vehicle Suspension Systems Incorporating the Same," US8210547B2, 2012.
[21] I. Guðmundsson, "A Feasibility Study of Magnetorheological Elastomers for a Potential Application in Prosthetic Devices," School of Engineering and Natural Science, 2011.
[22] P. D. Rodenbeck, "Magneto-Rheological Elastomer Wheel Assemblies with Dynamic Tire Pressure Control," US8176958B2, 2012.
[23] G. Du and X. Chen, "MEMS magnetometer based on magnetorheological elastomer," *Measurement*, vol. 45, no. 1, pp. 54–58, 2012.
[24] R. A. Ottaviani, J. C. Ulincy, and M. A. Golden, "Magnetorheological Nanocomposite Elastomer for Releasable Attachment Applications," US6877193B2, 2005.
[25] J. Thévenot, H. Oliveira, O. Sandre, and S. Lecommandoux, "Magnetic responsive polymer composite materials," *Chem. Soc. Rev.*, vol. 42, no. 17, pp. 7099–7116, 2013.
[26] Ubaidillah, S. A. Mazlan, J. Sutrisno, and H. Zamzuri, "Potential applications of magnetorheological elastomers," *Applied Mechanics and Materials*, vol. 663, pp. 695–699, 2014.
[27] N. A. A. Wahab, S. A. Mazlan, Ubaidillah, S. Kamaruddin, N. I. N. Ismail, S.-B. Choi, and A. H. R. Sharif, "Fabrication and investigation on field-dependent properties of natural rubber based magneto-rheological elastomer isolator," *Smart Mater. Struct.*, vol. 25, no. 10, 107002, 2016.
[28] A. M. A. Lerner and K. A. Cunefare, "Adaptable Vibration Absorber Employing a Magnetorheological Elastomer with Variable Gap Length and Methods and Systems Therefor," US7102474B2, 2006.
[29] X. L. Gong, Y. C. Fan, S. H. Xuan, Y. G. Xu, and C. Peng, "Control of the damping properties of magnetorheological elastomers by using polycaprolactone as a temperature-controlling component," *J. Ind. Eng. Chem. Res.*, vol. 51, pp. 6395–6403, 2012.
[30] G. J. Liao, X. L. Gong, S. H. Xuan, C. J. Kang, and L. H. Zong, "Development of a real-time tunable stiffness and damping vibration isolator based on magnetorheological elastomer," *J. Intell. Mater. Syst. Struct.*, vol. 23, no. 1, pp. 25–33, 2011.

[31] S. H. Eem, H. J. Jung, and J. H. Koo, "Modeling of magneto-rheological elastomers for harmonic shear," *IEEE Trans. Magn.*, vol. 48, no. 11, pp. 3080–3083, 2012.
[32] N. Hoang, N. Zhang, W. H. Li, and H. Du, "Development of a torsional dynamic absorber using a magnetorheological elastomer for vibration reduction of a powertrain test rig," *J. Intell. Mater. Syst. Struct.*, vol. 24, no. 16, pp. 2036–2044, 2013.
[33] Z. C. Ni, X. L. Gong, J. F. Li, and L. Chen, "Study on a dynamic stiffness-tuning absorber with squeeze-strain enhanced magnetorheological elastomer," *J. Intell. Mater. Syst. Struct.*, vol. 20, no. 10, pp. 1195–1202, 2009.
[34] G. H. Hitchcock, F. Gordaninejad, and A. Fuchs, "Controllable Magneto-Rheological Elastomer Vibration Isolator," US7086507B2, 2006.
[35] A. Fuchs, F. Gordaninejad, G. H. Hitchcock, J. Elkins, and Q. Zhang, "Tunable Magneto-Rheological Elastomers and Processes for Their Manufacture," US7261834B2, 2007.
[36] G. Naganathan and S. L. Vieira, "Actively Controlled Impact Elements," US2004/0126565A1, 2004.
[37] T. Komatsuzaki and Y. Iwata, "Design of a real-time adaptively tuned dynamic vibration absorber with a variable stiffness property using magnetorheological elastomer," *Shock Vib.*, vol. 2015, pp. 1–11, 2015.
[38] J. Fu, P. Li, Y. Wang, G. Liao, and M. Yu, "Model-free fuzzy control of a magnetorheological elastomer vibration isolation system: analysis and experimental evaluation," *Smart Mater. Struct.*, vol. 25, no. 3, p. 035030, 2016.
[39] T. Komatsuzaki, T. Inoue, and O. Terashima, "A broadband frequency-tunable dynamic absorber for the vibration control of structures," *J. Phys. Conf. Ser.*, vol. 744, no. 1, p. 012167, 2016.
[40] X. Dong, M. Yu, C. Liao, and W. Chen, "A new variable stiffness absorber based on magneto-rheological elastomer," *Trans. Nonferrous Met. Soc. China*, vol. 19, no. SUPPL. 3, pp. s611–s615, 2009.
[41] J. Fu, M. Yu, X. M. Dong, and L. X. Zhu, "Magnetorheological elastomer and its application on impact buffer," *J. Phys. Conf. Ser.*, vol. 412, p. 012032, 2013.
[42] H. Du, W. Li, and N. Zhang, "Semi-active variable stiffness vibration control of vehicle seat suspension using an MR elastomer isolator," *Smart Mater. Struct.*, vol. 20, no. 10, p. 105003, 2011.
[43] M. Behrooz, X. Wang, and F. Gordaninejad, "Control of structures featuring a new MRE isolator system," in *Active and Passive Smart Structures and Integrated Systems*, 2012, vol. 8341, p. 83411I.
[44] Y. Li, J. Li, W. Li, and B. Samali, "Development and characterization of a magnetorheological elastomer based adaptive seismic isolator," *Smart Mater. Struct.*, vol. 22, no. 3, p. 035005, 2013.
[45] J. Li, Y. Li, W. Li, and B. Samali, "Development of adaptive seismic isolators for ultimate seismic protection of civil structures," in *Proceedings of 2013 SPIE*, 2013, vol. 8692, p. 86920H.

[46] S. Sun, J. Yang, H. Deng, H. Du, and W. Li, "Horizontal vibration reduction of a seat suspension using negative changing stiffness magnetorheological elastomer isolators," *Int. J. Veh. Des.*, vol. 68, no. 1/2/3, p. 104, 2015.

[47] S. Sun, J. Yang, W. Li, H. Deng, H. Du, and G. Alici, "Development of an MRE adaptive tuned vibration absorber with self-sensing capability," *Smart Mater. Struct.*, vol. 24, no. 9, p. 095012, 2015.

[48] S. Sun, H. Deng, J. Yang, W. Li, H. Du, G. Alici, and M. Nakano, "An adaptive tuned vibration absorber based on multilayered MR elastomers," *Smart Mater. Struct.*, vol. 24, no. 4, p. 045045, 2015.

[49] X. Gu, Y. Li, and J. Li, "Investigations on response time of magnetorheological elastomer isolator for real-time control implementation," *Smart Mater. Struct.*, vol. 25, no. 11, p. 11LT04, 2016.

[50] M. Przybylski, S. Sun, and W. Li, "Development and characterization of a multi-layer magnetorheological elastomer isolator based on a Halbach array," *Smart Mater. Struct.*, vol. 25, no. 10, p. 105015, 2016.

[51] W. Li, X. Zhang, and H. Du, "Development and simulation evaluation of a magnetorheological elastomer isolator for seat vibration control," *J. Intell. Mater. Syst. Struct.*, vol. 23, no. 9, pp. 1041–1048, 2012.

[52] Y. Li and Z. Ding, "Research on magnetorheological elastomers air spring based NR/SBR rubber," *Appl. Mech. Mater.*, vol. 219, no. 3, pp. 526–529, 2012.

[53] P. R. Marur, "Magneto-rheological Elastomer-based Vehicle Suspension," US2013/0087985A1, 2013.

[54] Ubaidillah, J. Sutrisno, A. Purwanto, and S. A. Mazlan, "Recent progress on magnetorheological solids: materials, fabrication, testing, and applications," *Adv. Eng. Mater.*, vol. 17, no. 5, pp. 563–597, 2015.

[55] N. Hoang, N. Zhang, and H. Du, "A dual adaptive tunable vibration absorber using MREs for vehicle powertrain vibration control," in *Active and Passive Smart Structures and Integrated Systems*, 2010, vol. 7643, 76430.

[56] N. Hoang, N. Zhang, and H. Du, "An adaptive tunable vibration absorber using a new magnetorheological elastomer for vehicular powertrain transient," *Smart Mater. Struct.*, vol. 20, no. 1, 015019, 2011.

[57] J. D. Carlson, W. Matthis, and J. R. Toscano, "Smart prosthetics based on magnetorheological fluids," in *Smart Structures and Materials: Active Materials: Industrial and Commercial Applications of Smart Structures Technologies*, 2001, vol. 4332, pp. 308–316.

[58] F. Thorsteinsson, I. Gudmundsson, and C. Lecomte, "Prosthetic and Orthotic Devices Having Magnetorheological Elastomer Spring with Controllable Stiffness," US2013/0060349A1, 2013.

[59] F. B. Guðmundsson, "*Preparation and characterization of a prototype magnetorheological elastomer for application in prosthetic devices*," M.Sc. thesis, University of Iceland, Reykjavik 2015.

[60] K. H. Gudmundsson, F. Jonsdottir, and F. Thorsteinsson, "A geometrical optimization of a magneto-rheological rotary brake in a prosthetic knee," *Smart Mater. Struct.*, vol. 19, no. 3, p. 035023, 2010.

[61] M. D. Christie, S. S. Sun, D. H. Ning, H. Du, S. W. Zhang, and W. H. Li, "A torsional MRE joint for a C-shaped robotic leg, "*Smart Mater. Struct.*, vol. 26, no. 1, p. 015002, 2017.

[62] K.-J. Huang, C.-K. Huang, and P.-C. Lin, "A simple running model with rolling contact and its role as a template for dynamic locomotion on a hexapod robot," *Bioinspiration Biomimetics*, vol. 9, no. 4, p. 046004, 2014.

[63] L. Yuan, S. Sun, Z. Pan, D. Ding, and W. Li, "Semi-active chatter reduction for robotic machining using magnetorheological elastomers (MREs)," in *2017 IEEE 7th Annual International Conference on CYBER Technology in Automation, Control, and Intelligent Systems (CYBER)*, 2017, pp. 349–354.

[64] R. Fuhrer, C. M. Schumacher, M. Zeltner, and W. J. Stark, "Soft iron/silicon composite tubes for magnetic peristaltic pumping: frequency-dependent pressure and volume flow," *Adv. Funct. Mater.*, vol. 23, no. 31, pp. 3845–3849, 2013.

[65] S.-Y. Tang, X. Zhang, S. Sun, *et al.*, "Versatile microfluidic platforms enabled by novel magnetorheological elastomer microactuators," *Adv. Funct. Mater.*, vol. 28, no. 8, p. 1705484, 2018.

[66] H. Böse, R. Rabindranath, and J. Ehrlich, "Soft magnetorheological elastomers as new actuators for valves," *J. Intell. Mater. Syst. Struct.*, vol. 23, no. 9, pp. 989–994, 2012.

[67] S. Bednarek, "The giant magnetostriction in ferromagnetic composites within an elastomer matrix," *Appl. Phys. A Mater. Sci. Process.*, vol. 68, no. 1, pp. 63–67, 1999.

[68] S. Bednarek, "The giant volumetric magnetostriction of ferromagnetic composites with elastomer matrix," *Int. J. Mod. Phys. B*, vol. 13, no. 24, pp. 865–878, 1999.

[69] X. J. Wang, F. Gordaninejad, M. Calgar, Y. M. Liu, J. Sutrisno, and A. Fuchs, "Sensing behavior of magnetorheological elastomers," *J. Mech. Des.*, vol. 131, pp. 1–6, 2009.

[70] I. Bica, "Magnetoresistor sensor with magnetorheological elastomers," *J. Ind. Eng. Chem.*, vol. 17, no. 1, pp. 83–89, 2011.

[71] I. Bica, Y. D. Liu, and H. J. Choi, "Magnetic field intensity effect on plane electric capacitor characteristics and viscoelasticity of magnetorheological elastomer," *Colloid Polym. Sci.*, vol. 290, no. 12, pp. 1115–1122, 2012.

[72] G. Ausanio, V. Iannotti, E. Ricciardi, and L. Lanotte, "Magneto-piezoresistance in magnetorheological elastomers for magnetic induction gradient or position sensors," *Sensors Actuators A Phys.*, vol. 205, pp. 235–239, 2014.

[73] L. Ding, S. Xuan, J. Feng, and X. Gong, "Magnetic/conductive composite fibre: a multifunctional strain sensor with magnetically driven property," *Composites, Part A*, vol. 100, pp. 97–105, 2017.

[74] T. Kawasetsu, T. Horii, H. Ishihara, and M. Asada, "Mexican-hat-like response in a flexible tactile sensor using a magnetorheological elastomer," *Sensors*, vol. 18, no. 2, p. 587, 2018.

Chapter 13

Development of a hybrid nonlinear vibration absorber working with MR elastomers

Weihua Li[1] and Shuaishuai Sun[1]

Absorbers have been commonly used in the engineering industry aiming at absorbing vibrational energy from machinery or structure [1–4]. The concept of vibration absorber was first proposed by Frahm [5]. Consisting of a mass and stiffness element with appropriate damping, the vibration absorber is a simple, economic, and reliable choice for vibration suppression. The working principle of an absorber is to transfer the vibrational energy of the controlled objects to the oscillators by equaling its natural frequency to the frequency of the disturbing vibration. Therefore, the effectiveness of an absorber lies in the natural frequency bandwidth that it achieves. With slightly widening the operation bandwidth, the performance improves drastically. To improve the functionality of a vibration absorber, broadening the operation bandwidth is an important issue, as the vibration sources in the practical circumstances are usually frequency variant or cover a large frequency range.

In order to enlarge the effective bandwidth of the absorber, active absorbers and semi-active absorbers have been developed [6,7]. However, the active devices become less practical due to their high requirements in terms of power, cost, and maintenance. To the contrary, semi-active absorbers become more widely accepted because of simplicity, robustness, and reliability. Magnetorheological elastomers (MREs), as a smart material whose stiffness is controllable ([8–11]), are currently prevalent as the involvement of MREs endows the absorber adaptability to be tuned in real time by changing the stiffness [12–15]. This semi-active property enables the absorber to adapt to vibration sources which are frequency variant. Deng developed an MRE absorber working in shear mode, whose natural frequency can vary from 55 to 82 Hz [16,17]. Subsequently, Gong's group applied a voice motor on MRE absorber to counteract the effect of damping so as to overcome the damping effect and increase the absorber's effectiveness [18,19]. As an effort to further improve and complete the MRE-based absorber, Sun *et al.* developed a multilayered MRE absorber which is much more capable of dealing with vibrations

[1]School of Mechanical, Materials, Mechatronic and Biomedical Engineering, University of Wollongong, Australia

with larger amplitudes [20,21]. The effective frequency bandwidth of the MRE absorber under certain constant currents, however, is still narrow although it can tune its natural frequency to track the excitation frequency. In other words, the MRE absorber can attenuate the vibration with single frequency well but cannot suppress the vibration with multiple frequencies desirably because the MRE absorber can match only one frequency at one time. So it is necessary to find a way to overcome this weakness of MRE absorber.

To address the limitations of the MRE absorber, involving nonlinear stiffness to the absorber is a promising method. The passive nonlinear absorber has been fully verified to be more effective and has wider effective frequency bandwidth than its linear counterpart. Roberson [22] did the pioneer research on an undamped nonlinear dynamic vibration absorber which composes a linear and a hardening cubic spring. The investigation results illustrate that the vibration suppression bandwidth of the nonlinear absorber is much wider than a linear one. This conclusion was experimentally verified by Arnold [23]. Following Roberson's research, Snowdon [24] further improved the performance of the nonlinear absorber by involving a fixed damping factor and a spring whose stiffness is proportional to excitation frequency. The results have confirmed that the improved nonlinear absorber performs better on reducing the resonance of the primary structure. Li applied the magnets to energy harvesting system, which can be used as an absorber as well. The experimental results verified that the magnet-system-induced nonlinearity can broaden the effective frequency range [25]. Many other excellent works on nonlinear absorber have been done in recent years [26,27] and have fully confirmed that the effective frequency bandwidth of a passive nonlinear absorber is wider than passive linear absorber. As a result, introducing nonlinearity to MRE absorber can realize adaptive nonlinear vibration absorber which will not only have wider working bandwidth under each fixed current but also can shift the resonance to adapt to the excitation variation. This is the motivation of this work.

In order to introduce nonlinearity to the MRE absorber, two opposite polar magnets were introduced to the linear MRE absorber. The combination of the MRE structure and the permanent magnets contributes to coexist of adaptability and nonlinearity. The detailed configuration of the absorber is shown in Section 13.1. Experimental results verifying the nonlinearity and adaptability are shown in Section 13.2. Section 13.3 evaluates the absorption effectiveness of the absorber. Section 13.4 summarizes this chapter.

13.1 Structure, working principle, and analyses of the MRE absorber

13.1.1 Structure

Based on the design of the linear multilayered MRE absorber, nonlinear MRE absorber is developed by introducing magnetic interaction. As is shown in Figure 13.1, it is mainly composed of two parts: the MRE structure in the left of the

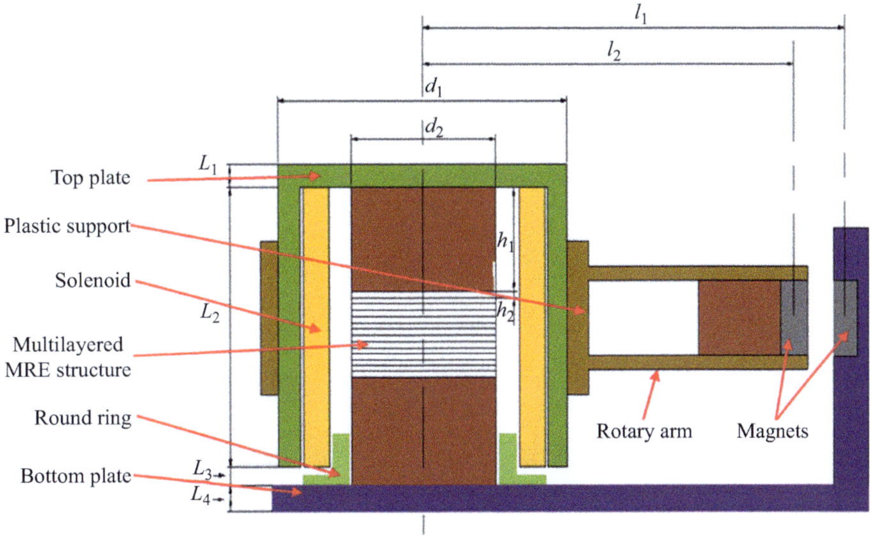

Figure 13.1 Structure of the hybrid MRE absorber

figure and the nonlinearity generation part in the right. The weight ratio of the iron particles, silicon rubber, and silicon oil used for MRE fabrication is 7:2:1. The MRE structure has a solenoid and a multilayered MRE inside the steel yoke, vertically. Surrounding the steel yoke, a plastically made supporter is fixed to the yoke as a connection to the rotary arm. The nonlinearity is generated by the relative motion of the two permanent magnets. One of the magnets is connected to the MRE structure through the rotary arm. The other is embedded into the plastic structure which is fixed to the bottom plate of the absorber. The two magnets generate attractive force between them. A round ring is mounted between the bottom plate and the solenoid to guarantee the MRE absorber works in an angular direction. The detailed size of the nonlinear MRE absorber is shown in Table 13.1. The m_1 and m_2 are the magnetic moments of the two magnets. The prototype of the hybrid MRE absorber is shown in Figure 13.2.

13.1.2 Working principle

Generally speaking, the proposed hybrid MRE absorber works in the angular direction. Its angular motion will be used to absorb the vibration energy from the control object. In terms of the MRE structure, the multilayered structure enables the MRE structure to have larger angular flexibility, which ensures the nonlinearity appearance [10,20]. Moreover, the presence of steel sheets enhances the magnetic induction of MRE and meantime strengthens the vertical support. Having had all these configurations ready, when the solenoid is powered, the whole structure can realize the adaptability because of the MRE's stiffness variability in response to the variations of magnetic field.

Table 13.1 Parameters of nonlinear MRE absorber

Parameters	Values	Parameters	Values
h_1	45 mm	L_1	5 mm
h_2	1 mm	L_2	100 mm
d_1	90 mm	L_3	5 mm
d_2	35 mm	L_4	5 mm
l_1	120 mm	coil	1,000 turns
m_1	0.258	l_2	110 mm
m_2	0.258	J	0.0151 kg m^2

Figure 13.2 The prototype of the hybrid MRE absorber

On the other hand, the pair of magnets is responsible for introducing nonlinearity to the system. The magnet-induced nonlinearity will significantly change the behavior and the performance of the absorber. As the permanent magnet m_1 was fixed to the bottom plate, it remains relatively motionless with bottom plate under external excitation while the magnet m_2 will move angularly around the center point of the MRE structure O shown in Figure 13.3 because of the angular flexibility of the MRE structure. The distance between the two magnets keeps changing, thereby generating a nonlinear magnetic force which in turn influences the movement of the MRE structure. In order to clearly demonstrate the nonlinear force generated by the two magnets, Figure 13.3 illustrates the dynamic interactions between the two magnets, where $\hat{r}$ is a unit vector parallel to the line joining the centers of the two dipoles, and $|r|$ is the distance between the centers of m_1 and m_2. m_1, m_2, and l_2 remain constant while a, b, and $\hat{r}$ change dynamically. The magnetic force can be calculated by (13.1) according to the classic dipole–dipole magnetic interaction [28]:

$$F = \frac{3\mu_0}{4\pi|r|^4}\Big((\hat{r}\times m_1)\times m_2 + (\hat{r}\times m_2)\times m_1 - 2\hat{r}(m_2\cdot m_1) + 5\hat{r}\left((\hat{r}\times m_1)\cdot(\hat{r}\times m_2)\right)\Big) \tag{13.1}$$

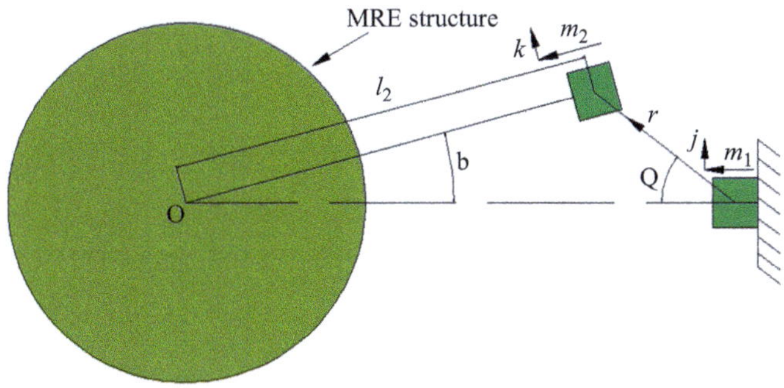

Figure 13.3 The magnetic interaction of the two magnets

According to the geometry relationship in Figure 13.3, the magnetic interaction force in unit vector k direction can be derived into

$$F = \frac{3\mu_0}{4\pi|r|^4}(|m_1||m_2|\sin(a) + |m_1||m_2|\sin(a+b)\cos(b) + 2|m_1||m_2|\sin(a+b)\cos(b) - 5|m_1||m_2|\sin(a+b)\sin(a)\sin(a+b)) \tag{13.2}$$

Based on the above expression, the relationship between the magnetic force and the relative angle b was simulated and presented in Figure 13.3. The parameters used in this simulation are based on the real size of the hybrid MRE absorber, shown in Table 13.1. It is seen that the relationship is nonlinear entirely; however, linearity exists in the certain region. The relationship between the force and the relative angle tends to be linear when the excitation is not fierce enough. Specifically, the AB segment shown in Figure 13.4 has a linear stiffness. When the excitation keeps increasing, the angle b will exceed AB segment and the nonlinearity will start to appear.

In summary, the working principle of the hybrid MRE absorber is the MRE structure provides adaptability for the hybrid MRE absorber allowing it to track the changing excitation frequency while the nonlinearity widens the effective frequency bandwidth of the hybrid MRE absorber under each constant current.

13.1.3 Analysis of the MRE absorber

In order to provide a fundamental resonance prediction for the MRE absorber design, theoretical derivation and analysis are provided as follows.

Based on reference [20], the increased shear modulus of MRE under different magnetic fields can be obtained by

$$\Delta G = 36\phi\mu_o\mu_p\beta^2\vec{H}_o^2\left(\frac{a}{d}\right)^3\zeta \tag{13.3}$$

where $\mu_p \approx 1,000$ and $\mu_l \approx 1$ are the relative permeability of the iron particles and silicon rubber, respectively, and μ_o is the vacuum permeability. $\beta = (\mu_p - \mu_l)/(\mu_p + 2\mu_l) \approx 1$, a is the average particle radius, ϕ is the volume

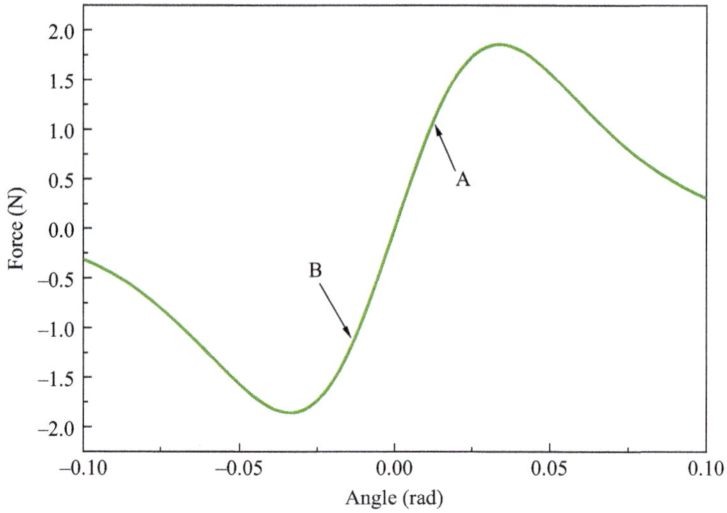

Figure 13.4 Relationship between the magnetic force and angle b

fraction, d is the particle distance before deflection, $\vec{H}_o$ is the intensity of the applied magnetic field, and $\zeta = \sum_{j=1}^{n} \frac{1}{j^3} \approx 1.202$.

The torsional stiffness of the ith MRE sheet can be written as

$$K_{ri} = \frac{GI_p}{L} = \frac{(G_o + \Delta G)I_p}{L} = \frac{\left(G_o + 36\,\phi\mu_o\mu_l\beta^2\vec{H}_o^2\left(\frac{a}{d}\right)^3\zeta\right)I_p}{L} \tag{13.4}$$

where G_o is the initial shear modulus of MRE, G is the overall shear modulus, and L is the thickness of the MRE sheet. I_p is the polar moment of inertia of the MRE layer and can be calculated by

$$I_p = \pi d_2^4/32 \tag{13.5}$$

where d_2 is the diameter of the MRE layer.

The torsional stiffness of the whole laminated MRE pillar (totally 10 layers MRE sheet) can be calculated by the following equation:

$$\frac{1}{k_{rw}} = \sum_{i=1}^{10}\frac{1}{k_{ri}} + \sum_{j=1}^{10}\frac{1}{k_{rj}} \tag{13.6}$$

where k_{rw} is the overall torsional stiffness of the laminated MRE pillar, and k_{rj} is the torsional stiffness of steel sheet. It is known that $k_{ri} \ll k_{rj}$; therefore k_{rw} can be simplified as

$$k_{rw} = \frac{k_{ri}}{10} \tag{13.7}$$

The linear torsional stiffness k_{rm} provided by the permanent magnets can be calculated based on AB segment shown in Figure 13.4, which is 8.7 N m/rad.

The natural frequency of the laminated MRE absorber in angular direction can be written as

$$f_r = \frac{1}{2\pi}\sqrt{\frac{k_{rm} + k_{rw}}{J}} \tag{13.8}$$

where J is the moment of inertia of the oscillator.

The magnetic field $\vec{H}_o$ in (13.3) can be calculated by the calculation method in [20]:

$$NI = B_{\mathrm{MRE}} S_{\mathrm{MRE}} R_m \tag{13.9}$$

$$B_{\mathrm{MRE}} = \mu_{\mathrm{MRE}} \vec{H}_o \tag{13.10}$$

$$R_m = R_1 + R_2 + R_3 + R_4 + R_5 \tag{13.11}$$

where NI is the magnetomotive force of coil, B_{MRE} is the magnetic flux density passing through MRE, and S_{MRE} is the flux area of MRE. $R_1, R_2, R_3, R_4,$ and R_5 are the magnetic resistances of ten MRE layers, two steel pillars, air gap, steel yoke, and top and bottom plates, respectively.

Thus, the frequency shift property of the innovative MRE absorber in angular direction can be calculated and is plotted in Figure 13.8.

13.2 Experiment test of the hybrid MRE absorber

13.2.1 Experiment setup

In this section, experimental tests were conducted to study the dynamic properties of the proposed nonlinear MRE absorber. Figure 13.5 shows the experimental setup. The absorber was fixed onto the vibration platform driven by a shaker (VTS, .VC 100-8). Two accelerometers (CA-YD-106) were used to measure the accelerations of the excitation and the oscillator, respectively. Please be noted the accelerometer for the oscillator is mounted on the rotary arm to measure the acceleration in angular direction. A DC power supply (Thurlby Thandar Instruments Ltd.) was used to provide current to the absorber solenoid. A data acquisition board was used as the interface between the hardware and the software and transferred the measured accelerations to the computer. The signal collection, recording, and processing systems were developed using the LabVIEW program.

When the shaking table is driven to vibrate horizontally, the MRE structure starts to rotate around its center, and as a result, the attached magnet will oscillate. Therefore, the distance between the two magnets will change and nonlinearity will appear. In this part, the influence of excitation amplitude on the MRE absorber was first investigated. Then the frequency shift performance under both linear case and nonlinear case were evaluated.

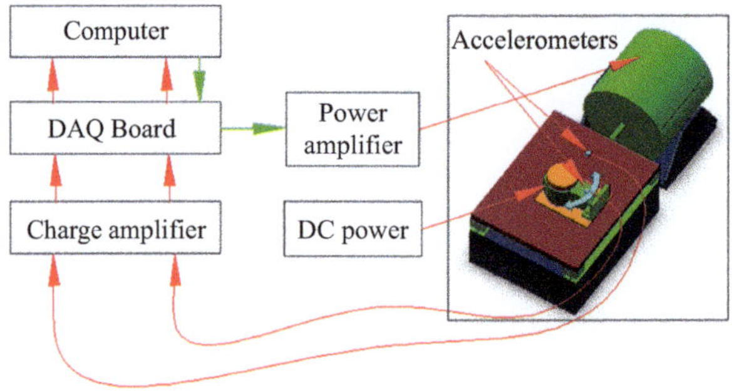

Figure 13.5 Experimental setup for characterizing the absorber

13.2.2 Testing result

13.2.2.1 Nonlinearity verification

In this amplitude-dependent investigation, harmonic sinusoidal signal with sweeping frequency from 3 to 5 Hz was used. The current to the coil is remained to be zero during this test. Two cases with different excitation amplitude were considered. It can be seen from Figure 13.6 that the acceleration response is linear when the excited amplitude is 1.5 m/s^2. In this linear case, the magnetic interaction works in linear AB range shown in Figure 13.4. The resonance frequency for this case is around 4 Hz. However, when the excited amplitude increases to 2.5 m/s^2, the acceleration response tends to be nonlinear and it can keep resonating for a wider frequency range. The abrupt increase/decrease of the acceleration demonstrates the nonlinearity. In this nonlinear working mode, the magnet interaction shown in Figure 13.4 will be beyond AB range and tend to be nonlinear. An obvious advantage can be observed from this phenomenon that the effective frequency bandwidth of the proposed nonlinear absorber will increase when the excitation becomes fiercer.

Another noteworthy finding from Figure 13.6 is that the acceleration response changes when a reverse sweep excitation signal was used. Reverse testing means the sweeping frequency decreases from 5 to 3 Hz. As observed from Figure 13.6, the acceleration response of the linear case under forward sweep matches well with that under reverse sweep. However, there is an obvious difference between the two curves of nonlinear case. It is seen that the acceleration under the reverse sweep excitation drops abruptly after it reaches the peak value and it covers a wider effective frequency bandwidth than that under forward sweep. The difference between the reverse test and forward test further verifies that the proposed MRE absorber performs nonlinearity.

13.2.2.2 Adaptability verification

The former paragraphs have introduced the appearance of nonlinearity because of the incorporation of the permanent magnets. Then in this part, adaptability of this

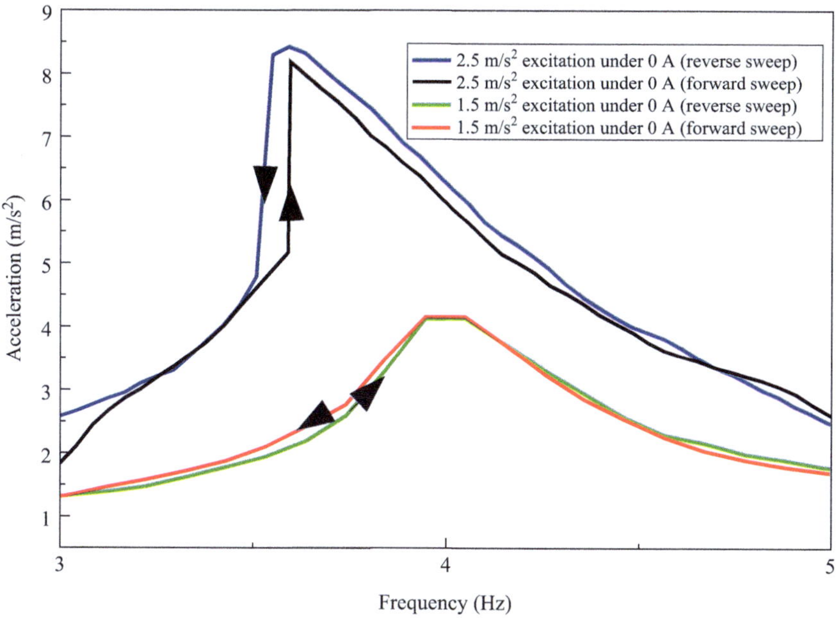

Figure 13.6 Performance of the absorber under different amplitudes

absorber working on both the linear mode and the nonlinear mode will be investigated. As mentioned earlier, it is the multilayered MRE structure that plays a significant role in making the adaptability possible. Figures 13.7 and 13.8 show the frequency shift responses for the linear mode and nonlinear mode, respectively.

In Figure 13.7, a sinusoidal signal with a frequency range sweeping from 3.5 to 5.5 Hz with the amplitude of 1.5 m/s^2 was used to drive the hybrid MRE absorber. The acceleration response under various current levels from 0 to 4 A was collected. It can be seen that the responses are all linear. The resonance frequency increases from 3.9 to 4.9 Hz as the current grows from 0 to 4 A, which demonstrates the adaptability under the linear working mode. The relationship between the current and natural frequency of the MRE absorber working in linear mode is plotted in Figure 13.8. From this figure, it can be seen that the model presented in Section 13.1.3 can predict the frequency shift property very well, which provides a fundamental design procedure for the MRE absorber.

Likewise, Figure 13.9 shows the frequency shift response of the absorber in the nonlinear working mode. The excited amplitude was 2.5 m/s^2 and the applied current was changed from 0 to 4 A with a step of 2 A. For each current level, the acceleration changes abruptly at the resonance frequency. Generally, the resonance frequency shifts from 3.5 to 4.6 Hz under the nonlinear working mode. In summary, both the linear working mode MRE absorber and the nonlinear working mode can be controlled by the current.

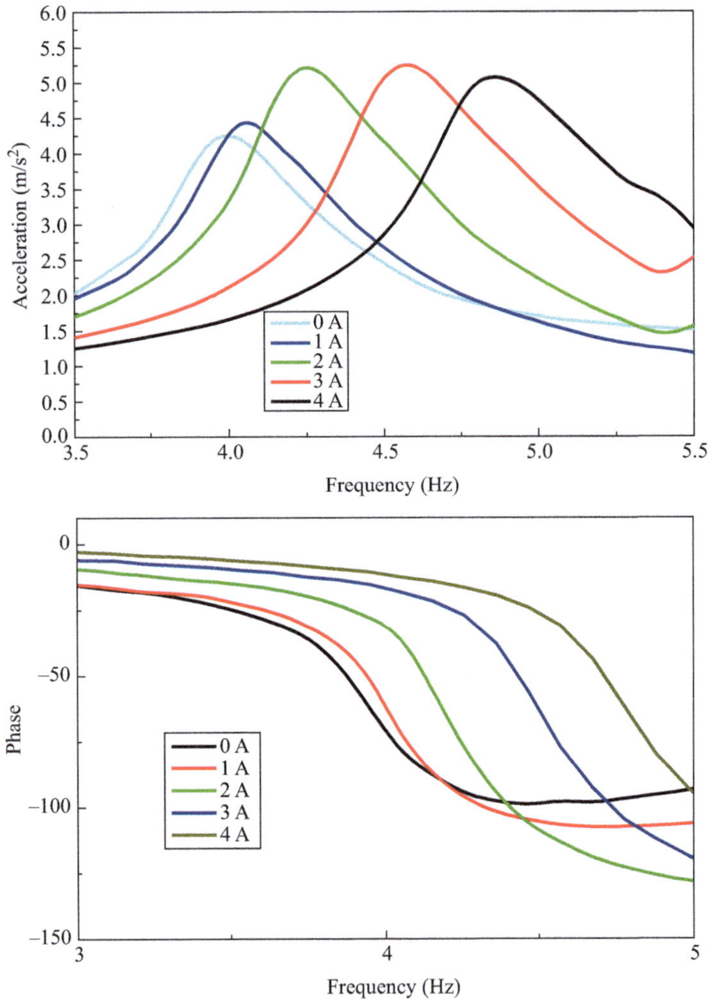

Figure 13.7 Frequency shift performance under 1.5 m/s²

13.3 Vibration reduction evaluation

Section 13.2 fully verified the nonlinearity and adaptive capability of the proposed MRE absorber. This section aims to evaluate its vibration reduction capability. Figure 13.10 shows the experimental setup. This testing system is similar to that introduced in Section 13.3 since the same hardware and software were used, with only one exception that a primary system as the control object was included in this system. This primary system is a mass spring system and consists a steel plate mounted on two linear bearings and four springs providing the lateral stiffness. The primary system was fixed to the shaking table. Then the nonlinear MRE absorber

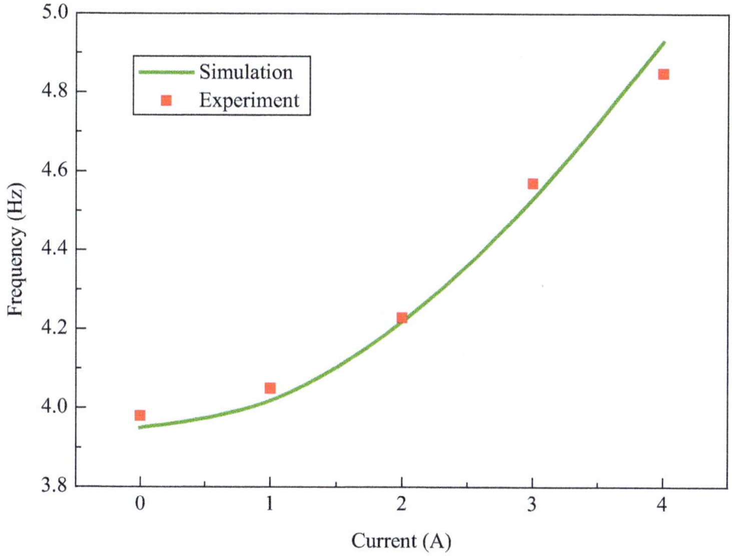

Figure 13.8 The relationship between the natural frequency and current under linear mode

was fixed on the top of the steel plate. Two accelerometers acquired the accelerations of the controlled plated and the excitation, respectively. The ratio of the object's acceleration to the excited acceleration was calculated and shown in the computer as transmissibility so as to evaluate the vibration reduction effectiveness. Unlike in the Section 13.3, desired scenarios demonstrated by the transmissibility would be such time when the value of transmissibility reaches the lowest. It is to say the object's acceleration would be furthest suppressed when the transmissibility reaches the minimum which is the optimal vibration absorption point of the absorber.

13.3.1 Performance of the MRE absorber under different amplitudes

The advantage of the nonlinear phenomenon in terms of the absorption ability is demonstrated in Figure 13.11. This figure shows the transmissibility responses under different excitation amplitudes with 4 A current, among which, one of them under the amplitude of 1 m/s^2 tends to be linear while the other two show nonlinear tendency. The sweep frequency is in reverse way from 5 to 3.5 Hz. It can be obviously observed that both linear and nonlinear working mode absorbers show good vibration absorption performance. An obvious difference of these two working modes, however, did exist as well. The transmissibility of the linear working mode absorber decreases from point a_1 to a_2, which means the absorber starts to work and absorb the vibration energy from the object. After that, the transmissibility increases to a high level gradually when the frequency exceeds the natural frequency of

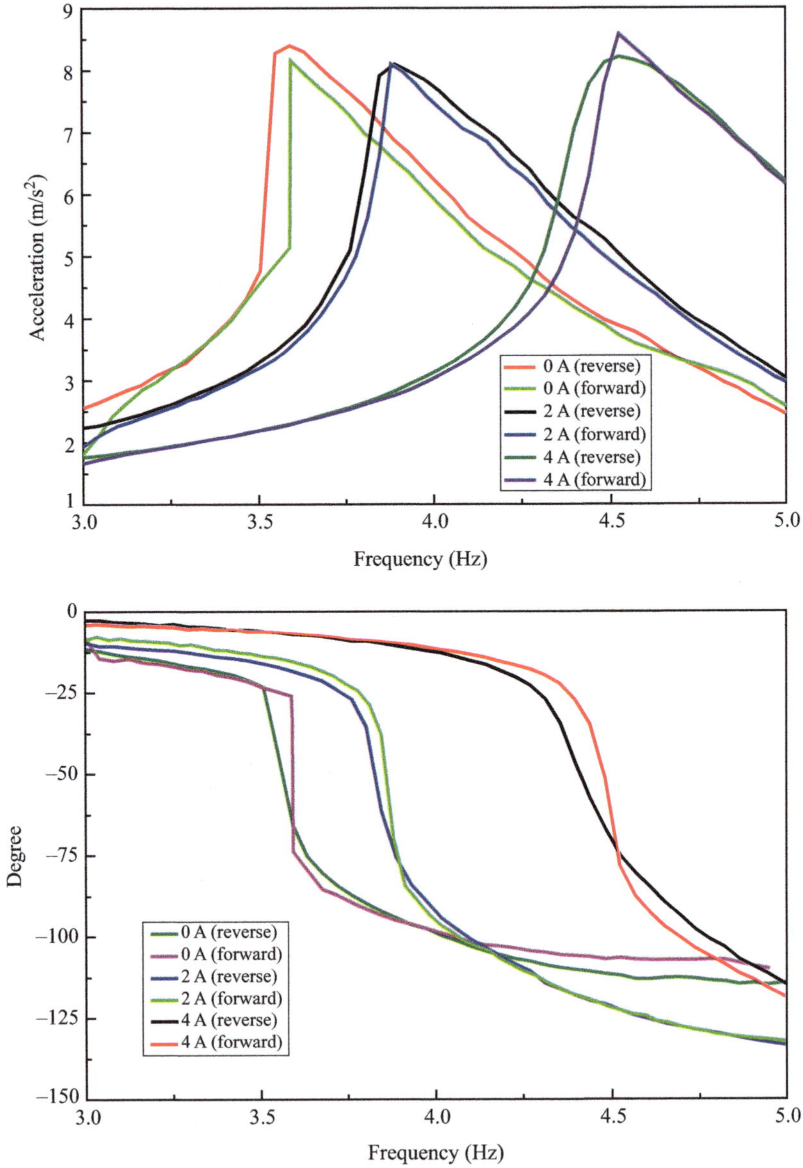

Figure 13.9 Frequency shift performance under 2.5 m/s² excitation

the absorber. In terms of the nonlinear working mode absorber, however, the oscillator will keep resonating until a much lower frequency and will sharply jump to a higher level from point b_1 to b_2 or from c_1 to c_2. The reason for this is that the oscillator working in linear mode starts and stops moving gradually while the

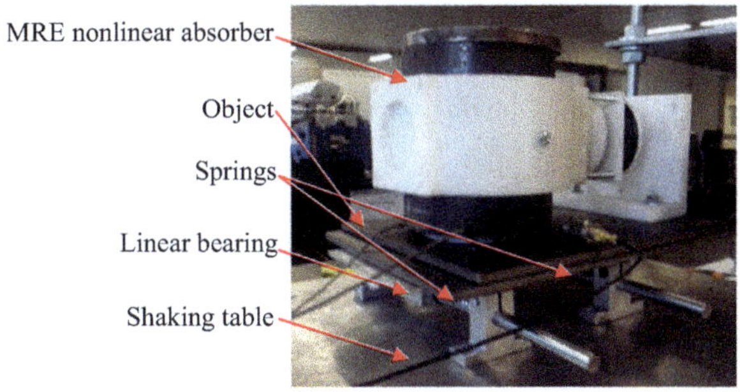

Figure 13.10 Vibration attenuation evaluation system

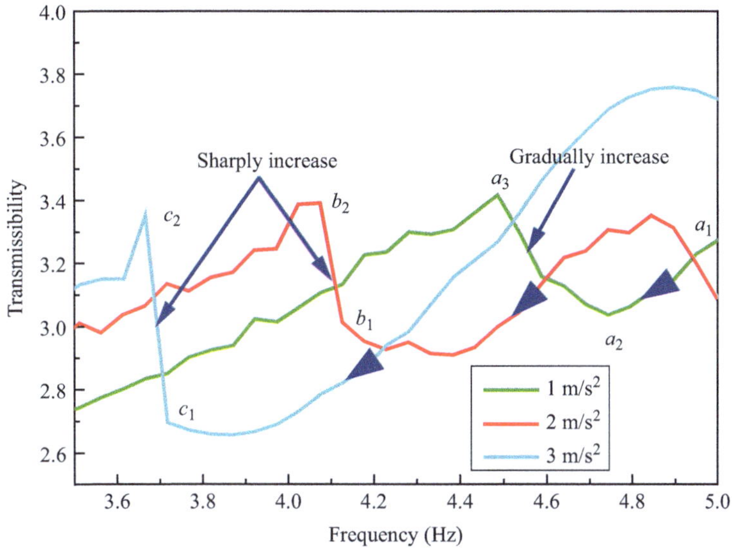

Figure 13.11 The vibration reduction performance under different amplitudes

nonlinear oscillator working in nonlinear mode begins the vibration gradually but stops sharply. This phenomenon can be verified in Figure 13.6. Another conclusion can be drawn from Figure 13.11 that the effective bandwidth of the absorber becomes wider when it works on nonlinear mode because the absorption frequency range in Figure 13.11 is wider when the excitation increases from 1m/s^2 to 2 and 3 m/s^2.

To conclude, the most innovative point of this proposed nonlinear MRE absorber is the synthesis of the controllability of MRE and the nonlinear characteristics of magnetic interaction with a goal to contribute to a more effective absorber. Its nonlinearity and adaptability have been verified through the experimental testing.

13.3.2 The performance comparison between passive absorber and controlled absorber

Apart from analyzing the impact of different amplitudes on the absorption effectiveness, absorption transmissibility under different current levels were also measured and analyzed. A series of absorption transmissibility under 0, 1, 2, and 3 A with 1.8 m/s^2 excitation were tested under reverse sweep frequency and are shown in Figure 13.12. It is seen that the absorber shows adaptability that the lowest point shifted to the right with respect to the x-axis. Obviously, the frequency shift is induced by the field dependence of MRE on the magnetic field. Another important observation to be noted is that the absorption transmissibility shows nonlinearity as all the curves increase abruptly after reaching the lowest point. As the lowest point indicates the moment when the absorber plays its utmost role in suppressing the vibration, it would be desirable if there is an optimal absorption transmissibility that connects all the lowest points. The semi-active control algorithm is actually responsible for controlling the stiffness of the MRE to track the lowest points. In the semi-active case, a short-time Fourier transform (STFT) control algorithm was used. The acquired acceleration signals were processed by the STFT algorithm to obtain the excitation frequency. This algorithm produces a current output by equaling the natural frequency of the absorber to the excitation frequency. Upon receiving the current signal, the absorber will adjust its stiffness to track the excitation frequency. As shown by the red line in Figure 13.12, the transmissibility of the controlled absorber remains at the lowest level through all the frequency range, which means the semi-active MRE absorber outperforms all passive absorbers.

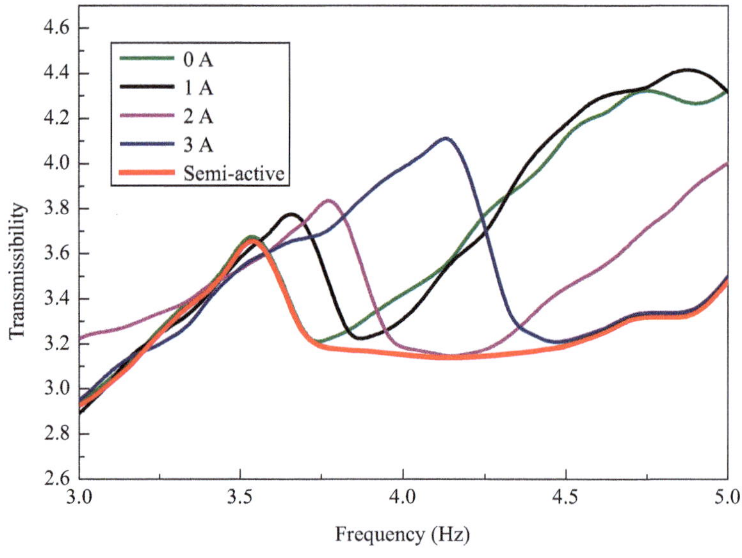

Figure 13.12 The comparison between passive absorber and semi-active absorber

13.4 Conclusion

This study proposes to incorporate magnetic nonlinearity into the linear MRE absorber, aiming at further broadening its operating bandwidth under each constant current. The test results under different amplitudes show that the absorber performs nonlinearity and its effective frequency bandwidth will be enlarged when it works on nonlinear mode. Then the adaptability for the linear case and the nonlinear case were both verified and evaluated. Afterwards, the absorption effectiveness of this nonlinear MRE absorber on vibration control was experimentally carried out. The vibration absorption evaluation also demonstrates that the nonlinear working mode absorber has larger effective bandwidth. And finally an STFT was used to control the nonlinear MRE absorber to track the excitation frequency variation. The test result demonstrates that the hybrid MRE absorber can also tune its stiffness to trace the excitation frequency and performs better than passive absorbers.

References

[1] Liu, K. & Liu, J. 2005. The damped dynamic vibration absorbers: revisited and new result. Journal of Sound and Vibration, 284, 1181–1189.

[2] Weber, F. 2014. Semi-active vibration absorber based on real-time controlled MR damper. Mechanical Systems and Signal Processing, 46, 272–288.

[3] Yang, Z., Qin, C., Rao, Z., Ta, N. & Gong, X. 2014. Design and analyses of axial semi-active dynamic vibration absorbers based on magnetorheological elastomers. Journal of Intelligent Material Systems and Structures, 25, 2199–2207.

[4] Lee, E., Nian, C. & Tarng, Y. 2001. Design of a dynamic vibration absorber against vibrations in turning operations. Journal of Materials Processing Technology, 108, 278–285.

[5] Fraham, H. 1911. Device for Damping Vibration of Bodies. U.S. Patent No. 989958.

[6] Davis, C. L. & Lesieutre, G. A. 2000. An actively tuned solid-state vibration absorber using capacitive shunting of piezoelectric stiffness. Journal of Sound and Vibration, 232, 601–617.

[7] Walsh, P. & Lamancusa, J. 1992. A variable stiffness vibration absorber for minimization of transient vibrations. Journal of Sound and Vibration, 158, 195–211.

[8] Xing, Z.-W., Yu, M., Fu, J., Wang, Y. & Zhao, L.-J. 2015. A laminated magnetorheological elastomer bearing prototype for seismic mitigation of bridge superstructures. Journal of Intelligent Material Systems and Structures, 1045389X15577654.

[9] Yang, J., Du, H., Li, W., LI, Y., LI, J., Sun, S. & Deng, H. 2013. Experimental study and modeling of a novel magnetorheological elastomer isolator. Smart Materials and Structures, 22, 117001.

[10] Yang, J., Sun, S., Du, H., Li, W., Alici, G. & Deng, H. 2014a. A novel magnetorheological elastomer isolator with negative changing stiffness for vibration reduction. Smart Materials and Structures, 23, 105023.

[11] Li, Y. & Li, J. 2015. Finite element design and analysis of adaptive base isolator utilizing laminated multiple magnetorheological elastomer layers. Journal of Intelligent Material Systems and Structures, 26, 1861–1870.

[12] Li, Y., Li, J., Li, W. & Du, H. 2014. A state-of-the-art review on magnetorheological elastomer devices. Smart Materials and Structures, 23, 123001.

[13] Popp, K. M., KR Ger, M., Hua Li, W., Zhang, X. Z. & Kosasih, P. B. 2010. MRE properties under shear and squeeze modes and applications. Journal of Intelligent Material Systems and Structures, 21, 1471–1477.

[14] Sun, S., Deng, H., Yang, J., Li, W., Du, H. & Alici, G. 2015a. Performance evaluation and comparison of magnetorheological elastomer absorbers working in shear and squeeze modes. Journal of Intelligent Material Systems and Structures, 26, 1757–1763.

[15] Ginder, J. M., Schlotter, W. F. & Nichols, M. E. Magnetorheological elastomers in tunable vibration absorbers. SPIE's 8th Annual International Symposium on Smart Structures and Materials, 2001. International Society for Optics and Photonics, 103–110.

[16] Deng, H.-X. & Gong, X.-L. 2008. Application of magnetorheological elastomer to vibration absorber. Communications in Nonlinear Science and Numerical Simulation, 13, 1938–1947.

[17] Deng, H.-X., Gong, X.-L. & Wang, L.-H. 2006. Development of an adaptive tuned vibration absorber with magnetorheological elastomer. Smart Materials and Structures, 15, N111.

[18] Liao, G., Gong, X., Kang, C. & Xuan, S. 2011. The design of an active-adaptive tuned vibration absorber based on magnetorheological elastomer and its vibration attenuation performance. Smart Materials and Structures, 20, 075015.

[19] Peng, C. & Gong, X. L. 2013. Active-adaptive vibration absorbers and its vibration attenuation performance. Applied Mechanics and Materials. Trans Tech Publ, 312, 262–267.

[20] Sun, S., Deng, H., Yang, J., *et al.* 2015. An adaptive tuned vibration absorber based on multilayered MR elastomers. Smart Materials and Structures, 24, 045045.

[21] Sun, S., Yang, J., Li, W., Deng, H., Du, H. & Alici, G. 2015. Development of an MRE adaptive tuned vibration absorber with self-sensing capability. Smart Materials and Structures, 24, 095012.

[22] Roberson, R. E. 1952. Synthesis of a nonlinear dynamic vibration absorber. Journal of the Franklin Institute, 254, 205–220.

[23] Arnold, F. 1955. Steady-state behavior of systems provided with nonlinear dynamic vibration absorbers. Journal of Applied Mechanics, 22, 487–492.

[24] Snowdon, J. 1959. Steady-state behavior of the dynamic absorber. The Journal of the Acoustical Society of America, 31, 1096–1103.

[25] Tang, L. & Yang, Y. 2012. A nonlinear piezoelectric energy harvester with magnetic oscillator. Applied Physics Letters, 101, 094102.

[26] Starosvetsky, Y. & Gendelman, O. 2008. Dynamics of a strongly nonlinear vibration absorber coupled to a harmonically excited two-degree-of-freedom system. Journal of Sound and Vibration, 312, 234–256.
[27] Gendelman, O. V., Gourdon, E. & Lamarque, C.-H. 2006. Quasiperiodic energy pumping in coupled oscillators under periodic forcing. Journal of Sound and Vibration, 294, 651–662.
[28] Levitt, M. 2001. Spin Dynamics: Basics of Nuclear Magnetic Resonance. Chichester: John Wiley & Sons.

Chapter 14

Development of smart base isolation system for civil structures utilising magnetorheological elastomer

Yancheng Li[1,2] and Jianchun Li[2]

This chapter details the development of a smart base isolation system for seismic protection of civil structures utilising magnetorheological (MR) elastomer (MRE). Critical components of the development of the system, i.e. adaptive base isolator, control design and shake table testing, are described. Finally, shake table testing is conducted to verify the effectiveness of the proposed smart-base isolation system.

14.1 Introduction

Base isolation, one of the most widely accepted and applied seismic protection techniques, mitigates the effects caused by hazardous earthquake events by decoupling the superstructure from ground motion [1]. Since it is the transverse wave of the seismic motion that jeopardise the structural elements and building contents [2], the base isolation system provides lateral flexibility to the building to lower the stiffness of the link between the building and ground. As a result, the natural frequencies of the isolated structure are shifted by base isolation system, and thus energy transmitted into the superstructure can be greatly reduced [3].

Extensive studies on traditional base isolation systems have bred a mature and diverse collection of base isolators [1], and each kind of isolator has its own scope of application and inherent shortcomings. For example, laminated rubber bearing (LRB) usually suffers from excessive base displacement under extreme earthquake inputs; Electricite-de-France base isolator may encounter potential hazard of unacceptable permanent horizontal displacement due to the lack of restoring force in the friction unit; friction pendulum system can only cope with certain types of earthquake due to its fixed isolation period is only correlated to the radius of the concave curvature. Therefore, practical base isolation design is achieved through the optimisation depending on designated superstructure, type of the soil or foundation, historical

[1]College of Civil Engineering, Nanjing Tech University, China
[2]School of Civil and Environmental Engineering, University of Technology Sydney, Australia

earthquake characteristics, etc. Such optimisation, on the other hand, compromises certain required parameters in order to seek for robust performance of the system. For instance, the near-fault seismic activities, which features intense long-period velocity waves (usually with an amplitude of 0.5 m/s and period range between 2 and 4 s) [4], can be detrimental and induce excessive displacement response in the conventionally base-isolated structures with a period in this range [5]. One approach to allocate large displacement is to use larger base isolator, quite commonly implemented in Japan, which is effective but not economical [6]. Another usual approach is to use high damping isolator, i.e. high damping rubber bearing and New Zealand (or lead core) rubber bearing or additional viscous dampers in the base isolation system. However, high damping system that can cope with large displacement may cause large acceleration response and be less effective under moderate earthquakes [7]. More accurately, the addition of damping, although reduces displacement, is at the expense of increasing the internal motion of superstructure (normally shown as inter-storey drift) as well as absolute accelerations. As a matter of fact, minimising the structural acceleration and displacement is recognised as an irreconcilable conflict, of which displacement indicates structural damage, while acceleration introduces condensed damage to the non-structural elements [2]. Hence, it is beneficial if there is a type of base isolator that can adjust itself real-timely for optimal performance at each time instant without compromising either or both of the acceleration and inter-storey drift responses. Additionally, audience of the classic elastomeric base isolation is hampered to low-to-medium rise, more rigid buildings, as a result of possible uplifting forces in the isolators when the superstructure is tall and slender. The reason behind such limitation lies in that, if the building is tall enough, the horizontal acceleration of each floor will produce inevitable overturing moment and thus potentially produce tension in the isolation system [8].

To address these issues, researchers started seeking solutions from perspective of 'smart' base isolation system combining passive base isolation system with controllable active or semi-active damping devices [9]. The most popular semi-active damping device is MR damper which features tunable viscous damping in the presence of a varying magnetic field [10]. In the disciplines of structural control, such isolation system should be categorised as hybrid control system. The hybrid control strategy has been proved to be effective in terms of seismic protection throughout comprehensive numerical and experimental testings [11–14]. However, the role of damping in seismic isolation has been comprehensively studied by Kelly [6] and results demonstrate that the use of supplementary dampers in seismic isolation is a misplaced effort and will cause undesirable side effects. It is well known that damping is able to primarily control vibration responses under the circumstances of steady-state resonance and free vibration stage [15]. Nevertheless, when confronted with impact load, which is particularly featured in near-fault earthquakes, not enough time is allowed for the damping to dissipate vibrational energy. Since the hysteresis nature of damping is not changed, it is worth questioning whether or not the semi-active or 'smart' damping can cope with the sudden change in external load or structure. Moreover, despite the fact that the supplementary damping may forcefully confine the base displacement of the passive base isolation system [16], high-frequency accelerations as well as increase of inter-storey drifts may be introduced to the superstructure by augmenting damping [17].

In this chapter, a new concept of smart base isolation system with controllable stiffness is proposed to resolve the inherent shortcomings of both traditional and hybrid base isolation systems. Following this, the details of the new adaptive base isolator are proposed and introduced. Nonlinear hysteresis modelling technique is used to portrait the characterisation of the base isolator. With the introduction of several structural control algorithms, a comprehensive shake table testing is conducted to demonstrate the feasibility and evaluate seismic protection performance of the innovative smart base isolation strategy.

14.2 Smart base isolation concept

Given the challenges facing traditional base isolation system as well as the hybrid base isolation system, one of the more desirable solutions is the truly smart base isolation system which can control the isolator's lateral stiffness in real time and hence decouple the superstructure from ground motion at any time instant according to the characteristics of the upcoming earthquake as well as the structural responses. In other words, the new type of smart base isolation endowing adaptability and intelligence directly to the base isolation itself has provided a completely different angle to address the aforementioned issues. Figure 14.1 compares the schematics of three types of base isolation systems, i.e. the traditional base isolation with passive damping, the hybrid base isolation which equips passive isolator with active or semi-active damper and the smart base isolation system with stiffness-controllable base isolator. As can be seen from the schematics, both

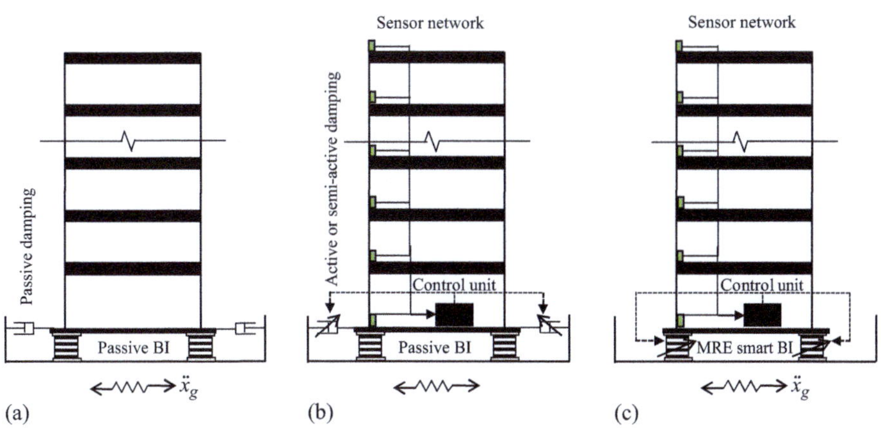

Figure 14.1 Schematic diagram of (a) traditional base isolation system, (b) hybrid isolation system combining passive base isolation with active or semi-active dampers, (c) smart base isolation system with MRE base isolators (BI, base isolation)

hybrid base isolation and smart base isolation system employ a sensing network, signal processor and control decision unit. However, the control effect is applied on the structure differently, i.e. in hybrid base isolation system, the control action is applied in the form of damping force while in smart base isolation system, while in the smart base isolation system, the control effect is realised by altering the lateral stiffness of the base isolator. By controlling the isolator's stiffness, the natural frequencies of the smart base isolation system are shifted away from dominant frequency components of the earthquake excitation, and hence a non-resonant state is achieved at every moment. Therefore, the proposed smart base isolation system can manage to attain a superior seismic protection performance.

Figure 14.2 illustrates the essential configuration of a smart base isolation system. As depicted in Figure 14.2, the realisation of the concept features four key parts:

1. Stiffness-controllable smart isolator: The stiffness controllable base isolator should be able to provide instantly controllable lateral stiffness to enable real-time decoupling from the ground motion. Hence, the proposed smart base isolation system should be categorised as a semi-active control system, and the adaptive isolator is the indispensable key to realise such concept.
2. Properly designed and representative structure model: As a testing bed, the scaled structure model should manage to represent typical low-to-mid-rise buildings. Meanwhile, the designated fundamental frequency should fall in the range of dominant frequency of common earthquakes to highlight the contracting seismic response between bare building and base-isolated structure.
3. Appropriate control algorithm: As known, the smart base isolator is a semi-active control device which influences the response of isolated structure by altering its lateral stiffness, so the controller should be developed to generate stiffness-changing command instead of simply calculating the desired control force. Moreover, due to the inherent hysteretic dynamics of the semi-active isolator, the control strategy should be able to bear or cope with the nonlinearity of the control system to avoid performance degradation or system instability.

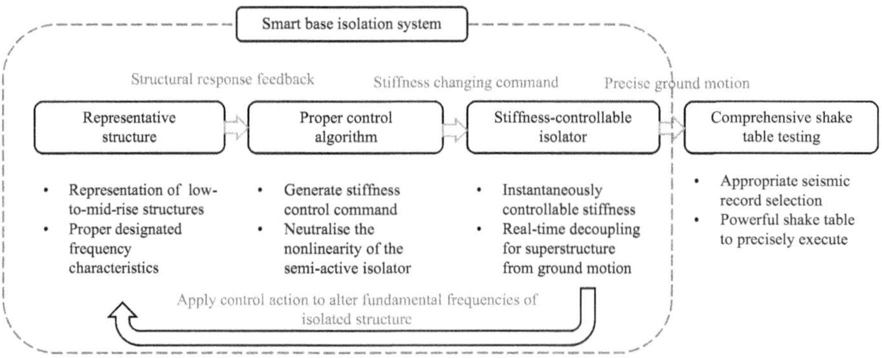

Figure 14.2 Essential configuration of a smart base isolation system

4. Comprehensive shake table testing: The selection of earthquake excitation collection should celebrate a diversity including both far- and near-fault earthquake to demonstrate the versatility of proposed smart base isolation system and the experiment should be conducted on a powerful shake table to ensure the earthquake inputs' intensity and accuracy.

This chapter focuses on experimentally exploring the feasibility to concept-prove the smart base isolation system. As previously discussed, the core element in the truly smart base isolation system is a base isolator possessing adaptive stiffness when subjected to the applied control current or voltage. In recent years, the employment of MRE in base isolator has shed light on the realisation of such novel isolators. Due to MR effect, MRE is a field-dependent material whose shear modulus can be controlled in real-time and reverse fashion [18]. By employing such material, the MRE base isolator acquires controllable stiffness in response to the applied magnetic field, which is fully consistent with the objective and requirement of a truly smart base isolation system. Hence, attracted by its promising performance, the development of the MRE base isolator has been intensively researched and several designs have been reported, including an MRE-based variable stiffness and damping isolator [19], highly adjustable laminated MRE isolator [20], MRE isolator with negative changing stiffness [21], etc.

In this research, the highly adjustable base isolator [20] is adopted for the development of MRE base isolation system due to its wide stiffness tuning range and high vertical loading capacity. Next, a scaled building structure is designed and manufactured as the testing bed, followed by system identification for control analysis and synthesis. To eliminate the vibration coupling and hence the disturbance on the movement in the direction of interest, the structure was designed as a pure shear building model in order to avoid other forms of building motion other than shear. With the integration of the MRE base isolation system, two control strategies, i.e. linear quadratic regulation (LQR) control with general regression neural network (GRNN) inverse model and genetic algorithm (GA) optimised fuzzy logic control, have been developed to exploit the uniqueness of the proposed system. Four benchmark earthquakes, El Centro 1940, Hachinohe 1968, Kobe (Sylmar) 1995 and Northridge 1994, are adopted as the ground motion, among which El Centro and Hachinohe records represent far-fault earthquakes while Kobe and Northridge records stand for near-fault earthquakes [22]. A shake table was utilised to reproduce the earthquake acceleration records without distortion. The seismic protection performance of proposed MRE base isolation system has been demonstrated through a comparative case study of fixed-base, passive-isolated, LQR-controlled base-isolated and fuzzy-logic-controlled base-isolated buildings.

14.3 Adaptive base isolator

14.3.1 Structure

The configuration of the novel MRE base isolator, as shown in Figure 14.3, incorporates the laminated structural design of traditional LRB. It consists of

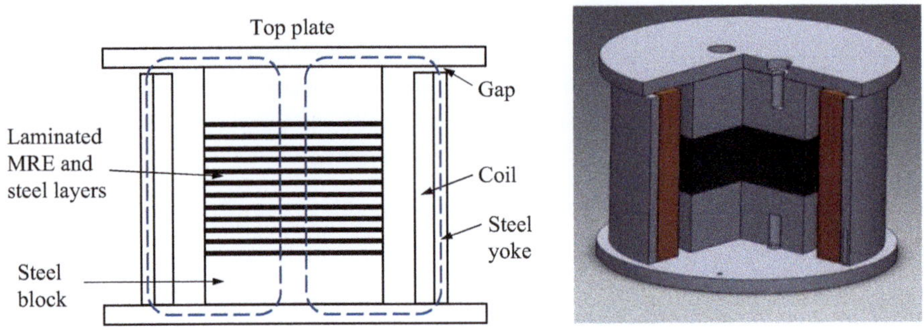

Figure 14.3 Structure and 3D illustration of the adaptive MRE base isolator

multilayer thin MRE sheets bonded onto multilayer thin steel plates. The laminated structure is essential for the seismic isolator used in civil engineering applications. It provides large vertical load capacity and stiffness and prevents lateral bulging and malfunctioning of the MRE rubber. The laminated structure allows high flexibility in horizontal direction by the shearing deformation of MRE sheets that can also be varied instantly under applied magnetic field. In this design, there are 24 layers of the steel sheet with thickness of 1 mm and 25 layers of MRE sheets with thickness of 1 mm being used. The diameter of the MRE and steel sheets is 120 mm.

In the design, the laminated bearing element is placed inside of a solenoid, which generates uniform magnetic field after it is energised with electric current [23–25]. The solenoid is made of electromagnetic coil and thin non-magnetic support as illustrated in Figure 14.3. The cylindrical shape non-magnetic support is made of epoxy material and has an inner diameter of 146 mm. The thickness of the thin wall is 2 mm. The cylindrical electromagnetic coil has an inner diameter of 150 mm and an outer diameter of 200 mm. The coil is firmly attached to the epoxy support. The diameter of the coil wire is 1.0 mm with a total winding number of 2,900 turns. The wire made of copper has electric resistance of 42.3 Ω. The space between the laminated MRE structure and the coil enables the MRE isolator to have a maximum deformation of 15 mm, equivalent to the maximum allowable shear strain of 60%.

The vertical load carrying capacity of the MRE isolator is 50 kg under maximum designed shear displacement (i.e. 15 mm) and zero magnetic field. When the shear displacement is less than 15 mm, the vertical loading capacity of the device will be greatly enhanced. As the loading capacity is a function of cross-sectional area, thickness and the shear modulus of MRE layer, it can be designed based on the requirements of applications.

14.3.2 Finite element analysis

Unlike the designs of other MRE devices, in this design, the MRE materials are placed inside the coil. According to the principle of magnetics, all the magnetic lines that enter any region must also leave that region. In a magnetic circuit, the area inside the coil is the region that all magnetic lines must pass through. Therefore, this placement guarantees uniform and strongest magnetic field given sufficient magnetic flux are provided. Laminated MRE structure with steel layer embedded serves as part of the magnetic core. The magnetic conductivity of the MRE material is fairly low due to the large volume fraction of rubber matrix, and hence to achieve saturated magnetic field for MRE material, considerable power consumption is expected. Laminated structure with high magnetic conductive steel layers between MRE layers improves its overall magnetic conductivity. Two solid steel blocks are added on the top and bottom of the laminated structure to further improve the permeability of the magnetic core. To form an enclosed path for the magnetic flux, two steel plates, one on the top and the other on the bottom, are designed to perform two roles as (1) creating paths for magnetic field and (2) being fixture to connect the device to the ground and the superstructure above [23]. In addition, an annular steel yoke is attached to the coil to complete the magnetic flux path. Gap between the top connection plate and the steel yoke allows lateral deformation of base isolator. Size of the gap is determined by considering the compression of MRE layers under maximum designated compressive loading and stability under maximum lateral deformation.

Finite element analysis was conducted using ANSYS Maxwell software. Magnetic property of the MRE material was obtained from the testing setup used in [26]. *H–B* curve of the material is shown in Figure 14.4. Other parts of the device, including annular yoke, steel plates, steel sheets and steel blocks, are all made of steel 1008. The magnetic property of the steel can be found in the software.

Figure 14.5 shows the magnetic flux density inside of the device when applied current is 3.0 A. It can be observed that the magnetic field inside of the MRE material is around 0.9 T and the distribution is uniform for all MRE materials. Li *et al.* [27] have demonstrated that the magnetic field in a solenoid reaches highest in the centre, while it gradually weakens when leaving the centre due to the dispersion of the field. Built on this, the laminated MRE and steel structure was designed in the centre of the magnetic core, while high permeable steel blocks were

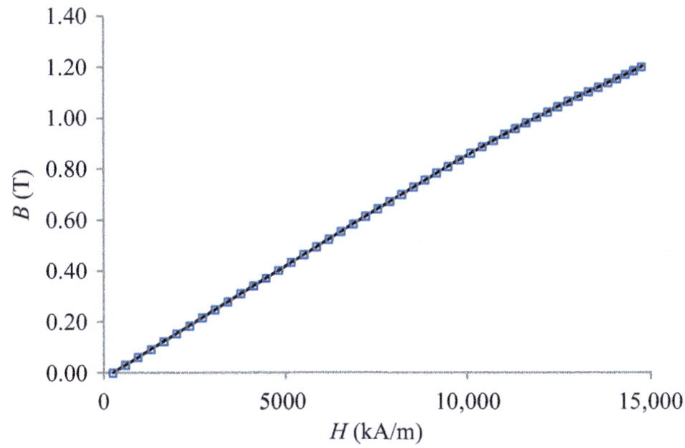

Figure 14.4 H–B curve, used in the analysis, tested using the equipment described in [26]

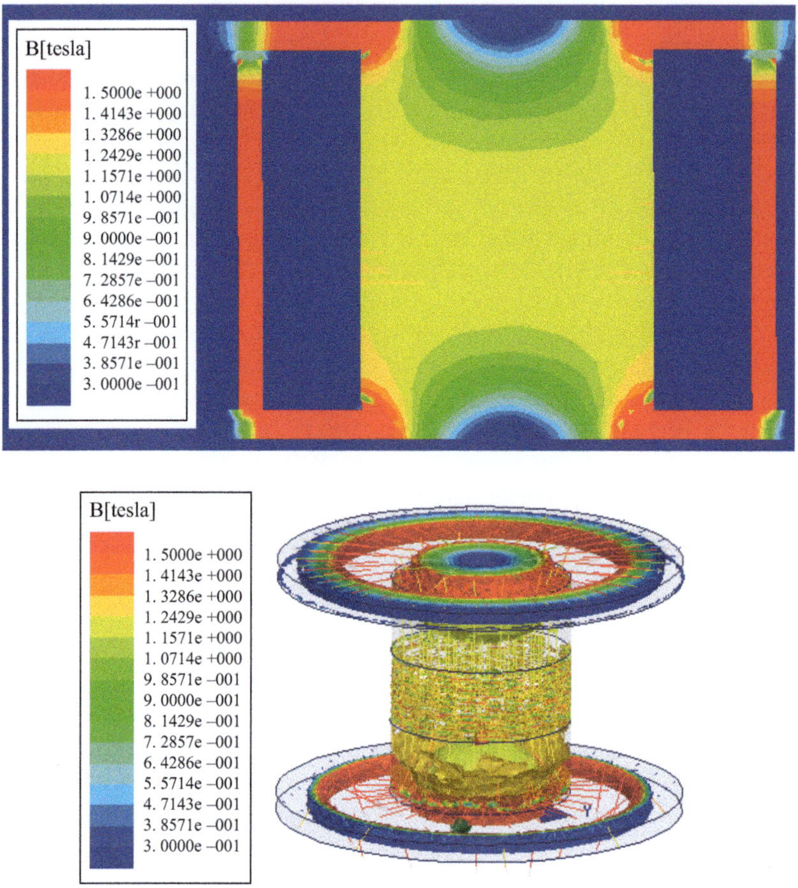

Figure 14.5 Magnetic flux density when current $I = 3.0$ A: 2D model and 3D model

attached to the laminated structure. This arrangement can shift the weakening magnetic field away from the MRE materials. Due to the better conductivity, the magnetic field inside other narrow magnetic paths, i.e. connecting plates and yoke, is much stronger than that in MRE material. This is not a surprise since the magnetic conductivity of MRE material is much lower than steel. However, laminated structure with alternative steel sheets and MRE sheets greatly improves the magnetic permeability of the structure as a whole.

14.3.3 Characterisation testing

A series of experimental tests were conducted to evaluate and characterise the performance of the MRE base isolator prototype utilising the experimental setup, shown in Figure 14.6. In experimental setup, the UTS shake table was used to provide horizontal loadings to the isolator either in the quasi-static or in dynamic mode.

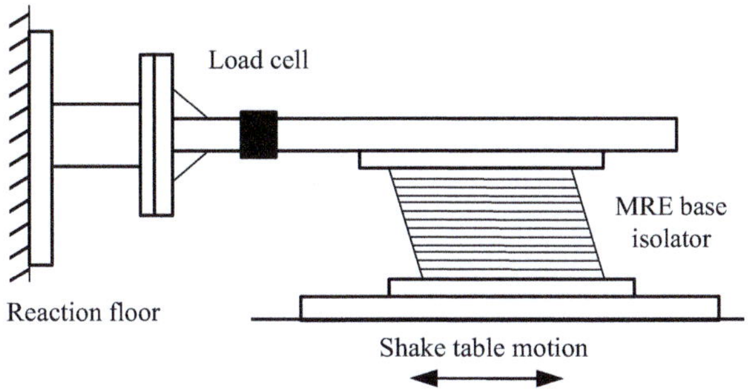

Figure 14.6 Sketch map of the experimental setup

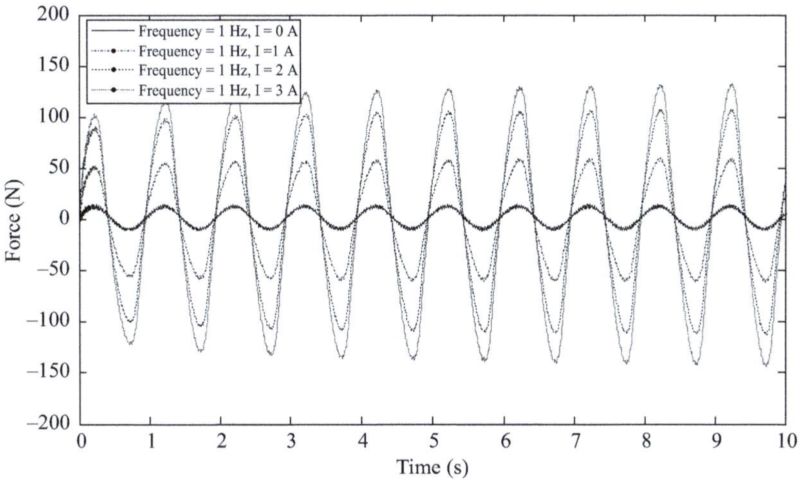

Figure 14.7 Force responses of the MRE base isolator at sinusoidal loading with $\Delta = 2$ mm and $f = 1.0$ Hz

The MRE base isolator was mounted on the shake table and moves along with the shake table motion. A load cell (Model No. STS-300-B10, Sun Scale INC) was installed to a fixed reaction rig to measure the lateral load applied to the isolators (Figure 14.7). During the test, the top plate of the MRE isolator and the load cell remained motionless, thus eliminating the undesired inertia force in the measurements. Displacement sensor which is built in the shake table (MTS, USA) provides the displacement measurement. A DC power supply (DC Power Conditioner, SOLA Electric, Division of SOLA Basic Australia) with capacity of 240 V and 5.3 A provides DC current to energise the magnetic coil. A slider (Type: S-260-10, Yamabishi Electric Co. Ltd, Tokyo, Japan) was used to adjust the applied current to the magnetic coil, also shown in Figure 14.10. A multimeter (Model No. 115, Fluke) was also used to monitor the current output from the slider during the testing.

For dynamic tests, various harmonic inputs under displacement control were chosen to load the base isolator. Three amplitudes (2, 4 and 8 mm, corresponding to 8%, 16% and 32% shear strain) were selected for the tests under various loading frequencies (0.1, 1, 2 and 4 Hz). For each loading case, four currents (0.0, 1.0, 2.0 and 3.0 A) were applied to examine performance of the device under different magnetic fields. At least 10 cycles were measured for each single loading case to ensure the stable performance of the device is captured for each case. In addition, each set of testing was repeated five times independently to ensure the accuracy and consistency of test results. It was therefore 240 tests in total that have been conducted. Since in the current design, the coil generates considerable amount of heat due to its high resistance (about 43 Ω), temperature control was also considered to maintain temperature at a relative constant level through different sets of tests. The sampling rate for the data acquisition (DAQ) was set at 256 Hz for capturing all test results including the dynamic tests.

Figures 14.8 and 14.9 show the force–displacement loops of the MRE base isolator at various sinusoidal loadings of three amplitudes at frequencies of 0.1 and 2.0 Hz, respectively. For each loading case, force increases with applied currents can be clearly observed. The measured force also increases with increase of the loading amplitude naturally. Since the slope of the force–displacement loops indicates the lateral stiffness of the MRE isolator, it can also be observed that at certain current range, the stiffness of the MRE base isolator gains considerable larger increase. Moreover, as the enclosed area of the loop represents the energy dissipation, or the damping equivalent, of the MRE isolator, it is noted that the damping of the MRE isolator increases when the applied current increases.

Table 14.1 lists the effective stiffness of the MRE base isolator under various loading conditions. The details for effective stiffness calculation can be found in [23,28]. The maximum effective stiffness is 66 kN/m and the maximum increase of the effective stiffness (i.e. when increasing the current from 0.0 to 3.0 A) is 1,630% from quasi-static tests.

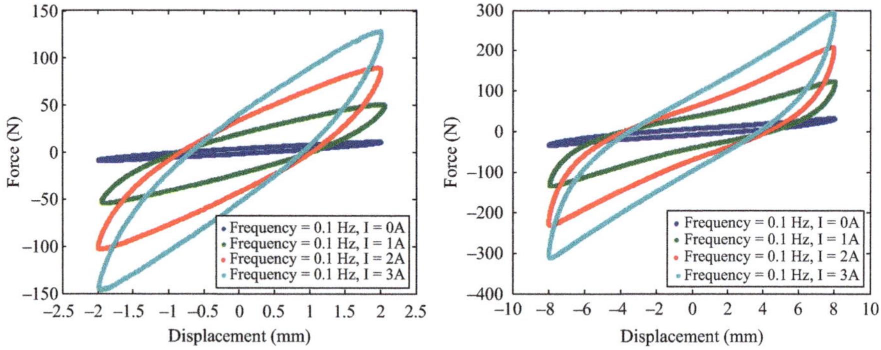

Figure 14.8 Force–displacement relationships of the MRE base isolator at quasi-static testing with frequency of 0.1 Hz ($\Delta = 2$ and 8 mm)

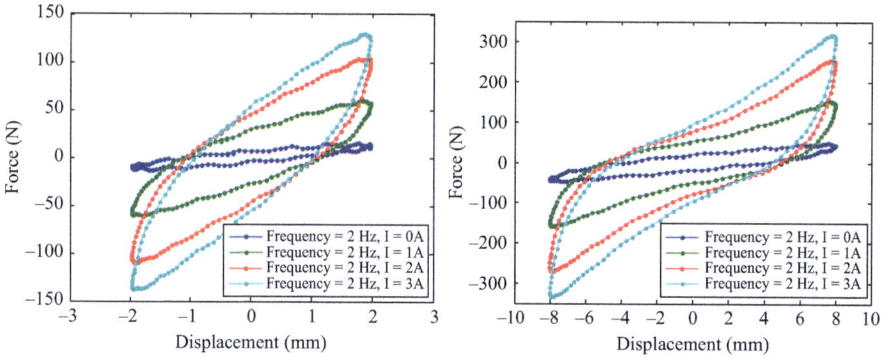

Figure 14.9 Force–displacement relationships of the MRE base isolator at quasi-static testing with frequency of 2 Hz ($\Delta = 2$ and 8 mm)

Table 14.1 Effective stiffness (kN/m) of the MRE base isolator under various loading conditions (amplitude and loading frequency)

Effective stiffness (normalised, kN/m)	**Amplitude = 2 mm**			**Amplitude = 4 mm**			**Amplitude = 8 mm**		
	0.1 Hz	**1.0 Hz**	**2.0 Hz**	**0.1 Hz**	**1.0 Hz**	**2.0 Hz**	**0.1 Hz**	**1.0 Hz**	**2.0 Hz**
0.0 A	3.64	4.96	5.80	3.63	4.69	5.33	3.72	4.62	5.24
1.0 A	25.58	27.13	27.52	19.35	20.72	20.87	15.99	17.72	17.81
2.0 A	45.73	52.22	50.64	33.73	38.15	37.53	27.32	31.44	31.15
3.0 A	62.98	66.13	65.02	46.64	48.74	47.88	36.41	39.31	39.29
Increase (0–3 A)	**1,630%**	**1,234%**	**1,022%**	**1,186%**	**939%**	**798%**	**878%**	**751%**	**650%**

14.4 Nonlinear hysteresis modelling

Establishing accurate model of the MRE base isolator is an essential task to reproduce the nonlinear device behaviour and to lay foundation for its control applications. Till now, several models have been proposed to describe the complex behaviour of the MRE base isolator, including parametric models and non-parametric models.

Regardless the type of the model, the modelling process normally involves two steps: first, the proposed model is trialled by standard sinusoidal loadings with various loading amplitudes, loading frequencies and applied currents. After this step, a generalised model is obtained which should include the linearisation process to establish the relationships between the model parameters and the applied currents. In the second step, the generalised model is to be tested by the output data of the device under randomised excitation and/or applied currents. This is due to the fact that the operating condition of the device is normally under stochastic loadings, and therefore the proposed model should be able to predict the behaviour of the device under such condition.

14.4.1 Parametric models

As shown in the force–displacement loops in Figure 14.9, it is clear that the parametric model should at least consist of stiffness and damping elements, representing by the inclines slope and the size of the hysteresis loops. For a model with stiffness and damping elements only, the hysteresis loop is an inclined ellipse shape. However, there is a nonlinear behaviour existing in the loop, and therefore the element with nonlinear hysteresis should be included in the model as well. To address this issue, several models have been proposed, such as Bouc–Wen model [29], strain-stiffening model [30–32], improved LuGre model [33], improved Dahl model [34].

Yang *et al.* [29] proposed a Bouc–Wen model to characterise the behaviour of MRE base isolator, Figure 14.10. This model consists of a spring, viscous dashpot and a hysteresis Bouc–Wen element in parallel, as following:

$$F = \alpha k_0 x + (1 - \alpha) k_0 z + c_0 \dot{x}$$

$$\dot{z} = A\dot{x} - \beta|\dot{x}||z|^{n-1}z - \gamma\dot{x}|z|^n$$

k_0 is the stiffness of the spring and c_0 represents the viscous coefficient, indicating the damping capacity of the system. The item of $c_0\dot{x}$ is a component of the total force. The rest part represents the restoring force as the summation of a linear component $\alpha k_0 x$ and a purely hysteretic component $(1 - \alpha)k_0 z$, in which $\alpha \in (0, 1)$ represents the linearity level of the hysteresis loops. A, n, β and γ, which are non-dimensional parameters, are responsible for the shape and the size of the hysteresis loops.

Yu *et al.* [33] proposed an improved LuGre model to characterise the behaviour of MRE base isolator (Figure 14.11). The model consists of a spring element, a viscous dashpot and a modified LuGre element connected in parallel, with mathematical expression is given by

$$F(t) = k_0 x(t) + c_0 \dot{x}(t) + \frac{\beta}{\alpha} y(t) + \frac{\varepsilon}{\alpha} \dot{y}(t) + f_0$$

$$\frac{1}{\alpha}\dot{y}(t) = \dot{x}(t) - |\dot{x}(t)|y(t)$$

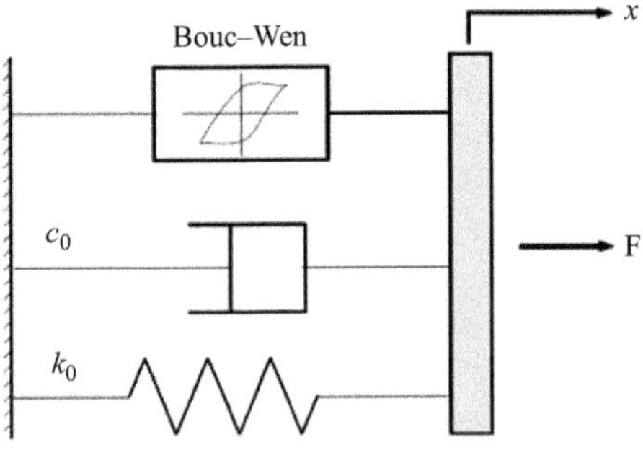

Figure 14.10 Bouc–Wen model for the MRE base isolator

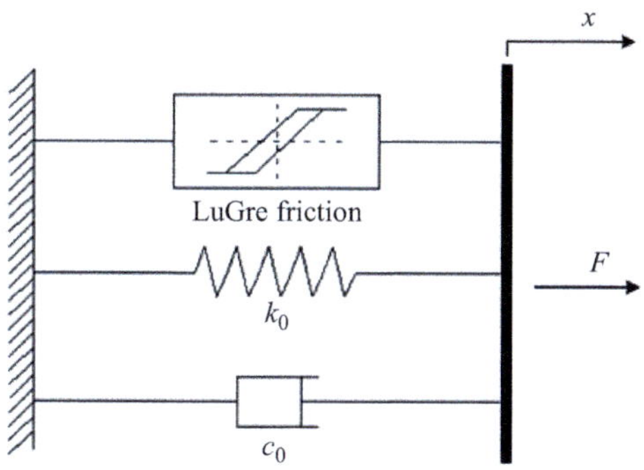

Figure 14.11 Improved LuGre friction model for the MRE base isolator

where $F(t)$ is the shear force of the model output at time t; $x(t)$ and $\dot{x}(t)$ denote the displacement and velocity of the device at time t, respectively; k_0, c_0 and f_0 are the spring stiffness, viscous damping and initial force of the device, respectively; y is an intermediate variable; α, β and ε are three non-dimensional parameters, which are used to control the shape and scale of the hysteretic loop.

Li and Li [30,31] have proposed a strain-stiffening model, Figure 14.12, with a hysteresis strain-stiffening element involved to characterise the behaviour of MRE isolator. The details of the model are as followings:

$$\begin{aligned} F &= k_1 y + az^3 \\ k_1 y &= k_0(x - y) + c_0(\dot{x} - \dot{y}) \\ az^3 &= c_1(\dot{x} - \dot{z}) \end{aligned}$$

where c_0, k_0, c_1, k_1 and a are the model parameters to be identified. This model contains a strain-stiffening element to portrait the nonlinear behaviour of the isolator. It also has a nominal displacement y in the standard solid model to be calculated, which may introduce extra computational efforts. To overcome this issue, a new strain-stiffening model has been proposed [35], as shown in Figure 14.13.

The mathematical expression of the proposed model is as follows:

$$F = k_0 x + c_0 \dot{x} + \alpha|x|x^3 + F_0$$

where k_0, c_0 and α are the model parameters as stiffness of the spring, damping of the dashpot and coefficient of the modified strain-stiffening element. F_0 is the offset of the shear force. Without the nominal displacement involved, the modified strain-stiffening model has simpler structure and only four model parameters are to be identified.

Table 14.2 compares the root mean square (RMS) errors and calculation time of different models after parameters are identified. The four models have similar RMS error, which show good model accuracy. Due to the variation of the model parameters included, the running time is quite different: Bouc–Wen model requires most computational resources while the improved strain-stiffening model demands

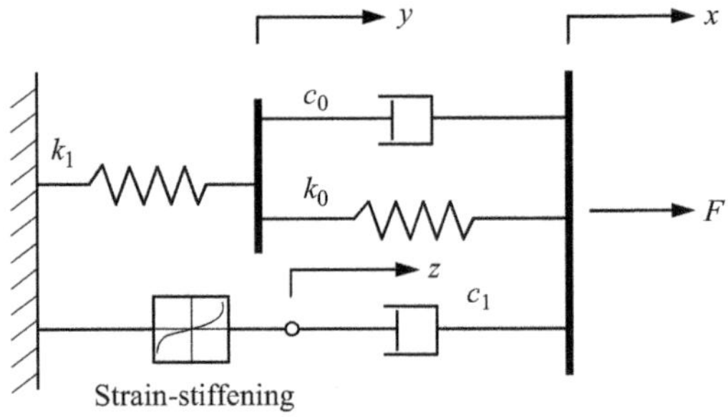

Figure 14.12 Strain-stiffening model for the MRE base isolator

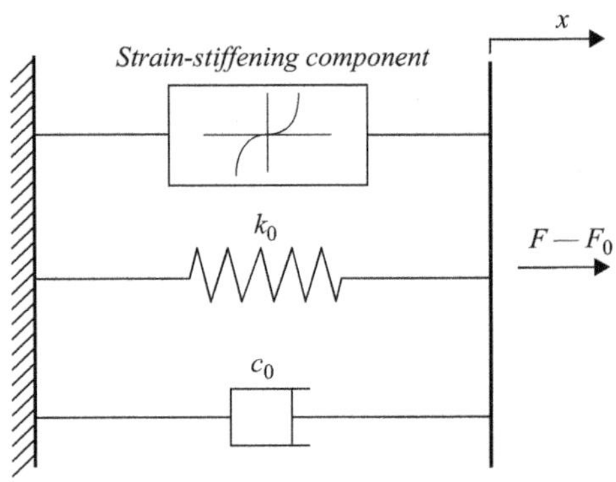

Figure 14.13 An improved strain-stiffening model for the MRE base isolator

Table 14.2 Model performance comparisons

Model	Number of model parameters	RMS error	Running time
Bouc–Wen	7	5.3194	168.4959
Improved LuGre friction	6	5.8728	151.6732
Strain-stiffening	5	5.6322	32.1143
Improved strain-stiffening	4	5.4935	2.7367

the least. The reason lies in the number of model parameters and the complexity of the nonlinear hysteresis elements. In terms of their applicability in the control design and analysis, Bouc–Wen model has proven popularity in this regard due to its explicit boundedness of the hysteresis part [36] while the strain-stiffening models have to face the challenge that they do not have boundedness in a pure mathematical point of view.

Figure 14.14 lists the comparison between the model prediction and the experimental data of the Bouc–Wen model, improved LuGre model, strain-stiffening model and improved strain-stiffening model. The comparison proves the

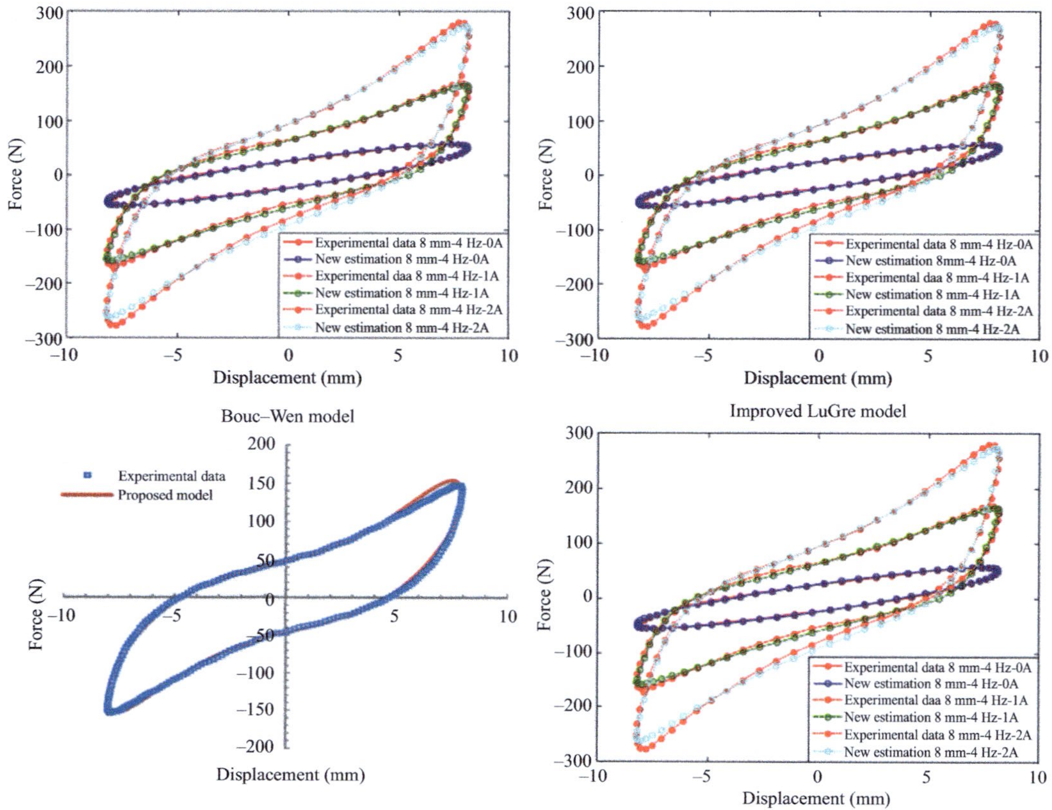

Figure 14.14 Comparison of the model prediction and measured data for four models

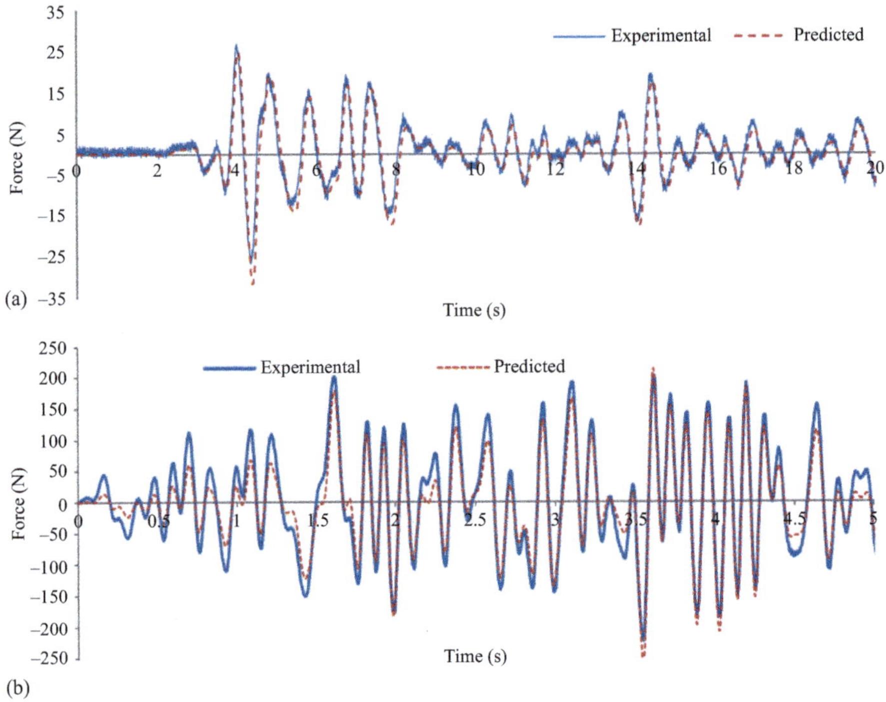

Figure 14.15 Examination of the strain-stiffening model under two random excitations: (a) comparison between the experimental data and generalized model under El-Centro earthquake input and (b) comparison between the experimental data and generalized model under random input

model accuracy of all the models for standard sinusoidal loadings. Figure 14.15 uses the strain stiffening model to illustrate the model performance under random loadings, such as El-Centro earthquake and a random input. The comparison between the predicted and measured data once again proves its model accuracy.

14.4.2 Non-parametric model

In the parametric models, all the model parameters possess physical meaning to explain the actual behaviour of the device. For example, the stiffness element refers to elastic behaviour of the device for which one can correlate to the inclined slope of the hysteresis loop. The damping coefficient links to the characteristic of viscous damping, usually represented by a dashpot, and the magnitude of the damping coefficient has direct relation with size of the enclosed loop. However, the model accuracy may be affected by limited choice of the model forms, and sometimes meaningless values of the coefficients may be identified based on the experimental data, such as negative stiffness and damping.

Non-parametric models differ from parametric models as the model structure is not specified *a priori* or not predefined but is instead determined from data. This feature adds flexibility and consequently accuracy into the model. In the non-parametric modelling process of MRE base isolator, the adopted methodology, such as neuro network, supporting vector, establishes the relationship between the motion of the isolator (e.g. displacement, velocity, loading frequency and applied current) and the force output. The preset form of the parametric model utilises current and previous states of responses to predict its output. The non-parametric model can capture more subtle aspects of the data and has higher accuracy and more freedom in the device modelling.

Several research has been reported to use non-parametric modelling technique to predict the nonlinear behaviour of the MRE base isolators. Koo *et al.* [37] developed a phenomenological model to capture dynamic behaviours, i.e. stress–strain relationship under compressive loadings, of MREs using artificial neural network (NN) method. Yu *et al.* [38] proposed a feedforward NN and ant colony algorithm to model the behaviour of laminated MRE base isolator. Leng *et al.* [39] established an artificial neuro network to model the behaviours of MRE isolator in shear–compression mixed mode. Yu *et al.* [40] utilised support vector regression (SVR) and particle swarm optimisation/fruit fly algorithm, respectively, to reproduce the behaviour of MRE base isolator.

Figure 14.16 illustrates the structure of a SVR-based model for MRE base isolator [40]. In this nonparametric model, the displacement and velocity of the device, as well as the current, are adopted as the inputs, while the shear force of the device is regarded as the output of the model. It is shown in Figure 14.17 that

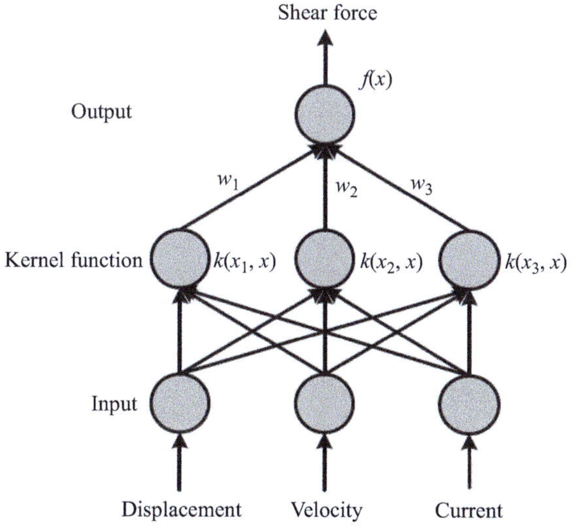

Figure 14.16 Structure of a support vector regression-based model

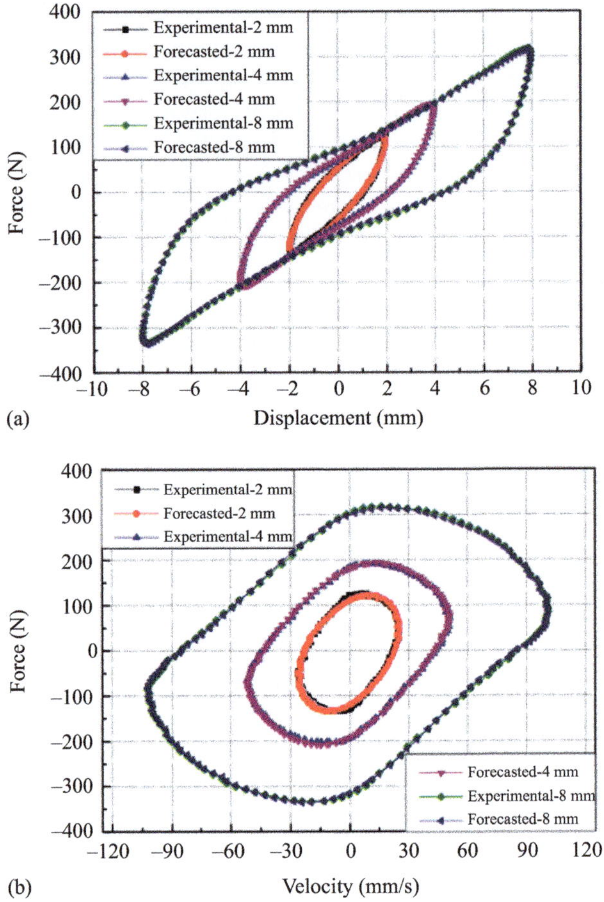

Figure 14.17 Comparisons between experimental and forecasted data with different loading amplitudes: (a) force–displacement responses and (b) force–velocity responses

the SVR model can well predict the behaviour of the MRE base isolator in all loading conditions, i.e. multiple loading frequencies and amplitudes.

Clearly, due to the flexibility, the non-parametric model performs better compared with parametric models. The non-parametric modelling technique is very popular to characterise the nonlinear behaviour of devices and structures. It is yet to see its popularity to emerge in the structural control application compared with parametric models.

14.5 Control algorithms

In this section, two control strategies adopted in the study of MRE base isolation system, i.e. LQR control with GRNN inverse model and GA optimised fuzzy logic control, are presented. Normally, the classic optimal controller like LQR calculates suitable control force based on the state-space function and system feedback. However, the semi-active MRE base isolator applies control effects indirectly on the isolated structure by altering the stiffness at base isolation level rather than generating control force directly. Hence, an inverse model is required to predict the control current or voltage on the isolator to generate desired control force. The inverse model is obtained by the training of GRNN. Moreover, the robustness of the control system can be significantly degraded by the hysteresis of the MRE base isolator. To this end, a fuzzy logic controller is developed and optimised by GA to cancel out the effect of the highly nonlinear dynamics of the isolator.

14.5.1 LQR control with GRNN inverse model

The flow diagram of the LQR control of MRE base isolation system is shown in Figure 14.18. As shown, the LQR controller calculates desired control force based on the system's response and then transmitted the force to the inverse model. Based on control force command, the inverse model generates desired control current. Due to current limitation of the MRE isolator, a saturation link is required to restrain the current within the range of 0–3 A. Considering the LQR method is a rather classic optimal control method, the formulae to calculate control gain is not discussed here.

After the LQR controller calculates the desired control force, an inverse model of the MRE isolator is employed to obtain the desired control current input. The inverse model is designed based on GRNN, whose schematic diagram is shown in Figure 14.19.

As can be observed in Figure 14.8, the GRNN consists of four layers, namely input layer, pattern layer, summation layer and output layer. In the case of MRE isolator, the input layer includes independent features chosen as displacement, velocity and control force of the isolator at present and previous time instant and the force at the present time instant. Each unit of pattern layer represents a training pattern. The output of GRNN, which is the control current in this case, calculates

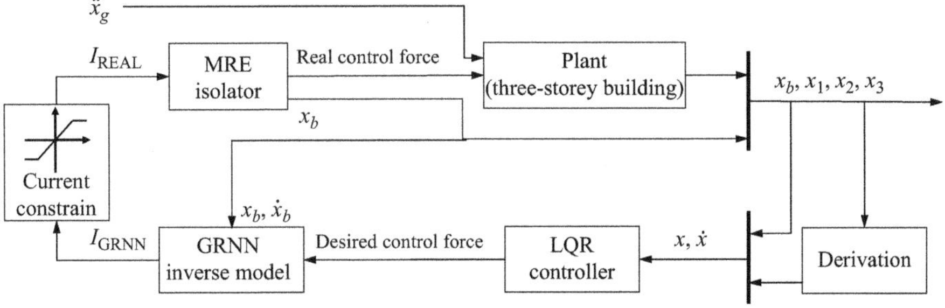

Figure 14.18 Schematics of LQR-controlled isolation system with GRNN inverse model

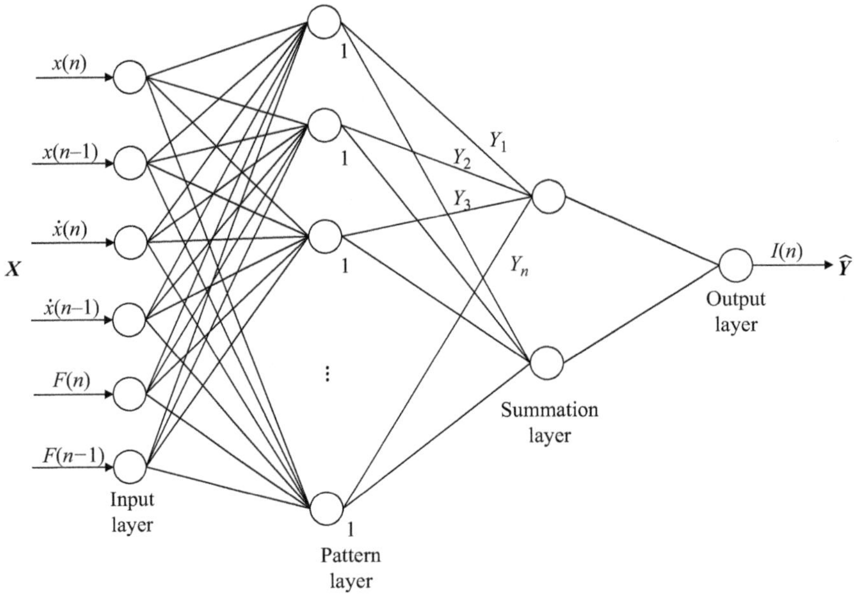

Figure 14.19 Schematic diagram of inverse model based on GRNN

the quotient of the two units of the summation layer. More details about GRNN can be found in [41–43].

14.5.2 GA optimised fuzzy-logic control

As mentioned previously, because of the high nonlinearity and uncertainty of MRE base-isolated structure, the uncertain and imprecise of the isolation system is a significant issue in real experimental applications. It is well known that the fuzzy logic control, which is not dependent on the synthesis and analysis of the mathematical control system [44], is quintessential for the control of MRE base isolation system and thus allows considerable nonlinearity and uncertainty of the input excitation, feedback signal and the controlled structure itself. The inputs and outputs of the fuzzy controller are described in linguistic directions and then connected by the fuzzy inferences of 'IF–THEN' rules.

Normally, a well-designed base isolation system should be able to achieve small base drift and structural acceleration simultaneously. Thus, the aim of the fuzzy logic controller designed in this study is to minimise the structural acceleration and base drift simultaneously [44]. Top floor acceleration $\ddot{x}_3$ and base level displacement are x_b then chosen to be the inputs of controller whose output is the control current of the MRE isolator. The responses can be divided into three types: (1) when the absolute acceleration is very large, the base isolation should be soft enough to release the superstructure and thus dissipate energy when the base drift is

Table 14.3 Optimised inference rule of the fuzzy logic controller

MF		Top floor acceleration $\ddot{x}_3$				
		PB	**PS**	**Z**	**NS**	**NB**
Base drift x_b	PB	M	B	S	S	M
	PM	S	B	S	B	S
	PS	S	S	B	B	S
	NS	S	S	B	S	S
	NM	S	B	S	S	S
	NB	M	B	S	B	B

The member functions (MF) of the input signals are abbreviated as follows: PB, big positive; PM, medium positive; PS, small positive; Z, zero; NS, small negative; NM, medium negative; NB, big negative, while the MFs of output signals are B, big; M, medium; S, small.

small and vice versa; (2) when the absolute acceleration is small, the stiffness of the base isolation should be positively correlated to absolute base drift, which ensures the stability of the building; (3) when the absolute acceleration is around zero, the control current should be small when the base drift is big and big when base drift is almost zero, which will ensure high base stiffness without the existence of seismic excitation.

Specific control law was obtained and optimised by a non-dominated sorting genetic algorithm type II (NSGA-II). Dynamic crowding distance (DCD) is introduced into the standard NSGA-II as a novel evaluation index to keep good diversity among the solutions. More details about NSGA-II with DCD can be found in [45]. The peak and RMS of top floor acceleration and base drift are adopted to construct the fitness function. The optimised inference rule of the fuzzy logic controller is shown in Table 14.3.

14.6 Experimental setup

14.6.1 Three-storey building

A three-storey building model is then selected as the testing bed of the MRE base isolation system. The schematic diagram and photos of the building model are shown in Figure 14.20. The testing building is a three-storey frame structure with a height of 1,200 mm, of which each floor is 400 mm high. For computation efficiency, it is ideal that the system can be simplified as a lumped mass model. Hence, this building model has been designed and manufactured as close to a pure shear model as possible. With such design, the likelihood of modal coupling of two directions is minimised and thus avoid the distortion of the building model. To achieve this objective, four identical steel strips, whose cross-section dimension is 34 mm × 4 mm, have been adopted as the columns of the structure

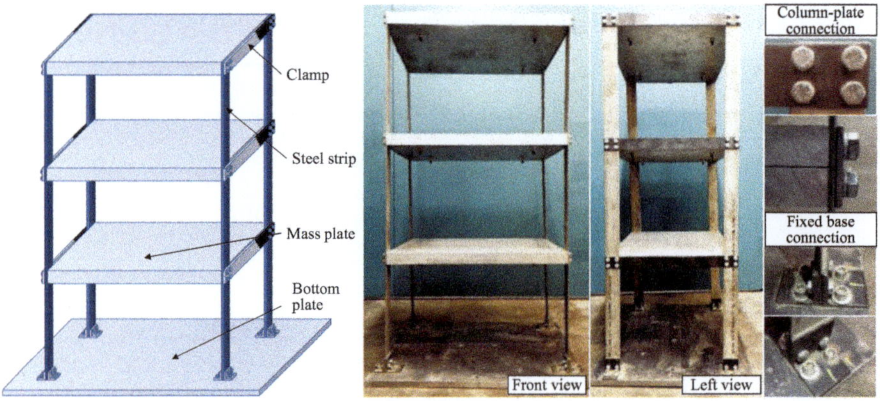

Figure 14.20 Schematic diagram and photos of the three-storey shear building model and connections in the structure

Table 14.4 Detailed designated parameters of each component

Component's name	Number	Material	Density (kg/mm^3)	Length (mm)	Width (mm)	Height (mm)	Weight (kg)
Mass plate	6	Aluminium	2.70×10^{-6}	600	450	20	14.58
Strip	4	Steel	7.85×10^{-6}	34	4	1,200	1.28
Clamp	6	Steel	7.85×10^{-6}	450	5	40	0.57
Bottom plate	1	Steel	7.85×10^{-6}	706	550	20	61

to provide low stiffness at the direction of the earthquake excitation and very high stiffness at the perpendicular direction. Slab of each level consists of two aluminium plates with the dimension of 600 mm × 450 mm × 20 mm bolted together with four countersunk head screws. At connections at each floor, a steel clamp is bolted on the top of the strip with the mass plate, endowing the connection with high rigidity to avoid the occurrence of torsional and rotational modes.

Detailed structural design, dimension and material of each component are shown in Table 14.4. To represent low-to-mid-rise buildings, the fundamental frequency of the structure is designated as 1.934 Hz, which also falls in the envelope of dominated frequency range of majority of natural earthquakes. With such frequency arrangement, the three-storey building is representative of the dynamic behaviour of real building model under earthquake attacks and will experience resonance and hence highlight the contract between seismic responses between bare and base-isolated building.

14.6.2 Isolation system assembly

The three-storey pure shear building model described in previous section is utilised as the testing bed. In isolation scenarios, two MRE base isolators are symmetrically mounted under the three-storey frame structure on the central axis of the structure's bottom plate. Hence, the equivalent mass of the isolation level is approximately 50 kg. In the passive isolation scenario, zero current is applied on the MRE isolator, which indicates the softest status of the isolator and thus the lowest corresponding natural frequency of the system. The photo of experimental setup is displayed in Figure 14.21.

14.6.3 Data acquisition and control system

A number of sensors are installed in this testing to measure the structure's movement feedback as well as the real-time current in the solenoid and magnetic field across the laminated MRE core of the isolator. Five Baumer laser distance sensors (Part No. OADM 20I4460/S14C) provide the measurements of shake table movement and relative displacement of each floor (only four are used in the fixed base building case). A sensor reference wall is built to hold the laser sensors precisely at the elevation of each floor. Sensors measuring second and third floors have a sensing span of 130 mm, while sensors measuring first and

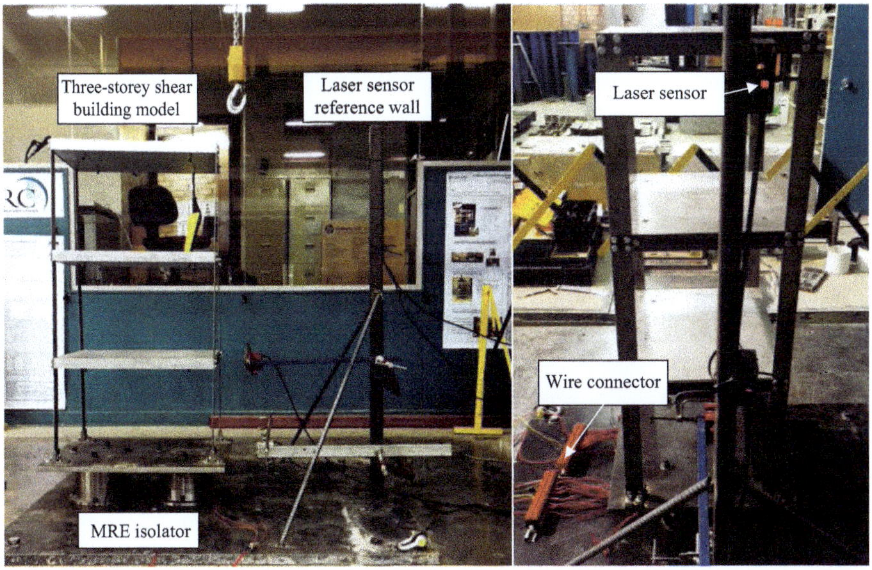

(a) (b)

Figure 14.21 Photo of experimental setup: (a) front view; (b) side view base floors as well as the table movement have a sensing span of 50 mm. Reason for this selection is that the bottom two floors and shake table feature relatively small displacement. Therefore, distance sensors with smaller measuring range can guarantee higher accuracy. Figure 14.22(a) shows the photos of employed laser sensor and the adapter between sensing system and the data-acquisition system. A Hall Effect current transducer (Part No. CSLA2CD) is utilised to monitor the real-time current, I in the solenoid of MRE isolator. The magnetic flux B across the MRE core is measured by a digital Hall Effect sensor IC (Part No. SS461A). Two wire connectors are used for flexible configuration of small coils in MRE isolators.

One of the essential equipment utilised in the control system is dSPACE Real-time PPC Controller Board (DS1104). This control board, based on MATLAB®/Simulink™ operational interface, is a software–hardware platform for semi-physical simulation and can be installed in virtually any PC with a free PCI or PCIe slot. With dSPACE controller board, the PC can be upgraded to a powerful development system for rapid control prototyping, which is crucial in the realisation of real-time control of the MRE base isolation system. The photo of dSPACE DS 1104 board is shown in Figure 14.22(b). In the experimental system, the role of controller board functions comes in functional variants. First of all, it was employed as DAQ system. There are eight A/D converters and eight D/A converters on the board. Second, it also functions as a real-time controller based on Simulink software by generating the PWM control signal governing MRE isolator at a switching frequency of 1,000 Hz. Seven out of eight A/D converters have been utilised to acquire feedback signals of five laser sensors, current transducer and Hall effect sensor. The control flow can be observed as follows: the structural and shake table movement (x_g, x_b, x_1, x_2, x_3) as

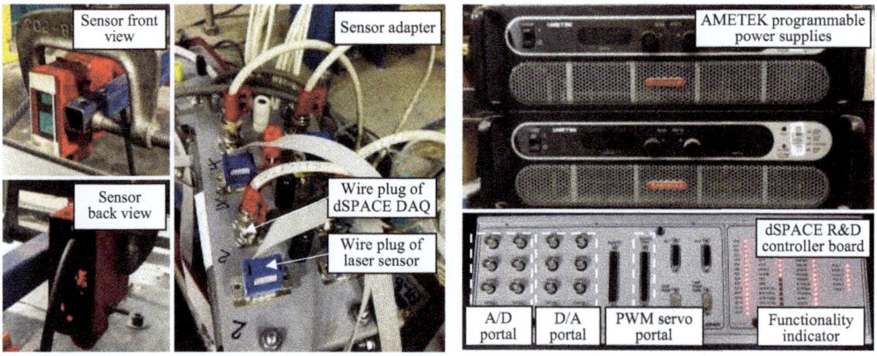

(a) (b)

Figure 14.22 Sensor installation, power supply and control system: (a) laser sensor and sensor adapter and (b) power supplies and data acquisition system with dSPACE well as current and magnetic field (I and B) in MRE isolator are measured and transmitted to DAQ system of dSPACE board; the inbuilt digital controller calculates desired control current and corresponding PWM signal; subsequently, the dSPACE board outputs command of duty cycle through PWM servo portal; according to the duty cycle, iGBT electrical switch aforementioned modulates the voltage applied on the MRE isolator. Two AMETEK programmable power supplies (Sorensen SG Series, 400 V/12 A) in Figure 14.22(b) have been employed to provide electricity.

14.6.4 Earthquake excitation

To appraise the seismic protection performance of the smart base isolation system under different control configurations, numerical and experimental testings are conducted under four natural acceleration records of historical earthquake events, namely Imperial Valley 1940, Kobe 1995, Tokachi-Oki 1968, Northridge 1994. Selected from the databases of Pacific Earthquake Engineering Research Centre (PEERC) and National Geophysical Data Centre (NOAA-NGDC) [22], records of Imperial Valley (station: El Centro Array) and Tokachi-Oki (station: Hachinohe) earthquakes are categorised as far-fault earthquake, which features wider frequency range but longer excitation time, while Kobe (station: KJMA) and Northridge (station: Sylmar) are classified as near-fault earthquakes which features waveforms containing large velocity pulses with lower frequency. Such selection consideration is capable of demonstrating the versatility of the proposed MRE base isolation system. The information of four earthquake records are summarised in Table 14.5 and the accelerograms, and pseudo-acceleration spectrum (damping ratio is assumed as 5%) of the four earthquakes are shown in Figure 14.23. From the time history records, long-period pulse-like waveforms can be clearly observed in Northridge earthquake. Meanwhile, pseudo-acceleration spectrums show that Kobe and Northridge earthquakes possess larger acceleration spectra when the structural period is larger than 1 s.

Table 14.5 Earthquake information

Record	Earthquake	Station	Magnitude	PGA (g)	PGV (cm/s)	Predominant freq. (Hz)
I-ELC180	Imperial Valley, 1940	El Centro Array	7.1	0.349	29.69	0.5–2.8
KJM000	Kobe, 1995	KJMA	7.2	0.834	40.65	0.6–2.7
Jap02.089	Tokachi-Oki, 1968	Hachinohe Harbor	7.9	0.229	22.71	0.2–3.2
SYL360	Northridge, 1994	Sylmar Olive View	6.8	0.843	64.68	0.5–2.5

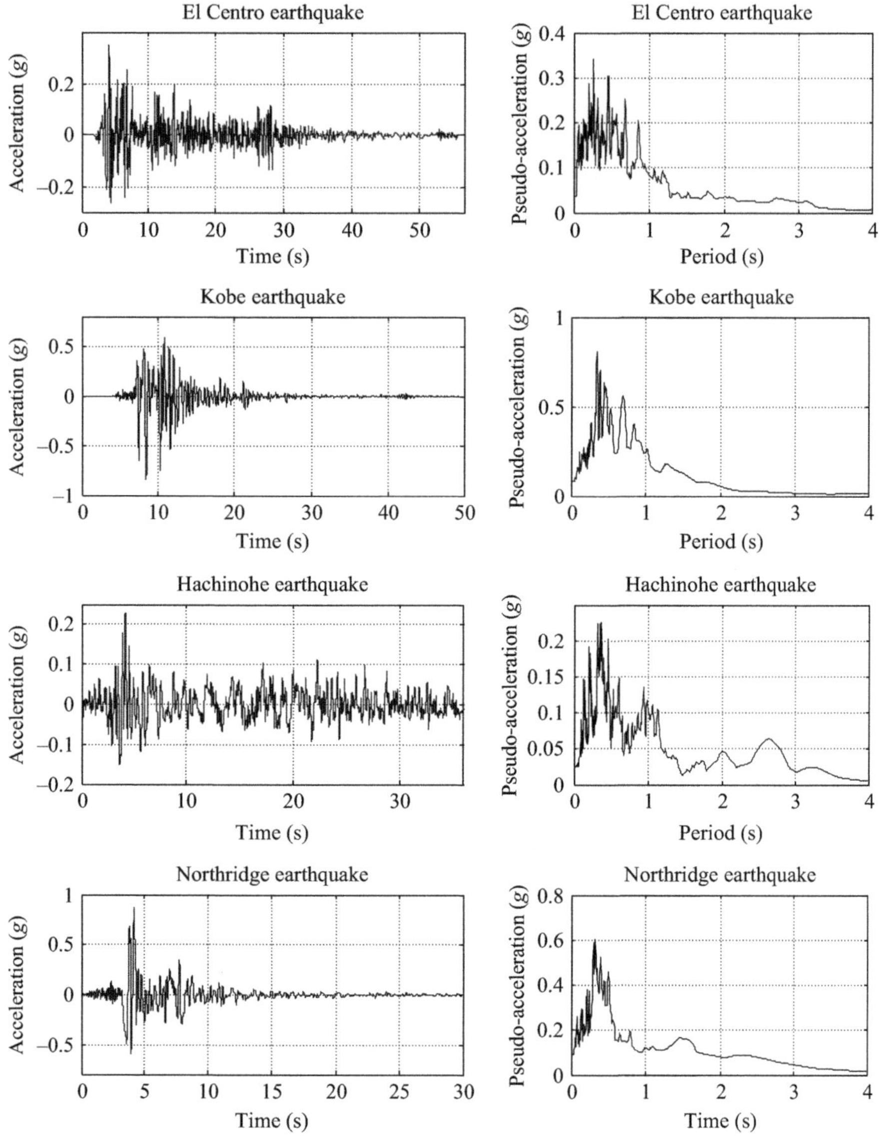

Figure 14.23 Earthquake time histories and pseudo-acceleration spectra (damping ratio = 5%)

14.7 Results and discussions

14.7.1 Time history

Figure 14.24 displays the time history of top-floor acceleration in the structural scenarios of fixed base building, passive isolated building, LQR and fuzzy-logic-controlled MRE isolation system under four earthquakes. As observed, the isolation

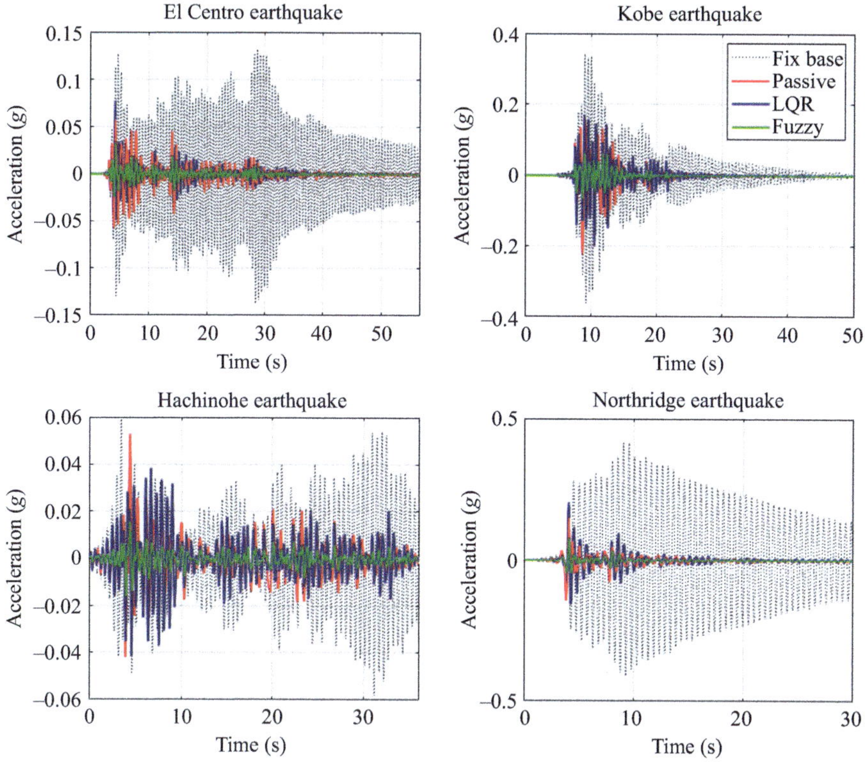

Figure 14.24 Time history of top floor acceleration of four structure scenarios

systems, no matter passive or controlled, are able to suppress the top floor acceleration effectively under different earthquakes. However, in Hachinohe earthquake, the passive isolation's response is close to that of the fixed base building during the first 6 s. In the latter part of all earthquakes, the fix building's behaviour can be recognised as a free vibration except in Hachinohe earthquake, under which the fixed base building is re-excited from around 12 s. It might be because, unlike other earthquakes, Hachinohe earthquake maintains a relatively high volatility until the end of the seismic event. Hence, the disturbance continues in the Hachinohe earthquake. Additionally, in the free vibration stage, the isolated structure's response can decay at a much faster speed than the fixed base building. It is because of the supplementary damping introduced into the structure when the isolation layer is integrated into the system. As for the controlled isolation systems, the fuzzy-logic-controlled base isolation system is able to significantly reduce the acceleration on the basis of passive controlled isolation system and maintain the acceleration at a low level during the entire time history. LQR-controlled isolation system, on contrast, does not have stable performance under different earthquakes. The acceleration response of LQR-controlled isolation system is smaller than but very close to the passive isolation system, and under Northridge earthquake, it even exceeds the passive acceleration response. In the end, in the free vibration stage, the vibration of structure presents a general period of 0.5 s, which is consistent with the conclusion drawn in last section that the major vibrational mode is the first mode whose natural frequency is about 2 Hz.

Figure 14.25 illustrates the comparison of base floor acceleration time history in the passive, LQR and fuzzy isolation scenarios. Significant reduction of base acceleration can be further achieved by both LQR and fuzzy-logic-controlled isolation systems compared to passive isolation system. However, the fuzzy logic control's performance is slightly superior to the LQR control under the first three earthquakes and much better under Northridge earthquake. Figure 14.26 shows the time history of base displacement under three isolation scenarios. Since excessive base displacement is one of the critical issues faced by passive base isolation, whether the smart semi-active controlled base isolation can resolve the challenge is of great interest. In the experiments, the passive base isolation system experienced hefty base displacement. Particularly in Kobe earthquake, the peak passive base displacement once reaches to 20 mm. Because of the bump between laminated core

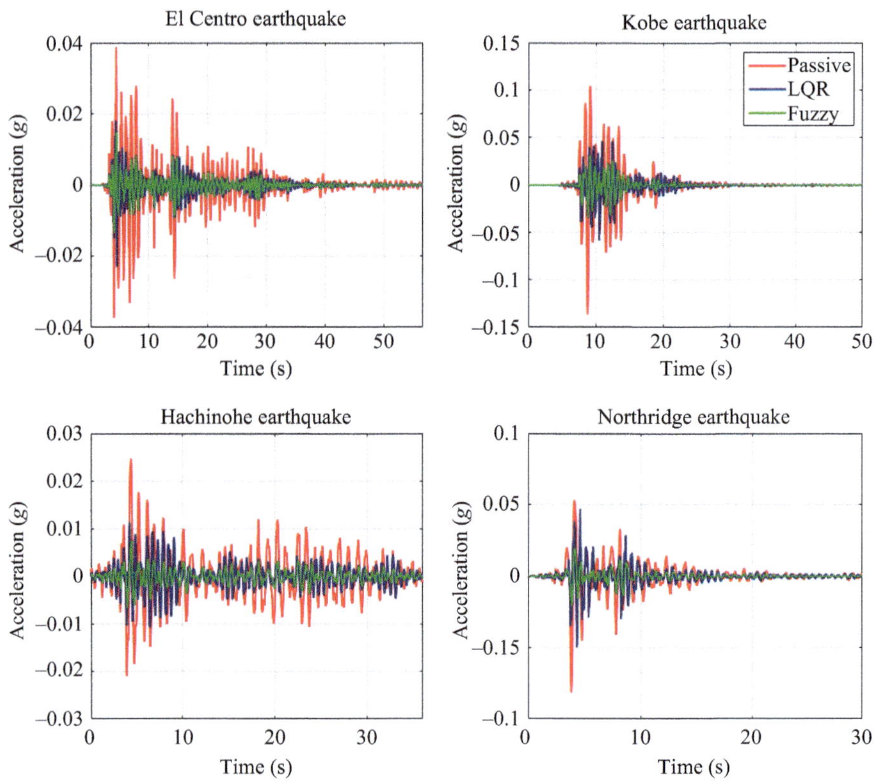

Figure 14.25 Time histories of base acceleration of passive, LQR and fuzzy-logic-controlled MRE base isolation

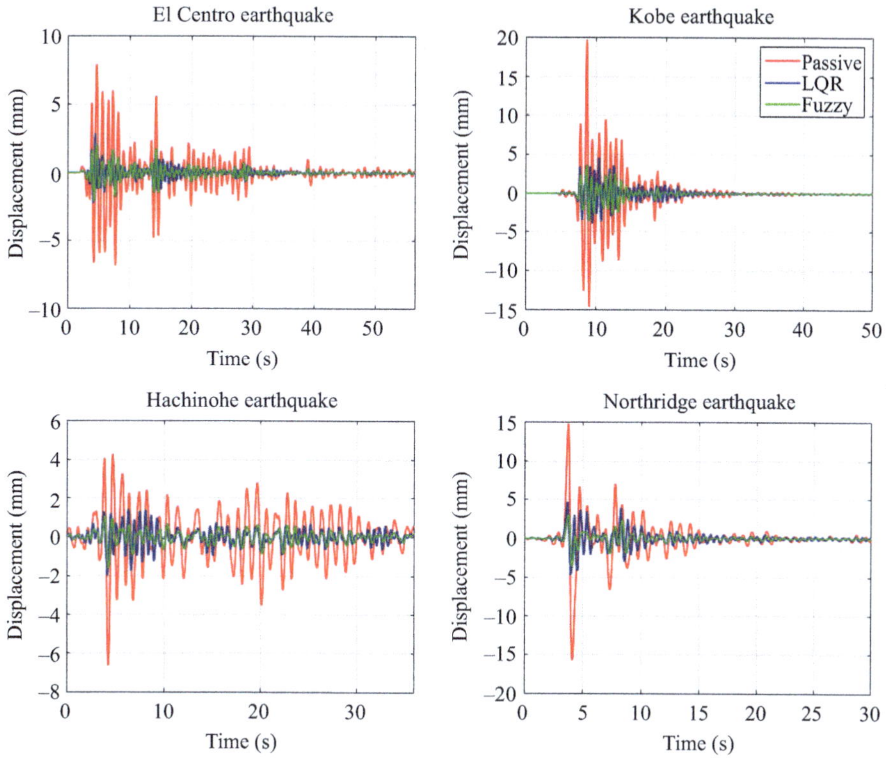

Figure 14.26 Time histories of base displacement of passive, LQR and fuzzy-logic-controlled MRE base isolation

and steel out wall, the peak acceleration also occurs at this time instant. Both of the LQR and fuzzy-logic-controlled MRE base isolation systems are able to significantly diminish the base displacement even under the most intense earthquake input. The displacements across the base isolation level are confined within an equitable range and considerably lower than the base deformation normally accepted in passive base isolation system [27], which shows promising capability of resolving the issue of excessive base movement.

14.7.2 Peek floor acceleration

Figures 14.27–14.30 illustrate the variations of peak value of inter-storey drift ratio, acceleration, displacement and floor shear over the elevation of the testing structure, in fixed base, passive isolation, LQR-controlled isolation and fuzzy logic isolation scenarios under four earthquake accelerograms. Under each earthquake, inter-storey drift ratios are obtained by dividing peak inter-storey drift with an inter-storey height of 400 mm; floor accelerations are the peak acceleration at each level throughout the whole time history expressed in the term of gravitational acceleration *g*; the last graph shows the ratio between floor shear force and the

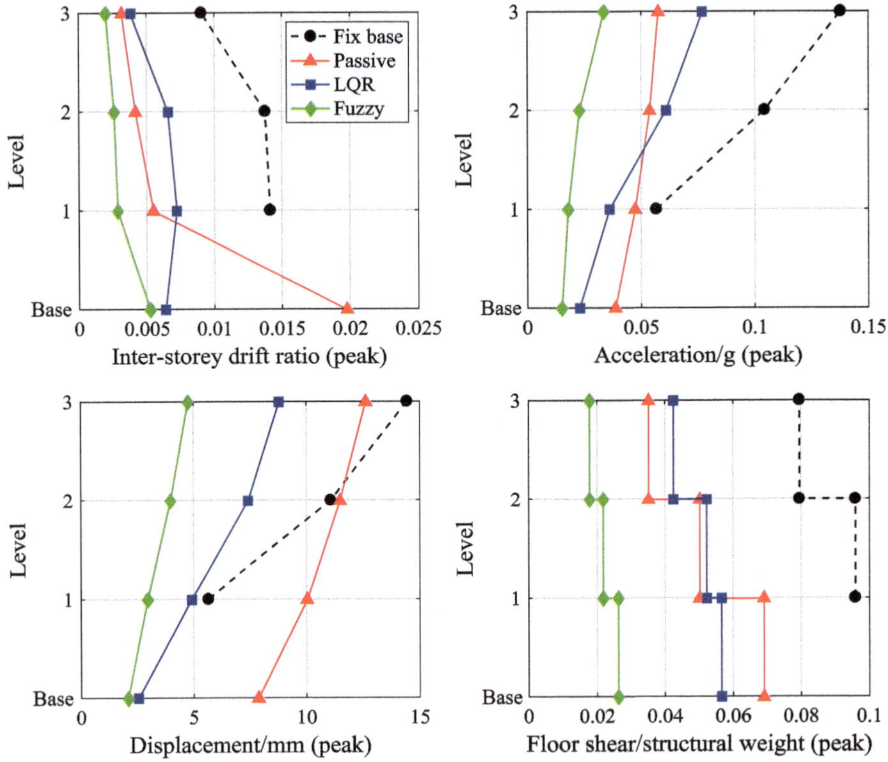

Figure 14.27 Peak floor responses of different isolation scenarios under El Centro earthquake (inter-storey drift ratio = inter-storey drift/storey height (0.4 m); structural weight = 912.57 N for fixed base building; 1,402.58 N for base-isolated building)

structural weight (912.57 N for fixed base building and 1,402.58 N for base-isolated building). The peak floor response of fixed base building is employed as the benchmark for protection effectiveness evaluation. When the listed value is greater than zero, it indicates a reduction in the corresponding index. In contrast, a value smaller than zero implies an amplification of the response.

Broadly, the reduction of peak floor responses achieved by controlled semi-active MRE base isolation system is rather pronounced, of which the fuzzy-logic-controlled isolation system attains the smallest monitored responses among all the isolation scenarios under any earthquake accelerograms. In contrast, the passive isolation system shows a great dependence on the characteristics of external excitations according to the responses, especially in terms of peak relative displacement. The LQR-controlled isolation system is able to achieve smaller evaluative indices values than the fixed base building. However, it does not show much superiority to the passive isolation system, and in some cases, its responses may even surpass those of passive isolation system.

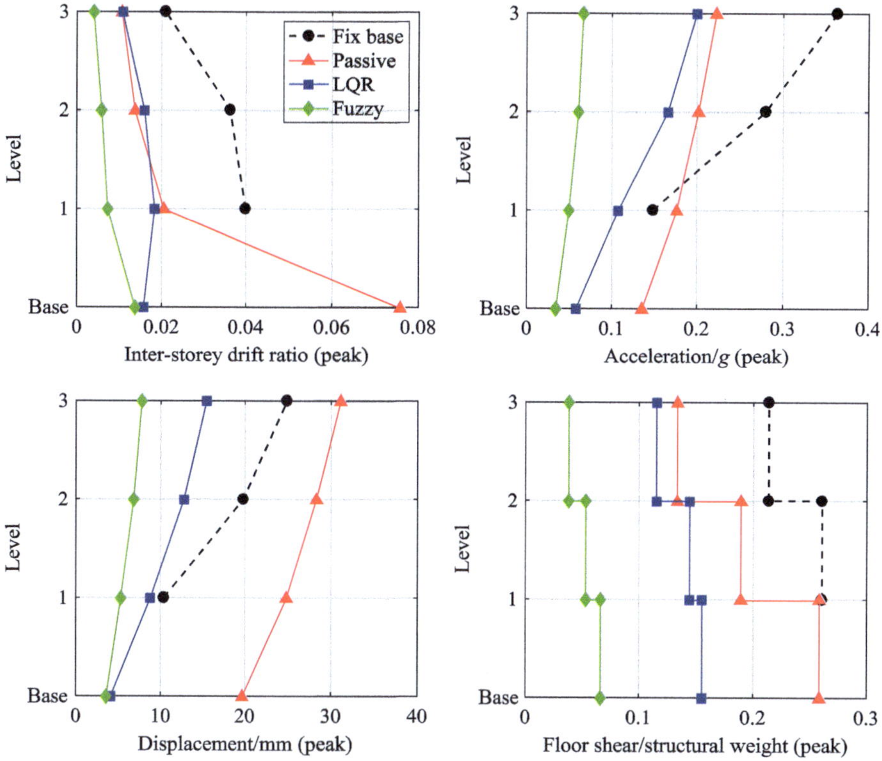

Figure 14.28 Peak floor responses of different isolation scenarios under Kobe earthquake (inter-storey drift ratio = inter-storey drift/storey height (0.4 m); structural weight = 912.57 N for fixed base building; 1,402.58 N for base-isolated building)

In the fixed building, the inter-storey drift ratio decreases as the structural height increases under all four earthquakes. Additionally, the inter-storey drifts of all three isolation scenarios (fuzzy, LQR and passive) are reduced, with respect to the fixed base building, on all levels except for the base level. The base inter-storey drift is actually equal to the base displacement and hence is not representative in the comparison of inter-storey drift. Moreover, the differences between inter-storey drift ratios of level 1–3 are not remarkable, which indicates that the superstructure of isolated building approaches rigid body motion through concentrating structural flexibility in the base isolation level. The LQR and passive-controlled isolation system shares similar inter-storey drift suppression performance and a reduction ranging from 48% to 74% are achieved under El Centro, Kobe and Northridge earthquakes. Under Hachinohe earthquake, both isolation systems could only reduce the inter-storey drift by around 30% at all levels. The fuzzy-logic-controlled isolation system, however, maintains a high level of drift reduction performance under all earthquakes, which ranges from 71% to 86%.

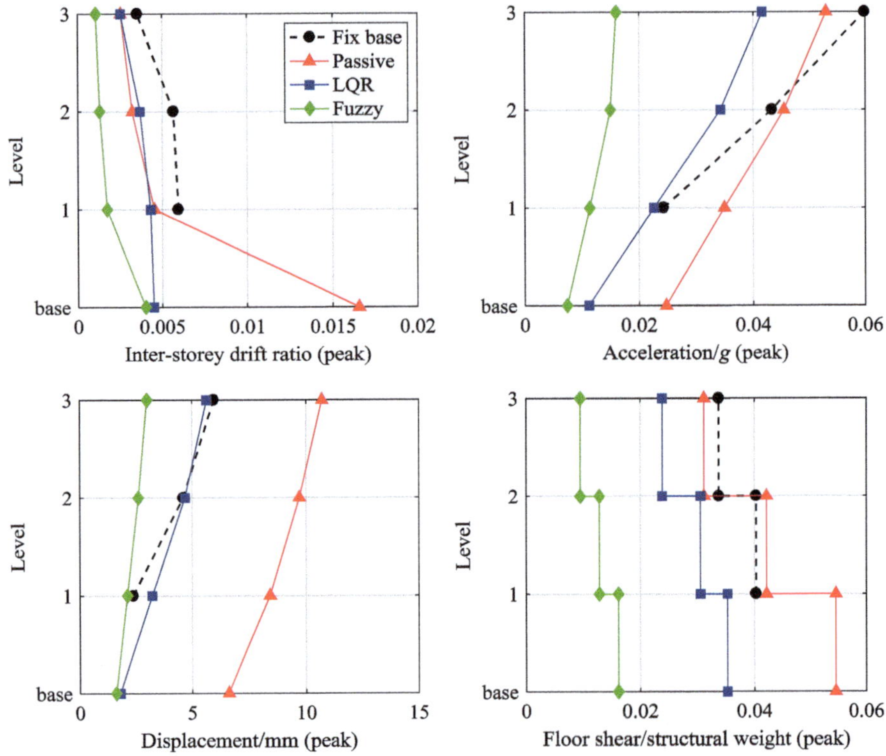

Figure 14.29 Peak floor responses of different isolation scenarios under Hachinohe earthquake (inter-storey drift ratio = inter-storey drift/storey height (0.4 m); structural weight = 912.57 N for fixed base building; 1,402.58 N for base-isolated building)

Both peak floor acceleration and relative displacement increases along the elevation of testing structure in all isolation scenarios and under all earthquake accelerograms. Such increase proves that the first mode (1.91 Hz) is the major vibrational mode participating throughout all four seismic events. The three isolation scenarios all managed to significantly reduce the peak floor acceleration under different earthquakes except for Hachinohe earthquake, where the passive isolation's acceleration response exceeds fixed base building at the first and second levels. Regarding to the reduction of peak floor acceleration, the passive, LQR and fuzzy-controlled isolation systems attained up to 58%, 44% and 78%, respectively, under El Centro earthquake; up to 39%, 45% and 82%, respectively, under Kobe earthquake; up to 64%, 51% and 81%, respectively, under Northridge earthquake. As for relative displacement, the passive isolation system shows a great reliance on the earthquake type in terms of suppression performance, i.e. under Kobe and Hachinohe earthquake, the passive isolation system's peak acceleration value is way larger than the fixed base building while under El Centro and Northridge

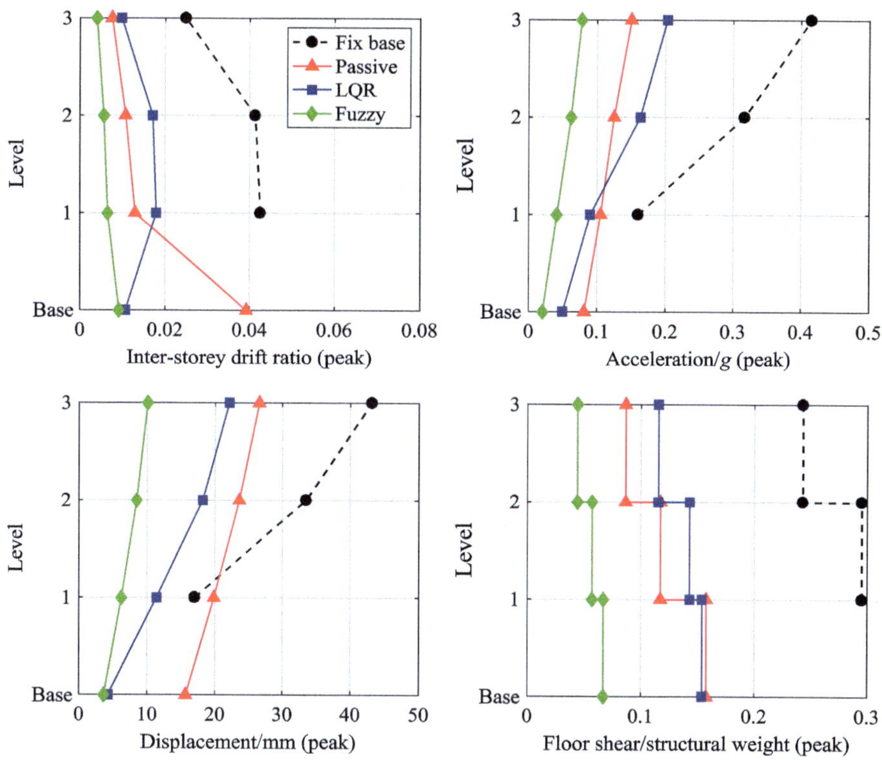

Figure 14.30 Peak floor responses of different isolation scenarios under Northridge earthquake (inter-storey drift ratio = inter-storey drift/ storey height (0.4 m); structural weight = 912.57 N for fixed base building; 1,402.58 N for base-isolated building)

earthquakes, the passive isolation system still realised considerable reduction (12% and 38%, respectively) of relative displacement at top floor. The LQR and fuzzy-logic-controlled base isolation system receive fairly close peak base displacement under all accelerograms: 2.5 and 2.1 mm under El Centro earthquake; 6.3 and 5.5 mm under Kobe earthquake; 1.8 and 1.6 mm under Hachinohe earthquake; 4.3 and 3.6 mm under Northridge earthquake; which are all far less than the base displacement of passive isolation system. Considering both inter-storey drift and base displacement graphs, although the LQR control shares similar inter-storey drift ratio as passive isolation system in the superstructure, the base displacement is well restrained by the LQR control, which shows one of the superiority of the controlled isolation system to the passive ones. The fuzzy-logic-controlled base isolation system once again achieved best relative displacement elimination performance among three isolation scenarios, with reductions ranging from 47% to 67% under El Centro earthquake; 49% to 69% under Kobe earthquake; 63% to 77% under Northridge earthquake. Under Hachinohe earthquake, a more pronounced performance dependence on the height is observed, with the reduction ranging from 11% to 49%.

The floor shear force indicates the vibration influence is applied on each level, and hence the smaller the floor shear force is, the less possible vibrational energy is transmitted into the superstructure. The three isolation scenarios all attained significant reduction in shear force at all floor under different accelerograms except for Hachinohe earthquake, where the shear force in passive isolation system surpasses that of the fixed base building between Level 2 and Level 1. The dependence of shear force reduction on the structural elevation is rather unpronounced under El Centro, Kobe and Northridge earthquakes. The passive, LQR and fuzzy logic isolation system brought about shear force reduction ranging from 48% to 60%, 45% to 47%, 77% to 78% under El Centro earthquake; 27% to 42%, 44% to 48%, 80% to 83% under Kobe earthquake 60% to 66%, 52% to 54%, 81% to 82% under Northridge earthquake. Nevertheless, the dependence is more pronounced under Hachinohe earthquake with a reduction ranging from −5% to 16% with passive isolation system; 24% to 34% with LQR controlled isolation system, 68% to 75% with fuzzy-logic-controlled isolation system.

14.7.3 Control force and applied current

The results discussed in this part are mainly regarding to the practical issues in the realisation of the semi-active MRE isolation system under LQR and fuzzy logic control. As explained in the section of numerical study, in order to realise the semi-active control with LQR controller, an inverse model of MRE isolator has to be employed to recurrent desired control force. Therefore, whether the force produced by MRE isolator can track the desired control force is one of the decisive factors of the control performance. Since it is hard to measure the shear force applied by MRE isolator on the structure without disturbing the system's movement, the force is calculated based on the real-time measurement of isolator's displacement and velocity as well as the stiffness and damping at the very time instant.

Figure 14.31 shows the comparison between desired control force and realised force by MRE isolator during four earthquakes. For better observation, it illustrates the control force during the representative 10 s of each ground accelerograms, during which relatively large control force is generated by the isolator (El Centro and Northridge earthquakes, 2–12 s; Kobe earthquake, 5–15 s; Hachinohe earthquake, first 10 s). As displayed in Figure 14.23, the MRE isolator is able to recurrent the trend of desired control force. Additionally, larger control force is required under Kobe and Northridge earthquakes around major seismic strikes to cope with the pulse-like wave, which distinguish the near-fault earthquakes from far-fault earthquakes. Furthermore, the MRE isolator is able to well recurrent the desired control force, especially when the control force is less than 4 kN, where the realised force perfectly matches the desired force. Nevertheless, the MRE isolator is reluctant to trace the desired control force around crests with a peak value larger than 5 kN, where the isolator's force normally outrides the desired force. Generally, however, in the proposed LQR control with inverse model, the MRE isolator can produce qualified control force to avoid unacceptable compromise in control performance.

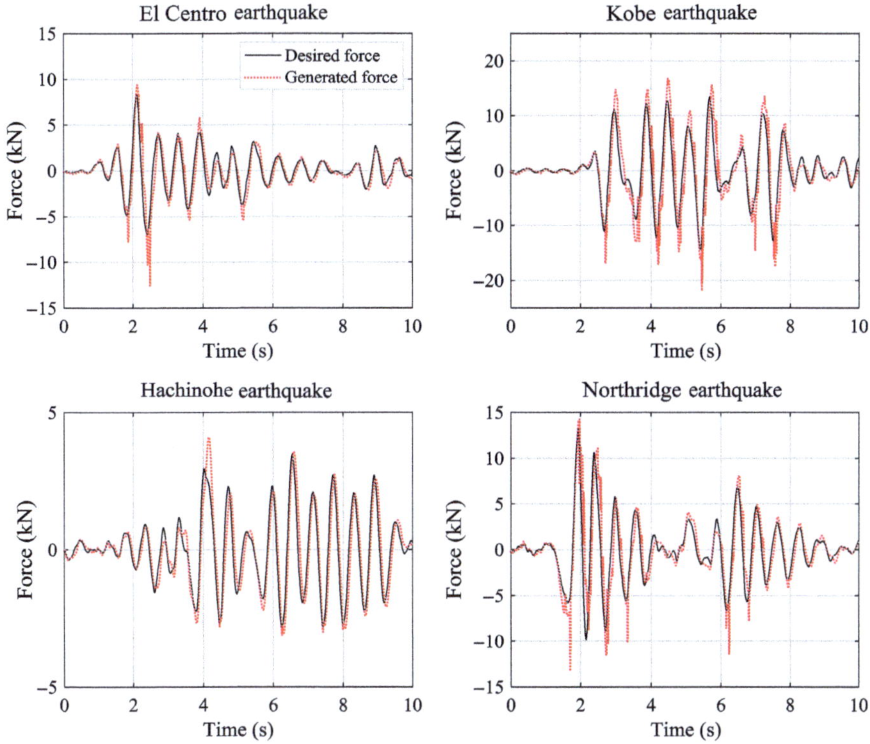

Figure 14.31 Comparison between desired force and realised force in LQR control (representative 10 s)

In the fuzzy logic control, the GA optimised fuzzy logic controller generates control current command based on the response feedback and the driving electrical circuits energise MRE isolator's solenoid accurately and immediately. As discussed before, overlong response time in MRE isolator's output regarding to the control current command will lead to degradation of control performance. Hence, whether the realised current can perfectly trace the control command is of great interest. In Figure 14.32, the measured current in the solenoid of MRE isolator is compared with the current command calculated by fuzzy logic controller in real-time. Under all four earthquakes, the initial current is 3 A to ensure the high stiffness of the semi-active base isolation system during normal time when there are no hazard excitation strikes. The realised current can accurately duplicate the current control command within reasonable response time except for some glitches in the signal due to overshoot, which happened occasionally. Through observation, the glitches normally occur at rising or falling edges of the signal and when the value of command current is jumping between maximum and minimum values frequently. Since the driving circuit is modulated by a PWM signal calculated by the PID controller, the frequent change in control command will lead to the PWM signal varying between 0 and 1 frequently, and thus during this period, the solenoid

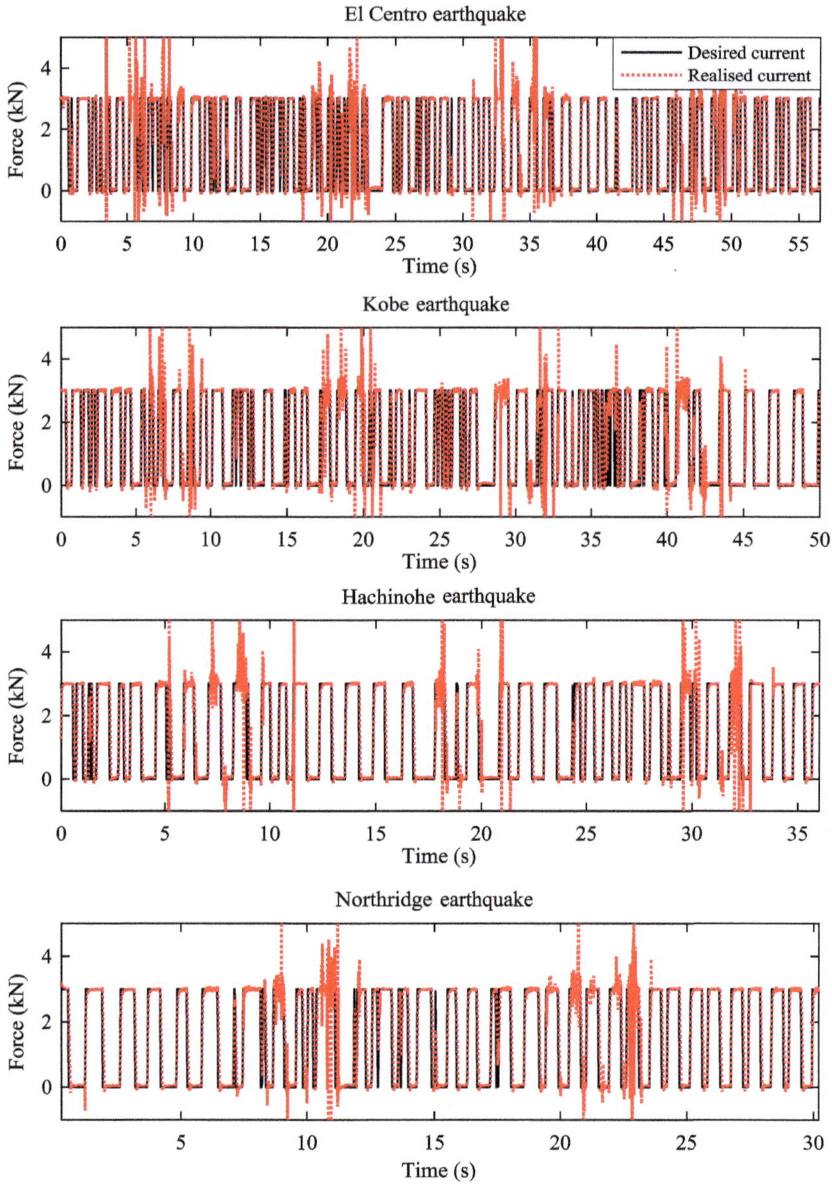

Figure 14.32 Comparison between desired force and realised current in GA-optimised fuzzy logic control

of MRE isolator will be overdriven with full voltage of the source constantly. Such frequent overdriving process is the major reason of signal glitches. Fortunately, the overshooting happens only occasionally and does not last for a long time, so such phenomena do not have a noticeable effect on control performance.

Acknowledgements

As the presented research was a team effort, the authors wish to acknowledge contributions from all members of their research team at Centre of Built Infrastructure Research, University of Technology Sydney, Australia. In particular, the authors wish to thank Dr Xiaoyu Gu for the work on the experimental realization of MRE adaptive isolation system, Dr Yang Yu for the work on MRE isolator modelling. The authors are also grateful to the helps from UTS laboratory staff, Peter Brown, Richard Turnell, Rami Haddad during the experimental research stage. In addition, the authors would like to acknowledge the financial supports on this research by Australian Research Council (DP150102636) and University of Technology Sydney in form of scholarship funding.

References

[1] Kelly JM. Aseismic Base Isolation. *The Shock and Vibration Digest* 1985 17 (7): 3–14.

[2] Kelly JM. Base Isolation: Linear Theory and Design. *Earthquake Spectra* 1990 6 (2): 223–44.

[3] Skinner RI, Robinson WH, and McVerry GH. *An Introduction to Seismic Isolation*. Chichester: John Wiley & Sons, 1993.

[4] Hall JF, Heaton TH, Halling MW and Wald DJ. Near-Source Ground Motion and Its Effects on Flexible Buildings. *Earthquake Spectra* 1995 11 (4): 569–605.

[5] Lu LY and Lin GL. Fuzzy Friction Controllers for Semi-Active Seismic Isolation Systems. *Journal of Intelligent Material Systems and Structures* 2009 20 (14): 1747–70.

[6] Kelly JM. The Role of Damping in Seismic Isolation. *Earthquake Engineering & Structural Dynamics* 1999 28 (1): 3–20.

[7] Jangid R and Kelly J. Base Isolation for Near-Fault Motions. *Earthquake Engineering & Structural Dynamics* 2001 30 (5): 691–707.

[8] Kelly J, Leitmann G, and Soldatos A. Robust Control of Base-Isolated Structures under Earthquake Excitation. *Journal of Optimization Theory and Applications* 1987 53 (2): 159–80.

[9] Yoshioka H, Ramallo J, and Spencer Jr B. "Smart" Base Isolation Strategies Employing Magnetorheological Dampers. *Journal of Engineering Mechanics* 2002 128 (5): 540–51.

[10] Chen L, Gong X, and Li W. Microstructures and Viscoelastic Properties of Anisotropic Magnetorheological Elastomers. *Smart Materials and Structures* 2007 16 (6): 2645.

[11] Yi F, Dyke SJ, Caicedo JM and Carlson JD. Experimental Verification of Multiinput Seismic Control Strategies for Smart Dampers. *Journal of Engineering Mechanics* 2001 127 (11): 1152–64.

[12] Dyke S, Spencer BF, Sain MK and Carlson JD. Modeling and Control of Magnetorheological Dampers for Seismic Response Reduction. *Smart Materials and Structures* 1996 5 (5): 565.

[13] Yoshida O and Dyke SJ. Seismic Control of a Nonlinear Benchmark Building Using Smart Dampers. *Journal of Engineering Mechanics* 2004 130 (4): 386–92.

[14] Nagarajaiah S and Narasimhan S. Smart Base-Isolated Benchmark Building. Part II: Phase I Sample Controllers for Linear Isolation Systems. *Structural Control and Health Monitoring* 2006 13 (2–3): 589–604.

[15] Crandall SH. The Role of Damping in Vibration Theory. *Journal of Sound and Vibration* 1970 11 (1): 3–18, IN1. Available at: https://onlinelibrary.wiley.com/doi/10.1002/eqe.4290220403

[16] Inaudi JA and Kelly JM. Hybrid Isolation Systems for Equipment Protection. *Earthquake Engineering & Structural Dynamics* 1993 22 (4): 297–313.

[17] Tsai HC and Kelly JM. Seismic Response of Heavily Damped Base Isolation Systems. *Earthquake Engineering & Structural Dynamics* 1993 22 (7): 633–45.

[18] Liao G, Gong XL, Xuan SH, Kang CJ and Zong LH. Development of a Real-Time Tunable Stiffness and Damping Vibration Isolator Based on Magnetorheological Elastomer. *Journal of Intelligent Material Systems and Structures* 2012 23 (1): 25–33.

[19] Behrooz M, Wang X, and Gordaninejad F. Modeling of a New Semi-Active/Passive Magnetorheological Elastomer Isolator. *Smart Materials and Structures* 2014 23 (4): 045013.

[20] Li Y, Li J, Tian T, and Li W. A Highly Adjustable Magnetorheological Elastomer Base Isolator for Applications of Real-Time Adaptive Control. *Smart Materials and Structures* 2013 22 (9): 095020.

[21] Yang J, Sun SS, Du HP, Li WH, Alici G and Deng HX. A Novel Magnetorheological Elastomer Isolator with Negative Changing Stiffness for Vibration Reduction. *Smart Materials and Structures* 2014 23 (10): 105023.

[22] K-Karamodin A and H-Kazemi H. Semi-Active Control of Structures Using Neuro-Predictive Algorithm for Mr Dampers. *Structural Control & Health Monitoring* 2010 17 (3): 237.

[23] Li Y, Li J, Li W, and Samali B. Development and Characterization of a Magnetorheological Elastomer Based Adaptive Seismic Isolator. *Smart Materials and Structures* 2013 22 (3): 035005.

[24] Li J, Li Y, Li W, and Samali B. Development of adaptive seismic isolators for ultimate seismic protection of civil structures, *Proc. SPIE 8692, Sensors and Smart Structures Technologies for Civil, Mechanical, and Aerospace Systems*, 2013, 86920H.

[25] Li Y, Li J, and Samali B. A novel adaptive base isolator utilising magnetorheological elastomer, *22nd Australasian Conference on the Mechanics of Structures and Materials*, Sydney, Australia, 11–14 December 2012, 763–7.

[26] Zeng J, Guo Y, Li Y, Zhu J, and Li J. Two-Dimensional Magnetic Property Measurement for Magneto-Rheological Elastomer. *Journal of Applied Physics* 2013 113 (17):17A919.

[27] Li Y, Li J, and Samali B. On the Magnetic Field and Temperature Monitoring of a Solenoid Coil for a Novel Magnetorheological Elastomer Base Isolator. *Journal of Physics: Conference Series* 2013 412 (1), 012033.

[28] AASHTO. *AASHTO Guide Specifications for Seismic Isolation Design*. Washington, DC, American Association of State Highway and Transportation Officials (AASHTO), 2010.

[29] Yang J, Du H, Li W, *et al.* Experimental Study and Modeling of a Novel Magnetorheological Elastomer Isolator. *Smart Materials and Structures* 2013 22 (11): 117001.

[30] Li Y and Li J. A Highly-Adjustable Base Isolator Utilizing Magnetorheological Elastomer: Experimental Testing and Modelling. *ASME Journal of Vibration and Acoustics* 2015 137 (01): 011009.

[31] Li Y and Li J. On Rate-Dependent Mechanical Model for Adaptive Magnetorheological Elastomer Base Isolator. *Smart Materials and Structures* 2017 26 (4): 045001.

[32] Yu Y, Li Y, and Li J. Parameter Identification of a Novel Strain Stiffening Model for Magnetorheological Elastomer Base Isolator Utilizing Enhanced Particle Swarm Optimization. *Journal of Intelligent Material Systems and Structures* 2015 26 (18): 2446–62.

[33] Yu Y, Li Y, and Li J. Parameter Identification and Sensitivity Analysis of an Improved LuGre Friction Model for Magnetorheological Elastomer Base Isolator. *Meccanica* 2015 50 (11): 2691–707.

[34] Yu Y, Li Y, and Li J. Parameter identification of an improved Dahl model for magnetorheological elastomer base isolator based on enhanced Genetic algorithm, *23rd Australasian Conference on the Mechanics of Structures and Materials (ACMSM23)*, Byron Bay, Australia, 9–12 December 2014.

[35] Yu Y, Li Y, Li J, and Gu X. A Novel Hysteresis Model for Dynamic Behaviour of Magnetorheological Elastomer Base Isolator. *Smart Materials and Structures* 2016 25 (5): 055029.

[36] Chen X, Li Y, Li J, and Gu X. A Novel Dual-Loop Adaptive Control for Minimizing Time Response Delay in Real-Time Structural Vibration Control with Magnetorheological (MR) Devices. *Smart Materials and Structures* 2018 27 (1): 015005.

[37] Koo JH, Khan F, Jang DD, and Jung HH. Dynamic Characterization and Modeling of Magneto-Rheological Elastomers under Compressive loadings. *Smart Materials & Structures* 2009 19 (11): 117002.

[38] Yu Y, Li Y, and Li J. Nonparametric Modeling of Magnetorheological Elastomer Base Isolator Based on Artificial Neural Network Optimized by Ant Colony Algorithm. *Journal of Intelligent Material Systems & Structures* 2015 26 (14): 1789–98.

[39] Leng D, Xu K, Ma Y, Liu G, and Sun L. Modeling the Behaviors of Magnetorheological Elastomer Isolator in Shear-Compression Mixed Mode Utilizing

Artificial Neural Network Optimized by Fuzzy Algorithm (ANNOFA). *Smart Materials and Structures* 2018 27 (11): 115026.
[40] Yu Y, Li Y, Li J, and Gu X. Self-Adaptive Step Fruit Fly Algorithm Optimized Support Vector Regression Model for Dynamic Response Prediction of Magnetorheological Elastomer Base Isolator. *Neurocomputing* 2016 211: 41–52.
[41] Specht DF. A General Regression Neural Network. *IEEE Transactions on Neural Networks* 1991 2 (6): 568–76.
[42] Gu X, Yu Y, Li Y, Li J, Askari M, and Samali B. An Experimental Study of Semi-Active Magnetorheological Elastomer Base Isolation System Using Optimal Neuro Fuzzy Logic Control. *Mechanical Systems and Signal Processing* 2019 119: 380–398.
[43] Gu X, Yu Y, Li J, and Li Y, Semi-Active Control of Magnetorheological Elastomer Base Isolation System Utilising Learning-Based Inverse Model. *Journal of Sound and Vibration* 2017 406: 346–62.
[44] Kim H-S and Roschke PN. Design of Fuzzy Logic Controller for Smart Base Isolation System Using Genetic Algorithm. *Engineering Structures* 2006 28 (1): 84–96.
[45] Symans MD and Kelly SW. Fuzzy Logic Control of Bridge Structures Using Intelligent Semi-Active Seismic Isolation Systems. *Earthquake Engineering & Structural Dynamics* 1999 28 (1): 37–60.

Chapter 15

The designation and mechanical properties of magnetorheological plastomer

Xinglong Gong[1], Jiaqi Xu[1] and Shouhu Xuan[1]

Magnetorheological (MR) plastomer (MRP) is a class of magneto-sensitive smart material that incorporates magnetic particles into a soft polymer matrix. The low cross-linking density matrix allows the magnetic particles to be entrapped without sedimentation. Moreover, the magnetic particles can be moved under the induction of the magnetic field, so the MRP shows well stability and exhibits high MR effect. This chapter provides a detailed introduction to MRPs, including preparation, mechanical and electrical properties, formation mechanism, and applications.

15.1 Introduction

MR material is a magneto-sensitive smart material composed of different matrices, magnetic particles, and additives. The magnetic particles within MR material can be magnetized under applying an external magnetic field, and the magnetic dipole force enhances the mechanical properties of the MR material. Therefore, the magneto-controllable properties of MR material (MR effect), in the aspect of rheological properties, electrical properties, and acoustic properties, are manifested. Due to the rapid response and high MR effect, MR material has been widely used in architecture [1,2], automobile [3,4], machinery [5,6] and so on.

As early as 1948, Rabinow invented the MR fluid (MRF) consisting of carbonyl iron microparticles and light machine oil [7]. Due to its quick and reversible response of rheology to magnetic field, the MRF attracted intense interest of many researchers [8–12]. However, the difference in density between the magnetic particles and the matrix caused the problem of particle sedimentation, which became a major problem for the development of MRF [13–15]. Jolly added iron powder particles to the elastic polymer matrix to solve the problem of particle sedimentation, and thus the MR elastomer (MRE) was obtained [16]. The elastic polymer matrix gave rise to high stability and large stiffness of the MRE [17–19]. To this end, MRE showed broad application in various fields such as isolator [20,21], absorber [22–24], damper [25],

[1]CAS Key Laboratory of Mechanical Behavior and Design of Materials, Department of Modern Mechanics, University of Science and Technology of China, Hefei, China

and sensor [26,27]. Unfortunately, the magnetic particles were tightly located within the polymer matrix during the MRE preparation, and then the MR effect was weak due to the immobility of particles. MRP was the intermediate system between MRE and MRF with rapid rheological response and oriented particle alignment. The matrix of MRP was soft polymer with viscoelastic properties. Due to the low density of the matrix, the magnetic particles could reassemble their aggregation structures to form more compact chains by applying the external magnetic field [28]. Because of the improving MR effect, much effort has been committed to develop high-performance MRP and investigate its unique mechanism.

In 1995, Shiga *et al.* first dispersed iron powder in a silicone gel matrix and tested its magneto-viscoelastic properties [29]. Since then, a variety of matrices, such as hydrogel matrix [30–33], polyurethane matrix [34–36], polymer matrix [37,38], paraffin matrix [39], grease matrix [40], and mixed matrix [41,42], have been applied for MRPs. Recently, our group reported a novel MRP, which was composed of magnetic particles and low cross-linked polymer matrix. Since the viscosity nature, the MRP can be molded into various shapes. The magnetic particles in MRP can move readily by the external magnetic-field-induced torques. The easy formation of the particle chains in MRP played a key role in the achieving of high MR effect.

In this chapter, after a brief introduction of the synthetic methods of the MRPs, we provided a detailed and focused overview of the mechanical and electrical properties of MRP, as well as the evolution of microstructure and the mechanism of the magnetic-field-induced properties. Finally, the applications of MRPs were presented.

15.2 Fabrication methods of MRPs

15.2.1 Hydrogel-like MRP

Hydrogels are three-dimensional cross-linking hydrophilic polymer networks containing a large amount of water. In the past years, they have received increasing interests in the fields of medicine, biology, and sensor because of their excellent biocompatibility and mechanical properties. The combination of magnetic particles and hydrogels can represent an intelligent hydrogel-like MRP that responds quickly to external magnetic fields. The preparation of hydrogel-like MRP is divided into two steps: the polymerization of hydrogel networks and the mixing of magnetic particles. In general, hydrogel networks mainly include carrageenan [43], agar [44], guar gum [45], and polymers and their derivatives (e.g., polyvinyl alcohol [33], polyethylene glycol and acrylamide [46], and polyacrylamide [47]). Magnetic particles are micro- and nano-sized irons and their oxides. The methods for synthesizing hydrogel-like MRP that have been reported can be broadly classified into blending method [48–50], *in-situ* precipitation method [51–53], freeze–thawing method [54,55], and chemical cross-linking method [56–58].

15.2.2 Polyurethane MRP

Polyurethane, which is composed of polyisocyanates and hydroxyl-terminated compounds, is a common matrix for MRPs. The mechanical properties can be adjusted by the ratio of soft and hard segments. Many research results have been obtained in the preparation of MRPs based on polyurethane. For example, Fuchs *et al.* synthesized the polyurethane matrix for the MRPs by using polyols and diphenylmethane diisocyanate (MDI) as raw materials [59,60]. MDI and castor oil were used as the main raw materials for the magnetic polyurethane matrix. In addition, polyurethane can also be synthesized from MDI and castor oil [61–63]. In 2011, our group developed a plasticine-like MR materials with plastic polyurethane matrix and carbonyl iron particles (CIPs) [64]. The molecular weight of polyurethane matrix was low, and it was synthesized by using toluene diisocyanate and polypropylene glycol (PPG-1000).

15.2.3 Shear-thickening MRP

Shear-thickening gels are smart materials whose mechanical properties are typically enhanced with external excitation of shear, compression, and tension [65]. Doping the magnetic additives into shear-thickening gel enables to exhibit a special magnetic response characteristic. For the synthesis of shear-thickening gel matrix, the boric acid and the dimethyl siloxane were raw materials, and the benzoyl peroxide was the vulcanizing agent [65,66]. Finally, the shear-thickening gel was made of 65% polydimethylsiloxane, 17% silica (crystalline quartz), 9% thixatrol ST (castor oil derivative), 4% dimethylsiloxane (hydroxy-terminated polymers with boric acid), 1% glycerin, and 1% titanium dioxide [67]. After introducing the magnetic particles into shear-thickening gel, the multifunctional magnetic polymer composite showed excellent MR behavior and shear-thickening effect, so it was named as shear-thickening MRP. Besides the plasticity, the external magnetic forces also can induce the shear-thickening MRP into various shapes. More importantly, when stimulated by shear frequency and magnetic field simultaneously, the mechanical properties of the shear-thickening MRP can greatly enhanced due to the synergistic effect [66].

15.2.4 Thermosensitive MRP

Finally, the real MR plasticine was prepared by dispersing CIPs in paraffin matrix and petroleum jelly. Equal amounts of paraffin wax and petroleum jelly were fully melted in a water bath until a transparent solution was formed. Then a certain amount of CIPs was added to the above system with a vigorous mechanical stirring. As soon as the particles and the matrix were well mixed, cooling the mixture to room temperature [39]. Since the paraffin matrix was a phase-changing matrix, this kind of MR material can be applied as MRP and MRF by changing the temperature. Similarly, the grease-based MR materials also exhibited such a thermosensitive property [40,68,69].

15.3 Mechanical properties

15.3.1 MR effect of MRP

MR property was one of the most important properties for evaluating MR materials. A commercial rheometer was commonly used to test the MR properties in rheological tests [63,64,70–73]. Different from the MRF and MRE, iron powder particles in MRP did not settle and were able to move under a magnetic field. Because the iron powder particles were movable, they aggregated to form chain-like structures along the magnetic field, which improved the MR effect [74]. Furthermore, the magnetic dipole moment between iron powder particles increased with increasing magnetic field density, which enhanced the storage modulus of MRP (Figure 15.1(a)). This magneto-induced storage modulus enhancement phenomenon occurred in many MRPs. Yang *et al.* observed that the storage modulus of magnetic polyurethane containing 80 wt% iron powder particles increased from 0.04 to 2.81 MPa with increasing the magnetic field from 0 to 1 T [63]. The storage modulus of MR material formed by dispersing triiron tetraoxide in hydrogel also increased as the magnetic field density and particle content increased [32]. At high magnetic field density, the restriction effect of the iron powder particles on soft segments of matrix decreased, resulted in a lower loss factor (Figure 15.1(b)). At different constant magnetic fields, the loss factor versus shear strain amplitude and frequency can be explained by the interaction between matrix and iron powder particles [63]. In addition, the magnetic interactions in isotropic MRP were smaller than that in anisotropic MRP and the interfacial slipping between the iron powder particles and the matrix in isotropic MRP was larger than that in anisotropic MRP, so the isotropic MRP exhibited a smaller

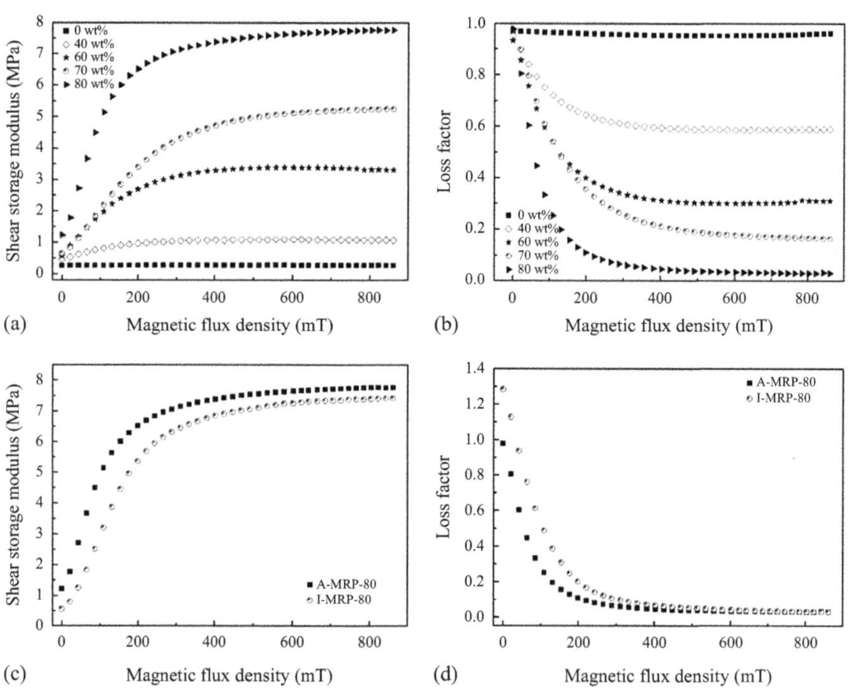

Figure 15.1 (a) Storage modulus and (b) loss factor of anisotropic MRP with different iron particle content under different magnetic flux density; (c) storage modulus; and (d) loss factor of MRP-80 (the MRP-x presents the weight percentage of iron particles) with different pretreatment process under different magnetic flux density [64]

storage modulus (Figure 15.1(c)) and a higher damping factor (Figure 15.1(d)) [28,64].

The MR properties were also dependent on the shear frequency. This phenomenon of storage modulus significantly increased with shear frequency and was found in shear-thickening MRPs [67,75,76]. Figure 15.2(a) showed that the storage modulus of shear-thickening MRP increased by four orders of magnitude through scanning frequency increasing from 0.1 to 100 Hz, which presented the typical shear-stiffening characteristic. In addition, the magnetic field promoted the aggregation of magnetic particles to form chain-like structures and enhanced the mechanical properties (Figure 15.2(b)).

Wang *et al.* demonstrated that when the shear frequency increased, "cross-links" induced by the B and O atoms in shear-thickening MRP attracted the polymer chains and impeded the movement of other chains, resulting in thickening of the matrix (Figure 15.3) [66]. The temperature dependence of the shear-thickening MRP was also discussed by Guo *et al.* [76]. The high temperature softened the matrix, causing a lower initial storage modulus and a less resistance to the arrangement of magnetic particles.

In particular, there have been many developments regarding the MR properties of modified MRPs. Malecki coated CIPs with the solgel silica layer to improve the adhesion between the hydrophilic magnetic particles and the hydrophobic polymer, which was advantageous for improving the stability and durability [77]. Yang replaced conventional (microsphere-based) CIPs with dendritic CIPs and flower-like CIPs to enhance the comprehensive performance of magnetic polyurethane [78,79]. Due to high strength and toughness, carbon materials such as graphite [80], carbon nanotubes, and carbon fibers [81,82] were doped in MRP to strengthen the mechanical properties. The MRP containing hard magnetic particles (NdFeB) exhibited a complex MR behavior different from that of the soft magnetic particles because of the hysteresis characteristics and particle shape of the NdFeB [83].

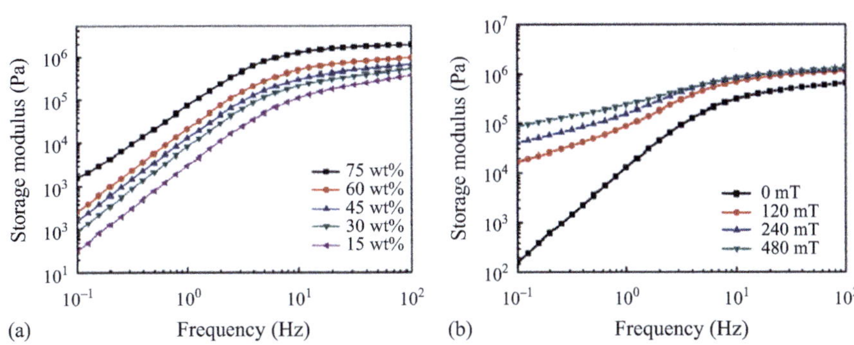

Figure 15.2 (a) Frequency sweeping tests of the shear-thickening MRP with different content of iron particles without magnetic field and (b) storage modulus of shear-thickening MRP-45 versus sweeping frequency in different magnetic field [65]

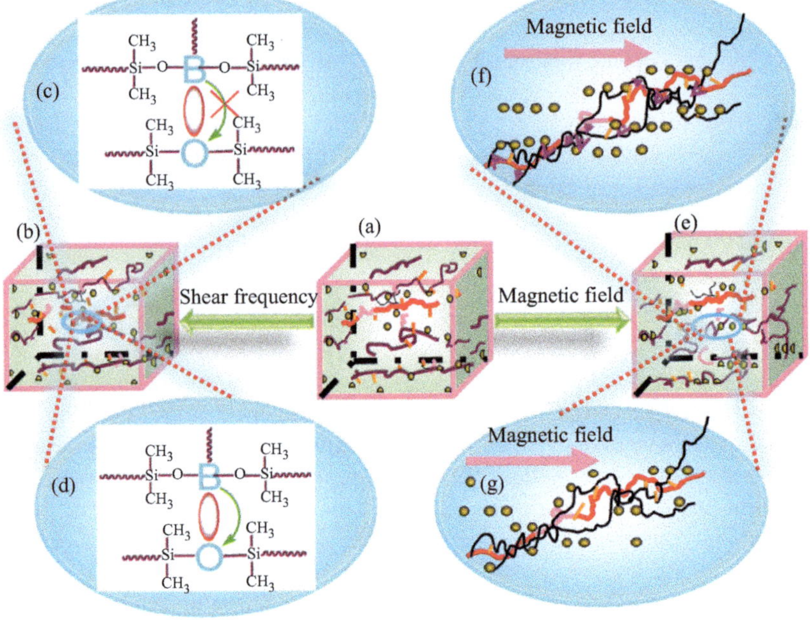

Figure 15.3 The changes in the microstructure of multifunctional polymer composite samples [66]

Interestingly, magnetically sensitive smart hydrogels of polyacrylamide in combination with barium ferrite nanoparticles (BaM) or BaM and graphene oxide exhibited a negative MR effect [84].

15.3.2 Dynamic mechanical properties of MRP

Considering the potential application of MRPs in vibration control, damping, and body armor, it was necessary to investigate their mechanical properties under high strain rates. According to the 1-D elastic wave propagation theory, the dynamic stress, the dynamic strain, and the strain rate can be expressed as

$$\sigma_s = \frac{E_b A_b}{A_s}\varepsilon_t \qquad \varepsilon_s = \frac{2C_b}{l_s}\int_0^t (\varepsilon_i - \varepsilon_t)d\tau \qquad \dot{\varepsilon_s} = \frac{2C_b}{l_s}(\varepsilon_i - \varepsilon_t)$$

where σ_s, ε_s, and $\dot{\varepsilon_s}$ are the dynamic stress, the dynamic strain, and the strain rate, respectively; ε_i and ε_t are the incident pulse and transmitted pulse, respectively. A_b and A_s are the cross-sectional areas of the bars and the samples, respectively and E_b is the elastic modulus of the bar material. C_b is the propagation velocity of the elastic wave in the bar and l_s is the length of the sample [75].

A modified split Hopkinson pressure bar (SHPB) system consisted of a striker bar, an incident bar, a transmission bar, an absorber bar, and a magnetic device (Figure 15.4) was used to study the dynamic mechanical properties of MRP [85].

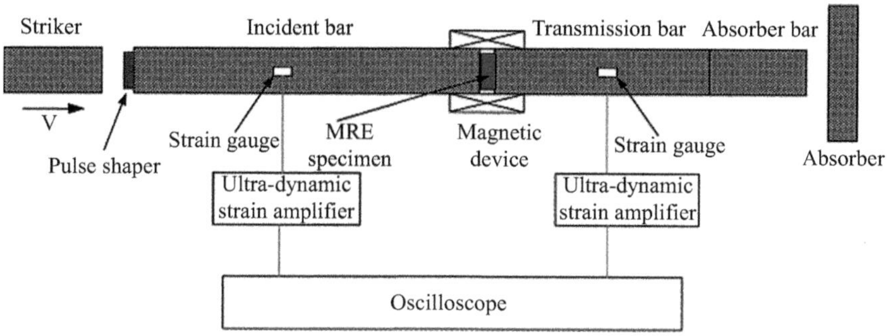

Figure 15.4 Schematic illustration of a modified SHPB setup [85]

By using the SHPB, Xu systematically investigated the rate-dependent and magnetic field-dependent dynamic mechanical properties of MRP at high strain rates [75]. As shown in Figure 15.5(a), stress increased with iron powder particle content and strain rate. On the one hand, the increase of iron powder particle content hardened the MRP, which was called particle reinforcement effect. On the other hand, due to the increasing strain rate, the matrix presented the thickening properties. The magnetic field density also enhanced the dynamic mechanical properties (Figure 15.5(b)). At lower strain rate, the agglomeration of polymer molecular chain inside the matrix was not significant, and the CIPs were easily assembled to form chains in a magnetic field. At a higher strain rate, the CIPs were tightly entangled by the polymer molecular chains, and they were difficult to form chain-like microstructures under the low magnetic flux density. Therefore, the increase rate of maximum stress at about 6,500 s^{-1} was smaller than the one at about 2,200 s^{-1} under a small magnetic field (Figure 15.5(c)). When magnetic flux density increased, the magnetic dipole force exceeded the matrix resistance and then the particles aggregated with the surrounding's to reform chain-like structures. The schematic diagram of the microstructure evolution was shown in Figure 15.5(d).

Similarly, the dynamic mechanical properties of shear-thickening MRP at high strain rates can be also studied by the SHPB. As the strain rate increased, it was proposed that shear-thickening MRP changed from the elastic state to the rubber state and the glass state (Figure 15.5(e)). In the elastic state and the rubber state, the cross bond blocked the movement of the molecular chain, but the chain segments can still rotate around a single bond, so the elastic modulus and the maximum stress continued to increase. When the strain rate was above 4,230 s^{-1}, the shear-thickening MRP went into a glassy state, the cross bonds were easily destroyed, and the chain segments cannot move nor rotate. The entangled molecular chains took shapes as molecular groups moving in a small range, and interaction with each other like crystal lattice [65]. Due to the recoverable cross-linking of B–O bonds, shear-thickening MRP exhibited a high self-healing performance. The impacted fragments were reshaped and tested at the same strain rate (Figure 15.5(f)). Both the maximum stress and the elastic modulus were very close to the initial sample which indicated that the shear-thickening MRP represented excellent repeatable mechanical properties (Figure 15.5(g) and (h)).

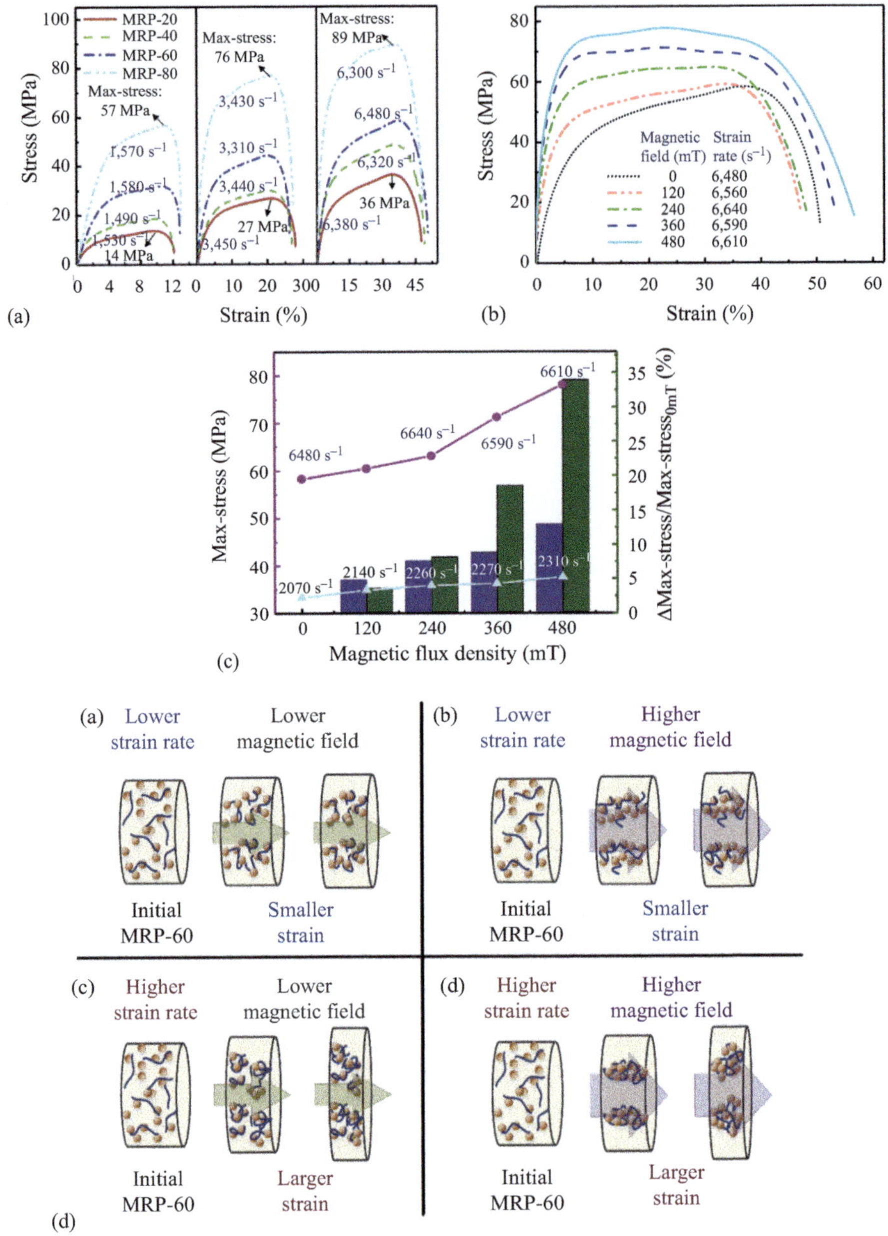

Figure 15.5 *(a) The stress–strain curves of MRPs at the different strain rates; (b) the dynamic compression stress–strain curves of MRP-60 with different magnetic field at the strain rate of about 6,500 s*$^{-1}$*; (c) the maximum stress and the increase rate of maximum stress of MRP-60; (d) the microstructure evolution of MRPs in dynamic compression process [75]; (e) stress–strain curves of the shear-thickening MRP-45 at different strain rate; (f) cracks appear in the specimens of shear-thickening MRP; stress–strain curves of primary specimens and remolded specimens at strain rate of about; (g) 2,650 s*$^{-1}$*, and (h) 5,500 s*$^{-1}$ *[65]*

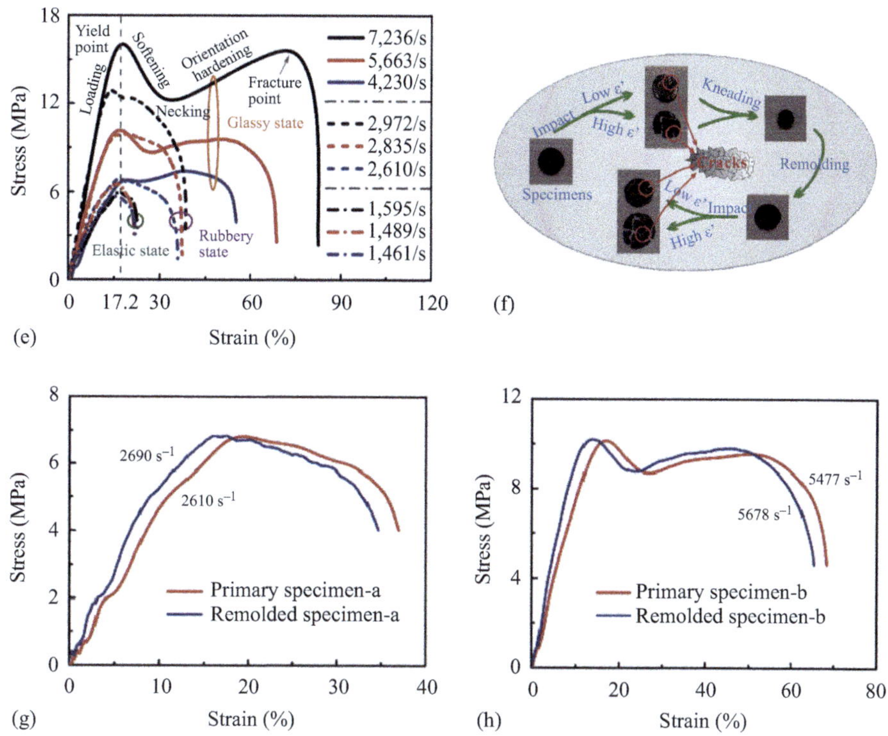

Figure 15.5 (Continued)

15.3.3 Magnetic–electric coupling properties of MRP

Since the iron powder particles were conductive, MRP also had unique electrical properties that can be controlled by magnetic field. The conductivity of MRP was attributed to the aligned distribution of particles. As illustrated in Figure 15.6(a), the spacing between adjacent particles and the thickness of the interface layer between the electrode and sample in anisotropic MRP were less than that in isotropic MRP, so the conductivity for anisotropic MRP was better than that for isotropic MRP. Moreover, the spacing between adjacent particles and the thickness of the interface would decrease as the magnetic field density increased, resulting in an increase of the conductivity for the MRP (Figure 15.6(b)). Therefore, the conductivity of MRP was also highly dependent on the direction of the particle chains (Figure 15.6(c)). Yu *et al.* reported that the ratio of castor oil and MDI, which were raw materials for the synthetic polyurethane matrix, affected the conductivity of the MR material [61]. In addition, the conductivity of magnetic polyurethane that replaced a part of CIPs with iron nanowires was also discussed [86]. In comparison to CIPs, iron nanowires had smaller saturation magnetization, larger residual magnetization, and coercivity (Figure 15.6(d)). As the content of iron nanowires increased, the conductivity increased first and then decreased (Figure 15.6(e)).

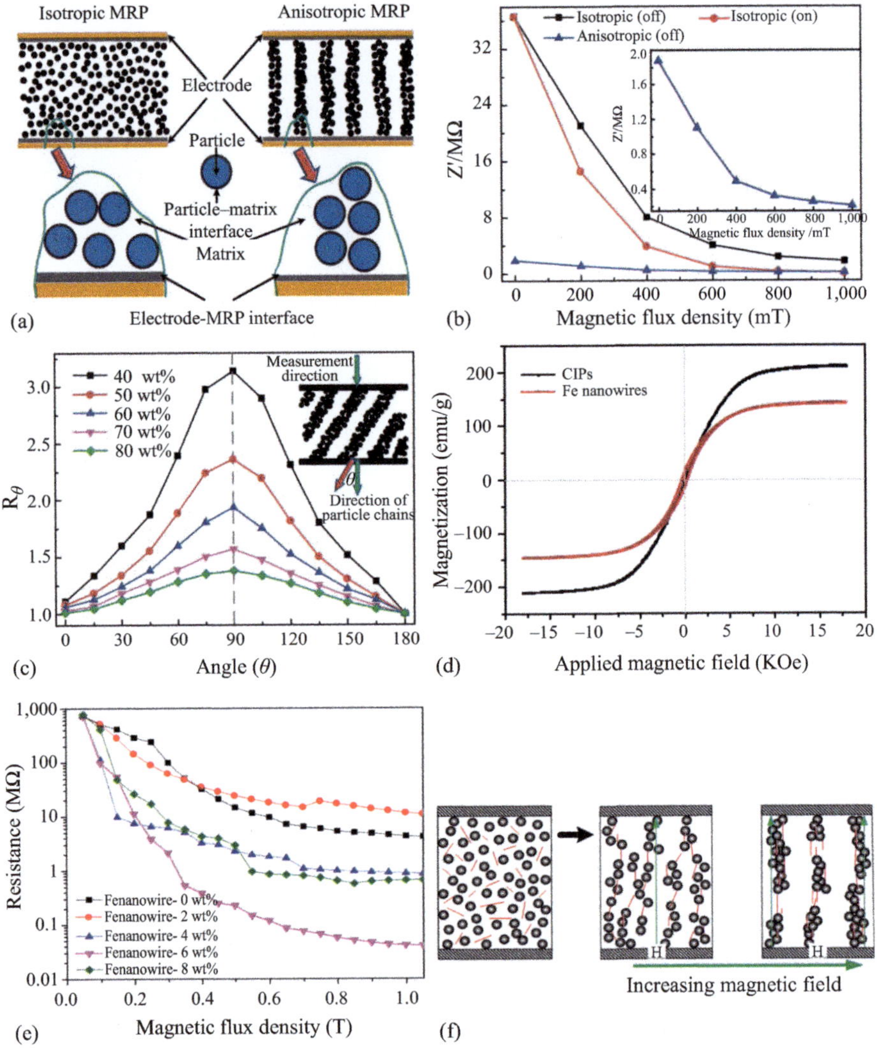

Figure 15.6 *(a) Schematic of MRP with different particle distributions and corresponding interfaces; (b) the resistive component of impedance spectroscopy at 0.1 Hz for MRP-80 with different preprocessing methods under different magnetic fields; (c) Z_0 ratio of anisotropic MRP with different particle contents in different particle chain directions [87]; (d) the magnetic hysteresis loops of the Fe nanowires and carbonyl iron particles at room temperature; (e) the resistance of resistors under different magnetic field; and (f) the sketch of the rearrangement of carbonyl iron particles with partial substitution by Fe nanowire under the magnetic field [86]*

The reason was that the viscosity of magnetic polyurethane increased with the content of iron nanowires, and high viscosity was not conducive to the formation of chain structures. In addition, the total weight of the iron nanowires and CIP content was defined. With increasing of the content of iron nanowires, the CIP content decreased. Then, the number of particle chains reduced, resulting in the decrease of the conductivity (Figure 15.6(f)).

In order to improve conductivity, Pang introduced graphite into MRP and proposed a particle–particle resistance model based on the dipole model (Figure 15.7(a)) [80]. The experimental results showed that the resistance of MRP with 15 wt% graphite doping decreased by four orders of magnitude than the resistance of non-doped MRP. Meanwhile, the conductivity of graphite-doped MRP was magnetically sensitive (Figure 15.7(b)). Carbon nanotubes and carbon fibers have also been applied to improve the electrical properties of MRP due to their excellent electroconductivities [81,88]. In the oscillating shear, the preconfiguration particle chains were tilted and fractured, and the conductive networks formed by carbon nanotubes and carbon fibers were also destroyed, resulting in an increase in electrical resistance and a decrease in the storage modulus. Under applying a magnetic field, adjacent particles reformed new chains; thus, the resistance decreased, and the

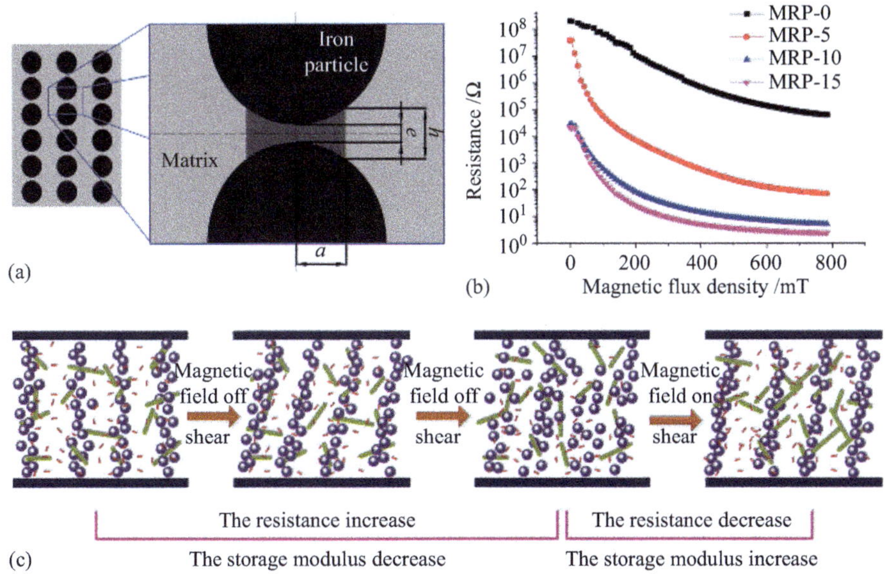

Figure 15.7 (a) Sketch of the particle–particle electrical resistance model; (b) resistance versus magnetic flux density of graphite-doped MRPs with different flake graphite powders contents [80]; and (c) the microstructure evolution of graphite-doped MRPs during the oscillatory shear. Purple balls represent the iron particles, green lines represent the carbon fibers, red lines represent the carbon nanotubes [81]

storage modulus increased accordingly (Figure 15.7(c)). Finally, a concept of electro-mechanical-magnetic coupling was proposed to investigate its unique properties.

15.4 Mechanism

The study of MR mechanisms is helpful to understand the mechanical properties and conducive to a better practical application of MR materials. The macro-mechanical behavior of MR materials is closely related to its microstructure. During the past decade, many researches have focused on establishing a micro-structure model to analyze the microstructure-dependent mechanical properties of MR materials. For example, Mohebi *et al.* performed microstructural modeling analysis on MRF and studied the evolution process of microparticle structure under an external magnetic field [89]. Heine *et al.* investigated the thermal conductivity of MRF under shear stress based on the self-consistent mean-field dipole model [90]. Keaveny and Maxey proposed a fixed-dipole model with a current density distribution function to simulate the magnetic model of particles [91]. Recently, Davis first established a simple analysis model of the MRE microstructure using the finite element method [92]. Later, Shen *et al.* proposed a magnetic dipole model to calculate the MRE magnetic shear modulus and established a mathematical model of the stress–strain relationship [93]. Subsequently, many microstructural models of MRE, such as finite-column mode [94], effective permeability model [18], and rectangular lattice model [95], were proposed. Most importantly, for MRPs, the three-dimensional dipole-spring model [96], coarse-grained model [97], continuum material model [98], were also developed, and they were proven to be effective for studying their mechanical properties.

The mechanical properties and mechanism of MRPs can be analyzed and deduced by creep and recovery experiments [100–105]. Xu performed MRP creep and recovery experiments to explain the deformation mechanism [99]. In the absence of a magnetic field, slipping between soft segments was the main-response mechanism of MRP to stress. Tensile, slipping, and curling of the soft section caused the strain to increase linearly with time, but the strain partially recovered after unloading (Figure 15.8(a)). When an external magnetic field was applied, the particle chains limited the motion of the soft segment. As the magnetic field increased, the creep strain decreased, the curvature of the creep curve and the delay elastic strain increased (Figure 15.8(a) and (b)). As shown in Figure 15.8(c), the proportion of the delayed elastic strain to the total strain increased from 11.6%, without magnetic field, to 32.7% at 930 mT magnetic field. Then, the physical state of MRP changed from viscoplastic to elastoplastic. Due to the magneto-induced restriction effect, the creep behavior of isotropic MRP at the initial stage was different from that of the anisotropic MRP (Figure 15.8(d)). This was because the anisotropic MRP had a larger restriction effect on the matrix and thus produced a smaller strain response at the same stress.

Particle-level dynamics simulation has been conducted by Liu *et al.* to further investigate the microscopic particulate structures of MRP [106]. The magnetic

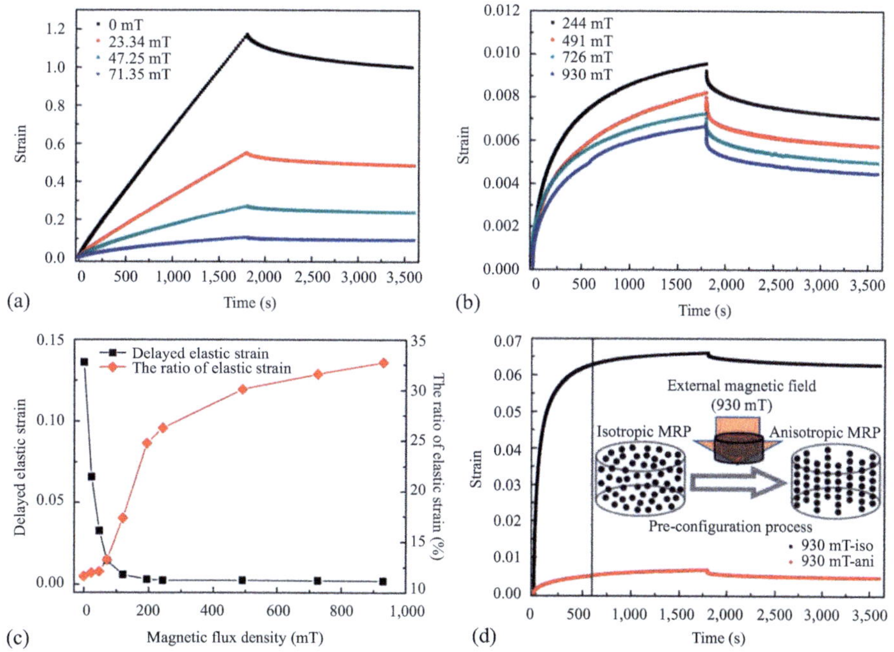

Figure 15.8 The creep and recovery curves of MRP under a constant stress of 2,000 Pa and (a) the low magnetic fields; (b) the high magnetic fields; (c) the delayed elastic strain and the ratio of elastic strain to the total strain under different magnetic field at the creep time of 1,800 s; (d) the creep and recovery curves for isotropic MRP and anisotropic MRP with a magnetic field [99]

interaction between the adjacent magnetic particles was described by a modified magnetic dipolar interaction force model. As shown in Figure 15.9(c), the iron powder particles were randomly distributed in the initial state (0 s). When a magnetic field was applied, a lot of short chain-like structures were quickly formed after 5 s. As time increased, the short chains attracted each other to form the long chain-like structures (20 s) and gradually formed stable column-like structures at 100 s. The variation of magnetic potential energy was shown in Figure 15.9(a). The larger the magnetic field density, the smaller the initial magnetic potential energy. Then, in the initial 10 s, the magnetic potential energy dropped down sharply. Meanwhile, the complex magnetomechanical coupling process proceeded. Later, the magnetic potential energy decreased to relatively stable values, and at last, more stable microscopic particulate structures were achieved. Figure 15.9(b) and (d) showed the evolution of microscopic particulate structures and the variation of magnetic potential energy under a rotating magnetic field, respectively. It was noticed that the chain-like structures changed with the direction of the magnetic field, and that the magnetic potential energy first

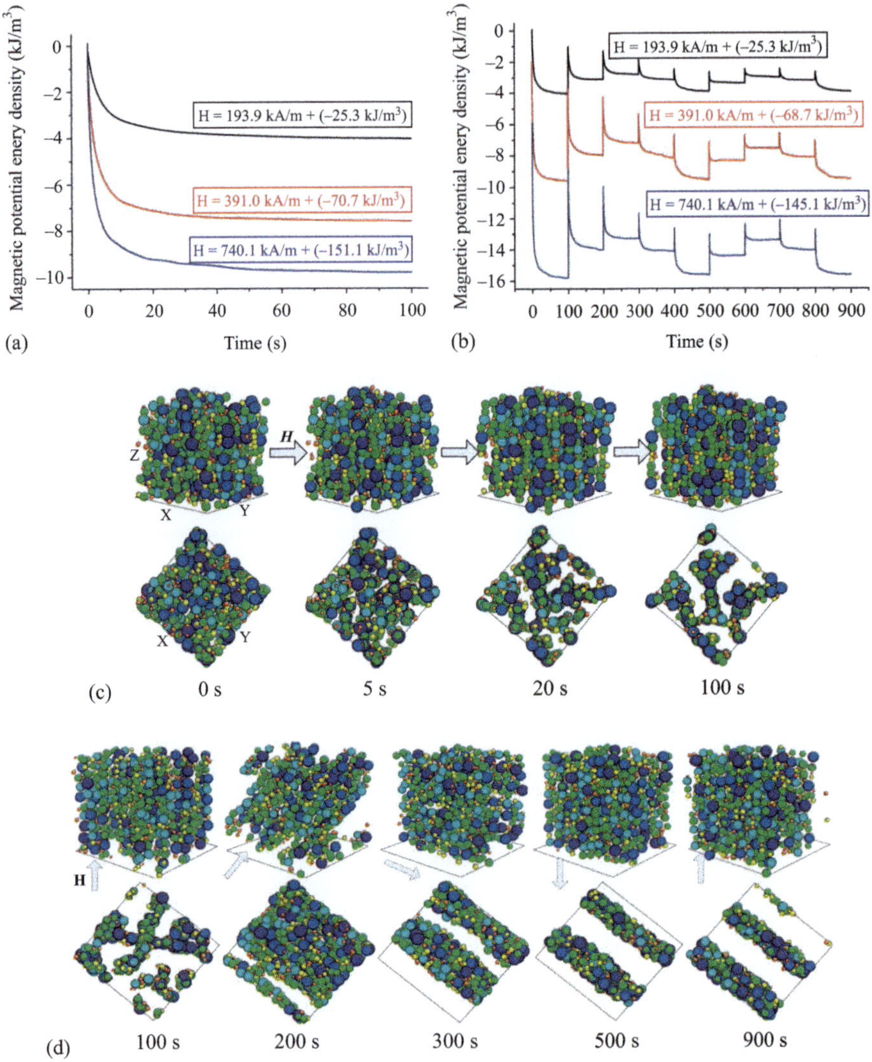

Figure 15.9 The magnetic potential energy of MRP-70 in (a) the different magnetic fields; (b) the stepwise in-plane rotating magnetic fields; (c) the evolution of magneto-induced microstructures with axonometric views and top views of MRP-70 in different magnetic fields; and (d) the magneto-induced microstructure of MRP-70 in a stepwise in-plane rotating magnetic field [106]

increased and then decreased at each magnetic field rotation. For each rotating step, the magnetic potential energy adhered to the principle of minimum potential energy. It can be seen that the aggregation process of particles was closely related to the magnetic potential energy.

15.5 Application

The MR absorber was a magnetic field-dependent absorber that rapidly and reversibly adjusted the damping force by changing the density of the magnetic field. MRF was commonly used as damping media for MR absorbers because of its rapid response and high MR effect [107–110]. However, the sealing and sedimentation also limited its practical application. With the high viscosity, good stability, and high MR effect, MRP has become a new damping media for MR damper. Recently, Pang designed a novel MR energy absorber (MREA) by using MRP as damping medium [111]. The schematic of the MREA with adjustable resistance gap was shown in Figure 15.10(a): (1) upper piston rod, (2) end cover, (3) coat, (4) exciting coil, (5) insulation lining, (6) piston, and (7) nether piston rod. The ModuLab material test system and drop hammer test system, respectively, tested the MR absorber's damping performance under low-speed compression and high-speed impact. By applying a current to the exciting coil, the resistance varied in a large range (Figure 15.10(b) and (c)) and was highly dependent on the speed of compression (Figure 15.10(d)). The shock absorbing performance of the MREA was further tested with drop hammers of different weights from different heights. The maximum force increased with increasing of the drop height (Figure 15.10(e)). With higher drop height, the velocity was larger and the viscose force was bigger, then the maximum force increased. Moreover, with different drop mass weights, the velocities of the drop hammers were the same, so the front part of the force curves was roughly coincident (Figure 15.10(f)). Figure 15.10(g) showed the energy absorption ratio increased from 4.13 J/mm with 0 A current to 5.07 J/mm with 2 A current which increased by about 23%. The energy absorption increased almost linearly with displacements, which meant the energy absorption was uniform for the whole stroke (Figure 15.10(h)).

The MRPs can be further applied in vibration absorbers, actuators, and sensors. For example, Kim proposed a stiffness feedback control system for magnetic carrageenan for the design of tunable shock absorbers that adaptively tune to a time-varying disturbance in real time. The absorption system can automatically tune from 56 to 67 Hz in a 100 mT field to minimize vibration of the main system [112]. Shin proposed a tunable vibration absorber based on magnetic hydrogel that produced 379% of stiffness change or 115% of tuning frequency under 200 mT magnetic field. It implemented a wide range of tonal vibration suppression on the main system to protect any connected sensitive devices [113]. Fuhrer described the introduction of metal nanoparticles into the skeleton of hydrogel and applied it as an ultra-flexible magnetic actuator with high mechanical stability, elasticity, and shape memory effect [114]. Zrinyi *et al.* introduced colloidally sized magnetic particles into a chemically cross-linking polyvinyl alcohol hydrogel to create a new class of stimuli-responsive gels that mimic muscle contraction. The shape change and movement of the magnetic field-sensitive gels were smooth and gentle, similar to that observed in muscles, which provided attractive means of actuation as artificial muscles for biomechanics and bionic applications [115].

To further improve the magnetically dependent conductivity of plastic MR materials, many conductive additives were introduced. Uniquely, the graphite-doped MRP exhibited a typical magneto-controllable electrical switch (Figure 15.11(a)).

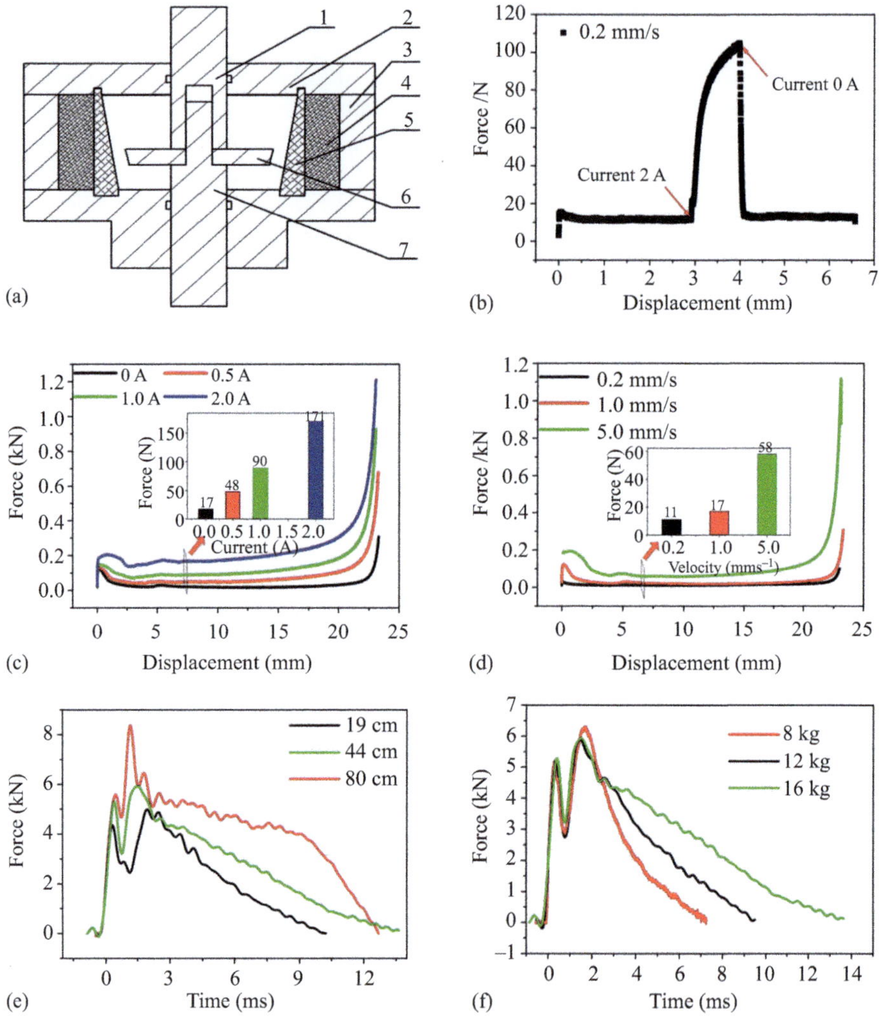

Figure 15.10 (a) Schematic diagram of the MREA; resistance force of the MREA under different current and different compression velocity (b) with a changing current (c) with different currents and a 1.0 mm/s compression velocity (d) under different compression velocity without current; the resistance force and energy absorption with different drop heights, different drop hammer weights and different current inputs: (e) and (h) different drop height, 8 kg drop mass, 0 A current input, (f) different drop mass weight, 44 cm drop height, 0 A current input, (g) different current input, 8 kg drop mass, 44 cm drop height [111]

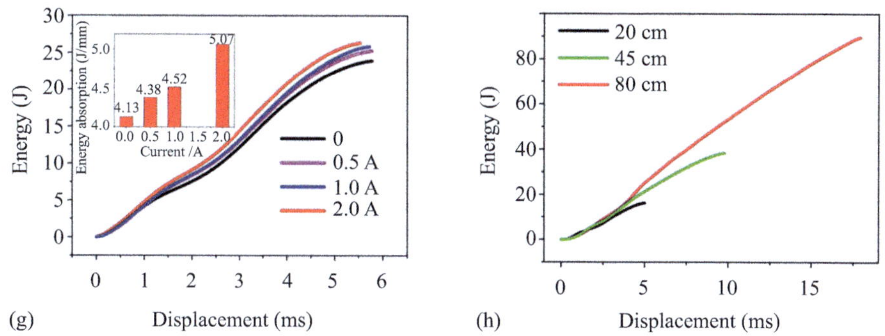

Figure 15.10 (Continued)

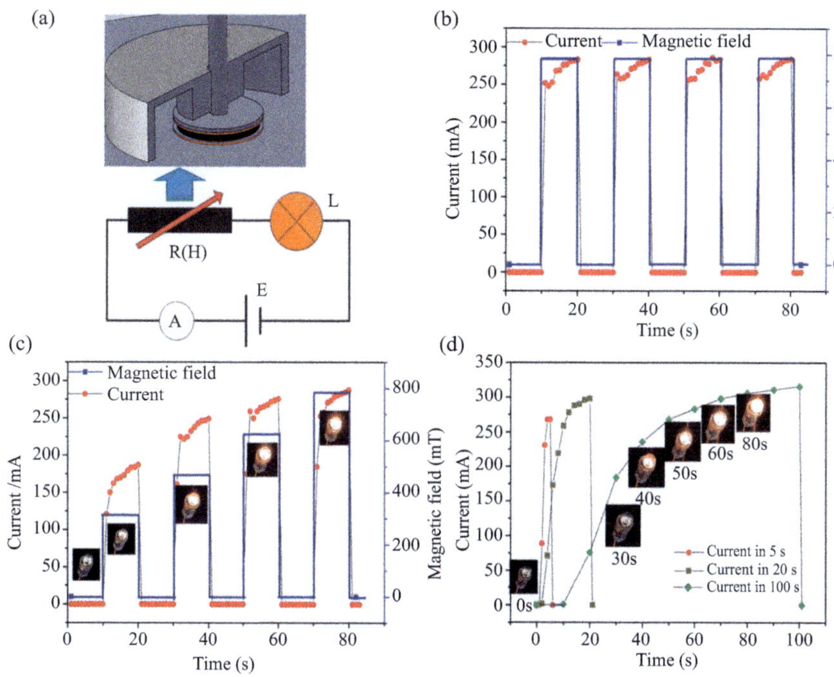

Figure 15.11 (a) Schematic illustration of devices in the experiment containing a magnetic on–off switch; conductive capability of sample MRP-15 and luminance of bulb change in terms of time under different magnetic field: (b) a square wave magnetic field; (c) a step wave magnetic field; and (d) a triangular wave magnetic field with different rising time [80]

Because the magnetic field changed the distribution of CIPs and graphite, the resistance decreased with the increase of the magnetic field, and the current increased accordingly. Similarly, once the magnetic field decreased, the resistance increased and the current decreased (Figure 15.11(b)). Different from other quick-response switch, this magneto-controllable switch can adjust the response time and sense current by tuning the magnetic field. As shown in Figure 15.11(c), the brightness of the bulb increased with the increase of the magnetic field density. Figure 15.11(d) showed the time dependence of the current by slowly increasing magnetic field. The current slowly increased to a maximum at different times and growth rates, which protected the electronic device from being destroyed by rapid and large currents.

15.6 Conclusion

As a kind of intelligent magneto-controllable material with good stability, and high MR effect, MRPs have attracted increasing attentions. In this chapter, we first introduced the preparations of the MRPs. According to the different matrices, the MRPs can be roughly classified into hydrogel-like MRP, polyurethane MRP, shear-thickening MRP, and thermosensitive MRP. Then, the excellent properties of MRPs, MR properties, dynamic mechanical properties at high strain rates, and magneto-electric coupling properties were described in detail. Due to the low cross-linking density of the polymer matrix, the magnetic particles can aggregate to form chain-like structures under a magnetic field, which improved the mechanical and electrical properties of MRPs. Furthermore, the evolution of microscopic particulate structures was also achieved by particle-level dynamics simulation. MRPs possessed excellent mechanical and electrical properties; thus they possessed broad applications in shock absorbers, brakes, sensors, and magneto-controllable switches. It is believed that more excellent properties of MRPs can be developed, and they will be applied to an expansive field for smart materials.

References

[1] P. Pan, D. Zhang, Y. Zeng, Y. Tsai, X. Nie, Development of a tunable friction pendulum system for semi-active control of building structures under earthquake ground motions, Earthquake Engineering & Structural Dynamics 47(8) (2018) 1706–1721.

[2] S. Sun, J. Yang, H. Du, *et al.*, Development of magnetorheological elastomers-based tuned mass damper for building protection from seismic events, Journal of Intelligent Material Systems and Structures 29(8) (2018) 1777–1789.

[3] J. Pan, J.-q. Zhang, Y.-g. Song, J. Zhang, Study on preparation and stability of special magnetorheological fluid for tracked vehicles, Journal of Academy of Armored Force Engineering 21(2) (2007) 59–61.

[4] A.L. Morales, A.J. Nieto, J.M. Chicharro, P. Pintado, A semi-active vehicle suspension based on pneumatic springs and magnetorheological dampers, Journal of Vibration and Control 24(4) (2018) 808–821.

[5] Q.H. Nguyen, S.B. Choi, J.K. Woo, Optimal design of magnetorheological fluid-based dampers for front-loaded washing machines, Proceedings of the Institution of Mechanical Engineers Part C-Journal of Mechanical Engineering Science 228(2) (2014) 294–306.

[6] Y. Zhou, H. Pan, Fuzzy semi-active control for seismic response reduction with magnetorheological dampers, Journal of Vibroengineering 16(4) (2014) 1640–1649.

[7] J. Rabinow, The magnetic fluid clutch, Electrical Engineering 67(12) (1948) 1167–1167.

[8] M.T. López-López, A. Gómez-Ramírez, J.D. Durán, F. González-Caballero, Preparation and characterization of iron-based magnetorheological fluids stabilized by addition of organoclay particles, Langmuir 24(14) (2008) 7076–7084.

[9] K. Shah, S.-B. Choi, Rheological properties of magnetorheological polishing fluid featuring plate-like iron particles, Smart Materials and Structures 23 (11) (2014) 117003.

[10] M.R. Jolly, J.W. Bender, J.D. Carlson, Properties and applications of commercial magnetorheological fluids, Journal of Intelligent Material Systems and Structures 10(1) (1999) 5–13.

[11] L. Shan, K. Chen, M. Zhou, X. Zhang, Y. Meng, Y. Tian, Shear history effect of magnetorheological fluids, Smart Materials and Structures 24(10) (2015) 105030.

[12] K.P. Hong, K.H. Song, M.W. Cho, S.H. Kwon, H.J. Choi, Magnetorheological properties and polishing characteristics of silica-coated carbonyl iron magnetorheological fluid, Journal of Intelligent Material Systems and Structures 29(1) (2018) 137–146.

[13] S. Gorodkin, W. Kordonski, E. Medvedeva, Z. Novikova, A. Shorey, S. Jacobs, A method and device for measurement of a sedimentation constant of magnetorheological fluids, Review of Scientific Instruments 71(6) (2000) 2476–2480.

[14] M. López-López, A. Zugaldia, F. González-Caballero, J. Durán, Sedimentation and redispersion phenomena in iron-based magnetorheological fluids, Journal of Rheology 50(4) (2006) 543–560.

[15] G.T. Ngatu, N.M. Wereley, Viscometric and sedimentation characterization of bidisperse magnetorheological fluids, IEEE Transactions on Magnetics 43(6) (2007) 2474–2476.

[16] M.R. Jolly, J.D. Carlson, B.C. Muñoz, T.A. Bullions, The magnetoviscoelastic response of elastomer composites consisting of ferrous particles embedded in a polymer matrix, Journal of Intelligent Material Systems and Structures 7(6) (1996) 613–622.

[17] W. Li, M. Nakano, Fabrication and characterization of PDMS based magnetorheological elastomers, Smart Materials and Structures 22(5) (2013) 055035.

[18] W. Li, X. Zhang, A study of the magnetorheological effect of bimodal particle based magnetorheological elastomers, Smart Materials and Structures 19(3) (2010) 035002.

[19] J. Wu, X. Gong, L. Chen, H. Xia, Z. Hu, Preparation and characterization of isotropic polyurethane magnetorheological elastomer through in situ polymerization, Journal of Applied Polymer Science 114(2) (2009) 901–910.
[20] M. Yu, L. Zhao, J. Fu, M. Zhu, Thermal effects on the laminated magnetorheological elastomer isolator, Smart Materials and Structures 25(11) (2016) 115039.
[21] R. Li, W. Mu, L. Zhang, X. Wang, Design and testing performance of a magneto-rheological elastomer isolator for a scaled bridge system, Journal of Intelligent Material Systems and Structures 29(2) (2018) 171–182.
[22] G. Liao, Y. Xu, F. Wei, R. Ge, Q. Wan, Investigation on the phase-based fuzzy logic controller for magnetorheological elastomer vibration absorber, Journal of Intelligent Material Systems and Structures 28(6) (2017) 728–739.
[23] S. Sun, H. Deng, J. Yang, *et al.*, An adaptive tuned vibration absorber based on multilayered MR elastomers, Smart Materials and Structures 24(4) (2015) 045045.
[24] F.-L. Xin, X.-X. Bai, L.-J. Qian, Principle, modeling, and control of a magnetorheological elastomer dynamic vibration absorber for powertrain mount systems of automobiles, Journal of Intelligent Material Systems and Structures 28(16) (2017) 2239–2254.
[25] W. Li, X. Wang, X. Zhang, Y. Zhou, Development and analysis of a variable stiffness damper using an MR bladder, Smart Materials and Structures 18(7) (2009) 074007.
[26] I. Bica, Magnetoresistor sensor with magnetorheological elastomers, Journal of Industrial and Engineering Chemistry 17(1) (2011) 83–89.
[27] T. Tian, W. Li, Y. Deng, Sensing capabilities of graphite based MR elastomers, Smart Materials and Structures 20(2) (2011) 025022.
[28] S. Xuan, Y. Xu, T. Liu, X. Gong, Recent progress on the magnetorheological plastomers, International Journal of Smart and Nano Materials 6(2) (2015) 135–148.
[29] T. Shiga, A. Okada, T. Kurauchi, Magnetroviscoelastic behavior of composite gels, Journal of Applied Polymer Science 58(4) (1995) 787–792.
[30] A.A. Abdeen, J. Lee, N.A. Bharadwaj, R.H. Ewoldt, K.A. Kilian, Temporal modulation of stem cell activity using magnetoactive hydrogels, Advanced Healthcare Materials 5(19) (2016) 2536–2544.
[31] Y. Pan, Y. Gao, J. Shi, L. Wang, B. Xu, A versatile supramolecular hydrogel of nitrilotriacetic acid (NTA) for binding metal ions and magnetorheological response, Journal of Materials Chemistry 21(19) (2011) 6804–6806.
[32] Y. Sun, Y. Wang, J. Yao, L. Gao, D.-s. Li, Y. Liu, Highly magnetic sensitivity of polymer nanocomposite hydrogels based on magnetic nanoparticles, Composites Science and Technology 141 (2017) 40–47.
[33] J. Wu, X. Gong, Y. Fan, H. Xia, Physically crosslinked poly(vinyl alcohol) hydrogels with magnetic field controlled modulus, Soft Matter 7(13) (2011) 6205–6212.

[34] E. Dohmen, M. Boisly, D. Borin, *et al.*, Advancing towards polyurethane-based magnetorheological composites, Advanced Engineering Materials 16 (10) (2014) 1270–1275.
[35] H. Grigoriew, L. Wiegart, A. Boczkowska, M. Mirkowska, Dynamic correlation in magnetorheological composite under magnetic field studied by XPCS, Solid State Communications 150(17–18) (2010) 840–843.
[36] L. Ge, S. Xuan, G. Liao, T. Yin, X. Gong, Stretchable polyurethane sponge reinforced magnetorheological material with enhanced mechanical properties, Smart Materials and Structures 24(3) (2015) 037001.
[37] H.-N. An, J. Groenewold, S.J. Picken, E. Mendes, Conformational changes of a single magnetic particle string within gels, Soft Matter 10(7) (2014) 997–1005.
[38] P.V. Rao, S. Maniprakash, S. Srinivasan, A. Srinivasa, Functional behavior of isotropic magnetorheological gels, Smart Materials and Structures 19(8) (2010) 085019.
[39] S. Xuan, Y. Zhang, Y. Zhou, W. Jiang, X. Gong, Magnetic Plasticine^{TM}: a versatile magnetorheological material, Journal of Materials Chemistry 22 (26) (2012) 13395–13400.
[40] N. Mohamad, S. Mazlan, S.-B. Choi, M. Nordin, The field-dependent rheological properties of magnetorheological grease based on carbonyl-iron-particles, Smart Materials and Structures 25(9) (2016) 095043.
[41] F. Guo, G. Lin Xiao, J. Yu Guo, The preparation and testing of rheological properties for single- and double-grading magnetorheological composite gels, Key Engineering Materials 730 (2017) 65–71.
[42] W. Zhang, X.L. Gong, S.H. Xuan, Y.G. Xu, High-performance hybrid magnetorheological materials: preparation and mechanical properties, Industrial & Engineering Chemistry Research 49(24) (2010) 12471–12476.
[43] T. Mitsumata, N. Abe, Magnetic-field sensitive gels with wide modulation of dynamic modulus, Chemistry Letters 38(9) (2009) 922–923.
[44] D. Diaz-Bleis, C. Vales-Pinzon, Y. Freile-Pelegrin, J.J. Alvarado-Gil, Thermal characterization of magnetically aligned carbonyl iron/agar composites, Carbohydrate Polymers 99 (2014) 84–90.
[45] A. Tiraferri, K.L. Chen, R. Sethi, M. Elimelech, Reduced aggregation and sedimentation of zero-valent iron nanoparticles in the presence of guar gum, Journal of Colloid and Interface Science 324(1–2) (2008) 71–79.
[46] M. Bonini, S. Lenz, R. Giorgi, P. Baglioni, Nanomagnetic sponges for the cleaning of works of art, Langmuir 23(17) (2007) 8681–8685.
[47] R. Messing, N. Frickel, L. Belkoura, *et al.*, Cobalt ferrite nanoparticles as multifunctional cross-linkers in PAAm ferrohydrogels, Macromolecules 44 (8) (2011) 2990–2999.
[48] I.Y. Tóth, G. Veress, M. Szekeres, E. Illés, E. Tombácz, Magnetic hyaluronate hydrogels: preparation and characterization, Journal of Magnetism and Magnetic Materials 380 (2015) 175–180.
[49] J.I. Kim, C. Chun, B. Kim, *et al.*, Thermosensitive/magnetic poly(organophosphazene) hydrogel as a long-term magnetic resonance contrast platform, Biomaterials 33(1) (2012) 218–224.

[50] A. Brunsen, S. Utech, M. Maskos, W. Knoll, U. Jonas, Magnetic composite thin films of FexOy nanoparticles and photocrosslinked dextran hydrogels, Journal of Magnetism and Magnetic Materials 324(8) (2012) 1488–1497.

[51] W. Zhao, K. Odelius, U. Edlund, C. Zhao, A.-C. Albertsson, In situ synthesis of magnetic field-responsive hemicellulose hydrogels for drug delivery, Biomacromolecules 16(8) (2015) 2522–2528.

[52] W. Wang, Y. Liu, T. Li, M. Zhou, Heterogeneous Fenton catalytic degradation of phenol based on controlled release of magnetic nanoparticles, Chemical Engineering Journal 242 (2014) 1–9.

[53] M. Helminger, B. Wu, T. Kollmann, *et al.*, Synthesis and characterization of gelatin-based magnetic hydrogels, Advanced Functional Materials 24(21) (2014) 3187–3196.

[54] R. Hou, G. Zhang, G. Du, *et al.*, Magnetic nanohydroxyapatite/PVA composite hydrogels for promoted osteoblast adhesion and proliferation, Colloids and Surfaces B: Biointerfaces 103 (2013) 318–325.

[55] R. Hou, L. Nie, G. Du, X. Xiong, J. Fu, Natural polysaccharides promote chondrocyte adhesion and proliferation on magnetic nanoparticle/PVA composite hydrogels, Colloids and Surfaces B: Biointerfaces 132 (2015) 146–154.

[56] J. Zhu, Bioactive modification of poly (ethylene glycol) hydrogels for tissue engineering, Biomaterials 31(17) (2010) 4639–4656.

[57] S. van Berkum, P.D. Biewenga, S.P. Verkleij, *et al.*, Swelling enhanced remanent magnetization of hydrogels cross-linked with magnetic nanoparticles, Langmuir 31(1) (2014) 442–450.

[58] N.A. Burke, H.D. Stöver, F.P. Dawson, Magnetic nanocomposites: preparation and characterization of polymer-coated iron nanoparticles, Chemistry of Materials 14(11) (2002) 4752–4761.

[59] M.J. Wilson, A. Fuchs, F. Gordaninejad, Development and characterization of magnetorheological polymer gels, Journal of Applied Polymer Science 84 (14) (2002) 2733–2742.

[60] A. Fuchs, M. Xin, F. Gordaninejad, *et al.*, Development and characterization of hydrocarbon polyol polyurethane and silicone magnetorheological polymeric gels, Journal of Applied Polymer Science 92(2) (2004) 1176–1182.

[61] M. Yu, B. Ju, J. Fu, S. Liu, S.-B. Choi, Magnetoresistance characteristics of magnetorheological gel under a magnetic field, Industrial & Engineering Chemistry Research 53(12) (2014) 4704–4710.

[62] B. Ju, M. Yu, J. Fu, X. Zheng, S. Liu, Magnetic field-dependent normal force of magnetorheological gel, Industrial & Engineering Chemistry Research 52 (33) (2013) 11583–11589.

[63] P. Yang, M. Yu, J. Fu, S. Liu, S. Qi, M. Zhu, The damping behavior of magnetorheological gel based on polyurethane matrix, Polymer Composites 38(7) (2017) 1248–1258.

[64] Y. Xu, X. Gong, S. Xuan, W. Zhang, Y. Fan, A high-performance magnetorheological material: preparation, characterization and magnetic-mechanic coupling properties, Soft Matter 7(11) (2011) 5246–5254.

[65] Y. Wang, S. Wang, C. Xu, S. Xuan, W. Jiang, X. Gong, Dynamic behavior of magnetically responsive shear-stiffening gel under high strain rate, Composites Science and Technology 127 (2016) 169–176.
[66] S. Wang, W. Jiang, W. Jiang, *et al.*, Multifunctional polymer composite with excellent shear stiffening performance and magnetorheological effect, Journal of Materials Chemistry C 2(34) (2014) 7133–7140.
[67] N. Golinelli, A. Spaggiari, E. Dragoni, Mechanical behaviour of magnetic Silly Putty: viscoelastic and magnetorheological properties, Journal of Intelligent Material Systems and Structures 28(8) (2017) 953–960.
[68] B.O. Park, B.J. Park, M.J. Hato, H.J. Choi, Soft magnetic carbonyl iron microsphere dispersed in grease and its rheological characteristics under magnetic field, Colloid and Polymer Science 289(4) (2011) 381–386.
[69] H. Sahin, X. Wang, F. Gordaninejad, Temperature dependence of magneto-rheological materials, Journal of Intelligent Material Systems and Structures 20(18) (2009) 2215–2222.
[70] H. An, S.J. Picken, E. Mendes, Enhanced hardening of soft self-assembled copolymer gels under homogeneous magnetic fields, Soft Matter 6(18) (2010) 4497–4503.
[71] B. Wei, X. Gong, W. Jiang, L. Qin, Y. Fan, Study on the properties of magnetorheological gel based on polyurethane, Journal of Applied Polymer Science 118(5) (2010) 2765–2771.
[72] H. An, S.J. Picken, E. Mendes, Nonlinear rheological study of magneto responsive soft gels, Polymer 53(19) (2012) 4164–4170.
[73] T. Mitsumata, N. Abe, Giant and reversible magnetorheology of carrageenan/iron oxide magnetic gels, Smart Materials and Structures 20(12) (2011) 124003.
[74] X. Gong, Y. Xu, S. Xuan, C. Guo, L. Zong, W. Jiang, The investigation on the nonlinearity of plasticine-like magnetorheological material under oscillatory shear rheometry, Journal of Rheology 56(6) (2012) 1375–1391.
[75] J. Xu, P. Wang, H. Pang, *et al.*, The dynamic mechanical properties of magnetorheological plastomers under high strain rate, Composites Science and Technology 159 (2018) 50–58.
[76] F. Guo, C.-b. Du, G.-j. Yu, R.-p. Li, The static and dynamic mechanical properties of magnetorheological Silly Putty, Advances in Materials Science and Engineering 2016 (2016) 1–11.
[77] P. Malecki, M. Krolewicz, J. Krzak, J. Kaleta, J. Piglowski, Dynamic mechanical analysis of magnetorheological composites containing silica-coated carbonyl iron powder, Journal of Intelligent Material Systems and Structures 26(14) (2015) 1899–1905.
[78] P. Yang, M. Yu, H. Luo, J. Fu, H. Qu, Y. Xie, Improved rheological properties of dimorphic magnetorheological gels based on flower-like carbonyl iron particles, Applied Surface Science 416 (2017) 772–780.
[79] P. Yang, M. Yu, J. Fu, H. Luo, Rheological properties of dimorphic magnetorheological gels mixed dendritic carbonyl iron powder, Journal of Intelligent Material Systems and Structures 29(1) (2018) 12–23.

[80] H. Pang, S. Xuan, T. Liu, X. Gong, Magnetic field dependent electro-conductivity of the graphite doped magnetorheological plastomers, Soft Matter 11(34) (2015) 6893–6902.

[81] J. Xu, S. Xuan, H. Pang, X. Gong, The strengthening effect of 1D carbon materials on magnetorheological plastomers: mechanical properties and conductivity, Smart Materials and Structures 26(3) (2017) 035044.

[82] P. Yang, M. Yu, J. Fu, Ni-coated multi-walled carbon nanotubes enhanced the magnetorheological performance of magnetorheological gel, Journal of Nanoparticle Research 18(3) (2016) 61.

[83] W. Zhao, H. Pang, X. Gong, Novel magnetorheological plastomer filled with NdFeB particles: preparation, characterization, and magnetic–mechanic coupling properties, Industrial & Engineering Chemistry Research 56(31) (2017) 8857–8863.

[84] B. Mordina, R.K. Tiwari, D.K. Setua, A. Sharma, Impact of graphene oxide on the magnetorheological behaviour of $BaFe_{12}O_{19}$ nanoparticles filled polyacrylamide hydrogel, Polymer 97 (2016) 258–272.

[85] G. Liao, X. Gong, S. Xuan, Magnetic field-induced compressive property of magnetorheological elastomer under high strain rate, Industrial & Engineering Chemistry Research 52(25) (2013) 8445–8453.

[86] M. Yu, H. Luo, J. Fu, P. Yang, The field-dependent conductivity of dimorphic magnetorheological gel incorporated with iron nanowire, Journal of Intelligent Material Systems and Structures 29(1) (2018) 24–31.

[87] Y. Xu, X. Gong, T. Liu, S. Xuan, Magneto-induced microstructure characterization of magnetorheological plastomers using impedance spectroscopy, Soft Matter 9(32) (2013) 7701–7709.

[88] H. Qu, M. Yu, J. Fu, P. Yang, Y. Liu, Enhanced field-dependent conductivity of magnetorheological gels with low-doped carbon nanotubes, Smart Materials and Structures 26(10) (2017) 105026.

[89] M. Mohebi, N. Jamasbi, J. Liu, Simulation of the formation of nonequilibrium structures in magnetorheological fluids subject to an external magnetic field, Physical Review E 54(5) (1996) 5407.

[90] M.C. Heine, J. de Vicente, D.J. Klingenberg, Thermal transport in sheared electro- and magnetorheological fluids, Physics of Fluids 18(2) (2006) 023301.

[91] E.E. Keaveny, M.R. Maxey, Modeling the magnetic interactions between paramagnetic beads in magnetorheological fluids, Journal of Computational Physics 227(22) (2008) 9554–9571.

[92] L. Davis, Model of magnetorheological elastomers, Journal of Applied Physics 85(6) (1999) 3348–3351.

[93] Y. Shen, M.F. Golnaraghi, G. Heppler, Experimental research and modeling of magnetorheological elastomers, Journal of Intelligent Material Systems and Structures 15(1) (2004) 27–35.

[94] L. Chen, X. Gong, W. Li, Microstructures and viscoelastic properties of anisotropic magnetorheological elastomers, Smart Materials and Structures 16(6) (2007) 2645.

[95] D. Ivaneyko, V.P. Toshchevikov, M. Saphiannikova, G. Heinrich, Magneto-sensitive elastomers in a homogeneous magnetic field: a regular rectangular lattice model, Macromolecular Theory and Simulations 20(6) (2011) 411–424.
[96] G. Pessot, H. Löwen, A.M. Menzel, Dynamic elastic moduli in magnetic gels: normal modes and linear response, The Journal of Chemical Physics 145(10) (2016) 104904.
[97] M.A. Annunziata, A.M. Menzel, H. Löwen, Hardening transition in a one-dimensional model for ferrogels, The Journal of Chemical Physics 138(20) (2013) 204906.
[98] D. Szabo, M. Zrinyi, Nonhomogeneous, non-linear deformation of polymer gels swollen with magneto-rheological suspensions, International Journal of Modern Physics B 16(17–18) (2002) 2616–2621.
[99] Y. Xu, X. Gong, S. Xuan, X. Li, L. Qin, W. Jiang, Creep and recovery behaviors of magnetorheological plastomer and its magnetic-dependent properties, Soft Matter 8(32) (2012) 8483–8492.
[100] B. Baldewa, Y.M. Joshi, Delayed yielding in creep, time–stress superposition and effective time theory for a soft glass, Soft Matter 8(3) (2012) 789–796.
[101] V. Awasthi, Y.M. Joshi, Effect of temperature on aging and time–temperature superposition in nonergodic laponite suspensions, Soft Matter 5 (24) (2009) 4991–4996.
[102] X. Wu, R.E. Sallach, J.M. Caves, V.P. Conticello, E.L. Chaikof, Deformation responses of a physically cross-linked high molecular weight elastin-like protein polymer, Biomacromolecules 9(7) (2008) 1787–1794.
[103] R.A. Riggleman, K.S. Schweizer, J.J. de Pablo, Nonlinear creep in a polymer glass, Macromolecules 41(13) (2008) 4969–4977.
[104] H. Muenstedt, N. Katsikis, J. Kaschta, Rheological properties of poly (methyl methacrylate)/nanoclay composites as investigated by creep recovery in shear, Macromolecules 41(24) (2008) 9777–9783.
[105] M. Kraft, J. Meissner, J. Kaschta, Linear viscoelastic characterization of polymer melts with long relaxation times, Macromolecules 32(3) (1999) 751–757.
[106] T. Liu, X. Gong, Y. Xu, S. Xuan, W. Jiang, Simulation of magneto-induced rearrangeable microstructures of magnetorheological plastomers, Soft Matter 9(42) (2013) 10069–10080.
[107] S.-B. Choi, W. Li, M. Yu, H. Du, J. Fu, P.X. Do, State of the art of control schemes for smart systems featuring magneto-rheological materials, Smart Materials and Structures 25(4) (2016) 043001.
[108] X. Gong, X. Ruan, S. Xuan, Q. Yan, H. Deng, Magnetorheological damper working in squeeze mode, Advances in Mechanical Engineering 6 (2014) 410158.
[109] Z.-X. Li, L.-H. Xu, Performance tests and hysteresis model of MRF-04K damper, Journal of Structural Engineering 131(8) (2005) 1303–1306.
[110] Z. Xing, M. Yu, S. Sun, J. Fu, W. Li, A hybrid magnetorheological elastomer-fluid (MRE-F) isolation mount: development and experimental validation, Smart Materials and Structures 25(1) (2015) 015026.

[111] H. Pang, S. Xuan, C. Sun, X. Gong, A novel energy absorber based on magnetorheological gel, Smart Materials and Structures 26(10) (2017) 105017.
[112] S.Y. Yang, K.H. Cho, Y. Kim, *et al.*, High performance twisted and coiled soft actuator with spandex fiber for artificial muscles, Smart Materials and Structures 26(10) (2017) 105025.
[113] B.-C. Shin, J.-H. Yoon, Y.-K. Kim, K.-S. Kim, Note: Vibration suppression using tunable vibration absorber based on stiffness variable magnetorheological gel, Review of Scientific Instruments 86(10) (2015) 106106.
[114] R. Fuhrer, E.K. Athanassiou, N.A. Luechinger, W.J. Stark, Crosslinking metal nanoparticles into the polymer backbone of hydrogels enables preparation of soft, magnetic field-driven actuators with muscle-like flexibility, Small 5(3) (2009) 383–388.
[115] M. Zrinyi, D. Szabo, Muscular contraction mimiced by magnetic gels, International Journal of Modern Physics B 15(6–7) (2001) 557–563.

Index